Wireless Communication Systems: Design and Implementation

Wireless Communication Systems: Design and Implementation

Editor: Archie Rogers

NYRESEARCH
P R E S S

New York

Published by NY Research Press
118-35 Queens Blvd., Suite 400,
Forest Hills, NY 11375, USA
www.nyresearchpress.com

Wireless Communication Systems: Design and Implementation
Edited by Archie Rogers

International Standard Book Number: 978-1-63238-716-5 (Hardback)

Cataloging-in-Publication Data

Wireless communication systems : design and implementation / edited by Archie Rogers.
    p. cm.
Includes bibliographical references and index.
ISBN 978-1-63238-716-5
1. Wireless communication systems. 2. Wireless communication systems--Design and construction.
3. Telecommunication systems. I. Rogers, Archie.
TK5103.2 .W57 2020
621.384--dc23

Contents

Preface

The purpose of the book is to provide a glimpse into the dynamics and to present opinions and studies of some of the scientists engaged in the development of new ideas in the field from very different standpoints. This book will prove useful to students and researchers owing to its high content quality.

Wireless communication refers to the transfer of information or power between two or more points that are not directly connected by an electrical conductor. Such communication is achieved with the help of radio waves. These waves cover a wide range of distance, from a few meters in the case of Bluetooth to as far as millions of kilometers in the case of deep-space radio communications. Wireless communication can also be achieved via free-space optical communication, sonic waves and electromagnetic induction. Various portable, fixed and mobile applications allow such communication to be established. GPS units, satellite television, radio receivers, cordless telephones and broadcast television are examples of systems that operate on wireless technology. This book outlines the process and applications of wireless communications in detail. It is a valuable compilation of topics, ranging from the basic to the most complex advancements in this field. For someone with an interest and eye for detail, this book covers the most significant topics in the field of wireless communication.

At the end, I would like to appreciate all the efforts made by the authors in completing their chapters professionally. I express my deepest gratitude to all of them for contributing to this book by sharing their valuable works. A special thanks to my family and friends for their constant support in this journey.

Editor

Stochastic Geometry Analysis of Downlink Spectral and Energy Efficiency in Ultradense Heterogeneous Cellular Networks

Jiaqi Lei, Hongbin Chen ⓘ, and Feng Zhao

Key Laboratory of Cognitive Radio and Information Processing, Guilin University of Electronic Technology, Guilin 541004, China

Correspondence should be addressed to Hongbin Chen; chbscut@guet.edu.cn

Academic Editor: Jinglan Zhang

The energy efficiency (EE) is a key metric of ultradense heterogeneous cellular networks (HCNs). Earlier works on the EE analysis of ultradense HCNs by using the stochastic geometry tool only focused on the impact of the base station density ratio and ignored the function of different tiers. In this paper, a two-tier ultradense HCN with small-cell base stations (SBSs) and user equipments (UEs) densely deployed in a traditional macrocell network is considered. Firstly, the performance of the ultradense HCN in terms of the association probability, average link spectral efficiency (SE), average downlink throughput, and average EE is theoretically analyzed by using the stochastic geometry tool. Then, the problem of maximizing the average EE while meeting minimum requirements of the average link SE and average downlink throughput experienced by UEs in macrocell and small-cell tiers is formulated. As it is difficult to obtain the explicit expression of average EE, impacts of the SBS density ratio and signal-to-interference-plus-noise ratio (SINR) threshold on the network performance are investigated through numerical simulations. Simulation results validate the accuracy of theoretical results and demonstrate that the maximum value of average EE can be achieved by optimizing the SBS density ratio and the SINR threshold.

1. Introduction

With the rapid growth of smart terminals, the wireless data traffic presented an exponential growth in recent years, which demands a much higher link throughput. However, latest wireless communication technologies have brought the point-to-point link throughput close to its theoretical limit, and the traditional homogeneous macrocell network cannot meet the enormous request on mobile services via the cell splitting technique anymore. In order to overcome these issues and provide a significant network performance leap, HCNs with the coexistence of macrocells and small cells have been introduced in the LTE-advanced standardization [1–3]. But in early HCNs, small cells were recommended as a complement for macrocells and were deployed in specific areas, such as hotspot and indoor scenarios. With the advancement of massive multiple-input multiple-output [4] and millimeter wave communication technologies, small cells were suggested to be densely deployed in all cellular network deployment scenarios [5]. So, the ultradense HCN, which has a much higher SBS density than ever, was viewed as one of the most promising paradigms to meet the 1000x wireless traffic volume increment in the next decade [6–11]. However, along with the increase of SBS density, the energy consumption of ultradense HCNs will inevitably increase [9–11]. Hence, the EE will be a major concern for designing environment-friendly ultradense HCNs. In this paper, endeavors are made to develop an analytical framework for evaluating the average link SE, average downlink throughput, and average EE in a two-tier ultradense HCN, as well as providing guidelines for practical deployments.

1.1. Related Works and Motivation. In HCNs, SBSs are usually randomly deployed, which pose significant challenges to the theoretical analysis and simulation validation. Recent research works have shown that such randomly deployed cellular networks can be successfully analyzed by utilizing the stochastic geometry tool [12]. For instance, modeling HCNs by the spatial Poisson point process (PPP) can provide an effective and tractable approach to analyze the performance of HCNs in terms of the coverage probability [13], outage probability [14], and average ergodic rate [15]. Furthermore, Singh et al. [16] and Dhillon and Andrews [17] characterized the downlink rate distribution with the mean load approximation. Besides, there exist many other point processes that can capture the spatial distribution characteristics of cellular networks, such as the binomial point process [18], hard-core point process [19], Poisson cluster process [20], and so on. Nevertheless, the PPP model has the advantage of being more analytically tractable. Furthermore, abovementioned works modeled the spatial distribution of macrocell base stations (MBSs) by a random point process as well and ignored the function of different tiers or the relationship between them, which is impractical in actual situations. So, in this work, the spatial distribution of MBSs is modeled by a regular hexagonal structure, while the spatial distribution of SBSs is modeled by a PPP. The stochastic geometry tool and statistical average method are adopted to analyze the performance of the macrocell tier and small-cell tier, respectively.

On the other hand, universal frequency reuse is a key technology to meet the ever-increasing throughput demands, which is one of the main characteristics of HCNs [21–24]. That is, the available spectrum will be aggressively reused by all of the coexisting network tiers. However, in ultradense HCNs, the cross-tier interference between a macrocell and a small cell and the co-tier interference among small cells have become a severe problem. As a result, there has been a significant amount of research efforts on managing both cross-tier interference and co-tier interference in a two-tier HCN, which consists of a macrocell network overlaid with small cells [25, 26]. In [25], the authors proposed a spectrum partitioning approach to avoid the cross-tier interference between the macrocell tier and the small-cell tier by using the orthogonal spectrum allocation. However, under a sparse small-cell deployment setting, this approach is clearly inefficient, and a much higher area SE can be attained if the spectrum sharing is allowed [26]. Inspired by earlier works mentioned above, the spectrum partitioning approach is adopted in the two-tier ultradense HCN with ultradense SBS deployment, which can obtain a higher network SE while significantly simplifying the theoretical analysis.

With respect to the EE, on one hand, significant efforts have been made to investigate the impact of base station density on the EE of HCNs. For example, the impact of base station deployment strategies especially the impact of base station density on the EE of ultradense HCNs was investigated in [27]. In [28], the authors analyzed the minimum base station density with the service outage constraint to minimize the network energy cost in HCNs. On the other hand, some earlier works concentrated on the EE by using energy-saving technologies, such as traffic awareness [29, 30], collaborative transmission [31], and base station sleep scheduling [32]. In this work, we focus on the optimization of network parameters (SBS density ratio and SINR threshold) to maximize the average EE without considering these related technologies.

As far as we know, most earlier works that adopted random network models to analyze the performance of HCNs focused on the network EE with respect to the SBS density ratio. In this work, the optimization problem of maximizing the average EE while meeting the average link SE and average downlink throughput requirements in the two-tier ultradense HCN is considered, and impacts of the SBS density ratio and SINR threshold on the network performance are investigated.

1.2. Contributions and Paper Organization. In this work, we provide a comprehensive average EE analysis of a two-tier ultradense HCN. Compared to earlier works, main contributions of this paper are summarized as follows:

(1) Taking different functions of macrocell and small-cell tiers into account, in a two-tier ultradense HCN, a modified nearest association scheme is proposed, namely, a UE first decides whether to associate with its nearest SBS according to the received SINR. If its received SINR surpasses a threshold, the UE will associate with its nearest SBS. Otherwise, it will associate with its nearest MBS. Note that a UE associates with its nearest MBS without the limit of SINR threshold, since the macrocell tier is not interference-limited and the power of interference is controlled at a certain level which is comparable to the noise power.

(2) The association probability is used to measure the burden of macrocell and small-cell tiers, and the average link SE is used to measure the quality of communications. Besides, the average link SE and average downlink throughput are derived by using the statistical average method.

(3) To evaluate the average downlink throughput experienced by UEs in the small-cell tier, instead of using the average ergodic rate, the SINR threshold is taken into account, and impacts of the SBS density ratio and SINR threshold on the average EE are investigated. The aim is to provide a tractable approach to seek the optimum SBS density ratio and SINR threshold to maximize the average EE while meeting minimum requirements of the average link SE and average downlink throughput experienced by UEs in macrocell and small-cell tiers, which is meaningful to guide the actual deployment of ultradense HCNs.

The rest of this paper is organized as follows: in Section 2, the two-tier ultradense HCN model, channel model, power consumption model, and association scheme are described. In Section 3, the average link SE, average downlink throughput, and average EE are analytically derived, and the optimization problem is formulated. In Section 4, numerical simulation results are presented to validate theoretical ones. Finally, some concluding remarks are given in Section 5.

Notations: E[·] represents the expectation operator, I_x represents the cumulative interference from all the other SBSs, $L_{I_x}(·)$ represents the Laplace transform of a random variable I_x, P_m and P_s denote the transmit powers of the MBS and SBS, respectively, N_m represents the number of

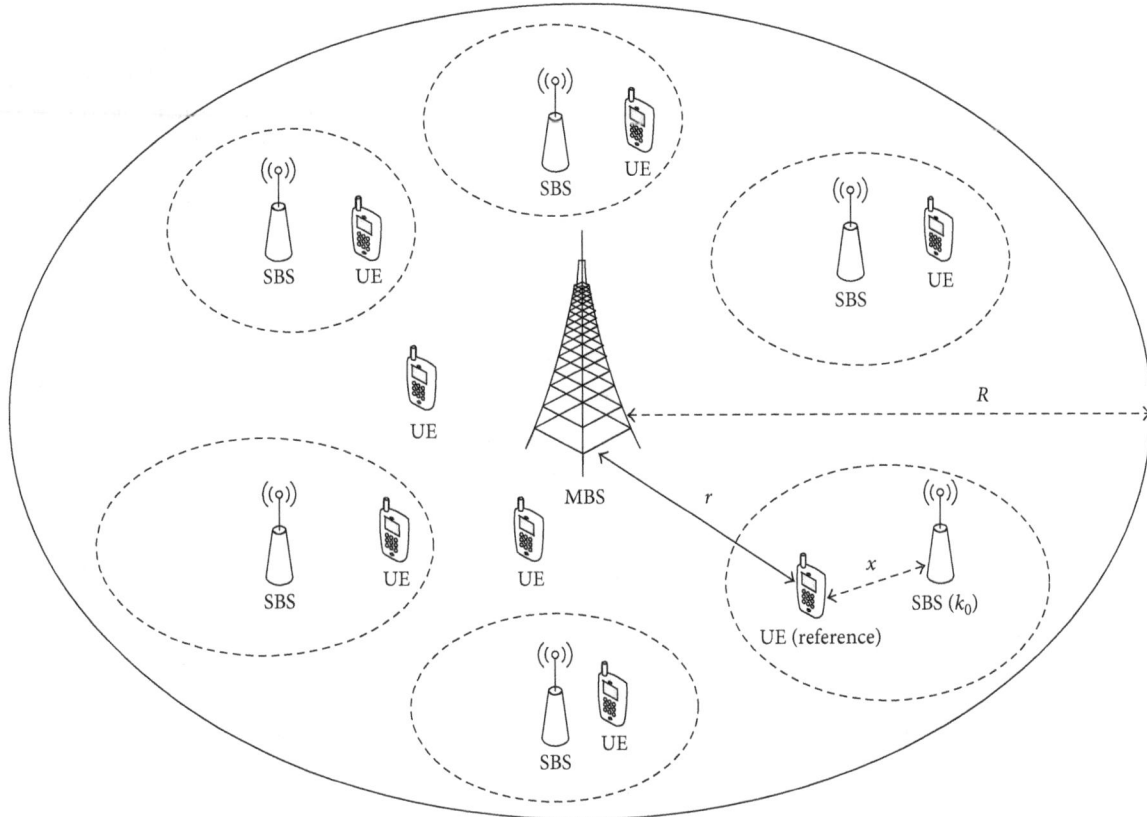

FIGURE 1: Illustration of a macrocell in the two-tier ultradense HCN.

UEs served by an MBS in the macrocell tier, N_s represents the number of UEs served by an SBS in the small-cell tier, the subscript x represents the distance between the reference UE and its nearest SBS (denoted by k_0), and subscripts m and s represent the MBS and SBS, respectively.

2. System Models

2.1. Ultradense HCN Model. The ultradense HCN consists of a traditional homogeneous macrocell network and an ultradense small-cell network, in which SBSs constitute the small-cell tier and their spatial distribution is modeled as a homogeneous PPP Φ_s with the intensity λ_s, while MBSs constitute the macrocell tier and their spatial distribution follows a regular hexagonal structure since they are aimed to provide seamless coverage for a large area. Note that, in earlier works on performance analysis of cellular networks, the hexagonal structure was commonly substituted by a circular one for tractability. In addition, UEs follow another independent homogeneous PPP Φ_u with the intensity λ_u. In order to avoid the cross-tier interference, the spectrum partitioning approach is adopted in the ultradense HCN. Moreover, because the analyzing process is the same for different macrocells and we try to obtain the average throughput in a tractable way, we pick out one macrocell with the radius R from the two-tier ultradense HCN to conduct the network performance analysis, as shown in Figure 1. In this macrocell, the MBS is located at the center, and the bandwidths W_1 and W_2 are dedicated for the MBS

and SBS, respectively. Without loss of generality, it is assumed that a reference UE is located within the range of the MBS with a radial distance r, and all base stations and UEs are assumed to be equipped with a single antenna.

Since the impact of SBS density on the average EE for a given UE density will be investigated, the SBS density is normalized to $\lambda_s = \rho \lambda_u$, where ρ is defined as the SBS density ratio. In this work, considering the feature of the ultradense HCN and referring to earlier works, let the value of ρ vary within 0.01~0.3. Note that ρ can take other values in different scenarios.

2.2. Channel Model. The path loss and fading are taken into account when modeling the wireless channel. The path loss exponent is denoted as $\alpha > 2$. The fading experienced by UEs is assumed to be Rayleigh fading with mean 1, and its power gain is denoted by h, which follows the exponential distribution, that is, $h \sim \exp(1)$.

In the macrocell tier, considering relevant techniques have been used to avoid the interference coming from adjacent MBSs, and in order to make the theoretical derivation tractable, it is assumed that the interference power coming from the other MBSs is comparable to the noise power σ^2, and the co-tier interference power $\xi^2 = \omega\sigma^2$ is the same all over the macrocell tier [33], where the parameter ω reflects the level of interference power in the macrocell tier. In the small-cell tier, the universal frequency reuse scheme is adopted to improve the network throughput. As such, except the serving SBS of the reference UE, all the other SBSs are potential interferers.

Therefore, the SINR of the reference UE at a distance r from its associated MBS can be expressed as

$$\gamma_1(r) = \frac{P_m h r^{-\alpha}}{\xi^2 + \sigma^2}, \qquad (1)$$

and the SINR of the reference UE at a distance x from its associated SBS can be expressed as

$$\gamma_2(x) = \frac{P_s h x^{-\alpha}}{I_x + \sigma^2}, \qquad (2)$$

where $I_x = \sum_{j \in \Phi_s, j \neq k_0} P_s h_j v_j^{-\alpha}$ denotes the cumulative interference from all the other SBSs, in which the jth SBS is at distance v_j from the reference UE.

2.3. Power Consumption Model. Generally speaking, the power consumption in a base station includes two parts: static power consumption and transmit power consumption. The static power consumption in a base station is independent of the transmit power, while the transmit power consumption scales with the traffic load. In this work, the situation where all the base stations are fully loaded is supposed, so it is reasonable to assume that the transmit power consumption in a base station is independent of the traffic load but scales with the average radiated power.

Therefore, in order to compute the average power consumption of the macrocell in the two-tier ultradense HCN, referring to [34], the power consumption in the MBS and the SBS can be simply written as linear expressions:

$$\begin{aligned} P_{m,tot} &= \Delta_m P_m + b_m, \\ P_{s,tot} &= \Delta_s P_s + b_s, \end{aligned} \qquad (3)$$

where the coefficients Δ_m and Δ_s, respectively, account for the power consumption that scales with the average radiated power due to amplifier and feeder losses. Static powers b_m and b_s, respectively, include ones used for signal processing, battery backup, and cooling. So, the average power consumption over the area covered by the macrocell in the two-tier ultradense HCN can be calculated as

$$\overline{P} = \frac{P_{m,tot} + \lambda_s \pi R^2 P_{s,tot}}{\pi R^2}. \qquad (4)$$

2.4. Association Scheme. In earlier works [13, 27], the spatial distribution of base stations in each tier of the HCN was modeled as an independent homogeneous PPP, and an SINR threshold was set to control the UE access to different tiers. Besides, both of them modeled the multitier network with a simple combination of single-tier networks, and the function of different tiers seems to be the same, which is impractical in an actual HCN. Furthermore, there are always some UEs dropped from the HCN, which means that the HCN cannot provide a seamless coverage.

In order to deal with problems mentioned above, in this work, a modified nearest association scheme is proposed, that is, a UE is covered by the small-cell tier when the SINR received from its nearest SBS is greater than the threshold β. Otherwise, it will be dropped from the small-cell tier and associate with the MBS.

Since UEs dropped from the small-cell tier will associate with the MBS, it is reasonable to use the association probability to characterize the burden of different tiers. The association probability is defined as $p = E_x[P\{\gamma_2(x) > \beta | x\}]$, in which $P\{\gamma_2(x) > \beta | x\}$ is the probability that a UE at distance x from its nearest SBS can achieve the target SINR β. It can be interpreted as (1) the average probability that a UE can achieve the target SINR β when associating with its nearest SBS and (2) the average fraction of UEs who at any time can achieve the target SINR β in the small-cell tier. According to [35], the association probability can be expressed as

$$\begin{aligned} p &= E_x[P\{\gamma_2(x) > \beta | x\}] \\ &= \int_{x>0} P\{\gamma_2(x) > \beta | x\} f(x) \, dx \\ &= 2\pi\lambda_s \int_0^\infty x L_{I_x}\left(\frac{\beta}{P_s x^{-\alpha}}\right) e^{-(\beta\sigma^2/P_s x^{-\alpha})} e^{-\lambda_s \pi x^2} \, dx, \end{aligned} \qquad (5)$$

where $L_{I_x}(s) = E[e^{-sI_x}] = \exp(-2\pi\lambda_s \int_x^\infty (1 - (1/(sP_s v^{-\alpha} + 1)))v dv)$ is the Laplace transform of the cumulative interference I_x and $f(x) = 2\pi\lambda_s x e^{-\lambda_s \pi x^2}$ is the probability density function of x.

3. Network Performance Analysis

In this section, taking the association probability into account, the average link SE, average downlink throughput experienced by UEs in macrocell and small-cell tiers, and average EE are derived. Then, an optimization problem is formulated to maximize the average EE while meeting the average link SE and average downlink throughput requirements.

According to (1), the SINR of a UE associating with the MBS is relevant to the distance r. So, the average link SE experienced by UEs in the macrocell tier can be expressed as

$$\begin{aligned} \tau_m &= E[\log_2(1 + \gamma_1(r))] \\ &= \frac{1}{\pi R^2} \int_0^{2\pi} \int_0^R \int_0^\infty \log_2\left(1 + \frac{P_m h r^{-\alpha}}{\sigma^2 + \xi^2}\right) e^{-h} r \, dh \, dr \, d\theta. \end{aligned} \qquad (6)$$

From (2) and taking the SINR threshold β into account, the average link SE experienced by UEs in the small-cell tier can be expressed as

$$\begin{aligned} \tau_s &= E[\log_2(1 + \gamma_2(x)) | \gamma_2(x) > \beta] \\ &= \log_2(1 + \beta) + \frac{1}{\ln(2)} \int_0^\infty A(x) \, dx, \end{aligned} \qquad (7)$$

where

$$A(x) = 2\pi\lambda_s \int_\beta^\infty x \frac{L_{I_x}(z/P_s x^{-\alpha})e^{-(z\sigma^2/P_s x^{-\alpha})}}{L_{I_x}(\beta/P_s x^{-\alpha})e^{-(\beta\sigma^2/P_s x^{-\alpha})}} \frac{1}{z+1}e^{-\lambda_s \pi x^2} dz.$$

(8)

Proof. Denote the random variable $\gamma_2(x)$ by Y. $E[\log_2(1 + \gamma_2(x))|\gamma_2(x) > \beta]$ can be calculated as follows [13]:

$$E_x\left[\int_0^\infty \log_2(1 + y)f_Y(y|Y > \beta)dy\right]$$

$$= \frac{1}{\ln(2)}E_x\left[\int_0^\infty \int_0^y \frac{1}{(z+1)}f_Y(y|Y > \beta)dz\,dy\right]$$

$$= \frac{1}{\ln(2)}E_x\left[\int_0^\infty \left(\int_z^\infty f_Y(y|Y > \beta)dy\right)\frac{1}{z+1}dz\right]$$

$$= \frac{1}{\ln(2)}E_x\left[\int_0^\infty \frac{P(Y > z|Y > \beta)}{z+1}dz\right]$$

$$= \frac{1}{\ln(2)}\int_0^\infty\left(\int_0^\infty \frac{P(Y > z|Y > \beta)}{z+1}dz\right)f(x)dx$$

$$= \frac{1}{\ln(2)}\int_0^\infty\left(\int_0^\infty \frac{P(Y > z|Y > \beta)}{z+1}dz\right)2\pi\lambda_s xe^{-\lambda_s \pi x^2}dx,$$

(9)

where the first step follows the integral transformation and the second step follows from changing the order of integration, while the conditional complementary cumulative density function (CCDF) of $Y = \gamma_2(x)$ is written as follows:

$$P(Y > z|Y > \beta)$$

$$= \frac{P(Y > z, Y > \beta)}{P(Y > \beta)}$$

$$= \frac{P(Y > \max(z, \beta))}{P(Y > \beta)}$$

(10)

$$= \begin{cases} \dfrac{L_{I_x}(z/P_s x^{-\alpha})e^{-(z\sigma^2/P_s x^{-\alpha})}}{L_{I_x}(\beta/P_s x^{-\alpha})e^{-(\beta\sigma^2/P_s x^{-\alpha})}}, & z > \beta \\ \\ 1, & \text{otherwise,} \end{cases}$$

where $P(Y > \beta) = P\{\gamma_2(x) > \beta|x\} = L_{I_x}(\beta/P_s x^{-\alpha})e^{-(\beta\sigma^2/P_s x^{-\alpha})}$.

It is clear that the average downlink throughput of a UE experienced in macrocell and small-cell tiers is mainly determined by the average number of UEs served by the MBS or its nearest SBS. According to the proposed association scheme, when the first interpretation of the association probability is adopted, that is, all the UEs have an equal probability to associate with the small-cell tier, referring to

[17], the spatial distribution of UEs that associate with the small-cell tier can be obtained by thinning the spatial distribution of all the UEs, which also follows a homogeneous PPP. In order to simplify the derivation, it is assumed that the spatial distribution of UEs served by the macrocell tier or the small-cell tier follows another independent homogeneous PPP, which means that the intensity of UEs served by the macrocell tier or the small-cell tier is identical in any area. Therefore, the average downlink throughput of UEs associating with different MBSs will be equal. Note that the independent thinning of the PPP of UEs that associate with the macrocell tier or the small-cell tier with a thinning probability is not considered. According to (5), in the macrocell with radius R, the average number of UEs served by the MBS is $E[N_m] = \pi R^2(1 - p)\lambda_u$. So, the average downlink throughput of a UE served by the MBS can be expressed as

$$C_m = \frac{W_1\tau_m}{E[N_m]}.$$

(11)

With respect to UEs served by the small-cell tier, taking the association probability into account, the equivalent UE density served by the small-cell tier is $\lambda_2 = p\lambda_u$. The average downlink throughput of a UE served by its nearest SBS can be expressed as

$$C_s = \frac{W_2\tau_s}{E[N_s]},$$

(12)

where $E[N_s] = 1 + 1.28(\lambda_2/\lambda_s)$ is the average number of UEs served by an SBS [16]. As mentioned above, the average downlink throughput of a UE in the two-tier ultradense HCN can be written as

$$\overline{C} = (1 - p)C_m + pC_s.$$

(13)

The average EE is defined as the ratio of average area throughput to average area power consumption, and the average link SE is defined as the weighted sum of the average link SE obtained in macrocell and small-cell tiers. According to (4–7) and (13), the average EE and average link SE can be, respectively, expressed as

$$\overline{\eta}_{EE} = \frac{\lambda_u\overline{C}}{\overline{P}},$$

(14)

$$\overline{\eta}_{SE} = (1 - p)\tau_m + p\tau_s.$$

(15)

Finally, we try to maximize the average EE while meeting the minimum requirements of the average link SE and average downlink throughput. The optimization problem can be formulated as

$$\max \quad \overline{\eta}_{EE}$$

$$\text{s.t.} \quad \overline{\eta}_{SE} \geq \tau$$

$$C_s \geq C_m \geq C_{\min},$$

(16)

where τ is the minimum average link SE requirement and $C_{\min}$ is the minimum requirement of the average downlink throughput experienced by UEs in macrocell and small-cell tiers. Note that (11), (12), (14), and (15) show that C_m, C_s, $\overline{\eta}_{EE}$,

TABLE 1: Simulation parameters.

Parameter	Notation	Value
Bandwidth	W_1 and W_2	10 MHz and 10 MHz
Path loss exponent	α	3
Noise power	σ^2	-100 dBm
Coverage radius of the MBS	R	500 m
UE intensity	λ_u	0.01 m^{-2}
Transmit power of the MBS	P_m	40 W
Transmit power of the SBS	P_s	2 W
Power parameters of the MBS	Δ_m and b_m	30 and 500 W
Power parameters of the SBS	Δ_s and b_s	2 and 10 W
Minimum average link SE requirement	τ	3 bps/Hz
Minimum average downlink throughput requirement	C_{min}	2×10^4 bps
Level of interference power in the macrocell tier	ω	10

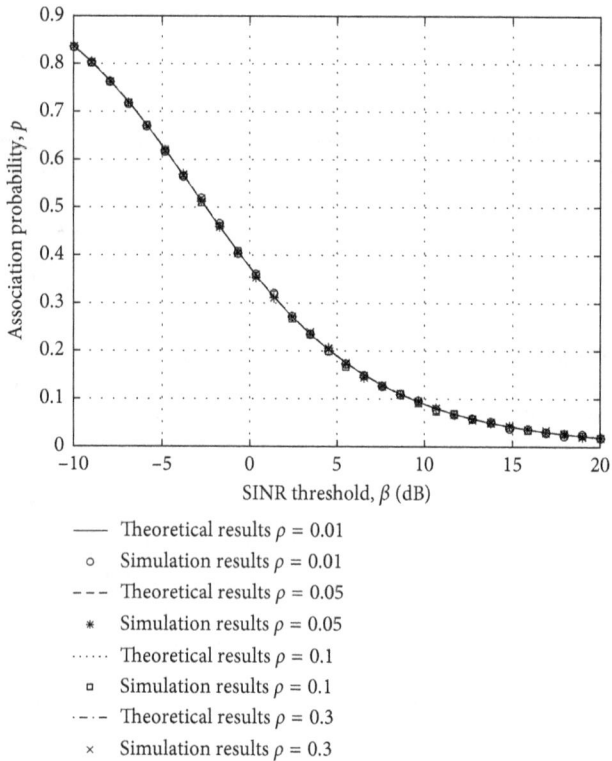

FIGURE 2: Association probability p versus SINR threshold β.

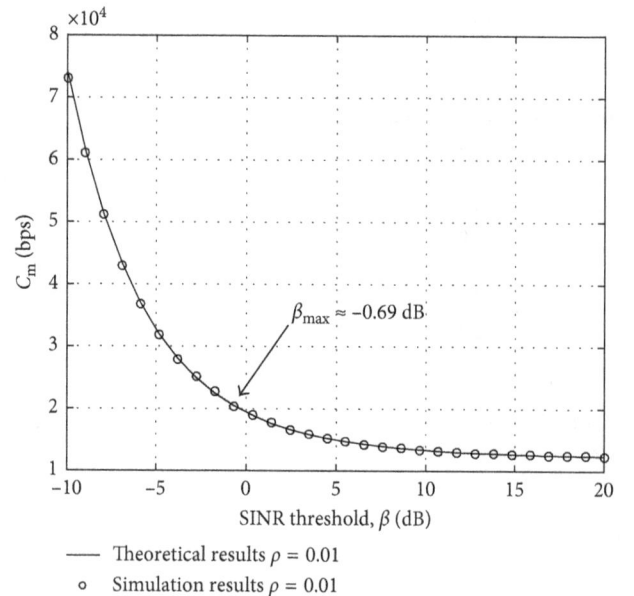

FIGURE 3: Average downlink throughput C_m of a UE served by the MBS versus SINR threshold β when the SBS density ratio $\rho = 0.01$.

and $\overline{\eta}_{SE}$ are functions of the SBS density ratio ρ and the SINR threshold β on condition that values of P_m, P_s, α, R, σ^2, ω, λ_u, W_1, and W_2 are all given. So, we can analyze impacts of ρ and β on $\overline{\eta}_{EE}$ and try to seek optimum values of ρ and β to maximize $\overline{\eta}_{EE}$ while meeting the requirements of $\overline{\eta}_{SE}$, C_m, and C_s. As it is difficult to obtain the explicit expression of $\overline{\eta}_{EE}$, in the following section, extensive numerical simulations are conducted to evaluate impacts of ρ and β on $\overline{\eta}_{EE}$.

4. Simulation Results

In the previous section, the average link SE, average downlink throughput, and average EE have been analytically derived. In this section, comparisons of Monte Carlo simulation results and theoretical results are presented to illustrate the performance of the two-tier ultradense HCN, especially the average EE. Due to differences among system models as well as considered factors, results are not compared with those in earlier works. Simulation parameters are listed in Table 1.

As a key intermediate parameter in the two-tier ultra-dense HCN, the association probability p has an immediate impact on C_m, C_s, and $\overline{C}$, which is a function of ρ and β when values of P_s, α, and σ^2 are given. So, impacts of ρ and β on p, C_m, C_s, and $\overline{C}$ are first investigated. Then, the values of P_m, P_s, α, R, σ^2, ω, λ_u, W_1, and W_2 are all set, and impacts of ρ and β on $\overline{\eta}_{SE}$ and $\overline{\eta}_{EE}$ are investigated. Note that analytical results are numerically calculated with (5) and (11)–(15).

4.1. Impacts of ρ and β on p, C_m, C_s, and $\overline{C}$. Figure 2 shows impacts of the SBS density ratio ρ and SINR threshold β on the association probability p. We can see that the SBS density ratio ρ has little effect on the association probability p, while the association probability p rapidly decreases with the increase of β. These results can be explained as follows:

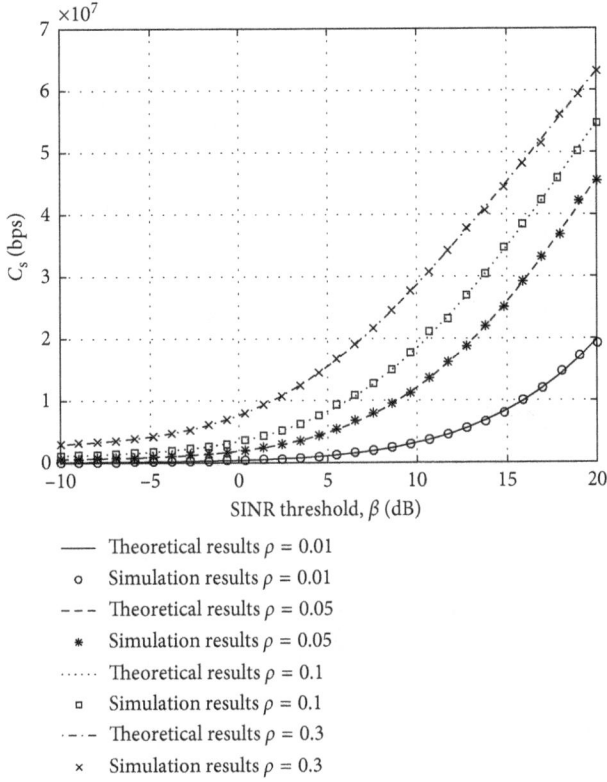

FIGURE 4: Average downlink throughput C_s of a UE served by its nearest SBS versus SINR threshold β.

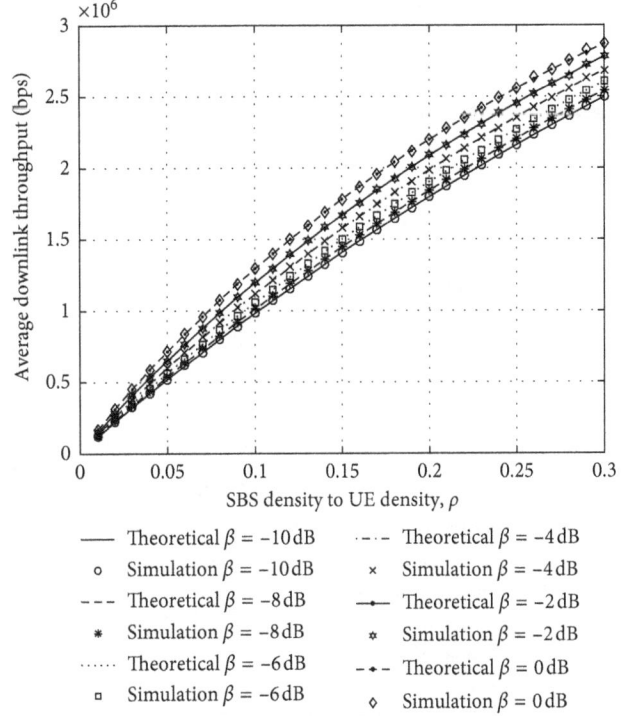

FIGURE 5: Average downlink throughput $\overline{C}$ versus SBS density ratio ρ.

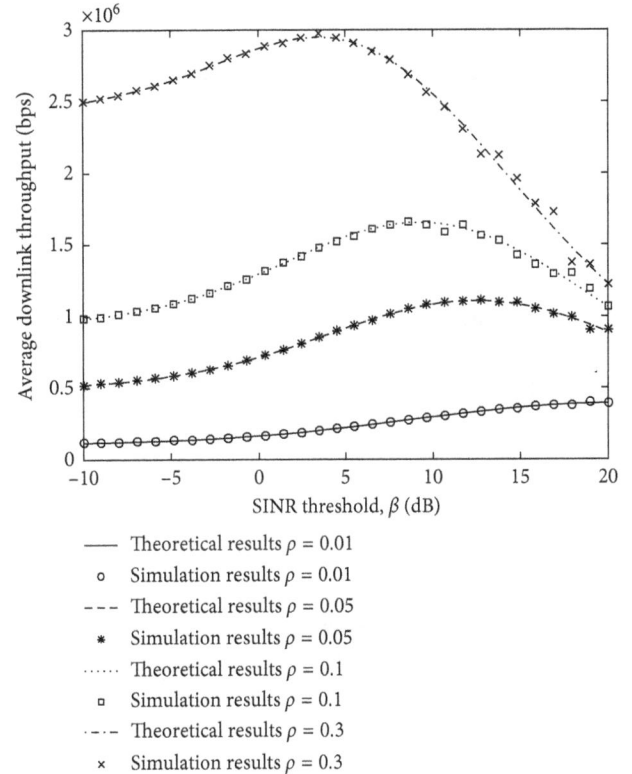

FIGURE 6: Average downlink throughput $\overline{C}$ versus SINR threshold β.

firstly, in the small-cell tier, the interference dominates the noise, which means that the noise power has little effect on the SINR [13]. Secondly, due to the random deployment of SBSs, the increase in signal power is exactly counterbalanced by the increase in interference power.

Since only the SINR threshold β has an obvious effect on the association probability p, according to (11), C_m is determined by β as well. So, we can observe the varying trend of the average downlink throughput of a UE served by the MBS, C_m versus β for any given ρ. Figure 3 depicts the impact of the SINR threshold β on C_m with $\rho = 0.01$. From this figure, we can observe that C_m rapidly decreases with the increase of β. It can be interpreted that more UEs drop from the small-cell tier and associate with the MBS due to the increase of β. With simulation parameters listed in Table 1, we can obtain the maximum value of β as $\beta_{\max} \approx -0.69$ dB.

Figure 4 depicts the impact of the SINR threshold β on the average downlink throughput C_s of a UE served by its nearest SBS for different values of ρ. From this figure, we can observe that C_s increases with β and ρ. The reason is that a UE served by its nearest SBS shares its serving base station with a smaller number of the other UEs and experiences a higher SINR.

Figures 5 and 6 show impacts of the SBS density ratio ρ and SINR threshold β on the average downlink throughput $\overline{C}$, respectively. We can see that simulation results coincide with theoretical results, which validates the expression of the average downlink throughput in (13). From Figure 5, we can observe that the average downlink throughput increases with ρ for any curve; it can be interpreted that less UEs are severed by its

nearest SBS when ρ increases, and it is sure to approach a certain limit when each small cell has only one active UE. From Figure 6, we can observe that if the SBS density ratio ρ is

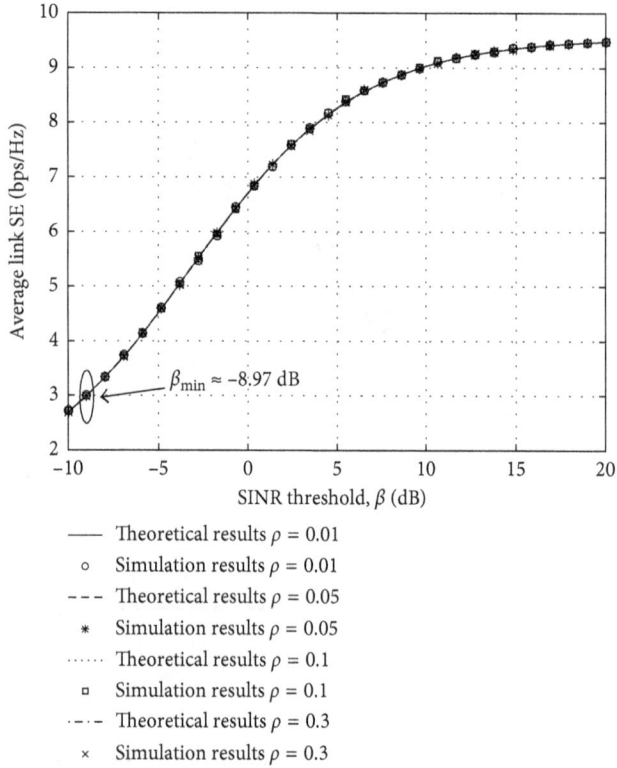

FIGURE 7: Average link SE $\overline{\eta}_{SE}$ versus SINR threshold β.

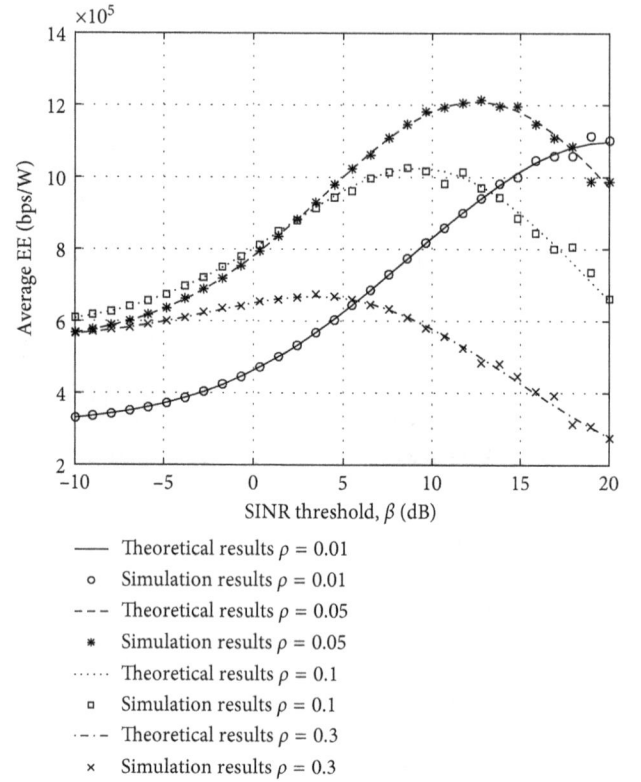

FIGURE 9: Average EE $\overline{\eta}_{EE}$ versus SINR threshold β.

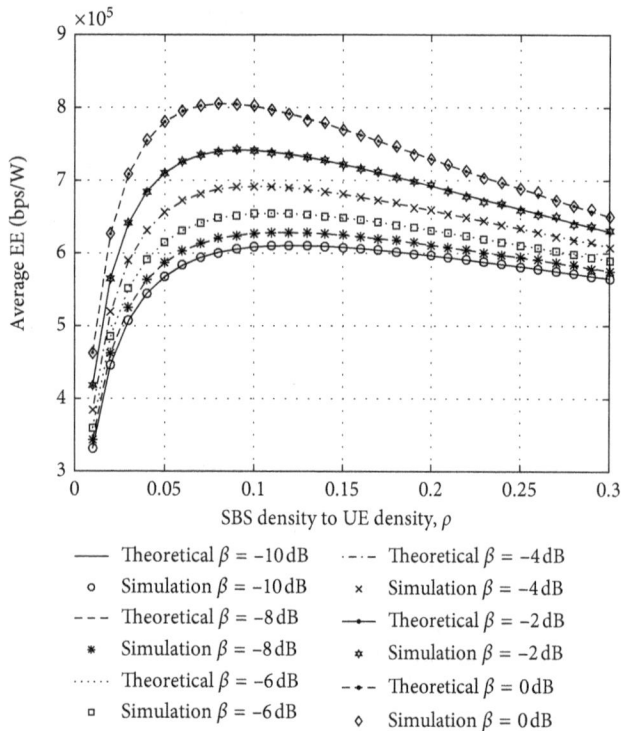

FIGURE 8: Average EE $\overline{\eta}_{EE}$ versus SBS density ratio ρ.

given, we can seek an optimum β to get the maximum average downlink throughput, which shows that β plays the role to adjust the traffic burden between macrocell and small-cell tiers.

4.2. Impacts of ρ and β on $\overline{\eta}_{SE}$ and $\overline{\eta}_{EE}$. Impacts of the SBS density ratio ρ and SINR threshold β on the average link SE $\overline{\eta}_{SE}$ and average EE $\overline{\eta}_{EE}$ are further simulated, which are shown in Figures 7–9. From Figure 7, we can see that the average link SE $\overline{\eta}_{SE}$ first increases and finally saturates with the increase of SINR threshold β, while the SBS density ratio ρ has little effect on $\overline{\eta}_{SE}$. Combining with results shown in Figure 6, we can conclude that a UE in the macrocell tier experiences a higher SINR, while in the microcell tier, a UE usually obtains a higher throughput. With the parameter τ listed in Table 1, we can obtain the minimum value of β as $\beta_{\min} \approx -8.97$ dB. From Figure 8, we can observe that the average EE $\overline{\eta}_{EE}$ first increases and then decreases with the increase of SBS density ratio ρ. It can be interpreted that the power consumption caused by the increment of SBSs surpasses the gain they bring to the throughput, which indicates that the SBS density ratio ρ should be carefully designed. From Figure 9, we can observe that there exists an optimum β to reach the maximum value of average EE for a given ρ. According to Figure 8 and considering the requirements of $\overline{\eta}_{SE}$, C_m, and C_s, we can obtain optimum values of the SBS density ratio and SINR threshold as $\rho \approx 0.08$ and $\beta \approx -0.69$ dB, respectively, to make the two-tier ultradense HCN most energy-efficient.

From above simulation results, we can find that $\beta > -0.69$ dB makes the average EE higher. However, it makes the throughput of UEs served by the MBS too low to bear. Besides, it is clear that further increasing ρ or β would not benefit the average EE. In order to further improve the average EE while meeting the minimum requirements of the average link SE and average downlink throughput in the two-tier ultradense HCN, more techniques like interference management and base station cooperation need to be incorporated, which will be conducted in our future work.

5. Conclusions

In this paper, impacts of the SBS density ratio ρ and SINR threshold β on the performance of a two-tier ultradense HCN have been studied by using the stochastic geometry tool. In addition to theoretically deriving expressions of the average link SE, average downlink throughput, and average EE, an optimization problem was formulated to maximize the average EE while meeting the average link SE and average downlink throughput requirements. Simulation results indicate that there exists an optimum value of ρ or β to maximize the average EE. But if one would like to achieve a high average EE while guaranteeing the performance of the two-tier ultradense HCN, such as meeting minimum requirements of the average link SE and average downlink throughput, both ρ and β should be carefully designed. From simulation results, we can observe that, with parameter settings listed in Table 1, the SBS density ratio as $\rho \approx 0.08$ and the SINR threshold as $\beta \approx -0.69$ dB can make the two-tier ultradense HCN most energy-efficient. In brief, this work is capable of providing theoretical insights for the energy-efficient planning of ultradense HCNs.

Acknowledgments

This work was supported by the National Natural Science Foundation of China (61671165 and 61471135), the Guangxi Natural Science Foundation (2015GXNSFBB139007 and 2016GXNSFGA380009), the Fund of Key Laboratory of Cognitive Radio and Information Processing (Guilin University of Electronic Technology), Ministry of Education, China, the Guangxi Key Laboratory of Wireless Wideband Communication and Signal Processing (CRKL160105 and CRKL170101), and the Innovation Project of GUET Graduate Education (2016YJCX91 and 2017YJCX27).

References

[1] A. Damnjanovic, J. Montojo, Y. Wei et al., "A survey on 3GPP heterogeneous networks," *IEEE Wireless Communications*, vol. 18, no. 3, pp. 10–21, 2011.

[2] D. Lopez-Perez, I. Guvenc, G. de la Roche, M. Kountouris, T. Q. S. Quek, and J. Zhang, "Enhanced intercell interference coordination challenges in heterogeneous networks," *IEEE Wireless Communications*, vol. 18, no. 3, pp. 22–30, 2011.

[3] A. Ghosh, N. Mangalvedhe, R. Ratasuk et al., "Heterogeneous cellular networks: from theory to practice," *IEEE Communications Magazine*, vol. 50, no. 6, pp. 54–64, 2012.

[4] W. Tan, M. Matthaiou, S. Jin, and X. Li, "Spectral efficiency of DFT-based processing hybrid architectures in massive MIMO," *IEEE Wireless Communications Letters*, vol. 6, no. 5, pp. 586–589, 2017.

[5] X. Ge, S. Tu, G. Mao, C. Wang, and T. Han, "5G ultra-dense cellular networks," *IEEE Wireless Communications*, vol. 23, no. 1, pp. 72–79, 2016.

[6] N. Bhushan, J. Li, D. Malladi et al., "Network densification: the dominant theme for wireless evolution into 5G," *IEEE Communications Magazine*, vol. 52, no. 2, pp. 82–89, 2014.

[7] I. Hwang, B. Song, and S. S. Soliman, "A holistic view on hyper-dense heterogeneous and small cell networks," *IEEE Communications Magazine*, vol. 51, no. 6, pp. 20–27, 2013.

[8] Z. Wang, B. Hu, X. Wang, and S. Chen, "Interference pricing in 5G ultra-dense small cell networks: a Stackelberg game approach," *IET Communications*, vol. 10, no. 15, pp. 1865–1872, 2016.

[9] J. Liu and S. Sun, "Energy efficiency analysis of cache-enabled cooperative dense small cell networks," *IET Communications*, vol. 11, no. 4, pp. 477–482, 2017.

[10] S. F. Yunas, M. Valkama, and J. Niemela, "Spectral and energy efficiency of ultra-dense networks under different deployment strategies," *IEEE Communications Magazine*, vol. 53, no. 1, pp. 90–100, 2015.

[11] C. Yang, J. Li, Q. Ni, A. Anpalagan, and M. Guizani, "Interference-aware energy efficiency maximization in 5G ultra-dense networks," *IEEE Transactions on Communications*, vol. 65, no. 2, pp. 728–739, 2017.

[12] H. ElSawy, E. Hossain, and M. Haenggi, "Stochastic geometry for modeling, analysis, and design of multi-tier and cognitive cellular wireless networks: a survey," *IEEE Communications Surveys & Tutorials*, vol. 15, no. 3, pp. 996–1019, 2013.

[13] H. S. Dhillon, R. K. Ganti, F. Baccelli, and J. G. Andrews, "Modeling and analysis of K-tier downlink heterogeneous cellular networks," *IEEE Journal on Selected Areas in Communications*, vol. 30, no. 3, pp. 550–560, 2012.

[14] H. Jo, Y. J. Sang, P. Xia, and J. G. Andrews, "Heterogeneous cellular networks with flexible cell association: a comprehensive downlink SINR analysis," *IEEE Transactions on Wireless Communications*, vol. 11, no. 10, pp. 3484–3495, 2012.

[15] M. Di Renzo, A. Guidotti, and G. E. Corazza, "Average rate of downlink heterogeneous cellular networks over generalized fading channels: a stochastic geometry approach," *IEEE Transactions on Communications*, vol. 61, no. 7, pp. 3050–3071, 2013.

[16] S. Singh, H. S. Dhillon, and J. G. Andrews, "Offloading in heterogeneous networks: modeling, analysis, and design insights," *IEEE Transactions on Wireless Communications*, vol. 12, no. 5, pp. 2484–2497, 2013.

[17] H. S. Dhillon and J. G. Andrews, "Downlink rate distribution in heterogeneous cellular networks under generalized cell selection," *IEEE Wireless Communications Letters*, vol. 3, no. 1, pp. 42–45, 2014.

[18] M. Afshang and H. S. Dhillon, "Fundamentals of modeling finite wireless networks using binomial point process," *IEEE Transactions on Wireless Communications*, vol. 16, no. 5, pp. 3355–3370, 2017.

[19] Y. Li, F. Baccelli, J. G. Andrews, T. D. Novlan, and J. Zhang, "Modeling and analyzing the coexistence of Wi-Fi and LTE in unlicensed spectrum," *IEEE Transactions on Wireless Communications*, vol. 15, no. 9, pp. 6310–6326, 2016.

[20] C. Saha, M. Afshang, and H. S. Dhillon, "3GPP-inspired HetNet model using Poisson cluster process: sum-product functionals and downlink coverage," *IEEE Transactions on Communications*, no. 99, p. 1, 2017.

[21] P. Lin, J. Zhang, Y. Chen, and Q. Zhang, "Macro-femto heterogeneous network deployment and management: from business models to technical solutions," *IEEE Wireless Communications*, vol. 18, no. 3, pp. 64–70, 2011.

[22] J. G. Andrews, H. Claussen, M. Dohler, S. Rangan, and M. C. Reed, "Femtocells: past, present, and future," *IEEE Journal on Selected Areas in Communications*, vol. 30, no. 3, pp. 497–508, 2012.

[23] S. Cheng, S. Lien, F. Chu, and K. Chen, "On exploiting cognitive radio to mitigate interference in macro/femto heterogeneous networks," *IEEE Wireless Communications*, vol. 18, no. 3, pp. 40–47, 2011.

[24] N. Saquib, E. Hossain, L. B. Le, and D. I. Kim, "Interference management in OFDMA femtocell networks: issues and approaches," *IEEE Wireless Communications*, vol. 19, no. 3, pp. 86–95, 2012.

[25] V. Chandrasekhar and J. G. Andrews, "Spectrum allocation in tiered cellular networks," *IEEE Transactions on Communications*, vol. 57, no. 10, pp. 3059–3068, 2009.

[26] W. C. Cheung, T. Q. S. Quek, and M. Kountouris, "Throughput optimization, spectrum allocation, and access control in two-tier femtocell networks," *IEEE Journal on Selected Areas in Communications*, vol. 30, no. 3, pp. 561–574, 2012.

[27] T. Zhang, J. Zhao, L. An, and D. Liu, "Energy efficiency of base station deployment in ultra dense HetNets: a stochastic geometry analysis," *IEEE Wireless Communications Letters*, vol. 5, no. 2, pp. 184–187, 2016.

[28] D. Cao, S. Zhou, and Z. Niu, "Optimal combination of base station densities for energy-efficient two-tier heterogeneous cellular networks," *IEEE Transactions on Wireless Communications*, vol. 12, no. 9, pp. 4350–4362, 2013.

[29] Z. Li, D. Grace, and P. Mitchell, "Traffic-aware cell management for green ultradense small-cell networks," *IEEE Transactions on Vehicular Technology*, vol. 66, no. 3, pp. 2600–2614, 2017.

[30] L. Xiang, X. Ge, C. Wang, F. Y. Li, and F. Reichert, "Energy efficiency evaluation of cellular networks based on spatial distributions of traffic load and power consumption," *IEEE Transactions on Wireless Communications*, vol. 12, no. 3, pp. 961–973, 2013.

[31] C. Yang, J. Li, and M. Guizani, "Cooperation for spectral and energy efficiency in ultra-dense small cell networks," *IEEE Wireless Communications*, vol. 23, no. 1, pp. 64–71, 2016.

[32] Y. S. Soh, T. Q. S. Quek, M. Kountouris, and H. Shin, "Energy efficient heterogeneous cellular networks," *IEEE Journal on Selected Areas in Communications*, vol. 31, no. 5, pp. 840–850, 2013.

[33] H. Chen, W. Chen, F. Zhao, L. Fan, and H. Zhang, "Stochastic geometry analysis of downlink energy efficiency for a relay deployment scheme in relay-assisted cellular networks," *Telecommunication Systems*, vol. 63, no. 2, pp. 263–273, 2016.

[34] J. Peng, P. Hong, and K. Xue, "Energy-aware cellular deployment strategy under coverage performance constraints," *IEEE Transactions on Wireless Communications*, vol. 14, no. 1, pp. 69–80, 2015.

[35] J. G. Andrews, F. Baccelli, and R. K. Ganti, "A tractable approach to coverage and rate in cellular networks," *IEEE Transactions on Communications*, vol. 59, no. 11, pp. 3122–3134, 2011.

Optimal Pricing for MDS-Coded Caching in Wireless D2D Networks

Tao Zhang,[1] **Lin Xiao** ⓘ**,**[1] **Dingcheng Yang** ⓘ**,**[1] **and Laurie Cuthbert**[2]

[1]*Information Engineering School, Nanchang University, Nanchang, China*
[2]*Information Systems Research Centre, Macau Polytechnic Institute, Rua de Luis Gonzaga Gomes, Macau*

Correspondence should be addressed to Lin Xiao; xiaolin@ncu.edu.cn

Academic Editor: Stefania Sardellitti

In this paper, we investigate the caching of files in mobile devices in a wireless D2D network using maximum distance separable (MDS) codes. The coded symbols of each file are stored in the mobile devices. To regulate the D2D communications and reduce the transmission cost among mobile devices, a price mechanism is used for D2D communications so that they are allowed to choose whether to transmit the requested coded symbol to the other device or not. The mobile device can cooperate with others in terms of the battery level and the reward that is given from the other device. If the mobile device fails to retrieve enough coded symbols, the BS will respond with the missing coded symbols. We derived the optimal payment after taking the battery level of devices into consideration, and the problem was formulated as an optimization problem, subject to a range of payment. Numerical results demonstrate that using the MDS-coded scheme can significantly reduce the cost of transmitting a file, and there exist an optimal number of coded symbols cached in the mobile device, which consumes minimal cost of transmitting a requested file.

1. Introduction

Mobile data traffic has increased dramatically in recent years [1]. In order to address this enormous growth of traffic, one of the most promising methods is wireless caching, that is, storing the files in the user devices [2] or other helper stations [3]. Devices and helper stations can store files during the off-peak hours and transmit the contents to the requesters during peak hours. In particular, helper stations, like small cells, can provide a better quality of experience for users, although the necessity of high-speed optical fiber backhaul makes this scheme significantly expensive. It is noteworthy that device-to-device (D2D) wireless communication is expected to play an indispensable role in the near future [4]. With the introduction of D2D communication, a device can communicate directly with other mobile devices in its vicinity. When a device sends a request for a file that cannot be satisfied by its local storage, any of its neighbors with that content precached in their local storage can respond to the request if they can establish a link. If, as is possible, the requested data cannot be retrieved by D2D

communication, then the BS will assist in providing the data. This is expected to improve the spectral efficiency, increase the network throughput, and reduce the power consumption for the whole system [5]. Unlike other network systems where the content is only stored in the infrastructure, the key property of D2D-caching networks is that the virtual cache capacity increases linearly with the number of devices in the wireless networks. This means that the aggregate cache capacity grows with the increase of the number of devices in the system.

However, in a practical scenario, the mobile devices may not always respond to other devices' requests even if they have stored the requested file. This is because the mobile device has to consider its own interest [6]. Although the power consumption of D2D communications is lower than that of cellular communication, we cannot expect mobile devices to use most of their energy on D2D communications. The battery level of mobile devices is an important factor related to be considered with D2D communications among devices: the limited battery capacity of the mobile device, necessitating frequent charging, has become the biggest complaint for users of smartphones. Hence, reducing the energy consumption of D2D communication is crucial.

Coded caching is one of the effective methods for energy-saving in wireless caching networks since it can significantly reduce the amount of data that is necessary to transmit over the channel [7, 8]. A promising coded scheme was maximum distance separable (MDS) code, which was first investigated in [9] where the authors expressed the wireless distributed caching optimization problem as a convex optimization. Then, the MDS-coded scheme was investigated for small cells, and the authors proposed a careful caching placement to minimize the backhaul in the system [10]. In [8], the download delay of wireless content delivery was studied with the use of the MDS-coded scheme, and it was shown that this scheme can reduce the delay greatly. However, in those works, the battery condition of the mobile devices was not considered; adding this consideration is key in this paper.

There is lack of a proper mechanism to regulate the caching devices whether to establish D2D links with other devices or not. For most of the existing studies about caching in D2D networks, it is assumed that if a mobile device sends a request for a file, other devices in the neighborhood will establish D2D links and deliver the content. In reality, this can hardly be possible. Each device in a wireless network will have a different battery level, and the previous works assume that even when a device is in a low battery condition, it is still likely that this device will help to deliver the requested file to others. This is obviously unreasonable because the power available to that device could easily be used up, which would be a failure of D2D communication. Under this circumstance, the battery level is clearly an important factor that needs to be considered when proposing a proper mechanism for the mobile device establishing D2D links. In [11], the authors took the battery levels of devices into consideration and proposed a price strategy to motivate the load sharing between mobile devices. Results showed that this strategy increased the average battery levels and reduced the communication outages. Chen et al. [12] investigated the relationship between the energy cost and the offloading ratio with the consideration of battery levels and proposed a user-centric protocol to control the transmission cost for D2D communications. Chen et al. [13] proposed an incentive mechanism to maximize the utility of D2D communications and minimize the cost of the BS; the authors formulated the conflict between mobile devices and BS as a Stackelberg game. Different from these works, we study the coded caching with a new consideration of battery levels in D2D communications.

The main contribution of this paper is summarized as follows:

(i) We focus on the caching of files in wireless D2D networks with the MDS-coded scheme and using a price mechanism to regulate the D2D communications with the consideration of battery levels.

(ii) We propose a file transmission protocol in the wireless D2D networks. Then, we derive the cost of transmitting a requested file in the MDS-coded

scheme and optimize the payment to the cache-helper device.

(iii) We investigate the performance of the MDS-coded scheme by measuring the effect of some key parameters. Then, we compare our approach with the uncoded scheme and no-caching scheme, and the results show that the MDS-coded scheme is better at power-saving.

The remainder of this paper is organized as follows: in Section 2, the MDS-coded scheme and the price model are introduced. The problem of minimizing the cost of transmitting a requested file was formulated with constraints in Section 3. Section 4 gives the numerical simulations to prove the advantages of the MDS-coded scheme with a price mechanism. Finally, conclusions are drawn in Section 5.

2. System Model

2.1. Network Model. As shown in Figure 1, we consider one cell in cellular networks where the BS serves a number of M mobile devices denoted by the set $M = \{1, 2, \ldots, M\}$. We assume that the location of mobile devices follows a homogeneous Poisson point process (PPP) with density λ. The mobile devices caching the coded symbols form a PPP with density $\rho\lambda$, which are referred to as storage nodes. The devices not caching any file form another PPP with density $(1 - \rho)\lambda$, which are referred to as regular nodes. Each node can send a request for the file, and each storage node can communicate and share contents with all its neighbors via D2D links.

The device sending requests for contents is referred to as the source device $i \in M$. When the source device sends a request for a file, the source device i can associate with other storage nodes in the range of D2D communications. We denote the set of storage devices with distance r from the source device i as its helping set $\mathcal{H}_i \subset M$. In the helping set, the storage nodes that are willing to help transmit the coded symbols to the source device i are referred to as the cache-helper devices.

The channel coefficient from the source device i to the cache-helper device $j \in \mathcal{H}_i$ and to the BS is denoted as h_{ij} and h_i, respectively, and the corresponding distances are denoted as r_{ij} and r_i. The channel coefficient follows the zero mean complex Gaussian distribution with unit variance. Then, the channel power gain g_i, $i \in I, I = \{ij, i\}$ is expressed as follows:

$$g_i = |h_i|^2 = n_i G_i, \tag{1}$$

where n_i is an exponential random variable with unit mean modeling the power envelope of the Rayleigh fading, and we denote

$$G_i = \begin{cases} n_i G_0 \left(\dfrac{r_i}{r_0}\right)^{-\alpha}, & r_k > r_0 \\ n_i G_0, & \text{otherwise}, \end{cases} \tag{2}$$

as the power attenuation at the distance r_k, where α is the path loss exponent and r_0 is the constant path loss at the

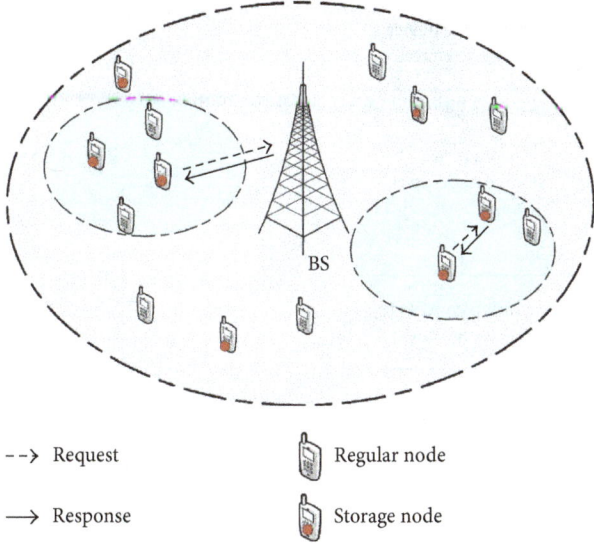

--> Request   Regular node

⟶ Response   Storage node

FIGURE 1: System model.

reference distance. Normally, D2D communications are of short range, so we can assume that

$$
G_i = \begin{cases} n_i G_0 \left(\dfrac{r_i}{r_0} \right)^{-\alpha}, & \text{BS communications} \\[2mm] n_i G_0, & \text{D2D communications.} \end{cases} \tag{3}
$$

For simplicity, a time-slotted system is considered in the system model, where coded symbols are transmitted in each time slot. For convenience, the number of coded symbols transmitted by a cache-helper device is normalized to unity in each time slot. When a mobile device initiates its data transmission, the coded symbols from the set $\{1, 2, \ldots, 2^{D_i}\}$ are sent, where D_i is the transmission rate in bits/sec/Hz. Thus, for a given transmission rate per coded symbol, the required energy for transmission with data rate D_i is

$$
E_i = \frac{\sigma^2}{g_i} \left(2^{D_i} - 1 \right), \tag{4}
$$

where σ^2 is the noise power at the receiver of the mobile device. When the cache-helper device needs to transmit more coded symbols, the transmission rate is larger, which consumes more energy.

2.2. Cache Model. The BS has access to a library of N files $\mathscr{L} = \{f_1, f_2, \ldots, f_N\}$. It is assumed that all contents have equal size B, and these contents are requested by mobile devices with the same probability. In the system, we adopt a deterministic allocation strategy, where contents are cached in the distributed storage nodes. Each content is partitioned into b fragments, and these fragments are used to create a coded symbols using the (a, b) MDS coding, like the Reed-Solomon (RS) codes [15]. The rate of the MDS erasure correcting code is denoted as $R = b/a$, and we compare the performance of different rates of the MDS coding in the paper. To simply the analysis, it is assumed that each mobile device caches coded symbols for each of the contents in $\mathscr{L}$,

and the coded symbols cached in each storage nodes are different. Overall, the total number of coded symbols cached at each storage device is $n_{cs} * N$, where $n_{cs} = a/m$ is the number of coded symbols for one content cached in the storage nodes. It is noted that a fragment is an uncoded part of a file and a coded symbol is a coded part of a file.

With the MDS coding, the response to a request from the mobile device can be via multiple D2D communications, as long as there are storage devices that are in the coverage of the source device willing to accept the payment and help to transmit the coded symbols. In practice, when the source device i has collected any subset of b coded packets, the source device i can recover the requested file with the MDS code (a, b). In order to make the system more efficient, we allow multiple D2D links to coexist in the cell if these D2D links are far apart in the space. Hence, the interference in this case can be neglected. A similar model is adopted in [16]. We also assume that when there are multiple D2D links from cache-helper devices in the coverage of the source device, the coded symbols of the requested file will be downloaded serially from the cache-helper devices to the source device.

In the off-peak time, the BS will cache the coded symbols of files in the storage nodes. Since the content allocation is executed by the BS, the BS has the information of coded symbols in the mobile devices. In addition, cache placement happens over a much larger timescale than scheduling and transmission. We assume that the static state of the content caching and the changes of cache updating over time are not considered for simplicity (similar assumptions have been made in previous literature, e.g., [3]).

2.3. Price Model. Each device's battery level is different from the others, and normally, the battery condition cannot be shared between mobile devices. Moreover, the mobile device has different valuations of the remaining energy in its battery when its battery is in different states. For example, the energy in a device is more valuable when the battery is insufficient. We assume that transmitting a coded symbol of the content consumes the same energy for all mobile devices.

We define ξ_j as the cost for the storage node $j \in \mathscr{H}_i$ transmitting a coded symbol to the source device i via D2D communications. The cost ξ_j is related to the battery level of the storage node, and the cost for each storage node is the function of the battery level B_j, that is,

$$
\xi_j = f(B_j), \tag{5}
$$

where $B_j \in [0, B_{\max}]$ is the battery level and $f : [0, B_{\max}] \to [0, \xi_{\max}]$ is a monotonically decreasing function of B_j. For simplicity, it is assumed that the maximum battery storage $B_{\max}$ is the same for each mobile device. If a storage node $j \in \mathscr{H}_i$ responds to the file request from the source device i, it will receive a price π_i for transmitting the requested content.

Due to advanced payment technologies, the reward can be in various forms, like currency or credit in a multimedia application. For the cache-helper device $j \in \mathscr{H}_i$, the benefit of being a cache-helper device is $\pi_i - \xi_j^{\text{D2D}} E_j^{\text{D2D}}$, where E_j^{D2D} is the energy consumed by the D2D communications for a cache-helper device. Furthermore, the cache-helper device has a reservation utility $\varepsilon > 0$, which means that the storage node will accept the request from the source device. Therefore, the utility of the cache-helper device $j \in \mathcal{H}_i$ for the source device i can be defined as

$$U_j = \begin{cases} \pi_i - \xi_j^{\text{D2D}} E_j^{\text{D2D}}, & \text{if } \pi_i - \xi_j^{\text{D2D}} E_j^{\text{D2D}} > \varepsilon \\ 0, & \text{otherwise.} \end{cases} \quad (6)$$

For the source device i, if there is a storage node accepting the price π_i, then the cost of the source device i is π_i. If not, the source device's request will be handled by the BS directly at the cost of $\xi^{\text{BS}} E^{\text{BS}}$, where ξ^{BS} is the cost of the BS transmitting the coded symbols to the source device. Hence, the cost of a source device i is

$$C_i = \begin{cases} \pi_i, & \pi_i - \xi_j^{\text{D2D}} E_j^{\text{D2D}} \geq \varepsilon \\ \xi^{\text{BS}} E^{\text{BS}}, & \text{otherwise.} \end{cases} \quad (7)$$

In order to ensure mutual benefit for both the source device and the cache-helper device in the collaboration, the price π_i should satisfy the following inequality:

$$\varepsilon \overset{(a)}{\leq} \pi_i \overset{(b)}{\leq} \xi^{\text{BS}} E^{\text{BS}} - \xi_j^{\text{D2D}} E_j^{\text{D2D}}, \quad (8)$$

where inequality (a) ensures the benefit that the cache-helper device deserves and inequality (b) makes sure that the D2D communication is more power-saving than the BS communication.

2.4. Transmission Protocol. With a given request from the source device, we can specify the content transmission protocol via D2D communications and BS communications. If a source device sends a request for a file, the protocol for obtaining the file is given as shown in Figure 2.

Response from the D2D communications is as follows:

(1) The source device will check if there are any storage nodes deployed in its neighbors.

(2) The source device submits the payment to the storage nodes. The storage nodes (if any) accept the payment and send the number of n_{cs} coded symbols of the requested file to the source device.

(3) When there is more than one storage node willing to respond to the source device's request, the source device can download the coded symbols serially from multiple nodes.

Response from the BS communications is as follows:

(1) If the source node has collected fewer than the b coded symbols from cache-helper devices, the file cannot be reconstructed. Hence, the source node will turn to the BS to finish the transmission of the requested file.

(2) If the source device cannot find any storage nodes that can help to transmit the requested file in its coverage of the D2D communications, the BS will respond to the source device's request directly.

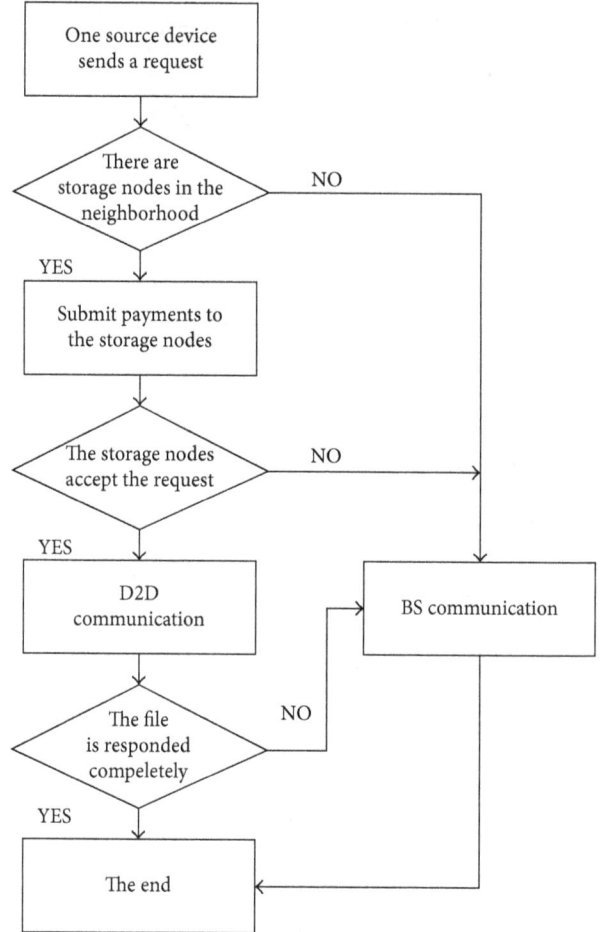

FIGURE 2: File transmission protocol.

3. Problem Formulation and Solutions

In this paper, in order to make the whole system more power-saving, our goal is to find the optimal payment for transmitting the coded symbols for contents in different (a, b) MDS-coded schemes. In this section, we first study the probability of storage nodes available when a request arrives from the source device. Then, we analyze the expected amount of data transmitted by D2D communications and by BS communications in the system. We formulate the price mechanism with the MDS-coded scheme in wireless D2D networks as an optimization problem.

Here, we study the probability that there are several storage nodes available when a request arrives. For later use, we denote $\mathcal{J}_k$ as the event that there are k storage nodes available in the coverage of the source device at the time of a request sent by a source device. We define $\mathcal{C}$ and $\bar{\mathcal{C}}$ as the events that the request is from a regular node and from a storage node. As we mentioned above, coded symbols of the contents in $\mathcal{X}$ are cached in the storage nodes. Therefore, the request for these contents can be responded via D2D communications.

Proposition 1: The probability that there are k mobile devices caching the coded symbols available in the coverage of the source device when the source device requests for a file from $\mathcal{X}$ is

$$q_k = \Pr\left(\mathcal{F}_k\right) = \left(\frac{M-m}{M} + \frac{k}{\pi r^2 \lambda}\frac{m}{M}\right)\frac{\left(\pi r^2 \lambda\right)^k}{k!}e^{-\pi r^2 \rho \lambda}, \quad k \geq 0.$$

(9)

Proof: The probability q_i is calculated by

$$q_k = \Pr\left(\mathcal{F}_k\right) = \Pr\left(\mathcal{F}_k \mid \bar{\mathscr{C}}\right)\Pr(\bar{\mathscr{C}}) + \Pr\left(\mathcal{F}_k \mid \mathscr{C}\right)\Pr(\mathscr{C}).$$

(10)

The number of storage nodes and regular nodes is m and $M - m$, respectively. The probability of a request coming from storage nodes and regular nodes is expressed as follows:

$$\Pr(\mathscr{C}) = \frac{m}{M}$$
$$\Pr(\bar{\mathscr{C}}) = \frac{M-m}{M}.$$

(11)

When a regular device sends a request for a file, the probability that having $k \geq 0$ storage nodes that have cached coded symbols of the content within the distance r is

$$F(k, r, \rho\lambda) = \Pr\left(\mathcal{F}_k \mid \bar{\mathscr{C}}\right) = \frac{\left(\pi r^2 \rho\lambda\right)^k}{k!}e^{-\pi r^2 \rho\lambda}.$$

(12)

Similarly, when a storage node sends a request, the probability of having k storage nodes in the radius r is

$$F(k-1, r, \rho\lambda) = \Pr\left(\mathcal{F}_k \mid \mathscr{C}\right) = \frac{\left(\pi r^2 \rho\lambda\right)^{k-1}}{(k-1)!}e^{-\pi r^2 \rho\lambda}.$$

(13)

Using (11)–(13), we get (9). Then, we consider the amount of data that is downloaded from the BS and cache-helper devices, respectively. For simplicity, we denote $\mathscr{A}_i$ as the event that the probability of the storage node $j \in \mathscr{H}_i$ accepts the payment π_i and helps the source device i to transmit the requested file. $\Pr(\mathscr{A}_i)$ is the probability that the storage node $j \in \mathscr{H}_i$ accepts the payment π_i and helps the source device i to transmit the requested file, which is expressed as

$$\Pr(\mathscr{A}_i) = \Pr\left(\pi_i - \xi_j^{\text{D2D}}E_j^{\text{D2D}} \geq \varepsilon\right).$$

(14)

Obviously, the storage node will help the source device to transmit the requested coded symbols only when the storage node can obtain benefit via the transmission for coded symbols. Provided that there are $k > 0$ storage nodes in the radius r of the source device i and each storage node can transmit the number of n_{cs} coded symbols, the amount of the coded symbols that are downloaded from the number of k storage nodes is denoted as

$$\gamma_i^{\text{D2D}} = \begin{cases} k \cdot \Pr(\mathscr{A}_i) \cdot n_{\text{cs}}, & \text{if } 0 < k \cdot \Pr(\mathscr{A}_i) \cdot n_{\text{cs}} < b \\ b, & \text{if } b \leq k \cdot \Pr(\mathscr{A}_i) \cdot n_{\text{cs}}. \end{cases}$$

(15)

Similarly, when a request comes from the source device i, the amount of the coded symbols that are downloaded from the BS is

$$\gamma_i^{\text{BS}} = \begin{cases} b - k \cdot \Pr(\mathscr{A}_i) \cdot n_{\text{cs}}, & \text{if } 0 < k \cdot \Pr(\mathscr{A}_i) \cdot n_{\text{cs}} < b \\ 0, & \text{if } b \leq k \cdot \Pr(\mathscr{A}_i) \cdot n_{\text{cs}}. \end{cases}$$

(16)

We formulate the optimization problem that minimizes the expected total cost of the source device i for receiving the requested file over the payment π_i as follows:

$$\min_{\pi_i} \sum_{k=1}^{m} q_k \cdot \left[\gamma_i^{\text{D2D}}\left(\pi_i + \xi_j^{\text{D2D}}E_j^{\text{D2D}}\right) + \gamma_i^{\text{BS}}\xi^{\text{BS}}E^{\text{BS}}\right]$$

(17)

$$\text{s.t. } \varepsilon \leq \pi_i \leq \xi^{\text{BS}}E^{\text{BS}} - \xi_j^{\text{D2D}}E_j^{\text{D2D}}.$$

The total cost includes the cost in D2D communications and BS communications. The constraint is the range of payment and has been explained in (8).

Next, we study the probability of the storage node $j \in \mathscr{H}_i$ willing to help the source device i to transmit the coded symbols. For simplicity, it is assumed that the relation between the content cost of a cache-helper device and its battery level B_k in (5) follows a linear function, and a similar assumption is adopted in [11]:

$$\xi_j = \xi_{\max}\left(1 - \frac{B_j}{B_{\max}}\right).$$

(18)

It is also assumed that the battery levels follow a uniform distribution. This means that the content cost ξ_j is uniformly distributed as

$$\xi_j \sim B_j \sim \mathscr{U}\left[0, B_{\max}\right].$$

(19)

Therefore, the probability of the cache-helper device $j \in \mathscr{H}_i$ helping the source device i transmit the requested file is

$$\begin{aligned} \Pr(\mathscr{A}_i) &= \Pr\left(\pi_i - \xi_j^{\text{D2D}}E_j^{\text{D2D}} \geq \varepsilon\right) \\ &= \Pr\left(\frac{\xi_j}{\eta_j} \leq \frac{G_0\left(\pi_i - \varepsilon\right)}{\sigma^2\left(2^{D_i} - 1\right)}\right) \\ &= \Pr\left(\eta_j \geq \frac{\xi_j}{w_i}\right) \\ &= \frac{1}{\xi_{\max}}\int_0^{\xi_{\max}}\int_{\xi_j/w_i}^{\infty} e^{-\eta_j}\,d\eta_j\,d\xi_j \\ &= \frac{w_i}{\xi_{\max}}\left(1 - e^{-\left(\xi_{\max}/w_i\right)}\right), \end{aligned}$$

(20)

where

$$w_i = \frac{G_0\left(\pi_i - \varepsilon\right)}{\sigma^2\left(2^{D_i} - 1\right)}.$$

(21)

The transmission rate D_i is expressed as

$$D_i = n_{\text{cs}}s_{\text{cs}},$$

(22)

where s_{cs} is the size of the coded symbol. With the result of (17) and (20), the objective function (23) can be expressed as

$$\min_{\pi_i} \sum_{k=1}^{m} q_k \cdot \left[\min\left(k \cdot \frac{w_i}{\xi_{\max}} \left(1 - e^{-(\xi_{\max}/w_i)} \right) \cdot n_{\mathrm{cs}}, b \right) \left(\pi_i + \xi_i^{\mathrm{D2D}} E_j^{\mathrm{D2D}} \right) + \max\left(b - k \cdot \frac{w_i}{\xi_{\max}} \left(1 - e^{-(\xi_{\max}/w_i)} \right) \cdot n_{\mathrm{cs}}, 0 \right) \xi^{\mathrm{BS}} E^{\mathrm{BS}} \right]$$

$$\text{s.t. } \varepsilon \le \pi_i \le \xi^{\mathrm{BS}} E^{\mathrm{BS}} - \xi_j^{\mathrm{D2D}} E_j^{\mathrm{D2D}}.$$

(23)

This problem is nonconvex as the first part of (23) is concave, but we can solve the problem by 1D search due to the fact that there is only one variable with a limited range.

4. Numerical Results

In this section, we evaluate the price of transmitting coded symbols and the cost of responding to a source device's request when using the MDS-coded scheme. The general simulation parameters are listed in Table 1, and the specific setup for different MDS-coded cases will be elaborated in the later section.

In Figure 3, we show the optimal price π_i for different numbers of coded symbols cached in the storage nodes when λ varies. We note that, for a given ξ^{BS} and the (a, b) MDS-coded scheme, the optimal price increases with the increase in the number of coded symbols. This is due to the fact that when more coded symbols of each file are cached in the device, it needs more energy to transmit the requested coded symbols to the source device. As shown in Figure 3, the optimal price of coded symbols is better when the value of λ is larger. This is reasonable because when there are more storage nodes, the probability that the source device receives help from storage devices is higher, which leads to a lower payment. We can also note that curves become steady when the number of symbols cached in the storage nodes is more than around 11. This is because when n_{cs} is large enough, the payment that a source device needs to pay to the cache-helper device will be sufficient if the source device wants to keep a good probability of receiving help from storage nodes. However, when the price becomes larger, the total cost of responding to the source device's request will be high, which is not optimal. Moreover, there is a limitation in the payment for transmitting coded symbols by D2D communications according to the expression in (8).

Figure 4 shows the optimal payment for different numbers of coded symbols cached in the storage nodes when ξ^{BS} varies. It is observed that when ξ^{BS} is higher, the price of coded symbols is higher. This is because the higher payment of coded symbols is expected to reach the optimal balance between the cost of transmitting coded symbols by D2D links and BS communications in this situation. Moreover, a larger ξ^{BS} gives a higher bound to the price. In Figure 5, we change different MDS-coded schemes in which the source device receives a different number of coded symbols to be enough to recover the requested file. It is shown that the payment of coded symbols is higher when the source device is required to receive more coded symbols to reconstruct the file. The reason is that it is more difficult to obtain the help

from the cache-helper device to transmit more coded symbols to the source device. Thus, the price is higher.

Figure 6 depicts the total cost of responding to the source device's request for a file when the price of coded symbols changes. It is noted that all curves decrease at the beginning and then start to increase until the end. This indicates that the number of coded symbols cached in the storage nodes has an effect on the cost of transmitting the requested file to the source device. When the number of coded symbols cached in the storage nodes is small, the source device is less likely to receive enough coded symbols by D2D communications. When n_{cs} is large, the probability of storage nodes willing to help to transmit the coded symbols is decreasing. Thus, the BS will respond to the devices' requests in a larger proportion in both cases, which results in a higher cost. It is also observed that the optimal number of coded symbols cached in the storage nodes is larger with the decrease of λ for curves. This is because when there are fewer storage nodes in the system, each storage node is expected to cache more coded symbols to help the source device get the requested file.

In Figure 7, we show the influence ξ^{BS} has on the cost of transmitting a requested file. It is observed that when the number of coded symbols cached in the storage nodes is less than 10, the curves are quite close, but for more symbols cached, the gap between different curves widens. This is because the price of coded symbols diverges, as can be seen from Figure 4. Hence, the cost of transmitting the coded file by D2D communications ($n_{\mathrm{cs}} : 10 \rightarrow 15$) varies notably, and the gap between curves is clear. In Figure 8, it shows when using different MDS coded schemes (the number of coded symbols that are required to recover the file is different), there will be different transmission costs. We can see that when b is larger, the cost of transmitting a file is higher, with the reason being explained above.

For a better comparison, we show the cost of responding to a source device's request in different schemes. In the optimal scheme, the information about the mobile devices' battery condition is shared to all; hence, the source device can select the storage nodes that have a high battery level as the cache-helper device. In the uncoded scheme, the file is cached in the mobile devices as a whole. In the no-caching scheme, there is no caching in the system, and the mobile devices' requests are responded by the BS. We set $\lambda = 0.03$, $b = 20$, and $\xi^{\mathrm{BS}} = 3$ in the MDS-coded scheme and the optimal scheme. As shown in Figure 9, the curve of the optimal scheme is close to the MDS-coded scheme. The uncoded scheme is above the MDS-coded scheme, which means that the cost of transmitting a file consumes more energy cost. This indicates

TABLE 1: General simulation setup.

Simulation Parameters	
Range of D2D communications	$r = 10\,\mathrm{m}$
Device reservation utility	$\varepsilon = 0.2$
Maximum battery level	$B_{\max} = 100\,\mathrm{J}$
Maximum D2D communication cost	$\xi_{\max} = 1$
Size of the file	$B = 75\,\mathrm{MB}$
Number of mobile devices	$M = 200$
Percentage of storage nodes	$\rho = 0.5$
Number of files in the catalog	$N = 500$
Noise power	$\sigma^2 = 110\,\mathrm{dBm}$
Path-loss exponent	$\alpha = 3.6$

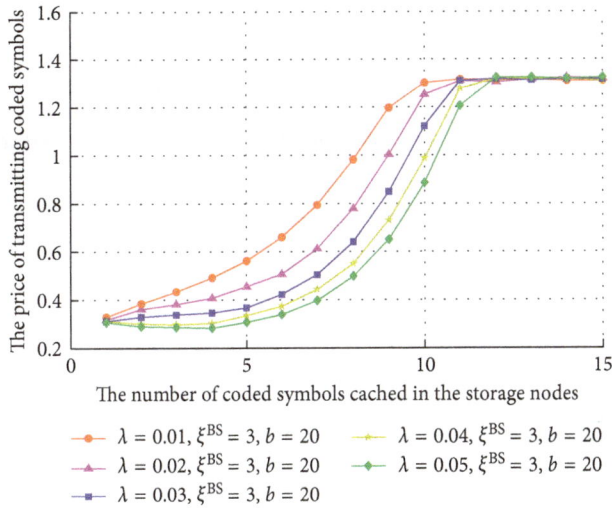

FIGURE 3: The optimal price of transmitting coded symbols, $\xi^{\mathrm{BS}} = 3$, and $b = 20$.

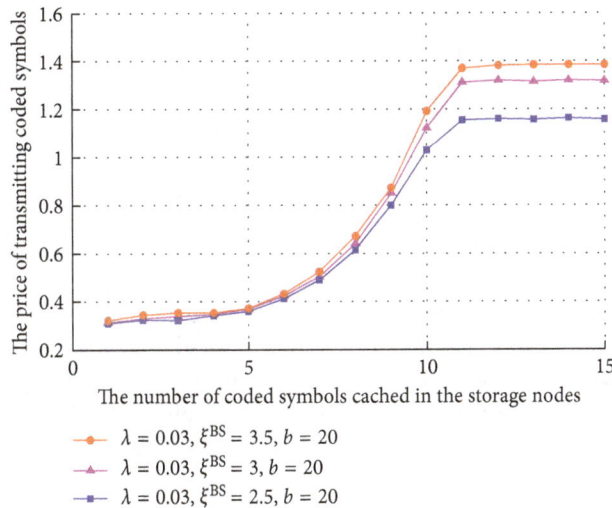

FIGURE 4: The optimal price of transmitting coded symbols, $\lambda = 0.03$, and $b = 20$.

that the MDS-coded scheme is better than the uncoded scheme and the no-caching scheme with the price mechanism.

At the beginning, the battery level of mobile devices is uniformly generated between the range $[0, B_{\max}]$. In order

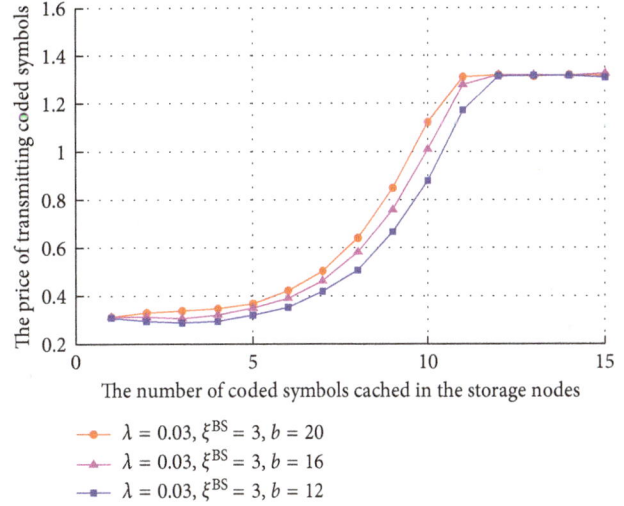

FIGURE 5: The optimal price of transmitting coded symbols, $\lambda = 0.03$, and $\xi^{\mathrm{BS}} = 3$.

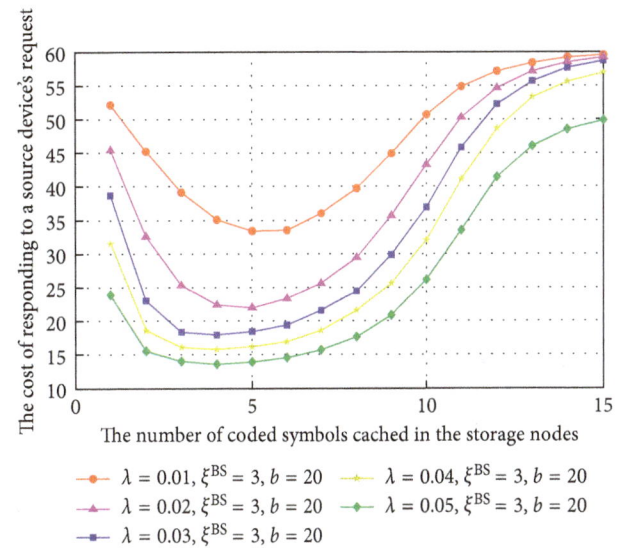

FIGURE 6: The cost of responding to a source device's request, $\xi^{\mathrm{BS}} = 3$, and $b = 20$.

to show the mobility of devices, the positions of mobile devices are regenerated within the area for each time slot. It is assumed that 10 requests are randomly generated by mobile devices in every time slot. In the setup, it is likely that there is more than one device that sends the request in a cluster at the same time slot. In order to avoid this situation, we regenerated the mobile devices' requests to avoid the interference caused by the D2D communication in a cluster. During a time slot of the system, if the battery level of the mobile device is 0, then the device is regarded as "dead" and it cannot provide any operation. In Figure 10, we can see that the curve of the MDS-coded scheme decreases faster than that of other schemes, which indicates that using the MDS-coded scheme can generate more D2D communications in the system. As for the optimal scheme, this curve is the same as the curve of the MDS-coded scheme. This is because

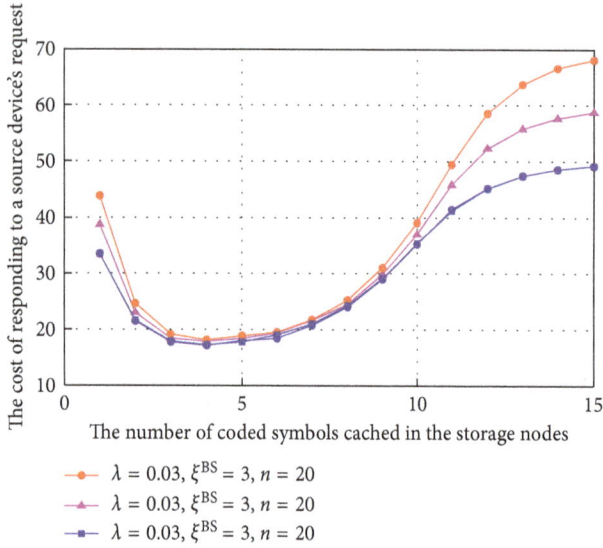

FIGURE 7: The cost of responding to a source device's request, $\lambda = 0.03$, and $b = 20$.

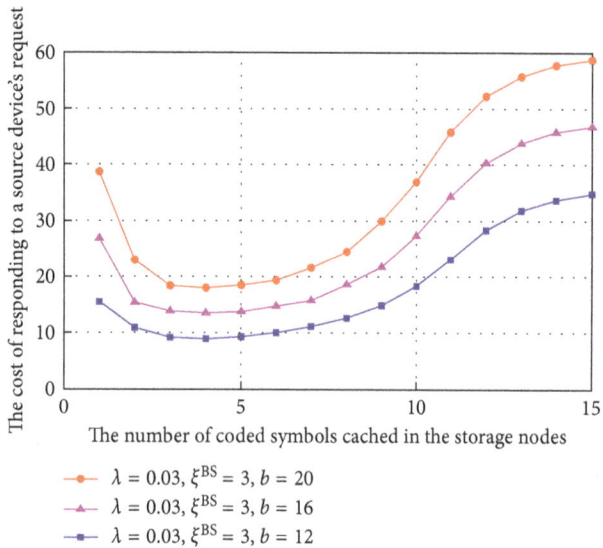

FIGURE 8: The cost of responding to a source device's request, $\lambda = 0.03$, and $\xi^{BS} = 3$.

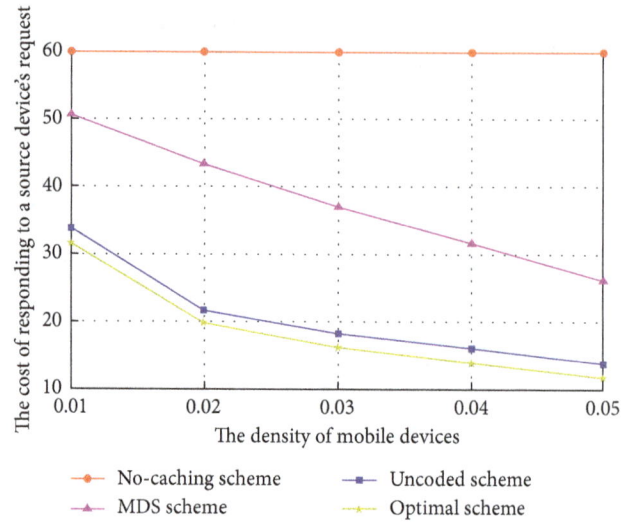

FIGURE 9: The cost of responding to a source device's request in different schemes.

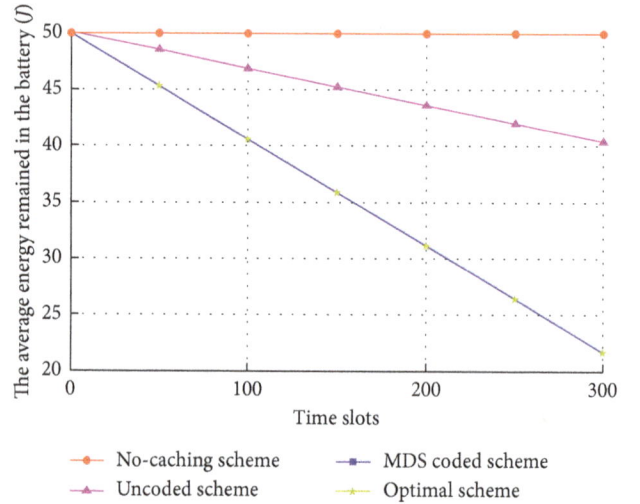

FIGURE 10: The average energy remained in the battery.

transmitting coded symbols consumes the same amount of energy for each storage node.

5. Conclusion

In this paper, we studied the problem of optimal payment for transmitting coded symbols from the cache-helper device to the source device with the MDS-coded scheme in wireless D2D networks. We derived the optimal payment after taking the battery level of devices into consideration, and the problem was formulated as a nonconvex problem. We then deeply investigated the performance of the optimal price and total cost in MDS-coded schemes by measuring the effect of the key parameters, such as the density of mobile devices and the number of coded symbols that are sufficient to recover

the requested file. With the optimal price of the MDS-coded scheme, it was shown that there existed an optimal number of coded symbols cached in the mobile devices, which can reduce the total cost of transmitting the requested file to the device. We also compared our scheme with other caching strategies, and numerical results showed that the MDS-coded scheme with a price mechanism could reduce the total cost of responding to a device's request for a file.

Acknowledgments

This work was supported in part by the National Natural Science Foundation of China (61703197, 61561032, 61461029, 41504026, and 61362009), the China/Jiangxi Postdoctoral Science Foundation Funded Project (2014MT561879 and 2014KY046), the Young Scientists Project Funding of Jiangxi Province (20162BCB23010 and 20153BCB23020), the Natural

Science Foundation of Jiangxi Province (20114ACE00200 and 20133ACB21007), and the Graduate Student Innovation Special Funds of Nanchang University (Grant no. cx2016265).

References

[1] CISCO, *Cisco Visual Networking Index: Global Mobile Data Traffic Forecast Update, 20152020 White Paper*, CISCO, San Jose, CA, USA, 2016.

[2] N. Golrezaei, M. Ji, A. F. Molisch, A. G. Dimakis, and G. Caire, "Device-to-device communications for wireless video delivery," in *Proceedings of the 46th Asilomar Conference on Signals, Systems, and Computers (ASILOMAR)*, pp. 930–933, Pacific Grove, CA, USA, November 2012.

[3] K. Shanmugam, N. Golrezaei, A. G. Dimakis, A. F. Molisch, and G. Caire, "FemtoCaching: wireless content delivery through distributed caching helpers," *IEEE Transactions on Information Theory*, vol. 59, no. 12, pp. 8402–8413, 2013.

[4] A. Asadi, Q. Wang, and V. Mancuso, "A survey on device-to-device communication in cellular networks," *IEEE Communications Surveys & Tutorials*, vol. 16, no. 4, pp. 1801–1819, 2014.

[5] Y. Chai, Q. Du, and P. Ren, "Partial time-frequency resource allocation for device-to-device communications underlaying cellular networks," in *Proceedings of the IEEE International Conference on Communications (ICC)*, Budapest, Hungary, June 2013.

[6] N. Golrezaei, A. F. Molisch, A. G. Dimakis, and G. Caire, "Femtocaching and device-to-device collaboration: a new architecture for wireless video distribution," *IEEE Communications Magazine*, vol. 51, no. 4, pp. 142–149, 2013.

[7] J. Llorca, A. M. Tulino, K. Guan, and D. C. Kilper, "Network-coded caching-aided multicast for efficient content delivery," in *Proceedings of the 2013 IEEE International Conference on Communications (ICC)*, pp. 3557–3562, Budapest, Hungary, June 2013.

[8] A. Piemontese and A. G. i Amat, "MDS-coded distributed storage for low delay wireless content delivery," in *Proceedings of the 2016 9th International Symposium on Turbo Codes and Iterative Information Processing (ISTC)*, pp. 320–324, Brest, France, September 2016.

[9] N. Golrezaei, K. Shanmugam, A. G. Dimakis, A. F. Molisch, and G. Caire, "Wireless video content delivery through coded distributed caching," in *Proceedings of the 2012 IEEE International Conference on Communications (ICC)*, pp. 2467–2472, Ottawa, ON, Canada, June 2012.

[10] V. Bioglio, F. Gabry, and I. Land, "Optimizing MDS codes for caching at the edge," in *Proceedings IEEE Global Communications Conference (GLOBECOM)*, San Diego, CA, USA, December 2015.

[11] Y. Guo, L. Duan, and R. Zhang, "Optimal pricing and load sharing for energy saving with cooperative communications," *IEEE Transactions on Wireless Communications*, vol. 15, no. 2, pp. 951–964, 2016.

[12] B. Chen, C. Yang, and A. F. Molisch, "Cache-enabled device-to-device communications: offloading gain and energy cost," *IEEE Transactions on Wireless Communications*, vol. 16, no. 7, pp. 4519–4536, 2017.

[13] Z. Chen, Y. Liu, B. Zhou, and M. Tao, "Caching incentive design in wireless D2D networks: a Stackelberg game approach," in *Proceedings of the 2016 IEEE International Conference on Communications (ICC)*, pp. 1–6, Kuala Lumpur, Malaysia, May 2016.

[14] L. Breslau, Pei Cao, Li Fan, G. Phillips, and S. Shenker, "Web caching and Zipf-like distributions: evidence and implications," in *Proceedings of Eighteenth Annual Joint Conference of the IEEE Computer and Communications Societies*, IEEE INFOCOM '99, vol. 1, pp. 126–134, New York, NY, USA, 1999.

[15] S. B. Wicker and V. K Bhargava, *Reed-Solomon Codes and Their Applications*, John Wiley & Sons, Hoboken, NJ, USA, 1999.

[16] N. Golrezaei, P. Mansourifard, A. Molisch, and A. Dimakis, "Base-station assisted device-to-device communications for high-throughput wireless video networks," *IEEE Transactions on Wireless Communications*, vol. 13, no. 7, pp. 3665–3676, 2014.

3

Energy-Neutral Communication Protocol for Living-Tree Bioenergy-Powered Wireless Sensor Network

Yin Wu ⓘ,[1] Bowen Li,[1] Yongjun Zhu,[2] and Wenbo Liu[2]

[1]*Department of Information Science and Technology, Nanjing Forestry University, No. 159, Long Pan Road, Nanjing 210037, China*
[2]*College of Automation, Nanjing University of Aeronautics and Astronautics, No. 29, Jiangjun Avenue, Nanjing 211106, China*

Correspondence should be addressed to Yin Wu; wuyin@njfu.edu.cn

Academic Editor: Carlos T. Calafate

The purpose of this paper is to represent a living-tree biological energy powered wireless sensor system and introduce a novel energy aware MAC protocol based on remaining energy level, energy harvesting status, and application requirements. Conventional wireless sensor network (WSN) cannot have an infinite lifetime without battery recharge or replacement. Energy harvesting (EH), from environmental energy sources, is a promising technology to provide sustainable powering for WSN. In this paper, a sensor network system has been developed which uses living-tree bioenergy as harvesting resource and super capacitor as energy storage. Moreover, by analyzing the power recharging, task arrangement, and energy consumption rate, a novel duty cycle-based energy-neutral MAC protocol is proposed. It dynamically optimizes each wireless sensor node's duty cycle to create a balanced, efficient, and continuous network. The scheme is implemented in a plant surface-mounted bioenergy power wireless sensor node system called PBN, which aims to monitoring the plant's growth parameters. The results show that the proposed MAC protocol can provide sustainable and reliable data transmission under ultralow and dynamic power inputs; it also significantly improves the latency and packet loss probability compared with other MAC protocols for EH-WSN.

1. Introduction

The limited available lifetime is a key bottleneck for most battery-powered wireless sensor networks (WSNs). Therefore, harvesting energy from the environment has been widely investigated to ensure the sustainability of WSN. As for this Energy Harvesting-WSN (EH-WSN), lots of studies have been carried out [1–3], the main research issue of energy consumption lies in two aspects: how to maximizing the harvested energy and how to maximizing the energy utilization efficiency, plus one research target: keep the EH-nodes stay in "energy neutral operation (ENO)" state [4]. Therefore, the hardware implementation and software communication protocols of EH-WSN are particularly attractive since they are just like the "body" and "nerve" of whole system. This paper proposes a novel MAC protocol with an energy harvesting prediction method to regulate the EH-node's work/sleep duty cycle based on the incoming power's changing status, residual energy level, and network

task requirements. The aim is to make the system continuously operational in practical monitoring situations by minimizing the energy wastage and increasing the energy efficiency.

Recently, there are several works in the literature in which the authors used the duty-cycling technique to optimize EH-WSN's performance [5–9]: Yoo et al. proposed two novel dynamic duty cycle scheduling schemes (called DSR and DSP) in order to reduce sleep latency, while achieving balanced energy consumption among sensor nodes in EH-WSN [5]; Le et al. designed an efficient wake-up variation reduction power manager (WVR-PM) for wireless nodes powered by periodic energy sources, an energy-efficient synchronized wake-up interval MAC protocol (SyWiM) has also been proposed to solve the timing offset and clock drift issues [6]; Liu et al. have proposed a load and energy balancing receiver-initiated duty cycle MAC protocol (LEB-MAC) [7], it outperformed RI-MAC and DSR protocols in following aspects: low receiver and sender duty cycle, high throughput, high fairness, and low end-to-end delay; Bouachir et al. presented the EAMP-AIDC protocol, an energy aware MAC protocol for EH-WSN based on individual duty cycle optimization [8]. It took into consideration nodes' residual energy and application and data requirements in order to define individual dynamic duty cycles so as to ensure continuous network operation; results showed that EAMP-AIDC protocol outperformed the IEEE 802.15.4 standard in terms of better energy consumption, increased survivability in energy savings, and in guaranteeing continuous operations. In particular, reference [9] proposed a duty cycle-based throughput optimal energy-neutral transmission work/sleep policy, then as well introduced a joint scheduling and routing protocol in multihop networks, aiming to provide network-wide packet communications under extremely limited node resources such as ultralow microbial fuel cell (MFC) power supply; results showed that it is able to provide sustainable and reliable data transmission under low and dynamic power inputs. With regard to EH-WSN of this specific kind, namely, using bioenergy as electrical power, researchers have also made quite a few contributions lately [10–13]: Zhang et al. compared two types of power management system (PMS) for MFC, charge pump capacitor converter type, and capacitor transformer converter type [10], They found that capacitor transformer converter type is recommended for ultralow MFC output and time-sensitive missions due to its wider input voltage range and shorter charging/discharging cycle; Erbay et al. proposed a PMS with dynamic maximum power point tracking capability; it could continuously detect the maximum power point (MPP) of the MFC and matches the load impedance of the PMS for maximum efficiency [11]; this PMS successfully powered a wireless temperature sensor that requires a voltage of 2.5 V and energy consumption of 85 mW, and it could transmit the sensor data every 7.5 min. Brunelli et al. presented the design of a battery-less monitoring system for plant health status, which exploits innovative Plant-MFC as joint power supply and biosensor for assessing the long-term health of the flora living in the surroundings [12], their bioelectrochemical system is used both as a power generator to supply the wireless embedded electronics and as a biosensor for estimating the status of the plant. Most recently Konstantopoulos et al. designed a self-powered battery-less electric potential wireless sensor that harvests near-maximum energy from the avocado plant itself and transmits the signal tens of meters away; it has a total power consumption of $10.6\,\mu W$ and could accommodate simultaneous operation of multiple plants [13].

In the above designs, ambient bioenergy can be harvested and stored in the batteries or super-capacitors. However, using EH-sensor nodes also meet many challenges, such as the trade-off between idle listening, overhearing, and control packet overhead; the trade-off between energy-neutral operation and long latency; the extremely low recharging speed due to the typical feature of ultralow power in bioenergy; and the energy conversion efficiency is always a research focus under the situation of entirely "self" power supply. Our work intends to address these problems by inducing three advanced approaches: (i) incorporating an energy harvesting prediction algorithm in the proposed MAC protocol; (ii) adopting an optimized self-adaptive work/sleep duty cycle mechanism for every EH-node; and (iii) executing a differentiated access priority that considers both energy sustainability [14] and task requirement.

By considering these issues, we set to implement a living-tree biological energy powered wireless sensor system and design an novel "energy neutral" MAC protocol based on EH-node's remaining energy level, energy harvesting status, and application requirements. The optimization purposes are to maximize the total number of sampled data under energy harvesting constraints, as well as minimize the network transfer delay. To the best of our knowledge, this is the first time to analyze MAC protocol optimization in WSN powered by bioenergy. The main jobs and innovations are as follows:

(1) We employ a prediction approach in the node's duty cycle adjustment. The nodes could change their duty cycle intervals with fully aware of the future available energy. This procedure notably improves the accuracy and promptness of timing sequence in the nodes' duty cycle.

(2) The EH-WSN equipped with our novel algorithm can adapt to the variations of bioenergy in environment effectively and sensitively.

(3) We evaluate and compare the performance of our proposed MAC protocol with some aforementioned schemes. Results show that it outperforms the others in terms of throughput and latency.

The rest of this paper is organized as follows. In Section 2, the sensor node architecture and hardware component design are described. Section 3 explains our system model and describes the unresolved problem. In Section 4, the proposed MAC protocol and its mathematical explanations for sustainable sensing in a bioenergy-powered WSN are presented. Section 5 discusses the simulation and practical experiment results. In Section 6, we conclude the paper with a brief summary.

2. PBN System Design

Nowadays WSN has become a research hotspot in forest information monitoring field because of its miniaturization and integration [15–17]. Certainly, it still faces some difficulties due to the complexity and particularity in forestry application environment, especially the power supply problem: it is very troublesome to replace the batteries as conventional WSN always takes batteries as the energy storage unit. Hence, harvesting local environmental energy to settle this power supply issue is imperative. However, the present studies in EH-WSN mainly focus on the use of solar energy, wind energy, vibration energy, thermal energy, and electromagnetic energy [18], which all have defects in a practical forestry environment. The intensity of light in the forest is relatively weak and cannot be harvested during rainy and cloudy days; vibration does not exist in the woods in most conditions; wind velocity is small near the ground surface and

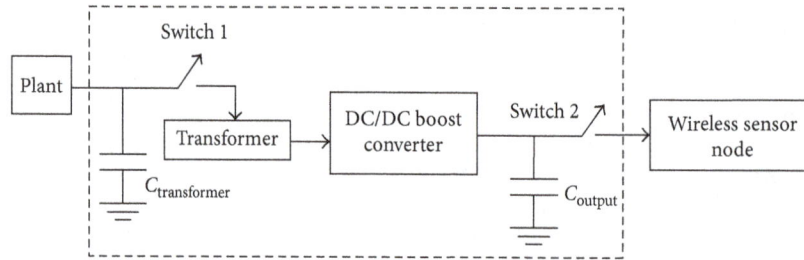

FIGURE 1: Schematic of proposed PBN.

is rather unstable; electromagnetic energy only exists close to the radiation emission station. Therefore, a more feasible energy harvesting method needs to be studied.

Recently, scientific explorations show that plants could become a potential source of bioenergy, and then this plant-bioenergy could be harvested to power small autonomous sensors (see above). Such EH-sensor does not need harsh working conditions and provides a new sight for EH-WSN in the forest. Accordingly, we conduct a research and design a prototype of plant surface-mounted bioenergy power wireless sensor node, with its block diagram shown in Figure 1.

The energy conversion circuit of PBN mainly consists of a voltage step-up converter. At the beginning, the first super-capacitor $C_{transformer}$ is charged by the low-voltage output of plant. Switch 1 is utilized to prevent the transformer and DC-DC converter drawing current from $C_{transformer}$ while it is being charged. Once the voltage of $C_{transformer}$ reaches the discharging voltage specified by the hardware, Switch 1 closes and $C_{transformer}$ works as the power source to drive the rest of the energy transfer circuits as well as the node. The voltage of $C_{transformer}$ is then amplified by the transformer. When $C_{transformer}$ discharges, the second super-capacitor C_{output} starts being charged. Once the voltage of $C_{transformer}$ drops below the charging voltage, Switch 1 opens and $C_{transformer}$ begins being charged again. This process iterates for a few times until the voltage of C_{output} reaches the required node voltage (such as 3.3 V for typical wireless sensors), when Switch 2 closes to power the node. The function of Switch 2 is to connect the node when C_{output} is fully charged and to prevent the load from drawing current when C_{output} is being charged.

We set the capacitance of $C_{transformer}$ to be 0.22 F, and it was charged to 350 mV and discharged to 100 mV. The average capacitor charging time was measured about 400 seconds. The super-capacitor C_{output} was selected as 2.5 F in order to transmit data packets once fully charged. When $C_{transformer}$ was charged to 350 mV, Switch 1 closed and began to charge C_{output}. It took about 2.5 h to charge C_{output} from 0 V to 3.3 V. Then Switch 2 closed and started driving the wireless sensor node. When the voltage of C_{output} decreased to 2.4 V, Switch 2 opened and C_{output} was then charged again. It took about 1 h to charge C_{output} from 2.4 V to 3.3 V. Actually we take the work/sleep mode toggle of wireless sensor node as Switch 2 because the power consumption in sleep mode is close to zero [19]. Hence, the charging procedure of C_{output} can be always running. Figure 2 just shows the voltage variations of super-capacitor charging

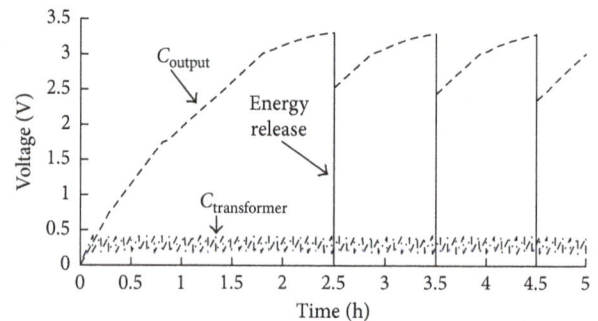

FIGURE 2: Changing statuses of super-capacitor voltage.

FIGURE 3: Illustration of proposed PBN.

process. In addition, the prototype of PBN is shown in Figure 3, an ultralow-power MSP430G2 is used as the control unit; a 315 MHz super-regeneration wireless transfer module of 10 mW power consumption is adopted as communication unit, with a photosensitive sensor to acquire the illumination of plant.

3. Problem Statement

The purpose of this work is to research and evaluate a MAC protocol that solves the problems associated with PBN, which is powered by ultralow and varying bioenergy. Every PBN has a fixed number (denoted as M) of data packets to send. The data packet, with a size regarded as Z, contains illumination information of each plant. Specifically, our work thinks of a sensor network that is made up of plentiful PBNs deployed in a multihop mode. One PBN can both generate data by itself and receive data packets from other nodes while it acts as a relay. The sink is considered to own an unlimited power supply which could process and store

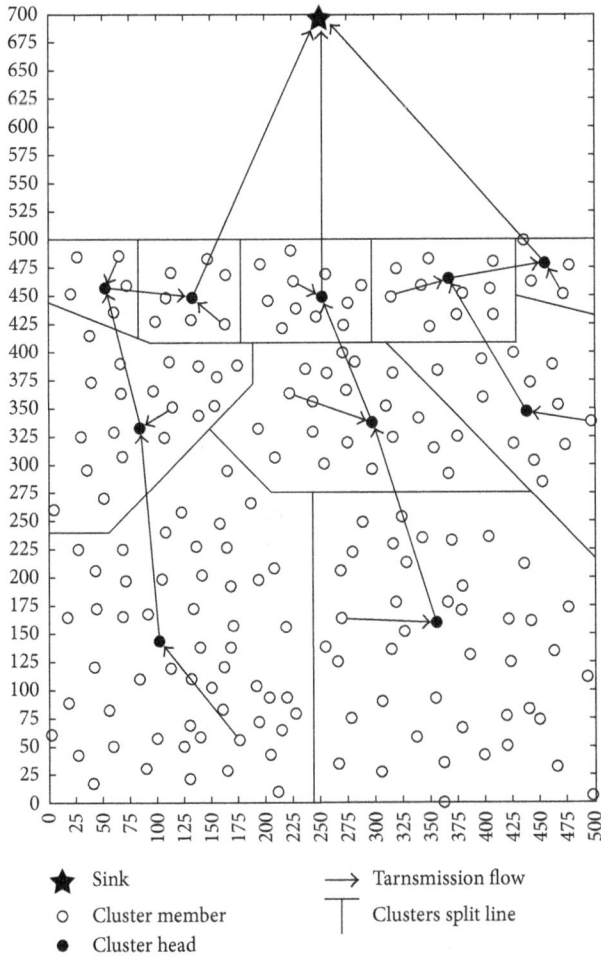

★ Sink	→ Tarnsmission flow
○ Cluster member	⊤ Clusters split line
● Cluster head	

FIGURE 4: Operation structure of EHGUR-OAPR.

sensor data from all the nodes. In consideration of practical application, we set the network's routing protocol as a clustering structure, and our former research achievement EHGUR-OAPR [14] is carried on. EHGUC-OAPR is an unequal clustering routing protocol: clusters that are closer to the sink have smaller size. Apparently, it can balance the energy consumption of the entire network and improve the data delivery ratio. Then, MAC protocol in such clustering routing path is easier to understand: it takes the responsibility of access control in intracommunication between cluster head and its members, along with the intercommunication between cluster heads and the sink. On the contrary, the elections of cluster heads, the partition of cluster members, and the routing paths to the sink have already been calculated by routing algorithm. A corresponding network diagram is shown in Figure 4.

Because of the time-changing and ultralow-power supply, PBNs should work in a duty cycle fashion. The ideal scenario is that nodes only wake up when there are data waiting for transmission; otherwise, it falls into deep-sleep. This procedure is mainly under the control of MAC layer. Following the literature in [9], the MAC protocols for EH-WSN can be classified into two categories: synchronous and asynchronous. The asynchronous type that adopts a receiver-initiated mechanism, which does not require any

clock synchronization between sensor nodes, has been proven to outperform the state-of-the-art of traditional sender-initiated protocols and the synchronous protocols. Figure 5 just shows a basic operation of receiver-initiated communication between a sender and a receiver.

As shown above, whenever a receiver wakes up, it sends a wake-up beacon (WUB) and then waits for an incoming packet. Meanwhile, whenever a transmitter has a packet to send, it opens an idle listening window to receive a WUB from its receiver. As soon as a WUB is received, the transmitter performs Clear Channel Assessment (CCA), Calculation before Transmission (CBT) and then, forwards its Data Packet to the receiver. If a packet is successfully received, the receiver sends an Acknowledgment packet (ACK). Finally, both transmitter and receiver turn into sleep mode for energy saving.

With regard to the PBN, energy harvesting rate of each node changes according to many factors: sap concentration of the plant, seasonal variation, day-night rhythm, electrode material, solar irradiation, ambient temperature condition, and so on. The maximum generated power could vary from approximately 800 nW to above 3000 nW throughout the day [21]. As a result, this long and quickly changing power property of PBN poses a unique challenge to the existing duty cycle-based MAC protocols: neighboring nodes may never be able to wake up at a common interval to transmit data. Therefore, protocols for PBNs must be clearly understand of each plant's power characteristic and set their communications in an individual independent optimization strategy. Only this can guarantee a successful and practical communication of such system.

4. Proposed MAC Protocol

In this section, we present a novel asynchronous duty-cycling energy-efficient MAC protocol called PB-MAC (plant-bioenergy MAC). PB-MAC achieves near-optimal energy efficiency both at receivers and at senders. In an optimal energy-efficient MAC protocol, when there is a packet to send, the sender and receiver wake up at the same time, transfer the packet reliably, and then both go to sleep again immediately. PB-MAC approaches this optimality in several ways: firstly, PB-MAC is a receiver-initiated protocol but introduces the use of an individual energy harvesting prediction-based sequence to control each node's wakeup times; this allows the senders accurately predict the time at which a receiver will wake up. Thus, PB-MAC reduces the duty cycle for receivers and senders both. Secondly, PB-MAC calculates the transmission energy consumption of every PBN to seek for an optimal "ENO" access policy that gains the maximum throughput. This means that if the cluster head's bioenergy cannot afford a full-course data collection of each cluster member, it should strive for an optimum option that maximizes the network data throughput. Finally, PB-MAC could adjust packet transmission delay based on PBN's priority according to the task requirement. In other words, if a member node has an urgent message, then PB-MAC will try to improve its access priority to optimize delivery latency.

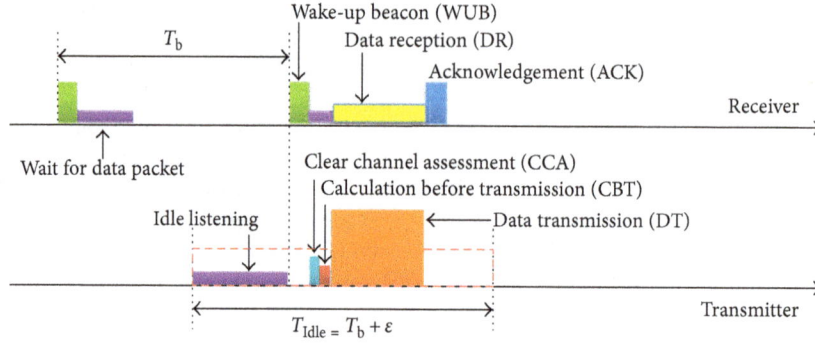

FIGURE 5: Basic communication scheme in RICER protocol [20].

The PB-MAC protocol operates in a time round structure, and the supposed round is divided into several time slots: each round has its own assignment individually, while the slots are used to apportion the working duty in a planned way at a detailed level. In this work, we use EH-round and EH-slot for the energy management arrangement, whereas the MAC-round and MAC-slot for communication protocol scheme. Both two are associated by the energy consumption calculations.

4.1. Predictive-Wakeup Mechanism. An analysis of the plant as an energy source was conducted in [22, 23] for Pachira and bigleaf maple trees, respectively. It is concluded that plant-bioenergy is heavily influenced by environmental sunlight and moisture, and it approximately follows a periodic sinusoidal variation of 24 hours. Thus, we carried out a bioenergy voltage testing experiment during 3 days on a Koelreuteria in both summer time and late autumn, as shown in Figure 6. It is obvious that the records coincide with a day and night variation rhythm, as well as comply with a seasonal change. So we can have a energy harvesting prediction algorithm based on Weather-Conditioned Selective Additive Decomposition model (WCSAD): We first set the weather conditions of every day into three categories: sunny, rainy, and mixed; then we proceed to forecast the weather that only belongs to the same sort. In addition, we combine both the season variations during all the year and weather changes in one day together to calculate the energy that could be harvested in next time EH-rounds.

Firstly, the PBN needs to record the harvested energy $E_{cal}(d, s)$ in time EH-round s of day d; hence, for the three categories there should be three matrixes $|E_{cal}(d, s)|_D$ that save D days data independently. The average harvested energy in three kinds is

$$\overline{E}_{sunny} = \frac{\sum_{i=1}^{D} \sum_{j=1}^{S} E_{sunny}(i, j)}{D},$$

$$\overline{E}_{rainy} = \frac{\sum_{i=1}^{D} \sum_{j=1}^{S} E_{rainy}(i, j)}{D}, \qquad (1)$$

$$\overline{E}_{mix} = \frac{\sum_{i=1}^{D} \sum_{j=1}^{S} E_{mix}(i, j)}{D},$$

where E_{sunny}, E_{rainy}, and E_{mix} are the harvested energy in sunny days, rainy days, and mixed days, respectively. Formula (1) should be updated every 24 hours to ensure the weather classification process. The detailed calculation rules are shown in Algorithm 1.

Here Diff (i, j) is a parameter concerns the seasonal changes:

$$\text{Diff}(i, j) = \beta \left[E_{cal}(i-1, j) - E_{cal}(i-2, j) \right] \\ + (1-\beta) \left[E_{cal}(i-2, j) - E_{cal}(i-3, j) \right], \qquad (2)$$

where β is just a tuning coefficient. After these corresponding classifications of former days, we can start to predict the up-coming energy, which would be harvested during time EH-round s in the present day. The prediction value $E_{est}(i, j)$ equals

$$E_{est}(i, j) = E_x(i, j) + \alpha \left[E_{cal}(i, j-1) - E_x(i, j-1) \right] \\ + (1-\alpha) \left[E_{cal}(i, j-2) - E_x(i, j-2) \right], \qquad (3)$$

where α is another adjustment coefficient. The latter two items in the above formula intend to represent the applicable harvesting energy's dynamic changes caused by different EH-rounds. $E_x(i, j)$ is a weighted sum of corresponding rounds' harvested energy in the past reference days that belong to the same weather kind: $E_x(i, j) = \gamma E_{x-1}(i, j) + \gamma^2 E_{x-2}(i, j)$. γ is also a debug parameter. $E_{x-1}(i, j)$ just shows the corresponding rounds' energy in the last time with the same weather kind, in turn $E_{x-2}(i, j)$ is the day before last time within the same weather type.

By default, we can take the present day's weather type as the same as yesterday. However, a detecting mechanism is also designed to test the default hypothesis: we set up four testing points in a single day, and we compute the average harvested energy during the time from the beginning of day to the test point moment, if the value has a great difference with the last day's data, we should regulate the weather type immediately.

4.2. Throughout Optimization Mechanism. Here, in this part, a duty cycle-based energy-neutral MAC protocol is proposed to accomplish an optimal throughput and appreciable quality of service. By studying on CSMA/CA protocols and

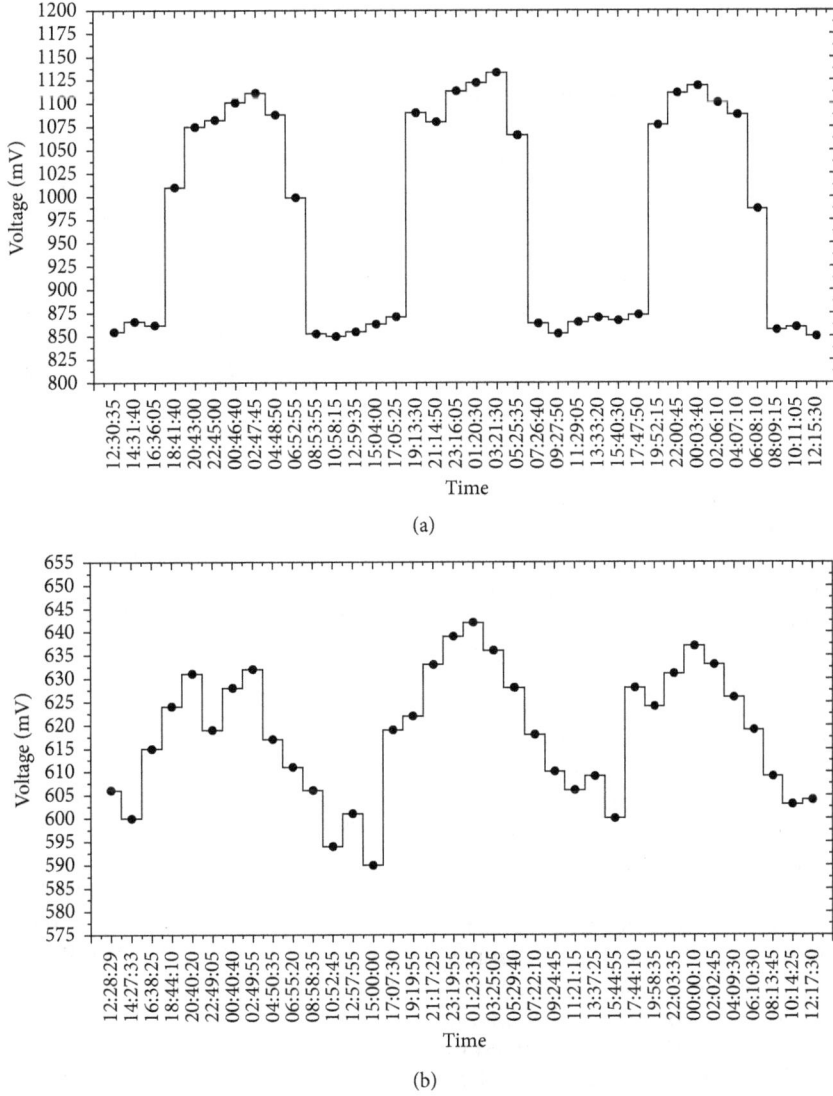

FIGURE 6: Bioelectric voltage variation of a Koelreuteria: (a) in summer time (b) in late autumn.

```
(1)  BEGIN
(2)      If (∑_{j=1}^{S} E(i, j) > (1 + Diff)E̅_rainy)
(3)          {day i is a rainy day, update E_rainy}
(4)      Else if (∑_{j=1}^{S} E(i, j) < (1 − Diff)E̅_sunny)
(5)          {day i is a sunny day, update E_sunny}
(6)      Else
(7)          {day i is a mixed day, update E_mix}
(8)  END
```

ALGORITHM 1: Harvested energy classification algorithm for different weather conditions.

TDMA protocols, we found that CSMA/CA protocol easily leads to serious conflicts, thus causing heavy package loss and TDMA protocol likely leads to large delay. Therefore, we try to combine CSMA/CA protocol and TDMA protocol together reasonably to design a MAC protocol that conforms to the PBN scenario. The hybrid TDMA/CSMA protocol, that is, PB-MAC, adopts TDMA mode at intracluster communication stage, while performs CSMA/CA mode at intercluster communication stage. The detailed time sequence strategy is described in the following section.

4.2.1. Initialization Phase. In this phase, all the PBNs should be set-up on the living trees and equipped with an initial energy $B_{initial}$ randomly. After that, every node transmits parameters such as its distances information and energy-harvesting rate directly to the sink and then the sink will run clustering algorithm to elect the cluster heads and members, receptively. Finally, this computational result will be downloaded to each node, namely, the routing procedure has been fulfilled [14].

Hence, the cluster member nodes should transmit messages to their corresponding cluster head to acknowledge the transmission arrangement, which is the next phase.

4.2.2. Intracluster Phase I. In this phase, the cluster members must establish communication connection with their cluster head by the CSMA/CA mode. Firstly, the member node i sends a request frame RTS before sending the data packet, and after receiving the RTS, the cluster head sends CTS to reply. Then node i sends its own information to the head node such as the ID number, amount of data it collects, energy harvesting rate, and predicting result. After receiving this packet, the head node should send an ACK packet that contains its wakeup sequence information to the node i. All the other nodes in the cluster will listen to the channel and wait until it is clear, then they would go on completing for the channel and repeat the process until all communications have completed.

4.2.3. Intracluster Phase II. By using the CSMA/CA mechanism, the cluster head collects all member nodes' preliminary information; thus, the intracluster transmission proceeds to the next scheduling access stage, which uses TDMA mechanism. We design an optimal access-scheduling algorithm to allocate the time MAC-slots for both maximum throughput and balanced network load (Algorithm 2).

In formula in Algorithm 2, $E_{M,n}$ means the maximum energy storage capacity of the node n, $E_{S,n}$ is the current stored energy in node n, $P_{EH,n} > 0$ is the individual harvesting power rate of node n, and σ, μ are appropriately chosen constants. We use $ST(n)$ to evaluate PBN's energy sustainability. The node with higher value is more patient than the node with lower value in transmitting packets; in other words, the low energy sustainability node is more desperate to transmit data before all the energy is discharged. So, in this context, the access probability can be determined to be inversely proportional to the energy sustainability.

Next, the cluster head should check if there is request to relay other head nodes' packets or transmit its own packet, as well as predict the forthcoming energy that can be harvested. Based on these above information, access allocation algorithm will calculate the number of nodes that could be visited:

$$E_{S,n} + E_{est}(i,\ j) = I[a(z)] \cdot E_{TX}(z) + I[b(z)] \cdot E_{RX}(z)$$
$$+ I[n1] \cdot E_{RX}(n1) + I[n2] \cdot E_{RX}(n2) + \cdots,$$
$$(4)$$

where $E_{TX}(z)$, $E_{RX}(z)$ are the energy consumptions in the cluster head node Z's data transmission and reception, respectively. $E_{RX}(ni)\,(i = 1, 2, 3, \ldots)$ is the data reception energy consumption for cluster member node i. $I(\cdot)$ is the indicator function and $a(z)$, $b(z)$ are the events that cluster head transmits and receives packets with other cluster heads; besides ni is the event that cluster member i transmits data to the head node. Apparently, the head node would obtain the exact number of executable nodes and broadcast the accurate access-scheduling time to them, thus performing the data communication in a TDMA mode. As for all the executable nodes, when they receive the broadcast message from the head node, they should analyze the schedule time based on their energy status: if matching with its work/sleep

duty cycle, the member node will send an ACK message back; in contrast, it will send a NACK message, and the head node should postpone to next one and adjust the scheduling plan.

Finally, the cluster head collects member nodes' packets and needs to decide whether to transmit these data forward or not. Here, we choose to use a random number generator with range from 0 to 1: if the generated number exceeds a specified threshold Ω, then it starts to prepare for the data transmission with next hop; otherwise, the head node sleeps immediately. Certainly here exists a head node-updating program, sink should collect every cluster nodes' energy information and compute an optimal election result. The whole operation diagram is shown below in Figure 7.

4.2.4. Intercluster Phase. After completing the data acquisition task in the cluster heads, it needs to decide how and when to transmit the packets to the sink. Based on the former phase, the head node could set the occasion of intercluster communication clearly. However, it should be noted that if the head node decides to transfer packet to the next hop, the actual sending process would take place in the next time MAC-round, and the current head node only transmits an appointment to the next hop at the present MAC-round. This procedure enhances the transmission reliability considerably and does not cause obvious time delay.

Therefore, all the cluster head nodes transmit packets accordingly based on the sequence of routing protocol. In addition, we set intercluster transmission in a sender-initiated way [24] to improve the energy efficiency, just as shown in Figure 8.

4.3. Priority Optimization Mechanism. In this section, we consider applications that require some PB nodes to transfer critical or urgent data to the sink as soon as possible from the time they are generated. The source node and all relay nodes between the source and destination should send these urgent data packets faster than normal data packets. Using an additional priority optimization mechanism, the described PB-MAC could specify these urgent data as high priority data packets and transmit these data packets first.

Here we modify the cluster head's access-scheduling algorithm with a packet priority value. In other words, time MAC-slots are allocated by both the energy sustainability of cluster member node and its packet priority; the new access order during the TDMA phase is determined by the sum of two parameters:

$$N_{Total} = \chi \cdot N_{ES} + (1 - \chi) \cdot N_P, \qquad (5)$$

where N_{Total} is the optimized sequence in the subsequent access allocation algorithm, χ is the coefficient range from 0 to 1, N_{ES} is the order of energy sustainability, and N_P is the order of packet priority. Using this priority control function, the sender nodes can notify their packet priority to the cluster head node. In addition, after receiving these senders'

(1) **BEGIN**

(2) Calculate member node n's energy sustainability $\mathrm{ST}(n) = (E_{M,n}/((P_{\mathrm{EH},n} + \sigma)\log\mu))(\mu^{\lambda(n)} - 1)$, where $\lambda(n) = (E_{M,n} - E_{S,n})/E_{M,n}$

(3) Judge if there exists inter-cluster communications on this cluster head node

(4) Predict the energy that can be harvested in this time round

(5) Sort the list of all the member nodes' $\mathrm{ST}(n)$ by ascending counts

(6) Allocate the time slot by the above list

(7) Compute the energy budget formulas; determine the total number of nodes that can be accessed

(8) Communicate with each cluster members according to the above result

(9) Decide whether to transmit the collected data to the next hop cluster head or not

(10) Decide whether to change the current cluster head node or not

(11) **END**

ALGORITHM 2: Access-scheduling optimization algorithm in the intracluster phase.

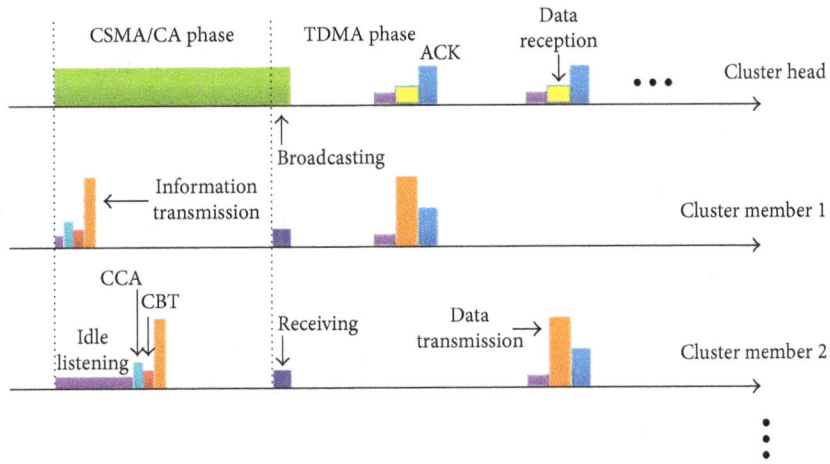

FIGURE 7: Intracluster work/sleep duty cycle diagram (note that the receiver and transmitter diagram are of very different time scales).

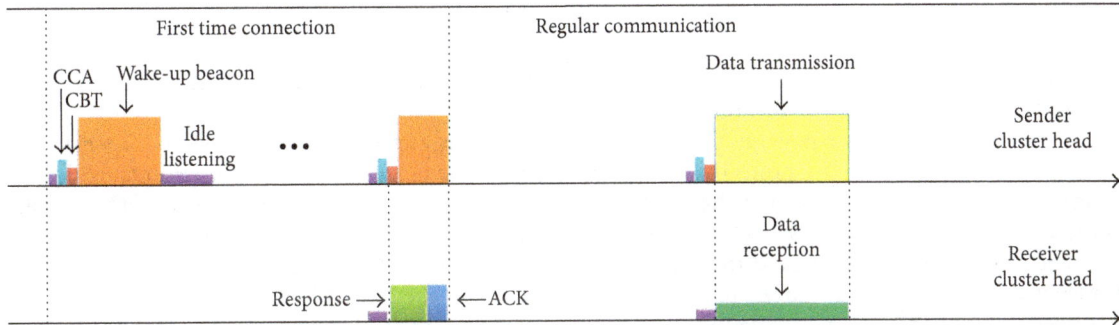

FIGURE 8: Intercluster work/sleep duty cycle diagram.

packets based on priority, the head node also must adjust its intercluster communication probability to transfer the data as soon as possible: it should start the intercluster transmission immediately and make an appointment with its next hop node.

As a result, PB-MAC could predict the harvested energy of next time round, calculate an optimal access sequence of the cluster member to keep all nodes under "ENO" state; maximize the data throughput based on the hybrid TDMA/CSMA transfer mode and minimize the transmission delay based on the priority technique. A detailed

duty cycle timing sequence along with the energy harvesting is described in Figure 9 for instance:

As shown above, all cluster nodes work in a duty cycle mode; they calculate the harvested power and turn down to sleep whenever the energy is insufficient. Due to the individual energy-harvesting rate, the working period of every member node may be different. They would follow the working sequence based on the proposed approach and carry out the data-sampling task as well. In turn, cluster heads take charge of gathering all the data and then send to the sink. Next, we will evaluate and compare the PB-MAC with some references.

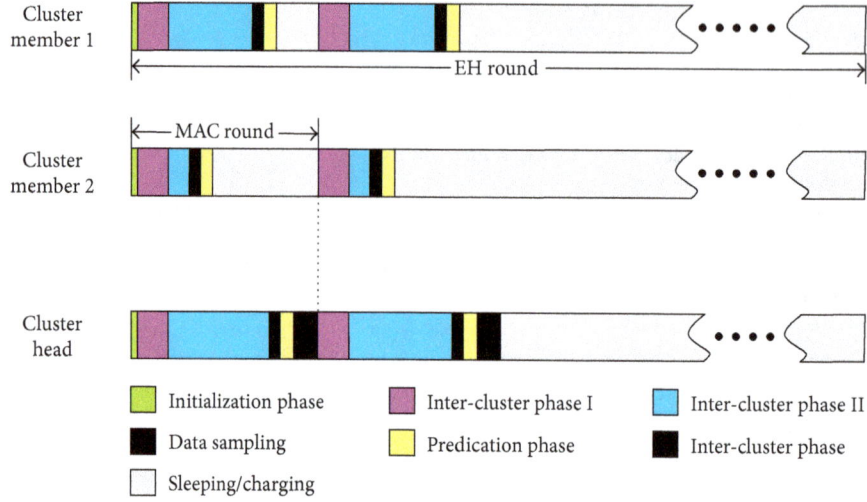

FIGURE 9: The overall duty cycle diagram of nodes in one cluster.

5. Performance of PB-MAC

To validate the proposed PB-MAC and evaluate its performance, we conduct simulation studies using OMNeT++. The simulated network is a 500 m × 500 m square area with total 200 randomly deployed nodes (Figure 4 shows the network topology). Key parameter values are listed in Table 1.

Here, we adopt the energy consumption model in [25]: $E_{TX}(k, d)$ means the transmission energy of transferring k bits for distance d and $E_{RX}(k)$ just means the reception energy of receiving k bits.

$$E_{TX}(k, d) = \begin{cases} E_{elec}k + E_{fs}kd^2, & \text{Intracluster,} \\ E_{elec}k + E_{mp}kd^4, & \text{Intercluster,} \end{cases} \quad (6)$$

$$E_{RX}(k) = E_{elec}k.$$

We compare PB-MAC against RI-MAC with DSP (denoted as RI-DSP) [5], EAMP-AIDC [8], and MFC-MAC [9], all operating under the plant-bioenergy-powered circumstance [13]. Firstly, we study the protocols using the same network settings as in Figure 4 and an initial energy level for all sensors to be 30%, we also assume that all sensors undertake equal power harvesting rate and there are three different scenarios, which are 0.25 mW, 0.75 mW and 1.25 mW, respectively. The results of a 24-hour simulation run are presented in Figure 10.

The results shown in Figure 10 are aimed at evaluating the functionality of PB-MAC. Since the initial energy level is set at 30%, there may be fewer nodes active, more network holes, and retransmissions. While PB-MAC shows the best overall performance, we note that MFC-MAC's probe-minimization mechanism was able to keep the senders' duty cycles short, saving significant energy. Longer sender duty cycles (Figure 10(b)) of RI-DSP and EAMP-AIDC manifest higher energy usage, more network holes as nodes need more sleep cycles, and consequently, higher node-to-sink delays (Figure 10(a)). All these protocols consume their energy storage quickly (Figure 10(c)) but will also replenish to different extent thereafter. Starting with

TABLE 1: Key parameter values.

Parameter	Value
Network monitoring area	(0, 0)~(500, 500)
Coordinate of sink	(250, 700)
σ, μ	0.1, 5
α, β, γ	0.5, 0.6, 0.62
D	5
EH-round	1 hour
MAC-round	12 second
Z	32 bit
E_{elec}	50 nJ/bit
E_{fs}	10 pJ/(bit·m^{-2})
E_{mp}	0.0013 pJ/(bit·m^{-4}s)
$E_{M,n}$	10 joule
χ	0.382
EH-slot	12 minute
MAC-slot	3 second
M	4

30% energy level puts many nodes in longer sleep cycles and leaves the network with dynamic routes, and this increases the packet loss probability among data flows. While both PB-MAC and MFC-MAC are able to achieve higher packet delivery ratio (PDR) than the other protocols, PB-MAC performs better and maintains 100% PDR under extremely limited energy harvesting conditions as shown in Figure 10(d). Therefore, in this particular plant-bioenergy scenario, the energy-harvesting rate is too low for the other three protocols, and only PB-MAC is able to continue operating and keep the energy level above 30% (note: for the convenience of analysis and comparison, all the prediction mechanisms in the above protocols are set full success due to the uniform energy harvesting rate, as well all data packets have the same priority).

Moreover, we adjust the position of sink (250, 700) along a straight line to the coordinate (250, 500), and we record two network quality indexes of PB-MAC likewise: the average node-to-sink latency and the node average remaining energy. The result is shown below in Figure 11. Here the energy-harvesting rate is assumed 0.75 mW. Obviously, we

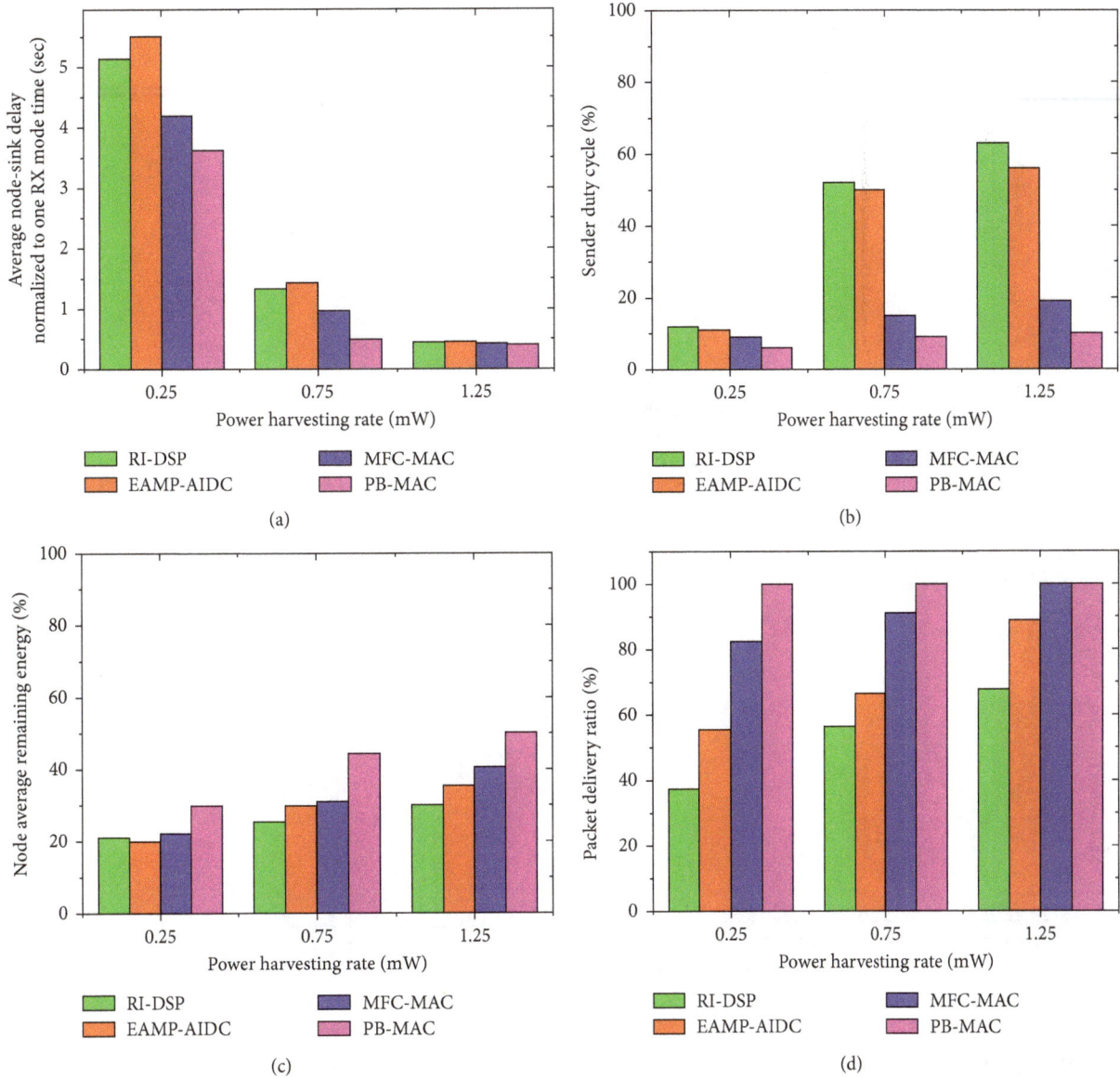

FIGURE 10: RI-DSP, EAMP-AIDC, MFC-MAC, and PB-MAC performance, with 30% initial energy level. (a) Average node-to-sink latency versus power harvesting rate, (b) sender duty cycle versus power harvesting rate, (c) node average remaining energy versus power harvesting rate, (d) packet delivery ratio versus power harvesting rate.

can see from the diagram that node average remaining energy gets greater when the sink moves towards the PBNs and the average node-to-sink latency decreases conversely. The reason is when the average distance between the sink and nodes becomes smaller, the transmission energy consumption also diminishes accordingly; meanwhile, the work/sleep duty cycle is optimized, and the improved operating time reduces the average latency.

Secondly, we use realistic data of our proposed PBNs measured at a mingled forest with area 50 m × 50 m (including Koelreuteria, cedar, birch, etc.) over 7 days in March 2018, starting at early morning, all 30 PBNs having an initial energy level of 60%. Nodes are placed in the field with a random uniform distribution while the sink is located at the center, and a random cluster member node has been

chosen to analysis specifically, and some of the related results are shown below:

As expected, the member nodes equipped with PB-MAC are able to harvest bioenergy systematically through the day and dissipate its energy supply rapidly. Nevertheless, by 17: 30 on the fifth day, the node starts to replenish its energy supply quickly (Figure 12(a)) and continue harvesting enough energy to operate: this is due to the precise operation of prediction algorithm. With the adequate storage, node is able to stay active longer, producing more data packets to the sink. This improves the data throughput and enhance the network performance in general (Figures 12(c) and 12(d)), especially based on the packets with randomly assigned priority (3 level), PDR is approximates to 100% over the latter testing period (Figure 12(b)).

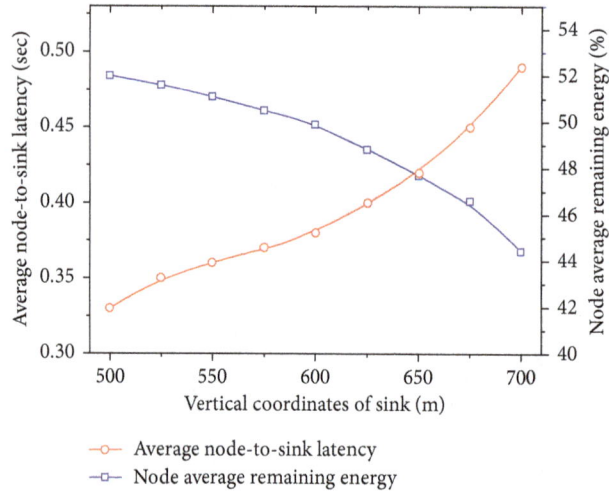

FIGURE 11: PB-MAC performance versus sink position changes.

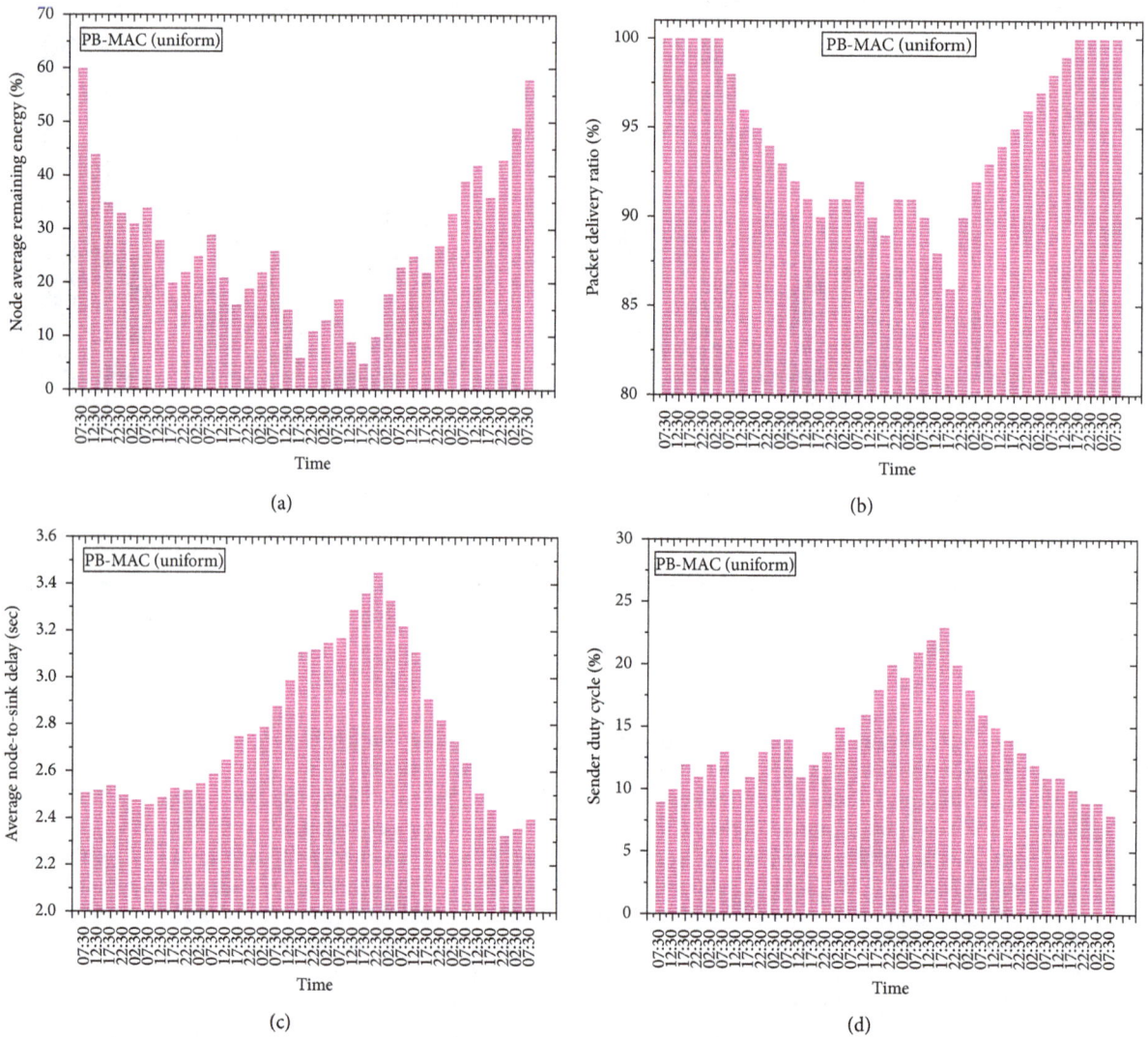

FIGURE 12: PB-MAC performance with 60% initial energy level (uniform deployment). (a) Nodes' average remaining energy versus time of 7 days, (b) packet delivery ratio versus time of 7 days, (c) average node-to-sink latency versus time of 7 days, (d) sender duty cycle versus time of 7 days.

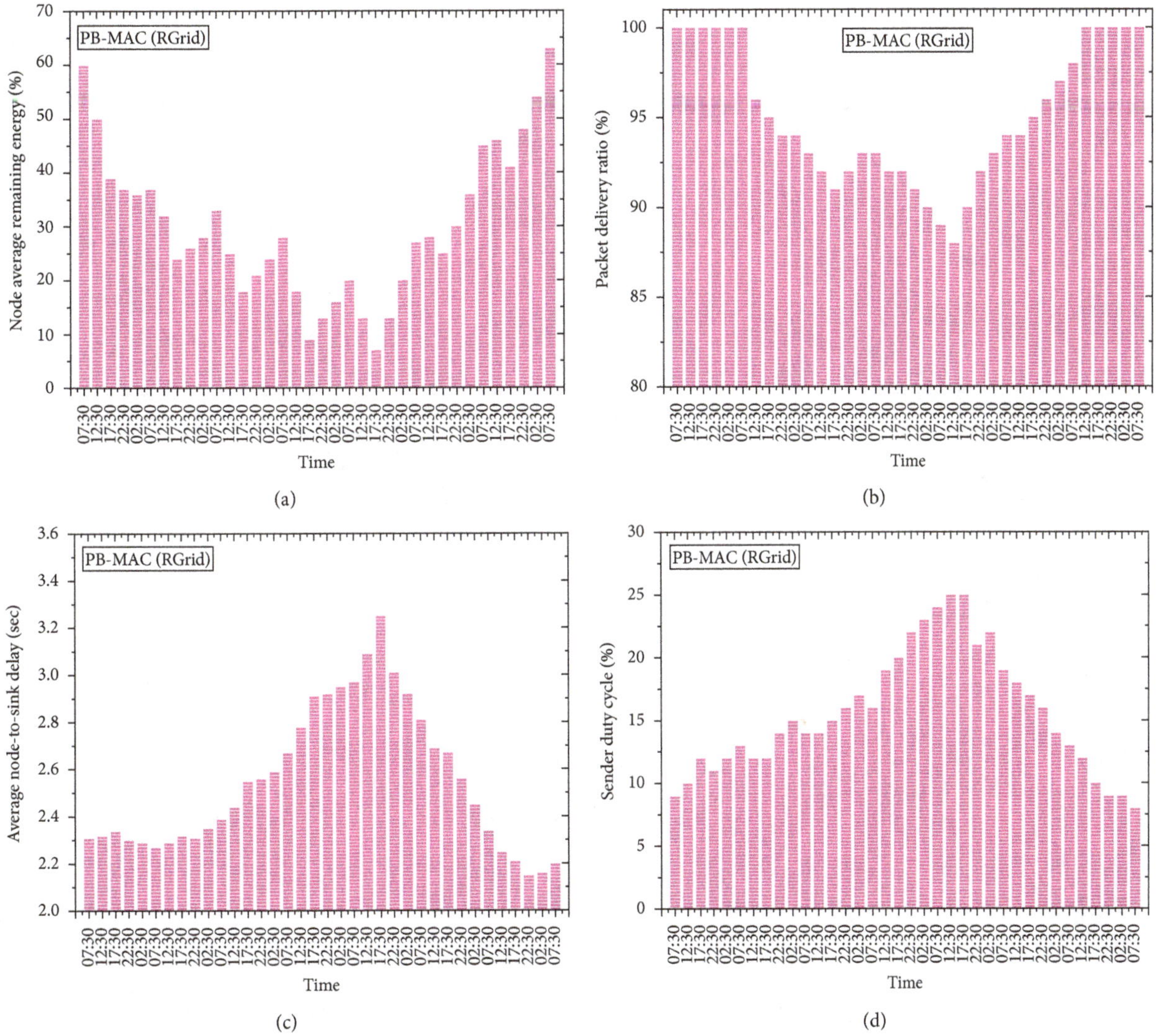

FIGURE 13: PB-MAC performance with 60% initial energy level (RGrid deployment). (a) Nodes' average remaining energy versus time of 7 days, (b) packet delivery ratio versus time of 7 days, (c) average node-to-sink latency versus time of 7 days, (d) sender duty cycle versus time of 7 days.

Furthermore, we modify the nodes deployment to have an additional similar testing of PB-MAC. The network topology structure is set up referring to article [26] and an RGrid-based deployment is employed: all PBNs are randomly divided into 8 grids that revolve around the sink grid in the center. We also carry out the proposed grid-based multihop routing scheme, and four experiment results of a random chosen cluster member node are given in Figure 13.

Apparently, PB-MAC in RGrid shows better performance than the PBNs with uniform distribution in Figure 12. Especially the average node-to-sink latency in Figure 13(c) improves significantly: the reason is that the scheduling of grid-based transmission is in a preset order; PB-MAC takes full advantage of this distinction and saves considerable waiting time of idle listening. This also helps to enhance the node's average remaining energy (Figure 13(a)). As well, the other two indicators further demonstrate that PB-MAC has superior adaptability and survivability for different network topologies.

Thirdly, we compare the four protocols using empirical data for network density performance on a cloudy day with rain in the morning. Each PBN starts with a 50% initial energy. The communication rate is 4 kB/s and the default packet size is 32 bytes. The experiment's result is shown in Figure 14.

As can be seen from the above diagram, data throughput rises along with the increased deployment of sensor nodes (Figure 14(a)). Particularly when the total number gets greater, average throughput of PB-MAC grows faster than others do. It is caused by the accurate dynamic adjustment of work/sleep duty cycle, which improves the working duration optimally and keeps all PBNs energy neutrally. On the other side, when the number of nodes increases, we find that the PBNs' average remaining energy decreases gradually and the node-to-sink latency increases (Figures 14(b) and 14(c)), the reason is that more network nodes would induce more relays and therefore, the more energy consumption and transmission delay. However, PB-MAC can adapt to the variation of

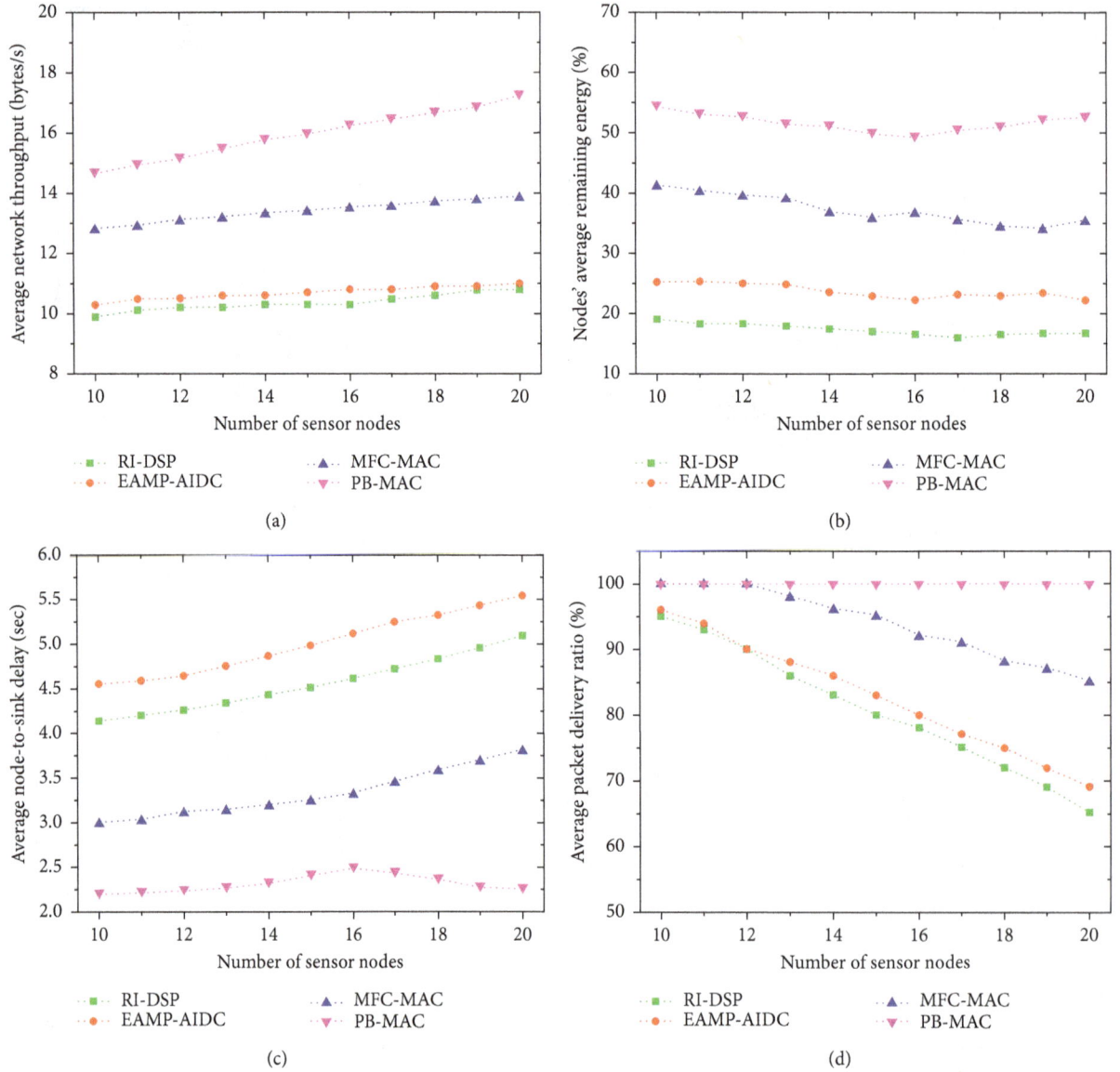

FIGURE 14: Node density performance comparison of four MAC protocols. (a) Average network throughput versus number of nodes, (b) nodes' average remaining energy versus number of nodes, (c) average node-to-sink latency versus number of nodes, (d) average packet delivery ratio versus number of nodes.

network topology and shows better scalability as for its high-efficient channel access-scheduling ability. In addition, PDR of PB-MAC always maintains 100% while the others drop drastically as seen in Figure 14(d). Consequently, it can be deduced that PB-MAC has high performance and robustness when the node density changes.

Finally, we conduct the statistical inference of our approach: the experiment on PB-MAC, RI-DSP, EAMP-AIDC, and MFC-MAC has carried out in the mingled forest over 7 days. The average remaining energy (ARE) of the proposed PBNs have computed, and it forms a bivariate normal variable when pair wise-comparison is done between PB-MAC and RI-DSP, or EAMP-AIDC, or MFC-MAC after every day. The pair of corresponding variables is in the form (x_i, y_i) where x_i is the ARE for PB-MAC and y_i is the ARE for the existing protocol as stated above; these are corelated

and should call for the use of paired T-test [27] for drawing statistical inference between the protocols being compared. The null and alternative hypotheses for the three cases are as follows:

(1) Null Hypothesis H_0: ($\text{ARE}_{\text{PB-MAC}} = \text{ARE}_{\text{RI-DSP}}$). Alternative Hypothesis H_1: ($\text{ARE}_{\text{PB-MAC}} > \text{ARE}_{\text{RI-DSP}}$).

(2) Null Hypothesis H_0': ($\text{ARE}_{\text{PB-MAC}} = \text{ARE}_{\text{EAMP-AIDC}}$). Alternative Hypothesis H_1': ($\text{ARE}_{\text{PB-MAC}} > \text{ARE}_{\text{EAMP-AIDC}}$).

(3) Null Hypothesis H_0'': ($\text{ARE}_{\text{PB-MAC}} = \text{ARE}_{\text{MFC-MAC}}$). Alternative Hypothesis H_1'': ($\text{ARE}_{\text{PB-MAC}} > \text{ARE}_{\text{MFC-MAC}}$).

The test statistic t with $n-1$ degrees of freedom is defined as: $t = D_{\text{avg}}/[S_{\text{d}}/\sqrt{(n-1)}]$, where D_{avg} and S_{d} denote

TABLE 2: Results of paired T-test.

Comparison of PB-MAC	T value	Significance level	95% confidence interval	
			Lower	Upper
RI-DSP	42.25	95%	44.21	48.65
EAMP-AIDC	33.21	95%	31.71	35.87
MFC-MAC	21.67	95%	6.59	7.91

the mean and standard deviation of the difference of ARE in two equal sized correlated large samples of size n. The 95% confidence limits for D_{avg} is $D_{\mathrm{avg}} \pm t_{0.05} * [S_d / \sqrt{(n-1)}]$. Here $t_{0.05}$ is the 5% point of the t-distribution on $n-1$ degrees of freedom. Let p indicate the probability of the calculated value for our test statistic t with $n-1$ degrees of freedom to obey the null hypothesis. A value of $p < 0.05$ indicates that H_0, H_0', and H_0'' is rejected at 5% significance level and hence H_1, H_1', and H_1'' be accepted at 95% confidence level.

Table 2 shows the result of paired T-test obtained by pair wise testing of PB-MAC with RI-DSP, EAMP-AIDC, and MFC-MAC, respectively. In all the cases $p < 0.05$ so H_0, H_0', and H_0'' is rejected at 5% significance level and H_1, H_1', and H_1'' is accepted at 95% confidence level. In addition, the lower and upper limits for the 95% confidence interval for D_{avg} are also listed. Hence, it can be convincingly concluded that PB-MAC outperforms RI-DSP, EAMP-AIDC, and MFC-MAC, and the results are statistically significant.

With all the scenarios studied, PB-MAC and MFC-MAC are able to achieve much outstanding performance with shorter sender duty cycles. Between them, the former even better satisfies the plant living environment. Therefore, we can conclude that while all protocols are able to operate under good circumstance, only PB-MAC is able to continue superior operating under extremely low energy harvesting conditions.

6. Conclusion and Future Work

In this paper, a novel MAC protocol for sustaining perpetual operation of PB-WSN is proposed. Some relevant researches have been conducted: First, an individual energy harvesting prediction algorithm is proposed to guarantee precise energy management. Second, an optimized self-adaptive work/sleep duty cycle mechanism is introduced, aiming to provide intracluster wide optimal packet communications under extremely limited node resources. Finally, a priority-based communication method is employed to improve the intercluster wide transmission delay. Numerical simulations and actual experiment results are evaluated to analyze the performance of the proposed algorithm. Parameters such as the average node-to-sink latency, the sender's duty cycle ratio, the node average remaining energy, and the packet delivery ratio have been analyzed and reviewed. Results show the effectiveness and efficiency of the proposed MAC protocol.

Future work will focus on the proposal of cross layer optimization problem for PB-WSN and consider cognitive radio network issues.

Acknowledgments

This work is partially supported by the National Natural Science Foundation of China (nos. 31700478 and 31670554). The authors also gratefully acknowledge the helpful comments and suggestions of the reviewers, which have improved the presentation.

References

[1] F. Ongaro, S. Saggini, and P. Mattavelli, "Li-ion battery-supercapacitor hybrid storage system for a long lifetime, photovoltaic based wireless sensor network," *IEEE Transactions on Power Electronics*, vol. 27, no. 9, pp. 3944–3952, 2012.

[2] S. Lee, B. Kwon, S. Lee, and A. C. Bovik, "BUCKET: scheduling of solar-powered sensor networks via cross-layer optimization," *IEEE Sensors Journal*, vol. 15, no. 3, pp. 1489–1503, 2015.

[3] M. Shin and I. Joe, "Energy management algorithm for solar-powered energy harvesting wireless sensor node for internet of things," *IET Communications*, vol. 10, no. 12, pp. 1508–1521, 2016.

[4] R. Rana, W. Hu, and C. T. Chou, "Optimal sampling strategy enabling energy-neutral operations at rechargeable wireless sensor networks," *IEEE Sensors Journal*, vol. 15, no. 1, pp. 201–208, 2015.

[5] H. Yoo, M. Shim, and D. Kim, "Dynamic duty cycle scheduling schemes for energy-harvesting wireless sensor networks," *IEEE Communications Letters*, vol. 16, no. 2, pp. 202–204, 2012.

[6] T. N. Le, A. Pegatoquet, O. Berder, and O. Sentieys, "Energy-efficient power manager and MAC protocol for multi-hop wireless sensor networks powered by periodic energy harvesting sources," *IEEE Sensors Journal*, vol. 15, no. 12, pp. 7208–7220, 2015.

[7] H.-I. Liu, W.-J. He, and W. K. G. Seah, "LEB-MAC: load and energy balancing MAC protocol for energy harvesting powered wireless sensor networks," in *Proceedings of the IEEE International Conference on Parallel and Distributed Systems*, pp. 584–591, Hsinchu, Taiwan, December 2014.

[8] O. Bouachir, A. Ben Mnaouer, F. Touati, and D. Crescini, "EAMP-AIDC-energy-aware MAC protocol with adaptive individual duty cycle for EH-WSN," in *Proceedings of the 13th International IEEE Wireless Communications and Mobile Computing Conference (IWCMC)*, pp. 2021–2028, Valencia, Spain, June 2017.

[9] F. Yang, K.-C. Wang, and Y. Huang, "Energy-neutral communication protocol for very low power microbial fuel cell based wireless sensor network," *IEEE Sensors Journal*, vol. 15, no. 4, pp. 2306–2315, 2015.

[10] D. Zhang, F. Yang, T. Shimotori, K.-C. Wang, and Y. Huang, "Performance evaluation of power management systems in microbial fuel cell-based energy harvesting applications for driving small electronic devices," *Journal of Power Sources*, vol. 217, pp. 65–71, 2012.

[11] C. Erbay, S. Carreon-Bautista, E. Sanchez-Sinencio, and A. Han, "High performance monolithic power management system with dynamic maximum power point tracking for microbial fuel cells," *Environmental Science and Technology*, vol. 48, no. 23, pp. 13992–13999, 2014.

[12] D. Brunelli, P. Tosato, and M. Rossi, "Flora health wireless monitoring with plant-microbial fuel cell," in *Proceedings of the 30th Anniversary Eurosensors Conference-Eurosensors*, pp. 1646–1650, Budapest, Hungary, September 2016.

[13] C. Konstantopoulos, E. Koutroulis, N. Mitianoudis, and A. Bletsas, "Converting a plant to a battery and wireless sensor with scatter radio and ultra-low cost," *IEEE Transactions on Instrumentation and Measurement*, vol. 65, no. 2, pp. 388–398, 2016.

[14] Y. Wu and W. Liu, "Routing protocol based on genetic algorithm for energy harvesting-wireless sensor networks," *IET Wireless Sensor Systems*, vol. 3, no. 2, pp. 112–118, 2013.

[15] Y. Liu, Y. He, M. Li, J. Wang, K. Liu, and X. Li, "Does wireless sensor network scale? A measurement study on GreenOrbs," *IEEE Transactions on Parallel and Distributed Systems*, vol. 24, no. 10, pp. 1983–1993, 2013.

[16] D. P. Smith, G. G. Messier, and M. W. Wasson, "Boreal forest low antenna height propagation measurements," *IEEE Transactions on Antennas and Propagation*, vol. 64, no. 9, pp. 4004–4011, 2016.

[17] F. Zhang, I. Joe, D. Gao, and Y. Liu, "An efficient multiple-copy routing in intermittently connected mobile networks," *International Journal of Future Generation Communication and Networking*, vol. 9, no. 5, pp. 207–218, 2016.

[18] N. Michelusi, L. Badia, R. Carli, L. Corradini, and M. Zorzi, "Energy management policies for harvesting-based wireless sensor devices with battery degradation," *IEEE Transactions on Communications*, vol. 61, no. 12, pp. 4934–4947, 2013.

[19] F. Yang, D. Zhang, T. Shimotori, K.-C. Wang, and Y. Huang, "Study of transformer-based power management system and its performance optimization for microbial fuel cells," *Journal of Power Sources*, vol. 205, pp. 86–92, 2012.

[20] E. Y. A. Lin, J. M. Rabaey, and A. Wolisz, "Power-efficient rendez-vous schemes for dense wireless sensor networks," in *Proceedings of the IEEE International Conference on Communications*, pp. 3769–3776, Paris, France, June 2004.

[21] Q. Ying, W. Yuan, and N. Hu, "Improving the efficiency of harvesting electricity from living trees," *Journal of Renewable and Sustainable Energy*, vol. 7, no. 6, p. 063108, 2015.

[22] A. Tanaka, T. Ishihara, F. Utsunomiya, and T. Douseki, "Wireless self-powered plant health-monitoring sensor system," in *Proceedings of the IEEE Sensors*, pp. 1–4, Taipei, Taiwan, October 2012.

[23] Z. Hao, G. Wang, W. Li, J. Zhang, and J. Kan, "Effects of electrode material on the voltage of a tree-based energy generator," *PLoS One*, vol. 10, no. 8, article 0136639, 2015.

[24] Y. Kim, C. W. Park, and T.-J. Lee, "MAC protocol for energy-harvesting users in cognitive radio networks," in *Proceedings of the 8th International Conference on Ubiquitous Information Management and Communication*, pp. 1–4, Siem Reap, Cambodia, January 2014.

[25] A. Goldsmith, *Wireless Communications*, Cambridge University Press, Cambridge, UK, 2005.

[26] A. Thakkar and K. Kotecha, "A new Bollinger band based energy efficient routing for clustered wireless sensor network," *Applied Soft Computing*, vol. 32, pp. 144–153, 2015.

[27] J. K. Sharma, *Fundamentals of Business Statistics*, Vikash Publishing House, Chennai, India, 2nd edition, 2014.

4

Efficient Maintenance of AODV Routes in the Vehicular Communication Environment with Sparsely Placed Road Side Units

Chanhyuk Cho and Sanghyun Ahn ⓘ

Department of Computer Science and Engineering, University of Seoul, Seoul, Republic of Korea

Correspondence should be addressed to Sanghyun Ahn; ahn@uos.ac.kr

Academic Editor: Jeongyeup Paek

Thanks to the vehicular communication network, vehicles on the road can communicate with other vehicles or nodes in the global Internet. In this study, we propose an enhanced routing mechanism based on AODV so that road side units (RSUs) can provide continuous services such as video streaming services to vehicles which may be intermittently located outside of the coverage areas of RSUs. In the highway environment with sparsely placed RSUs, the communications between RSUs and vehicles are frequently disconnected due to high vehicular speeds. To resolve this problem, both V2I and V2V communications are utilized. In order to reduce the route recovery time and the number of route failures in the sparsely placed RSU environment, backup routes are established through the vehicles with longer direct communication duration with the RSU. The backup route substitutes the main route upon route disconnection. Also, for the efficient handover to the next RSU, the route shortening mechanism is proposed. For the performance evaluation of the proposed mechanism, we carried out the NS-3-based simulations.

1. Introduction

The vehicular communication environment is composed of vehicles and road side units (RSUs) [1, 2]. A vehicle may act as a source, a destination, and/or a relay node and can communicate with other vehicles via vehicle-to-vehicle (V2V) communications and with nodes outside of the vehicular communication network through RSUs (i.e., vehicle-to-infrastructure (V2I) communications). RSUs form the vehicular communication infrastructure but require high installation and maintenance cost. Therefore, RSUs cannot be placed densely enough to cover all the areas. This implies that a vehicle may not have direct communication links with any RSUs at the interim area of two adjacent RSUs. In this case, for seamless data delivery services from RSUs, it is required to have a routing mechanism that can provide routes from RSUs to a vehicle via other vehicles. That is, both V2I and V2V communications are required. In this study, we specifically consider the *target* vehicular communication environment

with dispersedly placed RSUs from which steadily moving vehicles receive seamless data delivery services. The reason for assuming steadily moving vehicles is that the major focus of our study is not the communications between vehicles, but the communications between RSUs and vehicles.

So far, various types of routing protocols have been proposed for vehicular communications. We can categorize routing protocols for vehicular communications into topology-based and location-based (or geographic) routing protocols. In the topology-based routing protocol, the routes of the source and destination node pairs are maintained at nodes and are to be recovered in the case of route failures. Topology-based routing protocols are further categorized into reactive (or on-demand) and proactive routing protocols [3, 4]. Reactive routing protocols find a route for a source and destination node pair upon a route setup request, and proactive routing protocols configure routes prior to route requests. Due to frequent topology changes of vehicular networks, the reactive routing mechanism is preferred to the proactive one. On the other hand, geographic routing protocols do not maintain route information at nodes. Instead, each node determines the next-hop node for each data packet to be forwarded. In other words, each packet has to struggle to find its own way to the destination, which may result in significant delay and computing overhead. Also, nodes are required to periodically exchange geographic location information with other nodes, incurring significant communication overhead.

In this study, we focus on topology-based routing protocols, especially reactive routing protocols, in the target vehicular communication environment. Among the reactive routing protocols, we aim at applying the ad hoc on-demand distance vector (AODV) routing protocol [5] to the target vehicular communication environment. Even though AODV is originally proposed for mobile ad hoc networks (MANETs), AODV has been considered as a candidate routing protocol for the vehicular communication environment by many researchers [6–14]. Because AODV requires nodes to maintain routes, data packets can be delivered smoothly over the routes if they stay steadily.

Most of the previous work on applying AODV to the vehicular communication environment focuses on providing reliable routes in the harsh vehicular communication environment. To our knowledge, [14] is the only work related to applying AODV to the V2I environment by adopting the AODV+ [15]. However, M-AODV+ in [14] does not consider the quality of continuous data delivery services, such as video streaming services, from RSUs. Therefore, in this study, we propose a backup route mechanism in the coverage area of an RSU, an inter-RSU handover mechanism and a route shortening mechanism for the RSU to which a vehicle is heading. We assume a single-lane highway with steadily moving vehicles since our mechanism focuses on maintaining routes from/to RSUs. Because the vehicles on the route from an RSU to the destination vehicle can be on any lane in a multiple-lane road, the lane which the vehicle is on does not make a difference to our proposed mechanism. Thus, our mechanism is easily applicable to a multiple-lane road.

The rest of the study is organized as follows. In Section 2, we will describe the related work on the adaptation of AODV in the vehicular communication environment. Section 3 describes the detailed operation of our proposed mechanism. In Section 4, we evaluate the performance of our mechanism and analyze the simulation results. Finally, Section 5 concludes this study.

2. Related Work

There has been research done on applying AODV to the vehicular communication environment [6–14]. The previous studies [6–13] analyze the performance of AODV in vehicular ad hoc networks (VANETs) and/or try to figure out more reliable routes in dynamic vehicular communication environment. In [14], the authors propose M-AODV+ which is a modified version of AODV+ [15]. AODV+ allows AODV to be used for the communication between a MANET node and a node in the global Internet via a gateway. M-AODV+ extends AODV+ considering the frequently changing vehicular network environment so that the destination vehicle can communicate based on V2I and I2I routes. In M-AODV+, the proactive gateway (i.e., RSU) discovery mechanism of AODV+ is adopted and the mechanism of sharing the information of mobile nodes in the coverage area of an RSU with the other RSUs is proposed. However, M-AODV+ does not provide the way of continuous delivery of data from RSUs to the destination vehicle.

In [16], the implicit backup routing AODV (IBR-AODV) mechanism is proposed. In IBR-AODV, the neighboring nodes of a route become the backup nodes of the route. Each backup node overhears the ongoing transmissions on the route, and if it detects no ACK message for a transmitted data packet, it initiates the route recovery to the nodes on the route. Then, the nodes on the route include the backup node in their routing tables. Even though IBR-AODV quickly recovers route failures, it does not try to reduce route failures. Furthermore, IBR-AODV is not designed for the vehicular communication environment.

In [17], the authors propose a relay recovery route maintenance protocol that combines both proactive and reactive route recovery mechanisms based on AODV in MANET. A node which is the common neighbor of both the upstream and the downstream nodes of a link is chosen as a relay node. The relay node overhears transmissions from the upstream and the downstream nodes and detects possible link breaks by overhearing retransmissions from the upstream node. If the relay node detects the possibility of a link break, it sends a NOTICE message to the upstream node, and upon asserting a link break, the upstream node sends a CONFIRM message to the relay node. Then, the relay node becomes the new downstream node. Liang et al. [17] also propose a route shortening mechanism in which any node can initiate the route shortening procedure when it overhears a shorter route from its neighbors. However, the mechanisms in [17] are not suitable for the RSU-based vehicular communication environment.

3. RSU-Based AODV Route Maintenance

3.1. RSU-Based Vehicular Communication Environment. For the sake of convenience, we define the terminology related to RSUs. We call the RSU currently delivering data to the destination vehicle as the serving RSU of the vehicle. The serving RSU whose coverage area has been already passed by the destination vehicle is called the passed-by serving RSU of the vehicle, and the serving RSU to which the destination vehicle is heading is called the next serving RSU of the vehicle. The RSU which is not currently serving the destination vehicle heading to it is called the next RSU of the vehicle. The handover from the passed-by serving RSU to the next RSU occurs in the interim area of those two RSUs, and the next RSU becomes the next serving RSU. The next serving RSU of the destination vehicle becomes the current serving RSU once the vehicle moves into the coverage area of the RSU. Figure 1 shows the vehicular communication environment with two RSUs, RSU1 and RSU2.

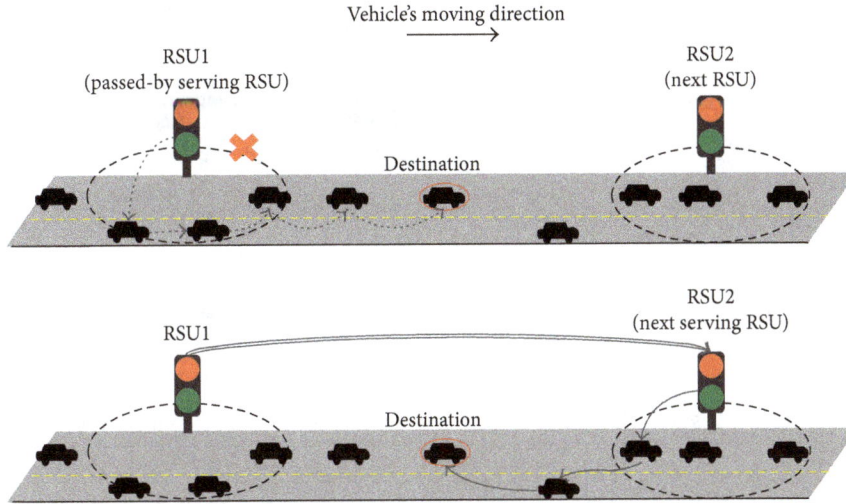

FIGURE 1: Vehicular communications based on V2I and V2V.

Because of the matter of cost, RSUs cannot be placed densely enough such that vehicles get communication services directly from RSUs. Therefore, we assume that the coverage areas of neighboring RSUs do not overlap.

3.2. Backup Route Mechanism. In the V2I communication environment, the passed-by serving RSU may detect frequent link disconnections to the 1-hop vehicle of the route to the destination vehicle because of vehicle movement. In this case, AODV recovers the broken link by making the RSU send a route error (RERR) message to the precursors. However, from the perspective of the vehicular network, the RSU does not have precursors, so the RSU has to discover a new route to the destination vehicle. This procedure is burdensome in the vehicular network with limited wireless link capacity. Figure 2 illustrates how AODV works in this situation.

We propose a backup route mechanism in the coverage area of the passed-by serving RSU to reduce the number of route recoveries. The procedure of establishing a backup route from the passed-by serving RSU is as follows:

(1) The RSU collects the location information of the vehicles in its own coverage area.

(2) Based on the location information, the RSU selects the vehicle Bn1, which is expected to stay in its coverage area the longest, as the 1-hop vehicle of the backup route.

(3) The RSU unicasts a backup RREQ (BRREQ) message with Gratuitous RREP flag = 1 to Bn1. The BRREQ message is a modified RREQ message with TTL = 1 and with the Destination IP Address = the address of Bn1.

(4) Once Bn1 receives the BRREQ message, it broadcasts an RREQ message to its one-hop neighbors.

(5) The vehicle receiving the RREQ message and having the route information to the destination vehicle sends back the Gratuitous RREP message to the RSU.

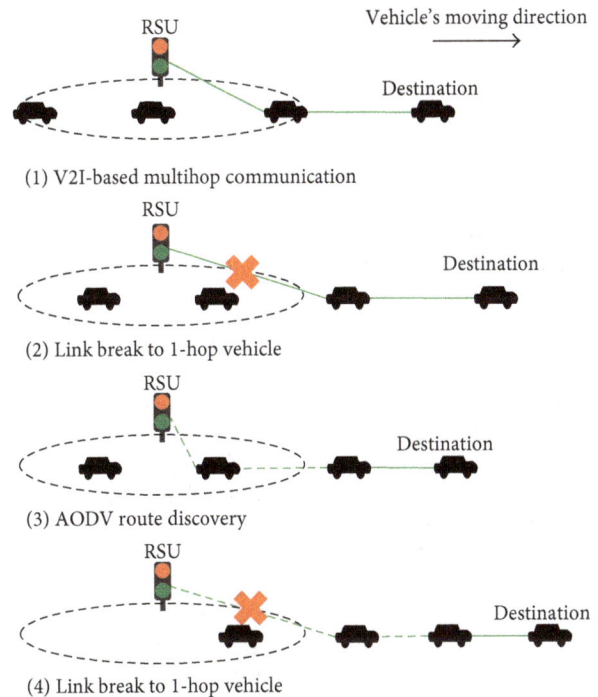

(1) V2I-based multihop communication

(2) Link break to 1-hop vehicle

(3) AODV route discovery

(4) Link break to 1-hop vehicle

FIGURE 2: The AODV route recovery procedure in the RSU-based V2I communication.

(6) Once the RSU receives the Gratuitous RREP message, it includes this backup route information in its routing table and sets the status of the backup route to inactive.

(7) Upon detecting the link disconnection to the 1-hop vehicle of the primary route, the RSU activates the backup route and deletes the main route.

Figure 3 shows how the passed-by serving RSU makes the backup route via the vehicle Bn1. We include the step numbers of the above-described procedure in the figure.

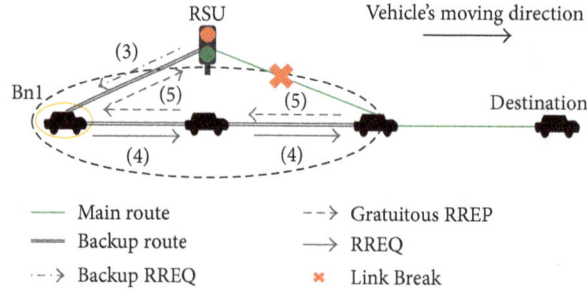

FIGURE 3: The backup route from RSU to the destination vehicle via Bn1.

FIGURE 4: The format of the RREQ message with the newly defined B and S flags.

For the BRREQ message, we include the backup route (B) flag in the RREQ message as shown in Figure 4.

3.3. Handover Mechanism.

The quality of delivery service from the passed-by serving RSU deteriorates as the destination vehicle moves away from the RSU because the route lengthens. Therefore, the passed-by serving RSU needs to decide when to handover to the next RSU. For that, the passed-by serving RSU acquires the speed information of the destination vehicle when the vehicle is in its coverage area and computes the average moving speed of the vehicle, V_{dest}. Once the destination vehicle leaves the RSU's coverage area, the RSU estimates the time, T_{mid}, when the destination vehicle is expected to arrive at the middle point of the passed-by serving RSU, R_{prev}, and the next RSU, R_{next}. For that, we assume that each RSU knows the locations of its neighboring RSUs:

$$T_{\text{mid}} = \frac{\overline{\text{Dist}\left(R_{\text{prev}}, R_{\text{next}}\right)}/2}{V_{\text{dest}}}, \quad (1)$$

where $\overline{\text{Dist}\left(R_{\text{prev}}, R_{\text{next}}\right)}$ means the distance between R_{prev} and R_{next}.

At T_{mid}, the passed-by serving RSU sends the handover request (HREQ) message with the address of the destination vehicle to the next RSU. When the next RSU receives this HREQ message, it establishes the route to the destination vehicle. Once the route is established, the next RSU becomes the next serving RSU and the next serving RSU sends back the handover done (HDONE) message back to the passed-by serving RSU and, at the same time, starts to deliver data to the destination vehicle. Then, the passed-by serving RSU stops delivering data to the destination vehicle. Thus, data are delivered to the destination vehicle through either the passed-by serving RSU or the next serving RSU. Figure 5 shows the procedure of the handover from RSU1 to RSU2.

3.4. Route Shortening Mechanism.

In Figure 6, we show the procedure of route shortening. As the destination vehicle approaches to the next serving RSU, the 1-hop vehicle, Sn1, of the next serving RSU notifies the next serving RSU of the address of the 2-hop vehicle, Sn2, and its own location and speed information. Sn1 knows the address of Sn2 because it is stored in the routing table of Sn1. The next serving RSU broadcasts a route shortening RREQ (SRREQ) message with Gratuitous RREP flag = 1, S flag = 1, and TTL = 1 and with the address of Sn2 as the Destination IP Address. Figure 4 shows the format of the SRREQ message which is the modified RREQ message with S flag = 1.

If the vehicle which is not Sn2 receives the SRREQ message, it discards the message. When Sn2 receives the SRREQ message, it replies back a Gratuitous RREP message to the next serving RSU. Once the RSU receives the Gratuitous RREQ message, it modifies the route information of the destination vehicle such that the next-hop node (i.e., the 1-hop node) to the destination vehicle is Sn2.

The SRREQ message is sent periodically by the next serving RSU at the period of T_{SRREQ} for the timely adjustment of the route:

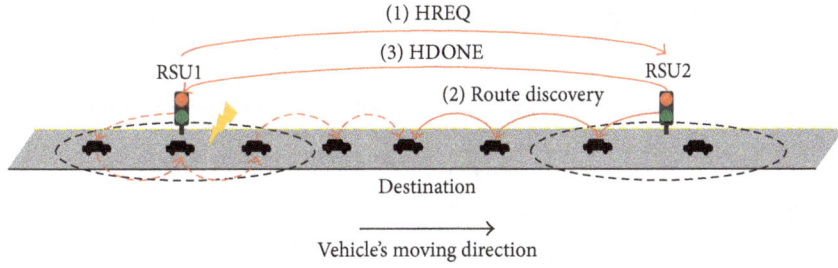

FIGURE 5: The handover procedure between the passed-by serving RSU, RSU1, and the next RSU, RSU2.

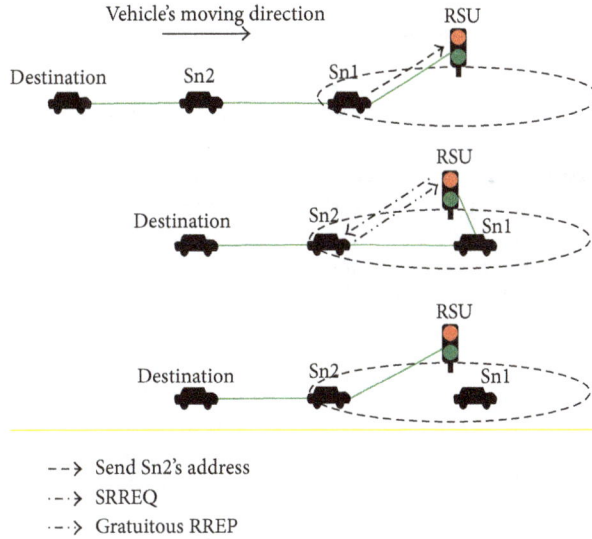

FIGURE 6: The route shortening procedure.

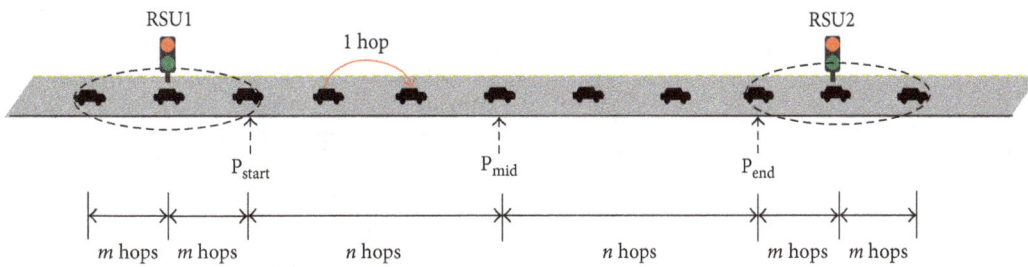

FIGURE 7: Simplified vehicular network for the control message overhead analysis.

$$T_{\text{SRREQ}} = \frac{R_{\text{vehicle}}}{N \times V_{\text{Sn1}}}, \qquad (2)$$

where N is the number of the SRREQ messages transmitted by the next serving RSU for the destination vehicle, V_{Sn1} is the speed of Sn1, and R_{vehicle} is the transmission range of a vehicle.

4. Performance Evaluation

4.1. Numerical Analysis for Performance Condition. To figure out the condition that our mechanism works better than AODV from the perspective of control message overhead,

we count the number of control messages such as RREQ, RREP, BRREQ, and SRREQ. For the simplicity of analysis, we use the simplified target vehicular network in Figure 7. In the simplified network, the intervehicular distance is the vehicular transmission range, in short, 1 hop. In this subsection, we use "hop" as the unit of the physical distance. The transmission range of each RSU is m hops and the distance between the boundaries of RSU1 and RSU2 is $2 \times n$ hops, where m and n are assumed to be integers greater than 0 for the sake of simplicity. We count the control messages generated while the destination vehicle moves from the start position P_{start} to the end position P_{end} as shown in Figure 7.

As for AODV, the number of control messages generated is C_{AODV}:

$$C_{AODV} = 2n \times (1 + (2m + 1) + 2), \qquad (3)$$

where "$2n$" is the number of route recoveries while the destination vehicle moves from P_{start} to P_{end}. For each hop the destination vehicle moves, "$(1 + (2m + 1) + 2)$" control messages are generated. In "$(1 + (2m + 1) + 2)$" of (3), the first term "1" is for the RREQ message generated by RSU1 to recover the route, the second term "$(2m + 1)$" is for the RREQ messages generated by the $(2m + 1)$ vehicles in the coverage area of RSU1, and the third term "2" is for the RREP messages generated by the 1-hop and 2-hop vehicles from RSU1.

The control message overhead of our mechanism, $C_{Proposed}$, is:

$$C_{Proposed} = \frac{n}{2m} \times 2 \times (1 + 2m) + 2 \times (1 + n) + 2 \times n, \qquad (4)$$

where "$n/2m$" is the number of proactive route recoveries while the destination vehicle moves from P_{start} to P_{end}. For each proactive route recovery, "$2 \times (1 + 2m)$" messages are generated to establish a backup route. Here, "1" is for the BRREQ message generated by RSU1 and "$2m$" is for the RREQ messages generated by the vehicles in the coverage area of RSU1 excluding the vehicle located at P_{start} and "2" is multiplied for the corresponding RREP messages. The second term "$2 \times (1 + n)$" is for the RREQ and the RREP messages generated during the new route discovery procedure initiated by RSU2 to the destination vehicle located at P_{mid}. And the third term "$2 \times n$" is for the SRREQ and the corresponding RREP messages generated during the route shortening of RSU2.

The condition that our mechanism outperforms AODV in terms of the control message overhead is $C_{Proposed} < C_{AODV}$, which results in the following condition:

$$n > \frac{m}{2 \times m^2 + m - 1}. \qquad (5)$$

With any m and n values, $m \geq 1$ and $n \geq 1$, the condition in (5) is satisfied. That is, our mechanism always outperforms AODV in terms of the control message overhead.

4.2. Simulation-Based Performance Evaluation. We carried out the NS-3 simulator-based performance evaluation [18]. The IEEE 802.11p is used as the MAC protocol for vehicular wireless communications. The simulation network is a 2.5 km single-lane highway with two RSUs and eleven vehicles.

The simulation parameters are given in Table 1. The transmission range of each vehicle is set to 130 m and that of an RSU is set to 250 m. The first RSU, RSU1, is placed at 500 m and the second RSU, RSU2, at 1800 m from the leftmost point, as shown in Figure 8(a). At first, the vehicles are placed such that the sixth vehicle is placed right under RSU1 as shown in Figure 8(a). Figure 8(b) shows the final status of the simulation network.

We evaluate and compare our mechanism with the AODV with or without our proposed handover mechanism.

TABLE 1: Simulation parameters.

Parameter	Value
Network size	100×2500 m
Simulation time	100 sec
Number of vehicles	11
Intervehicular distance	100 m
Vehicle transmission range	130 m
Vehicle velocity	60 km/h, 80 km/h, 100 km/h
Number of RSUs	2
Inter-RSU distance	1300 m
RSU transmission range	250 m
Traffic model	CBR (150 packets/sec)
Packet size	1024 bytes
MAC protocol	IEEE 802.11p

Through the simulations, we measured the packet delivery ratio and the packet delivery delay for the verification of the received quality of service at the destination vehicle.

Figures 9–11 are the simulation results of the case that the vehicular speed is 60 km/h. Figure 9 shows the number of the packets received at the destination vehicle for the previous 10-second time interval (i.e., the result of 45 seconds means the number of packets received during the 35- to 45-second interval). Because the handover occurs around 40 seconds after the simulation start, we can observe the performance of the backup mechanism for the first 40 seconds and, after that, the performance of the proposed route shortening mechanism. As Figure 9 shows, the proposed mechanism performs better than the other two mechanisms throughout the simulation. This indicates that our backup route and route shortening mechanisms perform well enough. Besides, after 60 seconds, AODV does not deliver any more data packets to the destination vehicle. Figure 10 shows the average packet delivery ratio during the simulation start to the point (i.e., the result of 45 seconds means the average packet delivery ratio during the 0- to 45-second interval). We can observe a significant performance improvement for the first 40-second interval thanks to the proposed backup route mechanism. In Figure 11, we can observe that our proposed mechanism shows the smallest delay all the time. After around 50 seconds, the delay of the AODV without the handover mechanism stays the same because no more packets are delivered.

Figures 12 and 13 show the performance results for various vehicular speeds. Figure 12 shows the average packet delivery ratio for the simulation time for the vehicular speeds of 60 km/h, 80 km/h, and 100 km/h. The performance of the AODV without the handover mechanism is the worst, and the proposed mechanism outperforms the AODV with the handover mechanism for all cases. The higher the vehicular speed is, the less the packets are delivered to the destination vehicle. Figure 13 is the graph showing the average packet delivery delay for the various vehicular speeds throughout the simulation. The AODV without the handover mechanism gives very large delay, and our mechanism gives slightly lower delay than the AODV with the handover mechanism. From these results, we can deduce that the backup route mechanism improves the packet delivery

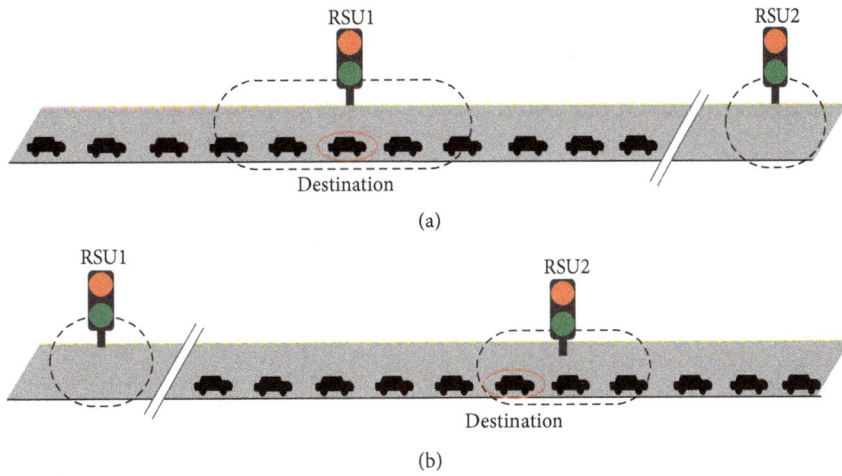

FIGURE 8: Simulation network: (a) initial status and (b) final status.

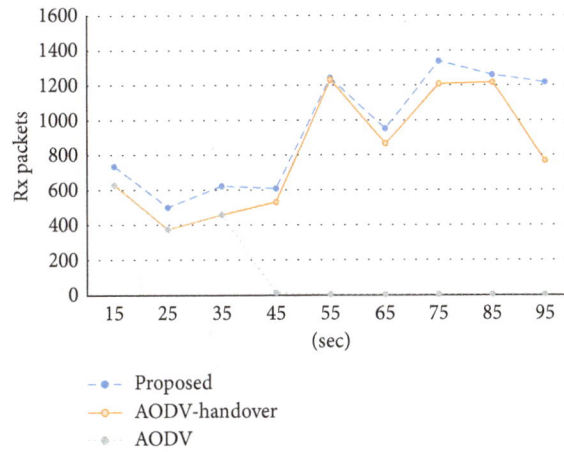

FIGURE 9: The number of received packets for the previous 10-second interval (for the vehicular speed of 60 km/h).

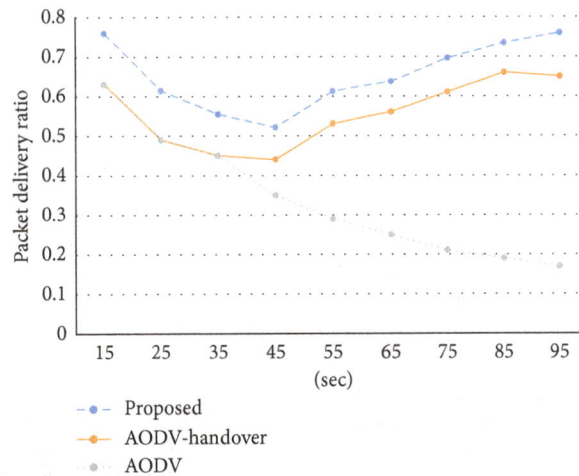

FIGURE 10: The average packet delivery ratio from the simulation start to that point (for the vehicular speed of 60 km/h).

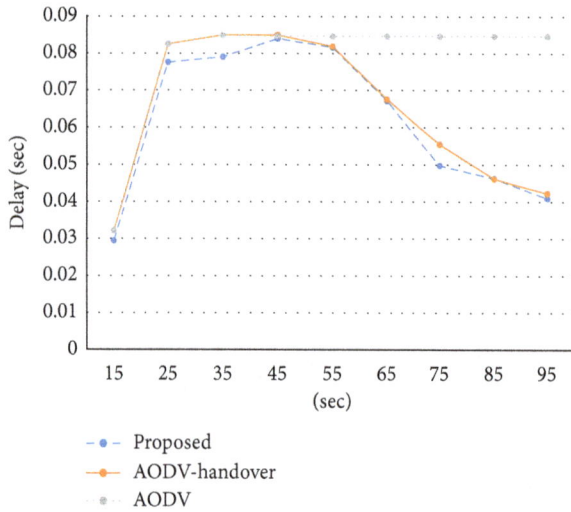

FIGURE 11: The average packet delivery delay for the previous 10-second interval (for the vehicular speed of 60 km/h).

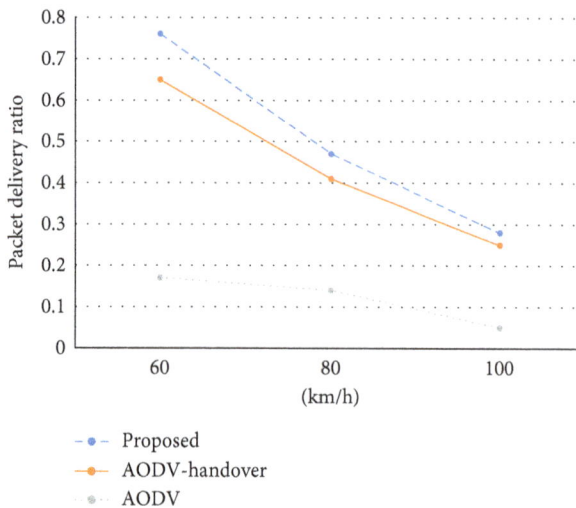

FIGURE 13: The average packet delivery delay for the various vehicular speeds (for the total simulation time).

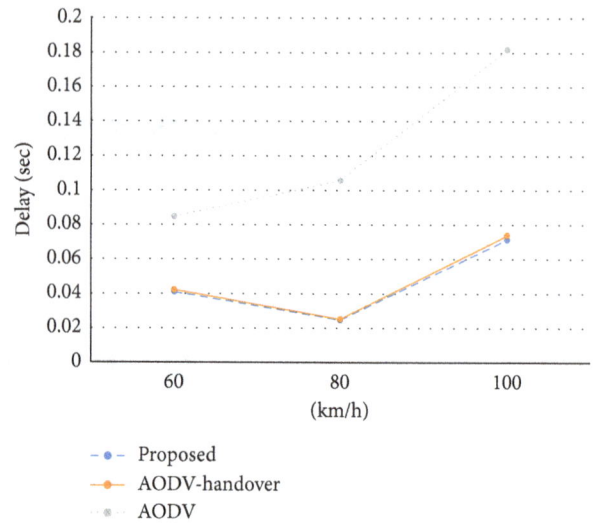

FIGURE 12: The average packet delivery ratio for the various vehicular speeds (for the total simulation time).

ratio but does not give significant impact on the delay performance.

5. Conclusion

AODV is a reactive routing protocol designed for MANET with mobile nodes. The vehicular communication network is similar to MANET, so AODV can be applied to the vehicular communication network. In this study, we tried to figure out how to apply AODV to the vehicular communication network with sparsely placed RSUs. By considering the constrained vehicular movement, we proposed the backup route mechanism in the coverage area of the passed-by serving RSU and the simple handover mechanism from the passed-by serving RSU to the next RSU and the route shortening mechanism in the coverage area of the next serving RSU. For that, we modified the RREQ message by

adding the B flag and the S flag. The performance of the proposed mechanism was verified by the NS-3-based simulations. Through the simulations, we showed that our mechanism improves the performance in terms of packet delivery ratio and delay compared with AODV with or without the handover mechanism.

Acknowledgments

This work was supported by the National Research Foundation of Korea (NRF) grant funded by the Korean government (MSIT) (no. NRF-2015R1A2A2A04005646) and the MSIT (Ministry of Science and ICT), Korea, under the ITRC (Information Technology Research Center) support program (IITP-2017-2015-0-00363) supervised by the IITP (Institute for Information and communications Technology Promotion).

References

[1] K. Abboud, H. A. Omar, and W. Zhuang, "Interworking of DSRC and cellular network technologies for V2X communications: a survey," *IEEE Transactions on Vehicular Technology*, vol. 65, no. 12, pp. 9457–9470, 2016.

[2] C. M. Silva, B. M. Masini, G. Ferrari, and I. Thibault, "A survey on infrastructure-based vehicular networks," *Mobile Information Systems*, vol. 2017, Article ID 6123868, 28 pages, 2017.

[3] R. Kumar and M. Dave, "A comparative study of various routing protocols in VANET," *International Journal of Computer Science Issues (IJCSI)*, vol. 8, no. 4, 2011.

[4] M. Barcia-Campos, D. G. Reina, S. L. Toral et al., "Performance evaluation of reactive routing protocols for VANETs in urban scenarios following good simulation practices," in *Proceedings of International Conference on Innovative Mobile and Internet Services in Ubiquitous Computing (IMIS)*, Fukuoka, Japan, July 2015.

[5] C. E. Perkins, E. M. Belding-Royer, and S. Das, *Ad Hoc on-Demand Distance Vector (AODV) Routing, IETF RFC 3561*, 2003.

[6] B. Ding, Z. Chen, Y. Wang, and H. Yu, "An improved AODV routing protocol for VANET," in *Proceedings of International Conference on Wireless Communications and Signal Processing (WCSP)*, Nanjing, China, October 2011.

[7] S. Harrabi, I. B. Jaafar, and K. Ghedira, "An enhanced AODV routing protocol for vehicular adhoc networks," in *Proceedings of BDAW*, Paris, France, November 2016.

[8] T. Kaur and A. K. Verma, "Simulation and analysis of AODV routing protocol in VANETs," *International Journal of Soft Computing and Engineering*, vol. 2, no. 3, 2012.

[9] A. Moravejosharieh, H. Modares, R. Salleh, and E. Mostajeran, *Performance Analysis of AODV, AOMDV, DSR, DSDV Routing Protocols in Vehicular Ad Hoc NetworkResearch Journal of Recent Sciences ISSN*, vol. 2277, p. 2502, 2013.

[10] P. Mittal and B. Yadav, "Improving AODV routing in vehicular ad hoc networks," *International Journal of Innovative Research in Computer and Communication Engineering*, vol. 4, no. 9, 2016.

[11] D. Rathi and R. R. Welekar, "Performance evaluation of AODV routing protocol in VANET with NS2," *International Journal of Interactive Multimedia and Artificial Intelligence*, vol. 4, no. 3, pp. 23–27, 2017.

[12] A. Datta, C. Chowdhury, and S. Neogy, "Ant-AODV-VANET: a bio-inspired routing protocol for VANET," in *Proceedings of International Conference on Emerging Research in Computing, Information, Communication and Applications (ERCICA)*, Bangalore, India, July2016.

[13] D. Zhu, G. Cui, and Z. Fu, "DT-AODV: an on-demand routing protocol based DTN in VANET," *Applied Mathematics and Information Science*, vol. 8, no. 6, pp. 2955–2963, 2014.

[14] J. Wantoro and I. W. Mustika, "M-AODV+: an extension of AODV+ routing protocol for supporting vehicle-to-vehicle communication in vehicular ad hoc networks," in *Proceedings of IEEE International Conference on Communication, Networks and Satellite (COMNETSAT)*, Jakarta, Indonesia, November 2014.

[15] A. A. Hamidian, *A Study of Internet Connectivity for Mobile Ad Hoc Networks in NS 2*, M.S. thesis, Lund University, Lund, Sweden, 2003.

[16] J. Jeon, K. Lee, and C. Kim, "Fast route recovery scheme for mobile ad hoc networks," in *Proceedings of International Conference on Information Networking (ICOIN)*, Kuala Lumpur, Malaysia, January 2011.

[17] Z. Liang, Y. Taenaka, T. Ogawa, and Y. Wakahara, "Proreactive route recovery with automatic route shortening in wireless ad hoc networks," in *Proceedings of International Symposium on Autonomous Decentralized Systems (ISADS)*, Tokyo, Japan, March 2011.

[18] *The Network Simulation, NS-3*, http://www.nsnam/org.

5

Reliable Flying IoT Networks for UAV Disaster Rescue Operations

Taemin Ahn,[1] **Jihoon Seok,**[2] **Inbok Lee,**[3] **and Junghee Han** ⓘ[4]

[1]*Samsung Electronics Co., Ltd., Yong-tong-gu, Suwon, Gyung-gi-do, Republic of Korea*
[2]*SK Hynix Co., Ltd., 2091 Gyeongchung-daero, Bubal-eup, Icheon-si, Gyeonggi-do, Republic of Korea*
[3]*Department of Software, Korea Aerospace University, 6 Hanggongdaehang-ro, Deokyang-gu, Goyang-si,*
Gyeonggi-do 412-791, Republic of Korea
[4]*Department of Telecommunication and Computer Engineering, Korea Aerospace University, 76 Hanggongdaehang-ro,*
Deokyang-gu, Goyang-si, Gyeonggi-do 412-791, Republic of Korea

Correspondence should be addressed to Junghee Han; junghee@kau.ac.kr

Academic Editor: Omprakash Gnawali

Recently, UAVs (unmanned air vehicles) have been developed with high performance, and hence, the range of system utilizing UAVs has also been widening. UAVs are even considered as connected mobile sensors and are claimed to be the future of IoT (Internet of Things). UAVs' mission fulfillment is relying on the efficiency and performance of communication in a FANET (Flying Ad hoc NETwork) environment where UAVs communicate with each other through an ad hoc network without infrastructure. Especially, for mission-critical applications such as disaster rescue operations, reliable and on-time transmission of rescue information is very critical. To develop the reliable FANETs, a realistic network simulation platform for UAV communication has become an important role. Motivated by this observation, this paper first presents a study on realistic FANET environment simulation platform. On top of the proposed platform, we also design a stable UAV communication protocol with high packet delivery and bounded end-to-end communication delay.

1. Introduction

Various types of UAV applications and services have been emerged, which range from commercial applications such as a real estate photography to public safety services. In the past, UAVs were used mainly in the field of defense including border surveillance and ground attack, but now they are used in various industrial fields such as logistics, IT, and agriculture. They have also been used in suppression, disaster relief, traffic monitoring, government departments, volcanic eruptions, and wildlife surveillance [1]. For example, Google's Rune Project, which uses airship-based UAVs to expand its wireless Internet penetration network, Amazon's next-generation delivery system using unmanned helicopters, Yamaha's industrial UAV, for use in agriculture, and AirDroid's personal drones that capture images and photographs are actively under development.

With emergence of smart and converged services for these autonomous UAVs, there has been rapid increase in the need for reliable connection among UAVs and control centers. However, the communication environment between UAVs changes frequently due to the fast movement of the nodes. In addition, since it moves in three-dimensional space, the surrounding topography changes frequently as well. This can cause frequent disconnection of the network link among UAVs and the infrastructure. To support reliable communication, reliable FANETs play important roles.

Inspired by this observation, this paper develops a new technology for reliable UAV networks. Main features of the proposed scheme are as follows:

(i) *We design and implement a realistic UAV simulation platform*: In order to develop a technology optimized for FANET environment, realistic network simulation is critical. To do this, we design a platform with combining a network simulator NS-3 and a robot simulator Gazebo together. Using these two tools, we develop a realistic UAV simulation platform that can realize and simulate UAV operations in the realistic FANET environment.

(ii) *We propose a reliable FANET routing methodology*: We examine problems of traditional routing schemes and we design a stable UAV communication algorithm with high packet delivery and bounded end-to-end communication delay. Specifically, the proposed routing algorithm has three main features: (1) hierarchical node clustering, (2) time division packet transmission, and (3) real-time estimation of transmission delay.

The rest of the paper is organized as follows: Section 2 introduces existing simulation platforms and routing protocols with discussing problems and limitations of existing platforms and routing protocols in FANETs. In Section 3, we introduce a design of the proposed simulation platform and a routing protocol developed for FANET environment. Section 4 explains the experimental procedure and the results for the performance evaluation of the proposed approach. Finally, Section 5 concludes this paper with discussion.

2. Related Works

2.1. Simulation Platforms. To develop reliable UAV systems and applications, it is important to be able to simulate UAV operations in a realistic way. So far, many network simulators have been developed, such as NS-2 [2], NS-3 [3], QualNet [4], and OMNet [5]. To use these simulators for mobile networks, either a mobility model like Gauss–Markov [6] is specified or trace files created by other software tools like SUMO [7] are given as inputs for the network simulators. Although SUMO software can produce very realistic motion patterns of nodes, there are some limitations that it can only make these patterns a priori, not on the fly. Also, it can only generate movement traces in 2D space. In contrast, Gauss–Markov mobility [6] model in NS-3 can make the movement of nodes in 3D space. However, a mobility model is based on statistical probability, which implies that previous movements of nodes influence the next movement decision in the model. Therefore, it has fundamental limitations to simulate UAV motions for disaster search and rescue operations. More importantly, these network-based simulators cannot reflect UAVs' realistic motions and behaviors affected by physical and mechanical features and its surrounding environment.

On the other hand, robot simulators like Gazebo [8] can control the physical movement of UAVs. Gazebo is a tool widely used for the development of robots with various physical characteristics, such as conveyor belts, unmanned probes, and line tracers. This tool can incorporate various sensor modules for these robots, and so we can make a UAV equipped with Wi-Fi signal detection sensor or a camera sensor. Also, it is possible to implement a collision avoidance algorithm inside Gazebo for controlling motion of the robot. Gazebo is an OpenGL-based simulation tool, and it can visualize the behavior of the robot as shown in Figure 1(b).

Overall, solely using the existing network simulators or robot simulators, we can test only either network performance or robot motion operations, not both considered. To handle this issue, this paper proposes a new FANET platform with combining the network simulator NS-3 and the robot simulator Gazebo together. To incorporate these two tools, we adopt ROS (Robot Operating System) [9] as a middleware. ROS provides various development tools as well as various functions for hardware abstraction, subdevice control, and interprocess message passing for software development. As shown in Figure 2, ROS handles a process that performs computation as a node. Each node sends and receives messages to the publisher-subscriber structure through a channel called *Topic*. Through this, it is possible to exchange information with nodes or to share desired information with a desired node.

2.2. Traditional Routing Protocols for MANETs. In this section, we introduce some of ad hoc routing protocols that are commonly used in MANET (Mobile Ad hoc NETwork) or VANET (Vehicular Ad hoc NETwork) environments and examine applicability of the existing protocols to UAV system in a FANET environment. An ad hoc routing protocol is specially designed to cope with link disconnection or topology change due to high mobility feature in an ad hoc network environment.

First, ad hoc routing protocol is a method to cope with link disconnection or topology change due to node mobility in ad hoc network environment where wireless communication is performed between nodes scattered without AP. There are three types of routing protocols: (1) table-driven proactive routing protocols, (2) on-demand reactive routing protocols, and (3) hybrid routing protocols. A proactive routing protocol is a method of preliminarily computing a route to create a routing table and updating it periodically by an algorithm. Typical proactive routing protocols are DSDV [10] and OLSR [11]. On the other hand, reactive routing protocols (e.g., DSR [12] and AODV [13]) are also called on-demand routing protocols. The reason for this is that, contrary to the proactive routing protocol, there is no operation if the packet transmission request does not occur, but when the packet transmission request is generated, the protocol for determining the forwarding path gets started only at that time. The hybrid routing protocol is a routing method that combines the advantages of proactive routing and reactive routing. Depending on the network conditions, proactive routing and reactive routing may be used in combination.

In addition, most VANET routing algorithms can be classified as either topology-based or geography-based. Due to its simplicity, geography-based routing has been proved to be more suited for VANET. The most critical issue in location-based approaches is that all nodes in VANET environment including source, destination, and forwarding nodes can move dynamically at a high speed. Hence, the positional information of destination node might need to be propagated by neighbors to all nodes at short intervals. Also, these geography-based algorithms have inherent limitations in the sense that packets cannot be delivered due to network disconnection or partition under low traffic density. To overcome these limitations, delay tolerant network- (DTN-) based methods have been proposed. In DTN protocols like vehicle-assisted data delivery (VADD) [14], a moving vehicle carries a packet and waits until a new vehicle moves into its vicinity.

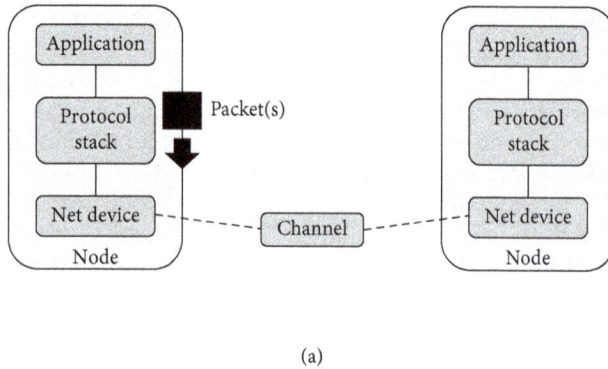

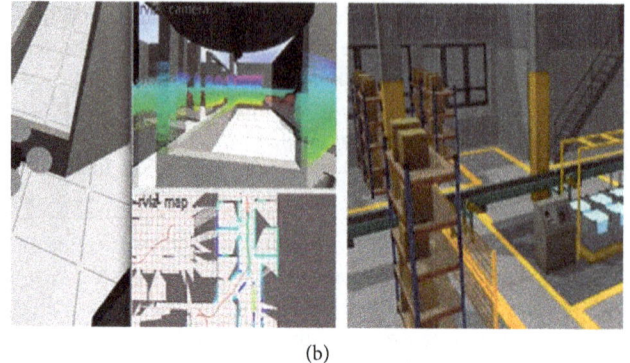

FIGURE 1: Existing network and robot simulators. (a) NS-3 structure [3]. (b) Gazebo [8].

3. Proposed Approach

Although various routing protocols have been developed so far, applying these protocols for FANETs is a challenging task due to high speed mobility of UAVs. Furthermore, the network simulation platform reflecting the UAV's physical features is not available yet, to the best of our knowledge. In what follows, this paper identifies important issues to solve for FANETs and proposes the integrative simulation platform. Then, we present the new routing protocol for FANETs with complementing the limitations of traditional methods.

3.1. Problem Statement. Based on the above observations, we identify a target scenario of UAV systems for disaster relief as follows:

(i) *Need for ad hoc flying networks*: if a disaster such as an earthquake, a landslide, or a tsunami occurs, it is not easy to use existing infrastructure in the area. Therefore, there is a need for a method of networking and communicating between UAVs without the help of an AP or a base station.

(ii) *Need for collaboration* among *UAVs*: even if the performance of the UAV improves day by day, there is a limit to finding a large number of survivors in a short period of time with only one high-performance UAV. Therefore, we assume a scenario in which a plurality of UAVs looks for survivors.

(iii) *Reliable data sharing among UAVs*: UAVs are fast enough to reach tens to hundreds of kilometers per hour and are not as dense as the nodes in MANET or VANET environments. In this environment, it is most important to guarantee packet transmission within a limited time in order to quickly and accurately transmit the message related to collision avoidance, flight control message, or survivor discovery information according to mission execution to other nodes.

(iv) *Need for UAVs with various sensors*: for effective disaster rescue operations with UAVs, various sensors such as camera sensor, Wi-Fi sensor, voice signal sensor are needed to detect survivors. Especially, in case of survivors buried because of disasters, it is

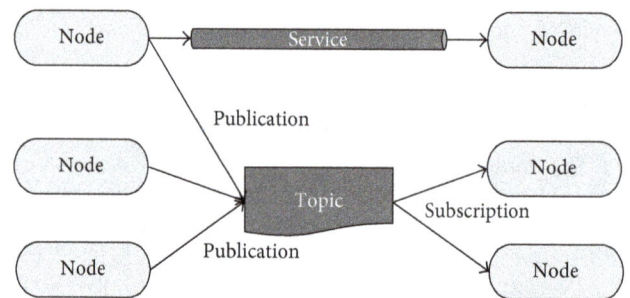

FIGURE 2: Robot Operating System (ROS) [9].

difficult to detect survivors only with camera sensors. Therefore, each UAV should be able to attach a Wi-Fi signal detection sensor or a voice signal detection sensor to detect the location of a survivor and perform rescue operation.

3.2. Development of FANET Simulation Platforms. As we mentioned before, it is important to be able to simulate UAV operations in a realistic way. Solely using the existing networks simulators or robot simulators, we can test only either network performance or robot motion operations, not both considered. To overcome this limitation, this paper proposes a holistic UAV simulation platform by integrating network simulators and robot simulators together. Specifically, we develop a new simulation platform to realistically implement the FANET environment as described in Figure 3. By adopting a robot simulator Gazebo [8], the proposed platform is able to represent the realization of UAV's mechanical movement such as cluster flight, collision avoidance, and 3-dimensional flight. Furthermore, the proposed platform with Gazebo can support operations of various sensors including a radar and a camera attached UAVs as well. On the other hand, we also integrate a conventional network simulator [2, 3] in the proposed platform to realistically analyze communication features among UAVs. Since multiple UAVs collaborate with each other to perform mission critical operations, successful on-time communications among UAVs are very essential. Using the NS-3 network simulator with Gazebo, we can accurately analyze communication performance of UAVs while simulating their flight motion and sensor operations at the same time.

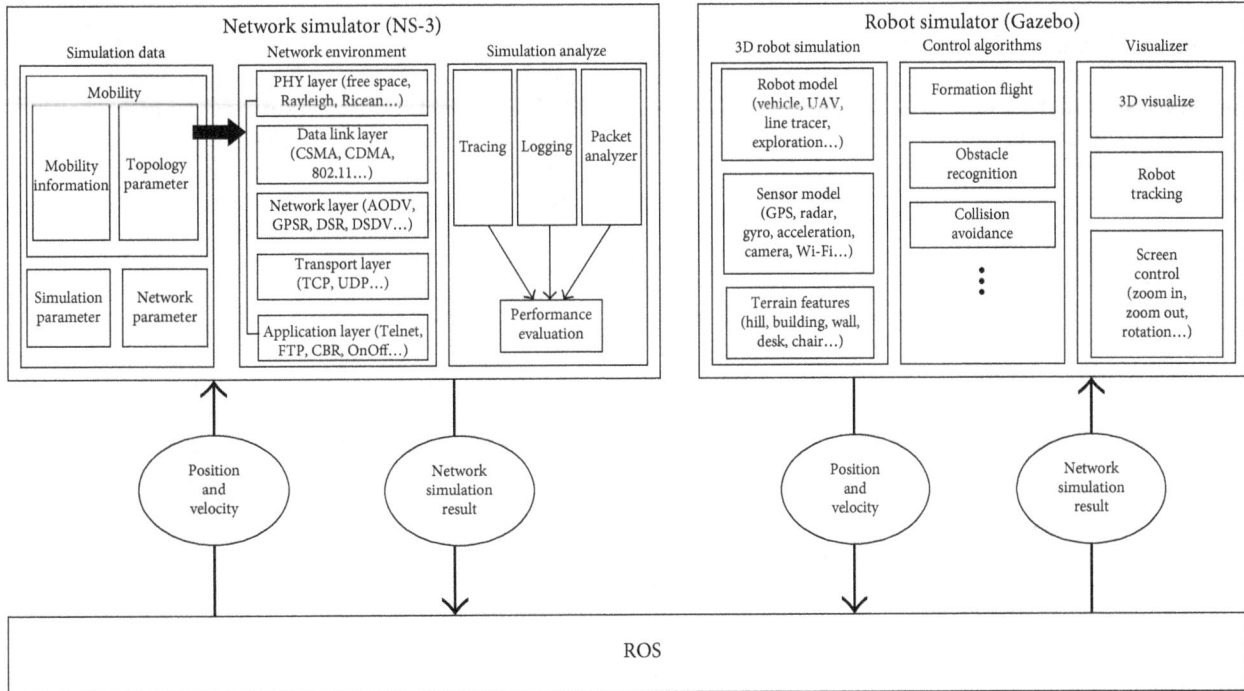

FIGURE 3: The proposed architecture of a FANET simulation platform.

To make both NS-3 and Gazebo interoperate with each other, we use a middleware called ROS [9], which is widely used in a robot control field. For example, Gazebo simulates a realistic robot motion in the simulator, making the same movement as a real UAV. UAV motion-related information is transmitted to the NS-3 network simulator through the ROS, which is illustrated as the right-most arrow line in Figure 3. In this platform, Gazebo transmits UAVs' current locations and velocities to NS-3. Based on this information, NS-3 can set up the nodes dynamically, each of which corresponds to each UAV in Gazebo. Specifically, initial locations of these nodes are determined based on this information from Gazebo, and these nodes move in NS-3 simulation with considering the velocities of UAVs transmitted from Gazebo on the fly. Overall, this proposed simulation platform not only enables the evaluation of the communication performance according to the realization of the UAV's mission in NS-3, but also realistically measures the UAV's performance according to the current communication state.

In order to integrate a network simulator NS-3 and a robotic simulator ROS, the two simulators must exchange information with each other. Specifically, NS-3 needs motion information of the UAVs created by the robot simulator, and the robot simulator requires the result of the communication from the network simulator. In the proposed platform, these two different processes in NS-3 and Gazebo communicate with each other via ROS interface to exchange information. In this system, a robot created by Gazebo becomes a subscriber node. ROS generates a publisher node to control the UAV and sends a UAV motion control message (twist) to the subscriber node. The physical movement information of the UAV created through this

process can be transmitted to the subscriber node of the network simulator through the publisher node of the robot simulator and conversely the mission execution message of the UAV generated at the publisher node of the network simulator can be transmitted to the subscriber node of the robot simulator (in the current implementation, these two processes are designed to update the data every 0.1 simulation second. The cycle can be adjusted depending on the volume of the exchanged data and degree of urgency) as described in Figure 3.

The proposed simulation platform is largely divided into a network simulator (NS-3), a robot simulator (Gazebo), and an interface part (ROS), each of which will be described in detail. NS-3 consists of three main parts: simulation data, network environment, and simulation analysis parts. The simulation data includes parameters related to the topology such as the size of the topology, the number of nodes, network layer parameters, simulation time, simulation environment setting, and finally mobility information related to node movement. Among them, the mobility information may be created by the mobility model. But the node in the proposed platform uses the ROS to acquire the motion information generated from the Gazebo instead of theoretical models. In addition, the result of the communication made by the realistic network simulation of the NS-3 is transferred to the robot simulator through the ROS. The 3D robot simulation part of the robot simulator is composed of various robot models such as UAVs and unmanned probes, as well as sensor models such as camera, radar, and GPS which are essential for the task of the robot. Furthermore, terrain features of 3D environment are also included. In addition, Gazebo incorporates an obstacle avoidance scheme with a visualizer part. The network simulation results received earlier in the NS-3 are used to determine the next movement of UAVs, and the next movement of the resulting UAV is sent to the NS-3 via the ROS in turn. In this way, two simulation tools interact to control the movement of the UAV according to the current communication situation and conversely analyze the communication performance according to the movement of the current UAV.

3.3. FANET Routing Protocol. Thanks to the proposed simulation platform in Section 3.2, we are able to realistically analyze the problems of existing routing protocols in FANETs. The proposed platform not only enables the evaluation of the communication performance in NS-3, but also realistically measures the UAV's performance reflecting physical environments provided by Gazebo tools. With this FANET simulator, we are able to design routing protocol optimized for the mission of UAV for disaster relief by realistically considering FANET environment. Also, we are able to effectively verify the performance of the proposed algorithm compared with traditional routing protocols based on the proposed simulation platform.

As mentioned in Section 2, both traditional reactive and proactive routing algorithms have their own drawbacks to be applied to FANETs. Since most information of UAVs is mission critical and so requires to be delivered (1) in high success ratio and (2) in time. To achieve these goals, the proposed algorithm adopts flooding-based routing methodology and modifies it with the following features:

(i) *Hierarchical node clustering*: the problems of flooding-based routing methods happen mainly due to the high volume of packets, which is inherent limitation of broadcasting or flooding based packet transmission. To ameliorate this problem, the proposed routing method proposes clustering-based mechanism so that it can reduce the amount of packet transmission.

(ii) *Time division packet transmission*: to avoid further packet congestion, we transmit packets in time division manner on top of clustering policy.

(iii) *Real-time estimation of end-to-end delay*: when constructing clustered FANETs, we perform real-time estimation of packet transmission delay. In FANET environments, on-time information transmission is really essential. With the proposed real-time estimation procedure, we are able to predict the worst case end-to-end delay for FANET communications and check the feasibility of a given set of clusters with a given time interval. Thanks to this procedure, we can guarantee the worst-case packet transmission delay so that we can provide reliable communication service in FANETs.

3.3.1. Hierarchical Node Clustering. The biggest problem of a traditional flooding mechanism [15] is that as the number of nodes increases, the packet overhead in the network exponentially increases. This causes collision between transmitted packets. In order to solve this problem, we propose a clustered network structure that hierarchically groups nodes—nodes are clustered in two layers, edge nodes and cluster heads. Hierarchical node clustering refers to a method of dividing a node into several groups and selecting a head node of each group and communicating through this node as shown in Figure 4. In this example, nodes in the overall topology are divided into four clusters, and each cluster has one cluster head node. Bold red solid arrows and dotted arrows indicate unicast transmission from edge nodes and flooding transmission among cluster heads, respectively. In particular, a drone on the network edge carrying out a disaster relief mission can send this information to each cluster head in a unicasting mechanism. When a cluster head drone receives unicast packets, it floods this packet to other cluster heads to share this information. In the traditional flooding mechanism, the edge node can trigger flooding while only cluster heads initiate flooding in the proposed clustering methodology. By doing this, we can significantly reduce the packet volumes on the networks and therefore can improve packet delivery rate. In Section 4, we verify this strategy using the proposed FANET platform.

3.3.2. Time Division Flooding. Although hierarchical node clustering makes it possible to reduce the total number of packets transmitted and received to a certain extent, if the number of clusters is large and/or the amount of data to be shared is huge, the proposed hierarchical method alone is not enough to support reliable FANET communication. Motivated by this observation, this paper incorporates additional feature, called time-division, in order to guarantee real-time and reliability of communication even when a large number of UAVs and drones perform many applications including collision avoidance and survivor search as assumed in this paper. Specifically, we divide the time axis of packet transmission into several time slots and allocate them to each cluster head to send packets only during their own time slots. The TDMA-based communication method of accessing the channel only to each time slot by dividing the time axis has already been studied extensively by various network protocols [16, 17]. For UAV and drone networks, a TDMA-based MAC layer protocol has been applied to a single-hop or multihop UAV network [16, 17] for improving the reliability of communication in FANETs. However, most of these TDMA algorithms are implemented in the MAC layer, which cause modification of existing MAC devices.

In this paper, we propose an algorithm that improves communication performance by adjusting the timing of sending flooding packets at the network layer, which can be compatible with existing MAC devices such as Wi-Fi and Zigbee. In the proposed protocol, each cluster head is assigned a certain time slot for packet transmission while edge nodes send unicast packets to their cluster head in CSMA/CA scheme. Figure 5 illustrates this idea with the time-axis representation of packet transmission at each cluster. In this example, we assume that five edge nodes (N_1, N_2, N_3, N_4, and, N_5) have detected the survivor's

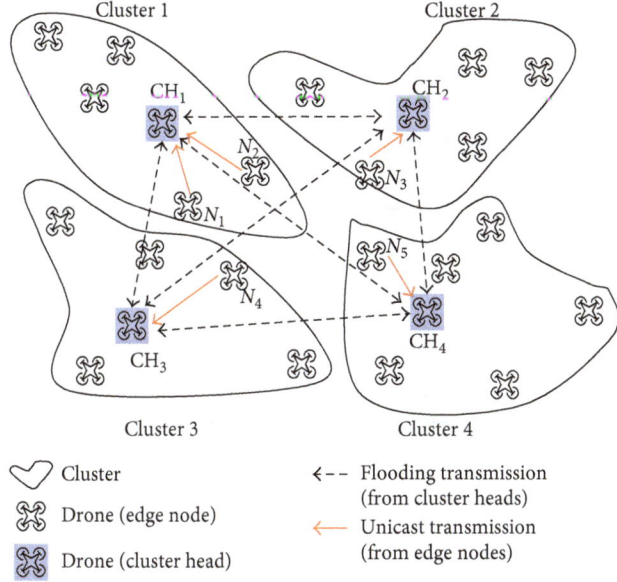

FIGURE 4: Hierarchical network structure.

signal, and they transmit this information to their own cluster heads as in the previous cluster network example of Figure 4.

As shown in this figure, unicast packets sent to the cluster head are not immediately forwarded. Instead, they are queued in the packet queue at corresponding cluster head node waiting for the time slot during which they can send the packets. For example, cluster$_1$ is assigned a time slot at t_3 and hence the packets received from N_1 and N_2 are queued and flooded at time t_3. Note that the packet from N_5 waits in the queue for a long time for a next time slot because its previous time slot has just passed before a packet arrives.

Overall, the proposed scheme with network-layer time division feature can improve packet delivery rate by avoiding collision among flooding packets without requiring modification of existing MAC layer devices.

3.3.3. Worst-Case Delay Estimation. So far, we have proposed hierarchical node clustering and time division schemes to reduce the number of packets and avoid collision of flooded packets. For mission critical disaster rescue operations of UAVs, it is also important to guarantee packet delivery within some time bounds (i.e., transmission deadline). Furthermore, since the moving speed of the UAV is as high as several tens to several hundred km/h, it is necessary to secure the packet transmission within the time limit in order to avoid collision between UAVs. For such applications, correctness of the systems depends on temporal behavior as well as functional correctness. We call such applications as real-time systems. Real-time system usually refers to hardware and software systems with time constraints. Specifically, real-time system requires that the system should response to events within specified deadlines [18, 19].

From this aspect, we now present a new methodology that guarantees real-time constraints of a target FANET system. Related to this timing delay issue, we answer the following three questions:

(i) *How to organize the cluster?* Depending on the hierarchy of clustered networks and the number of clusters, packet collision can be increased. Packet collision causes uncertainty of successful packet transmission and therefore makes it difficult to guarantee packet delivery time within deadlines. For example, as the number of clusters increases, the time slot interval of each cluster head becomes shorter, so that the risk of collision of the flooding packet sent by the head increases.

(ii) *How to assign time slots to cluster heads?* The second problem to answer is about parameters related to the time division packet transmission. The most important advantage of the time-division packet transmission is the fact that it can minimize packet collision and its periodicity can enable us to predict the worst transmission delay time. In this paper, we present the theoretical analysis for predicting worst case timing delay of packet transmission under the proposed hierarchical clustered networks.

Figure 6 illustrates the process for estimating the worst-case packet transmission delay time. In this example, the node N_1 generates the first packet q_1 at time t_1 and the second packet q_2 at t_2, and the cluster head (i.e., CH$_1$) floods q_1 and q_2 received from N_1 during its assigned time slot (i.e., t_3) (for simplicity of explanation, we assume propagation delay is zero in this example). These packets are received by CH$_2$ at t_3 assuming no propagation delay, and they have been queued waiting for the time slot for CH$_4$. Suppose the time slot for CH$_4$ is not long enough to accommodate these two packets. In this scenario, the packet q_1 might be flooded at t_4 first and then the second packet q_2 needs to wait for the next time slot at t_5. As a consequence, the destination node N_0 would receive q_1 and q_2 at t_4 and t_5, respectively.

$$\text{WD} = \sum_{\text{cluster } c \in \text{path}} \text{WD}_c. \tag{1}$$

In the scenario shown in Figure 6, the worst end-to-end transmission delay (i.e., WD) is determined by the sum of the worst-case transmission delay times of each cluster (i.e., WD$_c$) on the transmission path as shown in (1). If the packet generation period is p, the generation time of the qth packet is $(q-1) * p$. Suppose that the time required to transmit all q packets waiting for transmission is defined as $w(q)$. In this case, since the arrival time of the qth packet is $(q-1) * p$, the transmission delay time is calculated as shown in the following equation:

$$w(q) - (q-1) * p. \tag{2}$$

Therefore, since the worst transmission delay time in the corresponding hop is the largest one of the transmission delay time, WD$_c$ can be written as

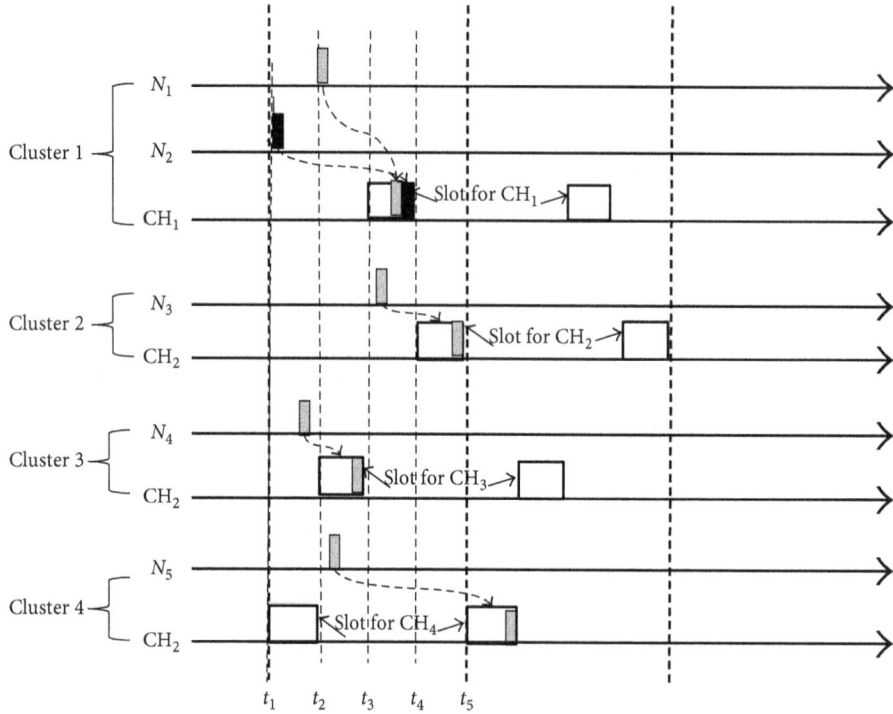

FIGURE 5: TDMA-based flooding at cluster heads.

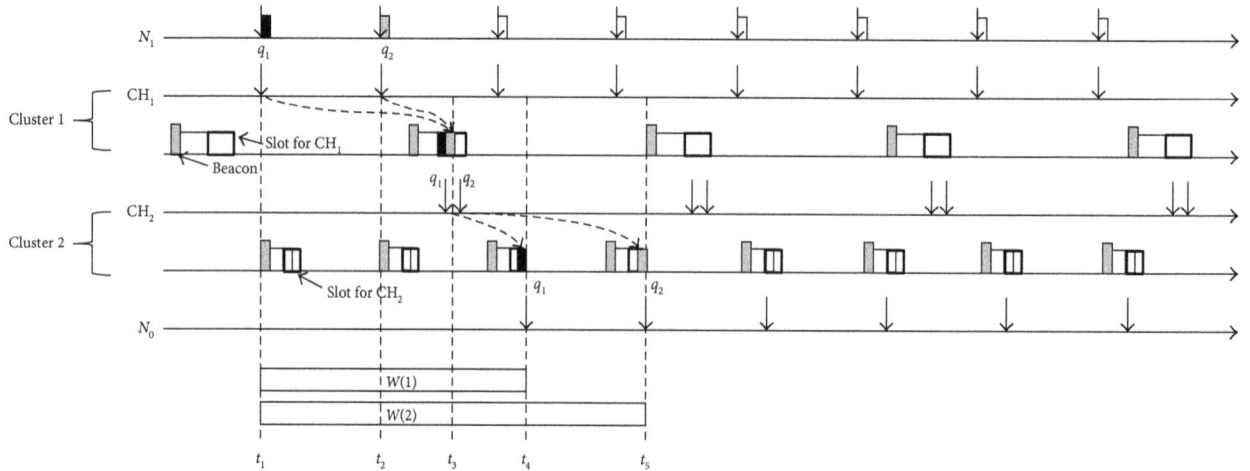

FIGURE 6: Worst-case delay estimation.

$$\text{WD}_c = \max_{q=1,2,\dots,Q}(w(q)-(q-1)*p), \qquad (3)$$

where Q is the first integer satisfying $w(q) \leq Q*p$, and the worst transmission delay time, WD_c can be predicted by knowing $w(q)$ using the following equation:

$$w(q) = \left\lceil \frac{q*b/R}{S} \right\rceil * p_c S = \frac{b/R}{p} * p_c. \qquad (4)$$

In (4), b/R means the time when b bits are sent to the rate R, that is, the time taken to send one packet. Therefore, S means the time taken to send all packets generated within one period, and p_c means the period for packet flooding of

cluster c. This equation is a variation of worst case busy period theorem [20]. In order to predict the worst-case transmission delay, it is necessary to know not only the packet size and the transmission rate of the link but also the period of packet generation in the application, the period of packet flooding in the cluster head, and the size of the time slot of each cluster head. In other words, the packet transmission period and the time slot allocated to each cluster head must be determined so that the worst transmission delay time can be calculated.

Using this procedure, we are able to predict the worst case end-to-end delay for FANET communications and check the feasibility of a given set of clusters with a given time interval. With this real-time feasibility check scheme, we will verify two-level hierarchical clustering which can improve PDR and end-to-end delay in the next section.

4. Evaluation

In this section, we develop the proposed FANET simulation platform with NS-3 and Gazebo tools to realistically evaluate network performance in the FANET environment as presented in Section 3.2. On top of the platform, we verify the FANET routing protocol proposed in Section 3.3. This paper conducted two types of evaluation. We first focused on the primitive communication performance of the proposed routing protocols by examining packet delivery ratio and delay in Section 4.3. In what follows, we also measured UAV's ability to successfully localize and approach survivors in Section 4.4.

4.1. Simulation Scenario. The simulation scenario for measuring the performance of the proposed routing protocol is shown in Figure 7. The figure is a screenshot from a Gazebo tool during the simulation. The red star represents the locations of two survivors and the red circle represents the cluster head UAV. Also, the white box in the center represents the leader node controlling all drone members. The entire nodes except the leader node are divided into four clusters as shown in Figure 7.

Nodes that have detected the survivors' Wi-Fi signals send this information to their cluster head node. Then, the cluster head node collects the received packets waiting for its time slot, and the head node floods them at once during the time slot. Any UAV node receiving flooded survivor detection information from three or more different nodes changes its flight direction to get near the survivors.

The parameters for the simulation are shown in Table 1. Reflecting the fact that the maximum Wi-Fi range is about 250 meters in general, the distance between UAV nodes is set to 200 meters for maximum coverage. The average moving speed of each UAV is 5~10 m/s, and 16 survivors are randomly distributed in the area.

4.2. Performance of Existing Mobile Routing Protocols. Before we introduce a new routing protocol, we first analyze limitations of the traditional routing protocols in FANET environments in more detail. For realistic simulations, we conduct evaluation with the above integrative simulation platform in Section 3.2. Figure 8 first shows the performance of a reactive routing protocol AODV [13] and a proactive routing protocol OLSR [11] (AODV and OLSR are well-known representative protocols among the various routing protocols introduced in Section 2).

As shown in Figure 8(a), for both AODV and OLSR, PDR (packet delivery rate) performance decreases dramatically as nodes move faster in FANET. Also, even at very low mobility rates, the network is still suffering from low PDR. Note that AODV shows better performance than OLSR. This is explained by the fact that AODV is an on-demand routing protocol, which builds a path for transmitting packets when a packet transmission request occurs.

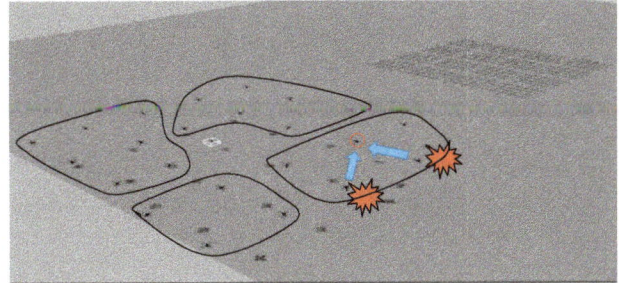

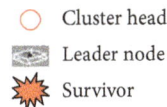

○ Cluster head

▱ Leader node

✴ Survivor

FIGURE 7: Simulation scenario.

TABLE 1: Simulation parameters [6, 8].

Parameter	Value
Channel	Wi-Fi
Application	On-off application
Traffic type	UDP
MAC protocol	802.11
Packet size	1024 (byte)
Data rate	DsssRate 11 Mbps
Topology size	2500×2500 (m^2)
Internode spacing	200 (m)
Simulation time	100 (s)
Number of nodes	45
Node movement speed	5~10 (m/s)
Number of survivors	16
Number of clusters	2, 4, 22, 44
Packet generation period	0.1(s), 1(s)
Packet flooding period	0.5(s), 1(s), 5(s)

This is trivial that OLSR is proactive and so it is more suitable for static networks. On the other hand, when the AODV routing protocol is used, the end-to-end delay time increases as the node moves faster. This is because the AODV, which is a reactive routing protocol, floods a large amount of control packets into the network every time the network topology changes, refinds the route, and finally transmits the data packet only after receiving the RREP packet. As a consequence, the frequent topology changes cause long end-to-end delay of the transmission packets.

Overall, the PDR and end-to-end delay performance of AODV and OLSR are not good enough to be applied in fast moving FANET environments. Note that most messages required for a UAV to operate are semibroadcast (or multicast) messages which all flying nodes in the topology should listen for inter-UAV collaboration. From this perspective, we believe that a flooding-like scheme that propagates information to all nodes in the topology might be more appropriate. However, a traditional flooding algorithm shows packet overhead problems as shown in Figure 9. As the number of source nodes that send packets is increased, PDR decreases due to packet overheads. Specifically, overhead causes an increase in the number of collisions among

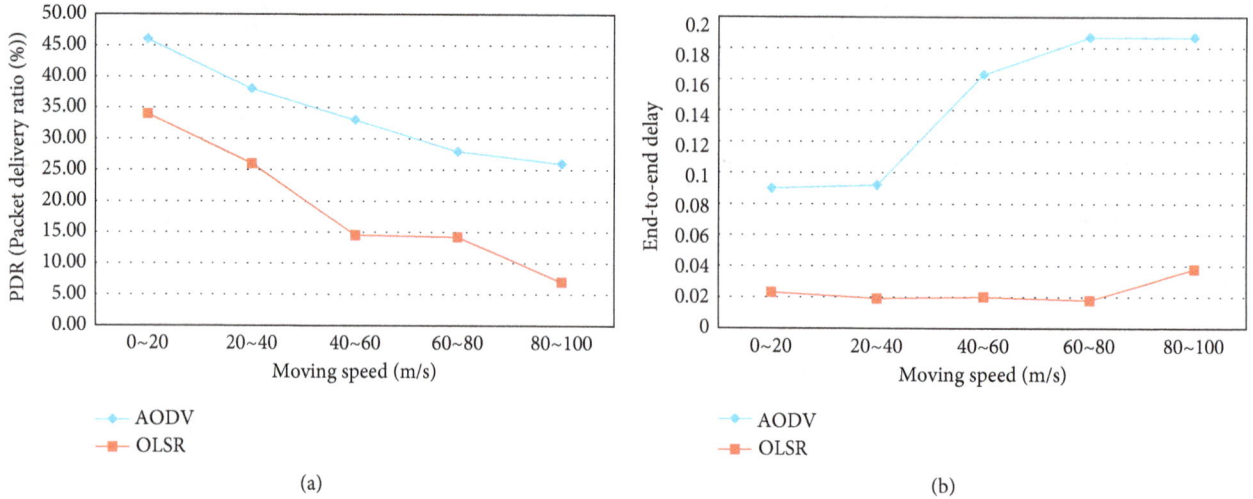

FIGURE 8: Limitations of existing MANET protocols in drone networks. (a) Packet delivery rate. (b) End-to-end delay.

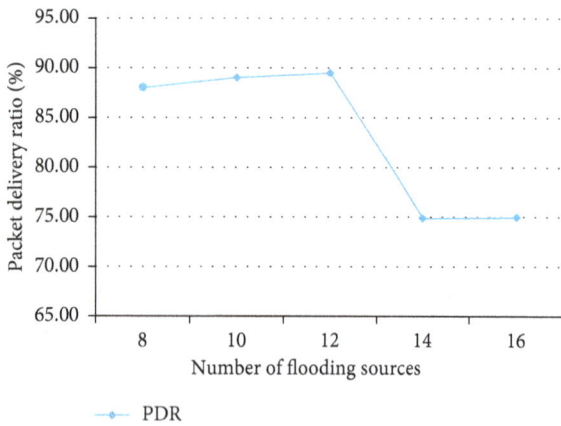

FIGURE 9: How about a flooding protocol?

packets, which means that the number of packet retransmissions in the MAC layer increases.

4.3. FANET Routing Performance.

As we mentioned before, we first examine performance of the proposed FANET routing mechanism in two metrics, PDR and end-to-end delay. The experiments are all conducted with the proposed simulation platform with Gazebo and NS-3. This implies that the provided results are reflecting communication performance with realistically considering UAVs' movement.

4.3.1. Packet Delivery Rate.

Figure 10 shows PDR (packet delivery rate) results according to the number of clusters and the number of flooding cycles, respectively. In this environment, the number of clusters has been varied to 2, 8, 22, and 44. Since the number of normal drones except a group leader is 44, in the case of 44-cluster scenario, each cluster has only one drone as a member and the only group member acts as a cluster head as well. In other words, all drones are cluster heads and so they all flood packets during the

assigned time slot without sending packet in unicast way at all.

From this experiment, we first observe that the more the clusters exist, the better the PDR is achieved. Theoretically, the smaller the number of clusters, the longer the time slot of one cluster head node, which means that the risk of collision between flooding packets flooding in the cluster head is reduced. However, the graph shows that the highest data rate is provided when the number of clusters is 22 or 44. The reason for this is explained by the fact that when the number of clusters is low, lots of unicast packets are sent to the same cluster head node. This causes these unicast packets collide with each other. Note that when the flooding rate is less than 1 s, the PDR is relatively low because the number of flooding packets is way too high and so they end up colliding with each other.

4.3.2. Transmission Delay.

Now, we examine the end-to-end delay of the proposed FANET routing protocol. Figure 11 shows the histogram of the packet transmission delay of the traditional flooding and the proposed algorithm with clustering and time-division features. In this experiment, the flooding period is 0.5 second with 8 clusters, and the application packet generation period is 0.1 second.

First thing we have noticed is that the average delay of the traditional flooding is better than the proposed routing scheme. However, the distribution of delays shown in Figure 11 shows a very important point. In the conventional flooding method, we found that some packets experience long delays as shown as a long tail in the left figure. This is explained by the fact that in the traditional flooding, packets are flooded as they are received and so many packets are subject to be retransmitted due to packet collision.

In contrast, the proposed algorithm shows that most packets are delivered within 0.5 seconds thanks to its TDMA feature. In other words, the proposed scheme is able to bind the transmission delay within a certain value. The worst-case delay guarantee is very important feature for mission-critical applications such as disaster rescue operations with time

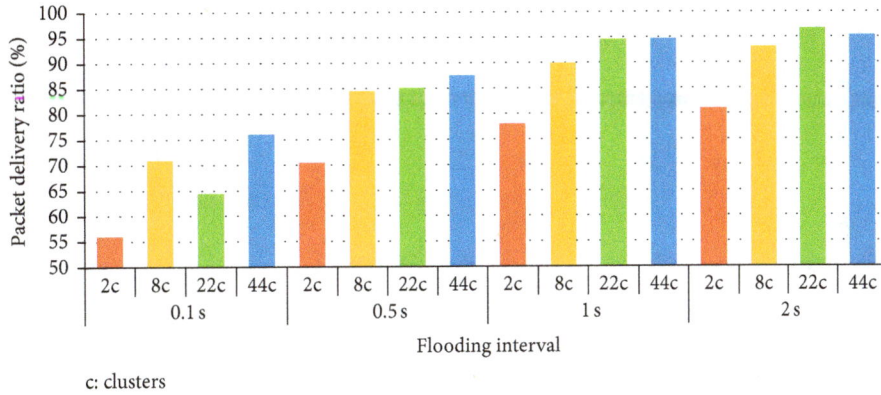

FIGURE 10: PDR (packet delivery rate) with 2, 4, 22, and 44 clusters for various flooding periods.

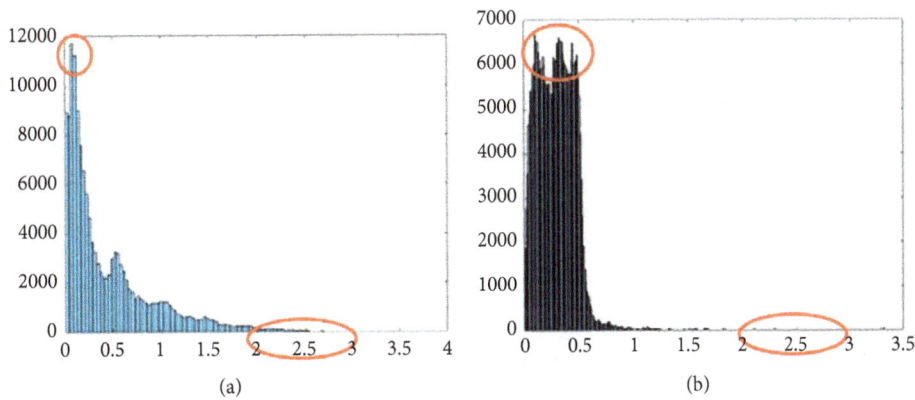

FIGURE 11: Transmission delay with original flooding scheme (a) and with the proposed algorithm (b).

constraints. Note that since the proposed TDMA transmission operation is performed at the network layer, not at MAC layer, it cannot completely guarantee that the packet is transmitted only in a predetermined time slot. This fact explains a few number of packets with long delays in the right graph (in order to guarantee hard deadline satisfaction, a MAC layer model supporting GTS (guaranteed time slot) such as Zigbee [17] should be incorporated with the proposed scheme).

4.4. UAV Mission Performance. In this section, we evaluate UAV mission performance by measuring how quickly each UAV approaches survivors. For this evaluation, we assume two types of applications running on UAVs, app_1 and app_2 as shown in Table 2. Recall that the main goal of the proposed rescue system is to locate survivors as quickly as possible in a stable communication environment. To do this, each UAV detecting a survivor's Wi-Fi signal is supposed to spread this information in order to make neighboring UAVs move near survivors in the proposed rescue scenario. Once these multiple UAVs arrive around the survivor, they cooperate with each other to accurately locate the exact survivor's location together using several localization metrics such as triangulation [21, 22]. So, the faster the time to approaching the survivor, the better the mission performance.

TABLE 2: Application parameters.

Parameter	app_1	app_2
Packet size (byte)	1024	512
Packet generation interval (sec)	1	0.1, 0.5, 1
Flooding interval (sec)	5	5
Number of UAVs	45	45

In this paper, we focus on how quickly each UAV can move near the survivor (localization algorithms are beyond this paper's scope).

Figure 12 shows a subset of our experimental results as a cumulative distribution function (CDF). The x axis and y axis represent the time to approach and the number of UAVs, respectively. From these results, we first observe the performance of the traditional flooding mechanism drops dramatically as the packet transmission rate increases, as shown in Figures 12(a)–12(c). This is because the network becomes more congested due to the faster period of packet transmission. Therefore, the probability that a neighbor node can successfully receive a packet is lowered and hence the time for the UAV to start approaching the survivor gets delayed. As an extreme example, when the packet generation cycle is 0.1 second, any node cannot receive app_2 packets and so no node approaches the survivor at all until the end of the

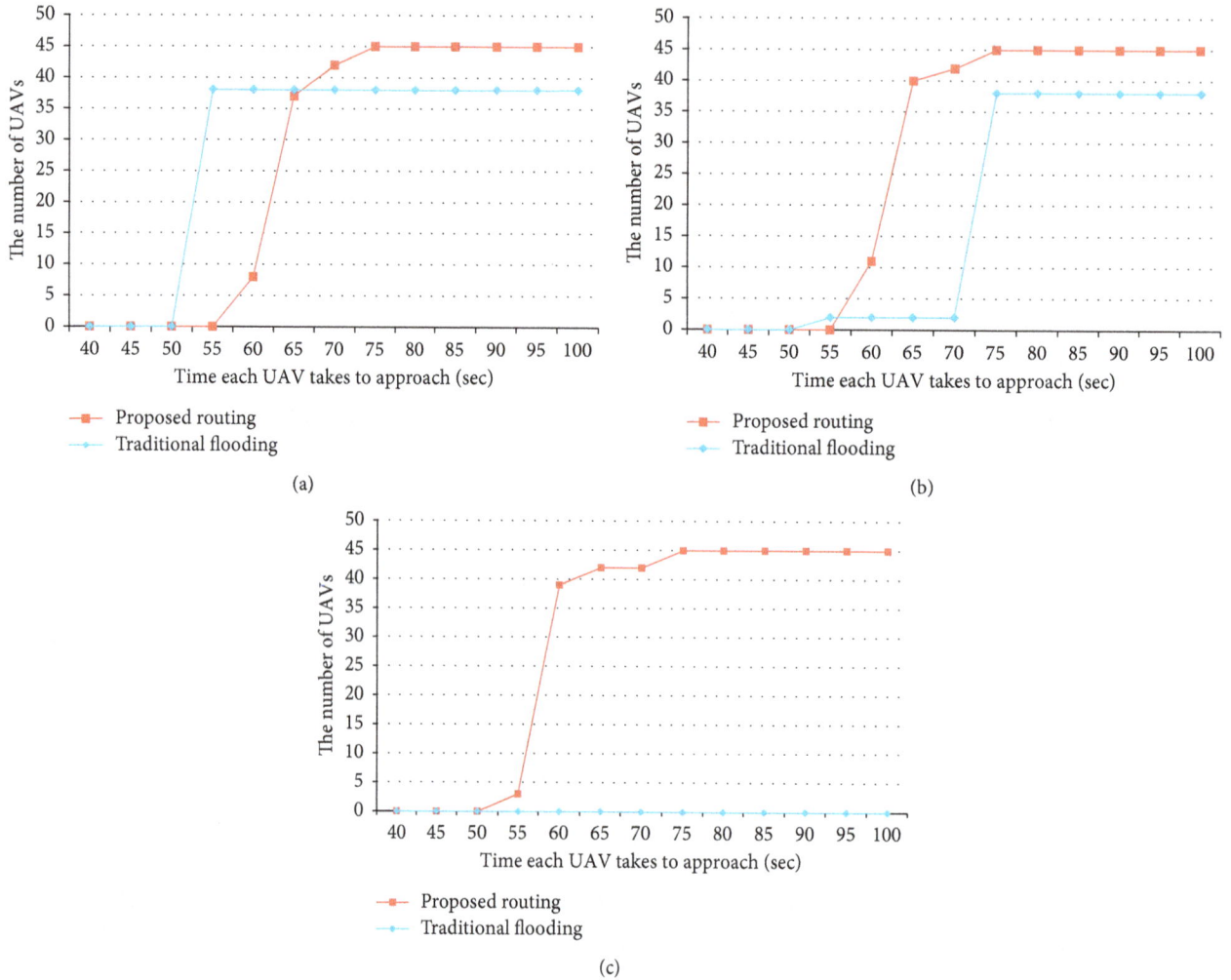

FIGURE 12: Mission performance comparison: (a) app$_2$'s transmission interval = 1 s, (b) app$_2$'s transmission interval = 0.5 s, and (c) app$_2$'s transmission interval = 0.1 s.

simulation. On the other hand, the performance of the proposed routing method is hardly affected by the application's packet interval. This is because the proposed algorithm reduces packet congestion through clustering and time-division transmission, as shown in the red-dot graph. In all of these three scenarios, we can see that most nodes start approaching the survivor in 50~70 seconds. One interesting thing is that in the first scenario with flooding interval 1 sec, the traditional flooding seems to provide better performance at the beginning than the proposed scheme. For example, after 50~55 seconds, around 38 UAVs approach the survivor in the traditional scheme whereas less than 5 UAVs move toward the survivor in the proposed method, as shown in Figure 12(a). This can be explained by the fact that when the packet generation cycle is 1 second, the network is not that congested at the beginning. So in this case, the traditional scheme can be beneficial because it floods packets as soon as it is ready. In contrast, the proposed method is supposed to wait for its time slot even when the network is not congested. As a consequence, transmission delay due to the time division transmission can affect the

mission performance of this simulation. However, as time goes, the flooding packet volume gets increased and so the network becomes more congested than before. In the latter part of the simulation, the proposed scheme ends up performing better than the traditional flooding.

5. Discussion

In this paper, we first propose a new simulation platform to overcome the disadvantages of existing network simulator and robot simulator and to create a realistic FANET environment for communication simulation. The proposed platform can realize not only realistic network simulation but also physical movement of robot and sensor operation by using network simulator and robot simulator together. Based on this simulation platform, we propose a new FANET routing protocol that overcomes the disadvantages of existing routing protocols. Since the existing unicast routing protocol is not suitable for carrying out the UAV mission, which requires the propagation of various information with time constraints, this paper develops the cluster-based flooding algorithm with a TDMA feature as well. In this paper, we verified that even the proposed simple two-level hierarchical clustering with real-time feasibility check methodology can improve PDR and end-to-end delay. We believe that more efficient clustering method will further improve overall performance. To do this, in the next study, we are going to analyze different clustering methodologies and develop the customized clustering method for FANETs.

Acknowledgments

This work was supported by the Ministry of Science, ICT and Future Planning (Award no. NRF-2015R1C1A2A01055444).

References

[1] H. Cai and Q. Geng, "Research on the development process and trend of unmanned aerial vehicle," in *Proceedings of the Vehicle International Industrial Informatics and Computer Engineering Conference*, Xi'an, Shaanxi, China, January 2015.

[2] M. Herrera, *NS2-Network Simulator*, Springer, Valparaiso, IN, USA, 2004.

[3] T. R. Henderson, M. Lacage, G. F. Riley, C. Dowell, and J. Kopena, "Network simulations with the ns-3 simulator," in *Proceedings of the SIGCOMM Demonstration*, Seattle, Washington, USA, August 2008.

[4] M. Subramanya Bhat, D. Shwetha, and J. T. Author Devaraju, "A performance study of proactive, reactive and hybrid routing protocols using QualNet simulator," *International Journal of Computer Applications*, vol. 28, no. 5, pp. 10–17, 2011.

[5] J. L. Kuo, C. H. Shih, and Y. C. Chen, "Performance analysis of real-time streaming under TCP and UDP in VANET via OMNET," in *Proceedings of the 13th International Conference on ITS Telecommunications (ITST)*, pp. 116–121, Tampere, Finland, November 2013.

[6] D. Broyles and A. Jabbar, "Design and analysis of a 3-D Gauss-Markov model for highly dynamic airborne networks," in *Proceedings of the International Telemetering Conference*, International Foundation for Telemetering, San Diego, CA, USA, October 2010.

[7] D. Krajzewicz, J. Erdmann, M. Behrisch, and L. Bieker, "Recent Development and Applications of SUMO-Simulation of Urban Mobility," *International Journal on Advances in Systems and Measurements*, vol. 5, no. 3, pp. 128–138, 2012.

[8] J. Meyer, A. Sendobry, S. Kohlbrecher, U. Klingauf, and O. V. Stryk, "Comprehensive simulation of quadrotor UAVs using ROS and gazebo," in *Proceedings of the International Conference on Simulation, Modeling, and Programming for Autonomous Robots*, pp. 400–411, Springer, Tsukuba, Japan, November 2012.

[9] M. Quigley, K. Conley, B. Gerkey et al., "ROS: an open-source robot operating system," in *Proceedings of the ICRA Workshop on Open Source Software*, vol. 3, no. 3, Kobe, Japan, May 2009.

[10] G. He, *Destination-Sequenced Distance Vector (DSDV) Protocol*, Networking Laboratory, Helsinki University of Technology, Espoo, Finland, 2002.

[11] T. Clausen and P. Jacquet, "Optimized link state routing protocol (OLSR)," Report No. RFC 3626, IETF, Fremont, CA, USA, 2003.

[12] D. B. Johnson, D. A. Maltz, and J. Broch, "DSR: the dynamic source routing protocol for multi-hop wireless Ad Hoc networks," *Ad Hoc Networking*, Addison-Wesley Longman Publishing Co., Inc., Boston, MA, USA, 2006.

[13] C. Perkins, E. Belding-Royer, and S. Das, "Ad hoc on-demand distance vector (AODV) routing," Report No. RFC 3561, IETF, Fremont, CA, USA, 2003.

[14] J. Zhao and G. Cao, "VADD: Vehicle-assisted data delivery in vehicular ad hoc networks," *IEEE Transactions on Vehicular Technology*, vol. 57, no. 3, pp. 1910–1922, 2008.

[15] J. K. Taek, M. Gerla, V. K. Varma, M. Barton, and T. R. Hsing, "Efficient flooding with passive clustering-an overhead-free selective forward mechanism for ad hoc/sensor networks," *Proceedings of the IEEE*, vol. 91, no. 8, pp. 1210–1220, 2003.

[16] D. T. Ho and S. Shimamoto, "Highly reliable communication protocol for WSN-UAV system employing TDMA and PFS scheme," in *Proceedings of the GLOBECOM Workshops (GC Wkshps)*, pp. 1320–1324, Houston, TX, USA, December 2011.

[17] P. Kinney, "Zigbee technology: wireless control that simply works," in *Proceedings of the Communications Design Conference*, vol. 2, pp. 1–7, San Jose, CA, USA, 2003.

[18] K. G. Shin and P. Ramanathan, "Real-time computing: a new discipline of computer science and engineering," *Proceedings of the IEEE*, vol. 82, no. 1, pp. 6–24.

[19] C. Liu and J. W. Layland, "Scheduling algorithms for multiprogramming in a hard real-time environment," *Journal of the ACM*, vol. 20, no. 1, pp. 46–61, 1973.

[20] K. Tindell and J. Clark, "Holistic schedulability for distributed hard realtime systems," *Microprocessing and Microprogramming*, vol. 40, no. 2-3, pp. 117–134, 1994.

[21] A. Savvides, C. Han, and M. B. Strivastava, "Dynamic fine-grained localization in Ad-Hoc net-works of sensors," in *Proceedings of the 7th Annual International Conference on Mobile ComPuting and Networking*, pp. 166–179, Rome, Italy, July 2001.

[22] K. F. Ssu, C. H. Ou, and H. C. Jiau, "Localization with mobile anchor points in wireless sensor net-works," *IEEE Transactions on Vehicular Technology*, vol. 54, no. 3, pp. 1187–1197, 2005.

6

Incentive Mechanism of Data Storage based on Blockchain for Wireless Sensor Networks

Yongjun Ren ⓘ,[1,2] Yepeng Liu ⓘ,[1,2] Sai Ji ⓘ,[1,2] Arun Kumar Sangaiah ⓘ,[3] and Jin Wang ⓘ[4]

[1]*School of Computer and Software, Nanjing University of Information Science & Technology, Nanjing, China*
[2]*Jiangsu Collaborative Innovation Center of Atmospheric Environment and Equipment Technology (CICAEET),*
 Nanjing University of Information Science & Technology, Nanjing, China
[3]*School of Computing Science and Engineering, Vellore Institute of Technology (VIT), Vellore, India*
[4]*School of Computer & Communication Engineering, Changsha University of Science & Technology, Changsha, China*

Correspondence should be addressed to Jin Wang; jinwang@csust.edu.cn

Academic Editor: Yuh-Shyan Chen

In this paper, the blockchain technology is utilized to build the first incentive mechanism of nodes as per data storage for wireless sensor networks (WSNs). In our system, the nodes storing the data are rewarded with digital money. The more the data stored by the node, the more the reward it achieves. Moreover, two blockchains are constructed. One is utilized to store data of each node and another is to control the access of data. In addition, our proposal adopts the provable data possession to replace the proof of work (PoW) in original bitcoins to carry out the mining and storage of new data blocks, which greatly reduces the computing power comparing to the PoW mechanism. Furthermore, the preserving hash functions are used to compare the stored data and the new data block. The new data can be stored in the node which is closest to the existing data, and only the different subblocks are stored. Thus, it can greatly save the storage space of network nodes.

1. Introduction

Wireless sensor network (WSN) has become very hot research topic recently in the field of microelectronics, communication, network, database, etc., because of its broad application prospects. It combines multiple technologies, such as sensing, computing, and wireless communication. The physical targets are monitored in real time through various types of microsensors, producing a large number of perceptual data at an unprecedented rate. Although the application scenarios and the deployment of hardware are different, the ultimate goal is to collect, transmit, and process the perceived data. Finally, users can achieve interesting information from the data [1, 2].

The wireless sensor network is a data-centric network. Therefore, the data storage of nodes is the fundamental problem in WSN, which should be solved. For the users, what they concerned are the perception of the data, rather than the sensor node itself and the networks they make up.

Furthermore, the wireless sensor networks support efficient and reliable data storage and access under the heterogeneous, unreliable environment. As the storage space and energy of each node are limited, how to effectively store data in the limited storage space has been an important research hot spot of data management in WSN.

The normal operations of WSN require the cooperation of network nodes. However, some network nodes may choose selfish behavior due to their limited resources, such as energy and storage space. If most network nodes take selfish behavior and do not forward packets, the entire network will not be able to provide normal service. Therefore, inciting selfish nodes to cooperate and ensuring the normal operation of the entire network are part of the important researches in WSN.

Traditionally, the solutions to the selfishness problem of nodes in WSN have based on the mechanisms of game theory and the mechanisms based on reputation. But the researches mainly focus on data transmission and packet forwarding. Moreover, now there is no specific incentive mechanism of data storage for nodes in WSN.

The storage capacity of nodes in WSN is limited, and the data storage capacity is also an important resource. This paper focuses on the incentive of data storage in WSN. In this paper, the blockchain technology is adopted to construct the first incentive mechanism of nodes' data storage in WSN. In our system, the data set which is storing every node is considered as a block of the blockchain. If the nodes store the data, they will be rewarded with digital money (bitcoins, etc.). Additionally, if the nodes store more data, they will attain more rewards. When mining and storing new data blocks in progress, we apply the provable data possession instead of the proof of work (PoW) in original bitcoins. The method can greatly reduce the computing power of the miners. Apart from this, comparing the existing data in nodes with the new data block, we can take advantage of the preserving hash functions. The node stores the new data, which is closest to the existing data, and only the distinct subblocks need to be stored. So, it greatly saves the storage space of network nodes.

The rest of this paper is organized as follows. Section 2 introduces the related works of data storage strategy in WSN and incentive mechanism. In Section 3, we analyze the existing problem of data storage in WSN. Section 4 presents the building blocks of our scheme based on the blockchain. And in Section 5, the incentive mechanism of data storage based on blockchain in WSN is proposed. Finally, Sections 6 and 7 present the discussion and conclusions of this paper, respectively.

2. Related Work

2.1. Data Storage Strategy in WSN. At present, there are three main ways of data storage in WSN: external storage, local storage, and data-centric storage [3].

2.1.1. External Storage. Sink node is a special kind of storage node, and its storage space and energy are not restricted and do not need to consume another node energy. Other nodes will send the collected data to the sink node, which will consume a lot of energy. If all the nodes in the network send data to the sink node, it will cause the network block and the nearby sink nodes will be invalid.

The LEACH protocol is proposed to collect data from the hierarchical sensor network, in which a subset of nodes is randomly selected as cluster heads, and the other nodes added different cluster according to the calculated distances between the nodes and the cluster heads. During a period, the nodes transmit data to the cluster heads, and the cluster heads process the data and then sends them to the sink nodes. The PEGASIS protocol [4] improved the LEACH protocol, in which the sensor network was organized into a chain structure. Each node receives and forwards data by its neighboring nodes. The sink nodes only select one other node to communicate with it. The data are aggregated in the process of forwarding from a node to the next node and eventually reaching the sink node. Thus, the consumed

energy in the PEGASIS protocol is less than that in LEACH. Wang et al. also proposed a new protocol [5], which is an improvement to the LEACH protocol. The protocol establishes the soft and hard threshold, which can dynamically adjust and compare the collected data to reduce unnecessary data transmission. When the node data are above the hard threshold, the data are transmitted and they are taken as a new hard threshold.

The storage strategy of external storage is mainly focused on data acquisition, ignoring the data storage ability of WSN and the demand of nodes for data.

2.1.2. Local Storage. In local storage, the data are stored in nodes of the network, which consumes little energy. The query commands are only sent to the other nodes. After the node receives the query and processes it, the result is passed to the sink node. So, queries consume longer delays.

The directed diffusion protocol stores the data collected by the nodes in the local nodes. The sink nodes achieve their information by broadcasting the "interest message" to the other network nodes. The node that received the message creates a gradient within the network, pointing to the sink node. The node establishes one or more paths to the sink nodes, doing flood search and performing data transmission. The geographic and energy-aware routing (GEAR) protocol [6] is the improvement of the directed diffusion protocol. In GEAR protocols, when a query message is sent in the target area, the propagation of the "interest message" is limited to the target area because of the geographical location, which avoids flooding in entire network and reduces the cost of routing.

The storage process of local storage strategy is simple. And the strategy focuses on data query processing and has less description of information, which leads to a lot of energy in the query process.

2.1.3. Data-Centric Storage. The data-centric storage is a hot research direction in recent years. It mainly studies how to store the perceived data of sensor nodes so as to ensure the high efficiency, stability, and real-time performance of the later query.

The concept of data-centric storage (DCS) is proposed by Meyfroyt et al., and the data storage algorithm GHT is designed based on the geographic information mapping table [7]. Its core idea is that data are stored according to their attributes, and a specific data are defined as an event. The sensor detects the data, hashes the event through a hash function, then achieves a geographic location, and saves the data to the nearest node based on the geographic information [8]. The algorithm is conducive to data query, which is only based on the query event attributes. And the use of mapping function can be found in the storage node, which avoids flooding. The disadvantage of the algorithm is the lack of efficient storage hot spot processing mechanism. When the data storage overloads, it cannot be transferred to another node. Moreover, accessing geographic information needs GPS and consumes system energy. In the data storage algorithm ARI [9], adaptive ring index structure is used to solve the hot spot problem of the DCS algorithm. And hash functions are utilized to hash a certain type of event to the event storage node. A ring is created around the event storage node, and events are dispersed and stored in the index nodes. In general, it is difficult to define clear demarcation of a wireless sensor network, which is not ideal for hot spot problems. In data storage algorithm of Reference [10], two-tier data storage structure is used to track the moving target of the mobile multisink node in the WSN. Data are transferred and stored through the creation of virtual grids in the algorithm. When the data collected by the grid storage nodes are queried, it is just needed to flood the request within the grid, which will save energy. In addition, some scholars have proposed a distributed index structure algorithm (DIFS) [11]. DIFS is an improvement of the TTDD algorithm. In DIFS, multilevel quadtree is constructed based on spatial decomposition technique and hash function, and the geography hash method is used as the index of data [12, 13]. The corresponding node stores the observed data through hash functions, and it can determine the range of the minimum number of index nodes by the query range.

2.2. Incentive Mechanism. At present, there are two main incentive mechanisms. One is based on game theory. The other is based on external incentives [14–20].

2.2.1. Incentive Mechanism Based on Game Theory. In the paper [14], the concept of multidomain wireless sensor network was first proposed, and the game theory was used to evaluate the impact of cooperative behavior. In the system, the participants in game analysis are the various individual wireless sensor networks, and it is assumed that each wireless sensor network has to make decisions: whether to help other networks to carry out data transfer and whether to request other networks to help its data transmission, which is the strategy of each participant in game analysis. On the basis of the above mechanism, the problem was continued to study the cooperative behavior among networks in multidomain wireless sensor networks [15]. The main differences in the game analysis are as follows: (1) the income function of the game is mainly expressed by the whole life cycle number of the network [16], rather than the calculation of the accumulated revenue of nodes and (2) the strategy of the sensor node is more intelligent. The choice of actions will be limited after many unsuccessful data transfers so that the network has minimal QoS guarantee. References [17, 18] analyzed the impact of different cooperation strategies on the life cycle of multidomain wireless sensor networks. The author proposes a linear design framework and uses the corresponding one-dimensional and two-dimensional linear models to assess the performance of different strategies. Based on the ideal conditions, the author adds various restrictions to observe the influence on the cooperation strategy. Simulation experiments have confirmed that cooperation can significantly extend the life cycle of the network. And under some special circumstances, some cooperative strategies can increase the life cycle to an order of magnitude.

2.2.2. Incentive Mechanism Based on External Incentives. In addition to the use of game theory to analyze the multidomain wireless sensor network, there are some researches of external incentive mechanisms. The main external incentive methods include virtual currency mechanism and honor incentive mechanism. In [19–21], an economic model of dynamic prices and incentive methods is proposed to study the cooperation in multidomain wireless sensor networks. And the proposed economic model and the traditional routing protocol AODV protocol [22] are merged into a hybrid protocol for simulation experiments. In the simulation experiment, the author compared the proposed NES method with other EES methods and PDM [23]. The experimental results confirm that the cooperation between the sensor networks will be enhanced, and the overall energy consumption in the network will be significantly reduced.

3. Problem Statement

The development of wireless sensor networks originated from military applications, such as battlefield monitoring. Nowadays, wireless sensor networks have been applied to many civilian applications, such as environmental and ecological monitoring, healthcare, home automation, and traffic control.

In the sensor network, nodes are deployed in a variety of ways within or around a perceived object. These nodes form a wireless network through self-organization method. And they can sense, collect, and process specific information in a cooperative way within the coverage area. Finally, it can realize the collection, processing, and analysis of any location information at any time. Each node of the sensor network is not only equipped with a radio transceiver but also a small microcontroller and an energy source (usually a battery), in addition to multiple sensors. The size of a single sensor node is as large as a shoe box, as small as dust. The size and complexity of the restrictions for sensor nodes determine the constraints of energy, storage, computing speed, and bandwidth. In large sensor networks, the sensor and network structure are different. Thus, the integration of heterogeneous networks often occurs in sensor networks. At the same time, the heterogeneous network structure also brings difficulty to data storage and sharing in WSN.

Moreover, the data storage capacity is also an important resource. But the storage capacity of nodes in a wireless sensor network is limited. Some network nodes give up storing data in order to save their own storage and energy resources, which are called selfish behavior. If the most network nodes behave selfishly and do not store data, then the entire network will not be able to provide normal service.

To solve the problem, we use incentive mechanisms based on blockchain to encourage network nodes to store data. The data storage based on the blockchain technology can not only provide the corresponding data storage function but also reward the digital currency to the network node that stores data. Therefore, data storage based on the blockchain technology in WSN is very suitable.

4. Building Blocks

4.1. Blockchain Technology. The blockchain system contains the following important components: underlying transaction data, distributed ledgers, important consensus mechanism, complete and reliable distributed P2P network, and distributed application on the network. And the framework is shown in Figure 1. The underlying data are organized into blocks, and each block is chained into a chain in the chronological order, which is called blockchain [24–26]. Each node of a fully distributed network stores a distributed ledger, that is, blockchain. The P2P protocol is used in the network to communicate with each other. All parties will reach agreement through consensus mechanisms. Advanced applications are generated based on these foundations. In the architecture, the nontampering blockchain data structure, the consensus mechanism in distributed network, the proof of work mechanism, and the increasingly flexible smart contracts are representative innovations [27, 28].

The underlying data are not stored in the blockchain. The raw data need further processing so that they can be written into the block. The underlying data are the most fundamental transaction records; the other data are only intended to encapsulate the message records. The network layer encapsulates the networking mode of the blockchain system, the message propagation protocol, and the data authentication mechanism. Combining with the practical application requirements and designing the specific propagation protocol and data verification mechanism, each node in the blockchain system can participate in the checksum accounting process of the block data. Only when the block data are verified by most nodes in the whole network, the block is recorded in the blockchain [29–31].

The PoW mechanism is an important innovation that closely integrates the functions of currency issuance, transaction payment, and verification. And the safety and decentric of the blockchain system are ensured through the competition of computing force. The core idea is to ensure the consistency of data and the security of the consensus by the computing force competition of distributed nodes. In the bitcoin system, the miners work together to solve a complex but easy-to-valid SHA-256 mathematical problems (i.e., mining) based on their respective computer forces. The nodes that solve the problem the fastest will get the right to account the block and bitcoin reward. The mathematical problem can be expressed as follows. Based on the current difficulty value, a suitable random number (Nonce) is sought so that the double SHA-256 hash of the metadata of the block header is less than or equal to the target hash value. However, the PoW consensus mechanism has a significant flaw: the waste of resources (such as electricity), caused by their strong computing power, has always been criticized by researchers [32–34].

The consensus process of the blockchain system realizes the data validation and accounting of shared blockchain ledgers by aggregating the computational power resources of large-scale consensus nodes, so it is essentially a task crowdsourcing process of consensus nodes. In the decentralized system, the consensus nodes themselves are selfish, and maximizing its own revenue is the fundamental goal of its participation in data validation and accounting. Therefore, it is necessary to design a reasonable and well-conceived mechanism of incentive and compatibility so that the individual rational behavior of the consensus node maximizing its own income is consistent with the overall goal of guaranteeing the safety and effectiveness of the decentralized blockchain system. The blockchain system integrates large-scale nodes and forms a stable consensus on the history of the blockchain by designing a modest economic incentive mechanism and integrating with the consensus process.

The contract layer is business logic and algorithm based on the blockchain virtual machine, which is the basis for realizing the flexible programming and operation data of the blockchain system. The smart contract has important significance to the blockchain system, which not only provides the programmable capabilities to the underlying data of the blockchain but also encapsulates the complex behavior of each node in the blockchain network. And it provides a convenient interface for building an upper application based on blockchain technology. Thus, blockchain technology with smart contract is extremely broad prospects.

4.2. PDP Mechanism. Provable data possession (PDP) mechanism is used to determine whether the data on the remote node are damaged (Figure 2). The PDP mechanism was first used in grid computing and P2P networks. He et al. constructed the PDP mechanism using RSA-signed homomorphic properties, but this mechanism requires that the entire file is represented by a large number, which results in high computational costs. Wang et al. proposed a probabilistic strategy to complete the integrity verification, using the homomorphic properties of the RSA signature mechanism to aggregate the evidence into a small value, greatly reducing the communication overhead of the protocol [35–38]. Wang et al. realized another mechanism that supports full dynamic operation of the PDP mechanism. It considers the use of the Merkle hash tree in order to ensure the correctness of the data block in position, and data block value ensures its correctness through the BLS signature mechanism [39–42]. In order to reduce the burden on the user, the mechanism also introduces an independent third party instead of the user to verify the integrity of outsourced data. In this article, this algorithm is used to replace the PoW mechanism in the original blockchain.

The PDP scheme is as follows. At first, encode M into M' so that each data block m_i of M' contains s data segments, that is, $m_i = (m_{i,1}, m_{i,2}, \ldots, m_{i,s})$. The metadata σ_i are calculated for each data block m_i as follows:

$$\sigma_i = \left(H(\text{name}||i) \times \prod_{j=1}^{s} u_j^{m_{i,j}} \right)^{\alpha}, \tag{1}$$

where α is the private key of the user and $u_j (1 \le j \le s)$ is randomly selected from the bilinear group G. Similar to the

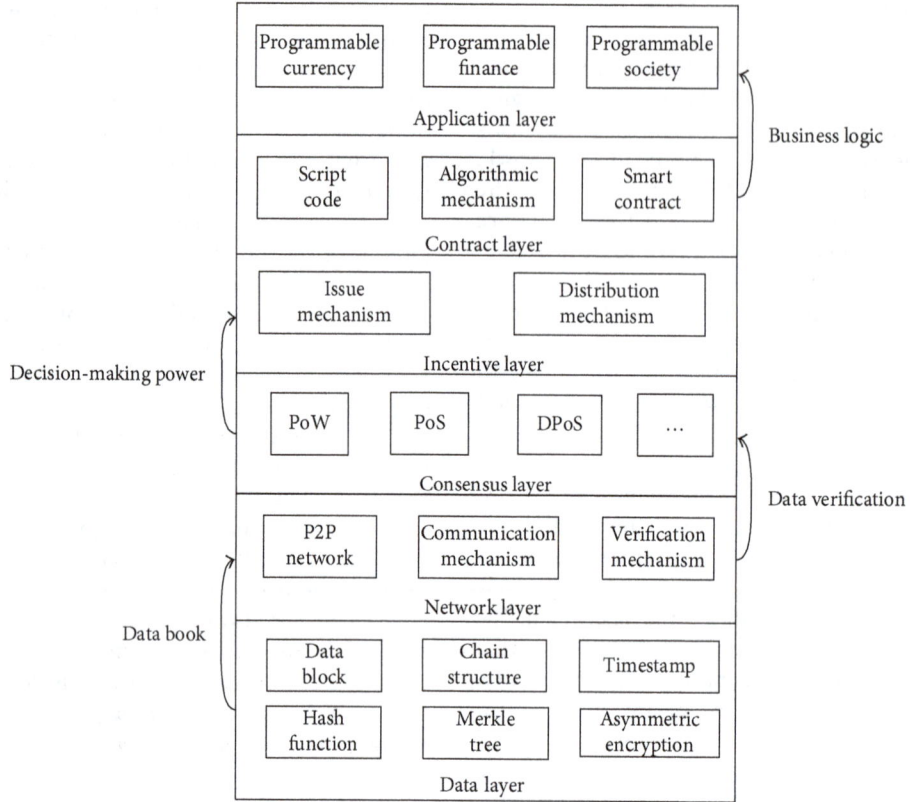

FIGURE 1: A basic framework of blockchain.

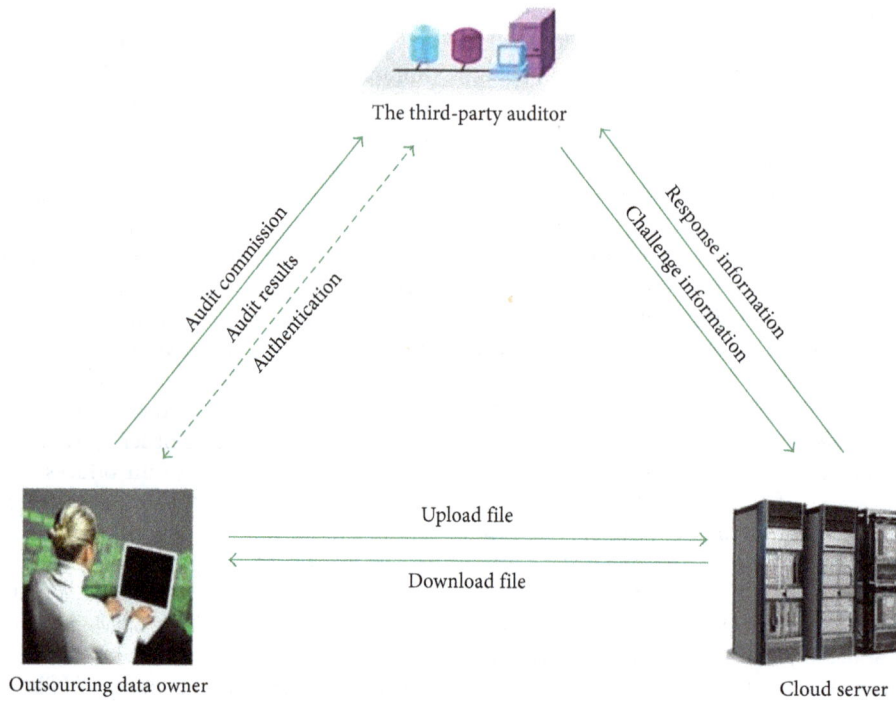

FIGURE 2: Provable data possession.

literature [4], the factor $(\prod_{j=1}^{s} u_j^{m_{i,j}})^{\alpha}$ contained in the metadata σ_i also supports the aggregation operation. So, the cloud storage server can generate the corresponding partial aggregation in the integrity verification phase. The algorithm also signs the data name, the number of data blocks, and the parameter u_j to obtain a tag of data r.

To verify the integrity of the outsourced data, a query challenge $C = \{(i, v_i)\}$ is submitted by the verifier, including the block number i which is randomly selected and the corresponding coefficient v_i. The cloud server calculates aggregated data blocks $\mu = (\mu_1, \mu_2, \ldots, \mu_n)$ and metadata σ as the proof, that is, (μ, σ), where $\prod_{(i,v_i) \in C} \sigma_i^{v_i}$ is the metadata, the aggregated data blocks is $\mu_i = \sum_{(i,v_i) \in C} v_i m_{i,j}$.

Verification is done by checking the following formula and performing two bilinear operations:

$$e(\sigma, g) = e\left(\prod_{(i,v_i) \in C} H(\text{name}||i)^{v_i} \times \prod_{j=1}^{s} u_j^{\mu_i}, v \right), \qquad (2)$$

where v is the user's public key corresponding to α.

In the above scheme, Shacham and Waters double the data for the first time so that each data segment $m_{i,j}$ corresponds to a data block of the aforementioned scheme. This segmentation strategy has the obvious advantage that by generating metadata for a set of data segments, the size of the processed data can be reduced, thereby reducing the storage costs of the cloud server.

5. Incentive Mechanism of Data Storage Based on Blockchain in Wireless Sensor Network

In this paper, the blockchain technology is utilized to build the first incentive mechanisms of nodes' data storage in WSN. In our system, the data set stored by every node is treated as a block of the blockchain. The nodes storing the data are rewarded with digital money (bitcoin, etc.). Moreover, the more the data stored by the node, the more the reward it achieves. Our proposal adopts the provable data possession to replace the proof of work (PoW) in original bitcoin to carry out the mining and storage of new data blocks. The method can greatly reduce the computing power by PoW mechanism. Furthermore, the preserving hash functions are used to compare the stored data and the new data block. Thus, the new data can be stored in the node which is closest to the existing data, and only the different subblocks are stored. So, it can greatly save the storage space of network nodes.

5.1. Blockchain of Data Storage for Sensor Node. The sensor network is often composed of multiple heterogeneous subsystems, and various network nodes have different capabilities in computing, energy, communication, and storage. In addition, the network nodes which using different types of sensors make the types of collected data varied. Therefore, the shared data storage mechanism should be adopted to realize the storage and management of the data in wireless sensor network. The blockchain has the advantage of decentralization. Moreover, the data storage based on decentering credit can be realized in the WSN, where the node does not need to be trusted using the encryption algorithm, time stamp, tree structure, consensus mechanism, and reward mechanism. Each network node can use the Merkle tree in the blockchain to store its data. The data of the nodes are stored in the leaves of the Merkle tree. Each stored datum can be a block, and all the data stored by the nodes are linked to form the data blockchain (Figure 3).

5.2. Trust Management of Network Node. In the system, the trust of network nodes is managed. When the network node is found to be fraud and with other behaviors, it is removed from the WSN network. We use the reputation system to manage the nodes in WSN. Once the network node is found cheating, it will be immediately excluded from the WSN.

In the system, the trust of the data initiator (node i) in the network to the data store (node j) can be obtained by calculating the number of success and failure of the node data storage in a certain period of time. After the kth data storage is successful, d_{ij}^k indicates the trust evaluation value of the data initiator node i to the data store node j; $\delta (0 \le \delta \le 1)$ is the time attenuation coefficient of the trust, which is used to reflect the influence degree of trust for the network node in data storage procession. The larger the weight of the recent score record, the greater the weight of the calculation of the trust value, as shown in the following equation:

$$d_{ij}^k = \sum_{m=1}^{k} \delta_m d_{ij}^m. \qquad (3)$$

In the system, the trust of data storage between node i and node j is divided into five levels according to the satisfaction degree, and 0, 0.25, 0.5, 0.75, and 1 are assigned in turn. The first level indicates that the data storage between the network node i and the network node j is failure, and the node i considers the node j is malicious. The second, third, fourth, and fifth levels of trusts are sequentially increased. The fifth level is the highest level, indicating that the data storage between the network node i and the network node j is successful, and that the node i fully trusts the node j. When there is a data storage relationship between the two nodes i and the node j, the degree of trust of node i to node j is calculated using Equation (1). When there is no direct transaction between the two nodes, use the following formula to calculate the average trust of the network as the recommended trust degree of node:

$$d_0 = \frac{\sum_{i=1}^{n} \sum_{j=1}^{n} \sum_{k=1}^{k} d_{ij}^k}{n^2 \sum_{k=1}^{k} k}. \qquad (4)$$

Most nodes in the network play dual role. One role is consumer, who is provided with storage service in the system. Another is the service provider, who provides storage service for other nodes. As a consumer, the trust evaluation of network nodes to other nodes is always considered accurate and deterministic. Therefore, the node modifies the data in the table with minimal possibility. Even if making a recommendation for a particular node, it does not make sense. In addition, it is safe to locally store the relevant calculated data of the trust value. As a service provider, it is the object to be evaluated. Any node i in the network cannot know the storage node which stores its reputation information, which avoids the possibility of the node to raise its reputation.

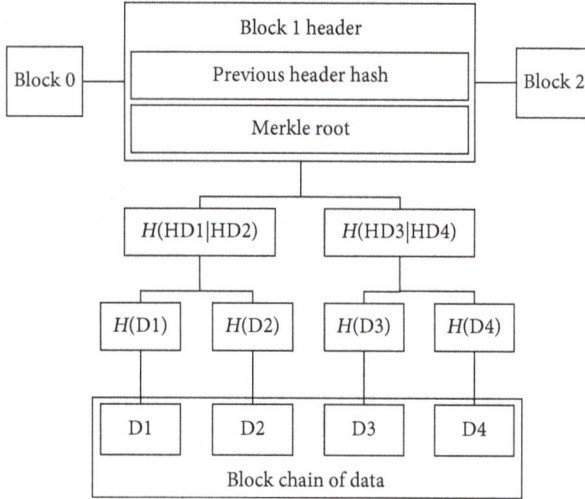

FIGURE 3: Data blockchain of the network node.

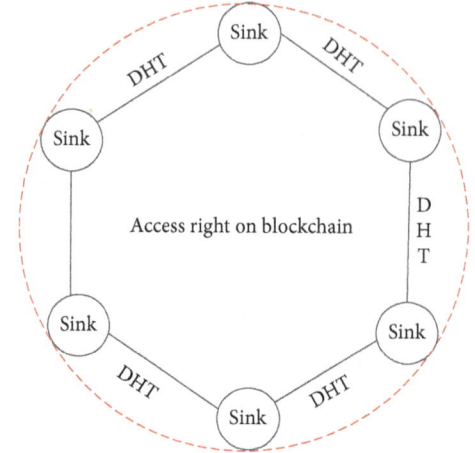

FIGURE 4: Access control based on blockchain.

5.3. Access Control Based on Blockchain. We use blockchain to securely store access right to the data stored in the sink nodes. The data owner, the data visitor, and some additional metadata are included in the signed storage transaction. Each data block is set to access rights and is restricted in time. The data owner can extend or revoke the right to access the data. For any data retrieval request, another node first checks the access rights record through the corresponding distributed hash table (DHT). Theoretically, malicious nodes can share data without permission. Since the access rights of the data are monitored, unauthorized data access will be detected. In addition, if the malicious node is detected, it will be removed out of the network. Therefore, the possibility of such insecure data access is very small. It is shown in Figure 4.

Below we build a block-based DHT for distributed storage and management of index data. DHT is a huge hash table, which is shared by a large number of nodes. Each sink node is assigned to a hash block that belongs to itself and becomes the manager of the hash block. Through the hash function, any data can be mapped to a 160-bit hash value, and the network nodes are mapped to a space. DHT can adapt to the dynamic join and exit sink nodes and has the characteristics of balance and query accuracy. We use the DHT algorithm based on Chord network; through the SHA series hash function, the data are mapped to 160-bit hash value. For chord structure, we use the predecessor list positioning to improve the positioning fault tolerance, by selecting the node to reduce the positioning delay. That is, in the positioning process to select the next jump, those nodes which are a small delay and closer to the other nodes are selected in the bottom of the logistics. Take the above predecessor's search function as an example. Assuming that a node m returns to the predecessor list of the node n, in addition to the node location information, there is a delay of each node to m. Based on these delays, the node n evaluates each node in the list and selects the node that it considers the most reasonable.

5.4. Mining and Incentive Mechanism Based on PDP for Data Storage of Node in WSN. There is a significant flaw of the PoW consensus mechanism in traditional blockchain technology, which requires a lot of computation and causes serious waste of resources (such as electricity). That has always been criticized by academics and industry. In order to solve the problem, the PDP mechanism is used to replace the PoW mechanism to construct the mining and incentive mechanism for data storage of node in the resource-constrained WSN.

5.4.1. Scheme Description. A new data block, which will be stored in the sensor network, is broadcast. And each network node then calculates the challenge of PDP for the data block. If the PDP is verified correctly, the new data block will be stored by the node, and the node will receive a reward for storing the data block as a result, that is, a unit of the digital currency. The proposed scheme is as follows.

(1) A new data block $M = \{m_1, m_2, \ldots, m_n\}$ which will be stored. The public key of the data publisher is (g^x, u), and the private key is x; H_1 is a preserving hash function, and data publisher computes $\{H_1(m_i)\}$ and generates the authenticator $\sigma_i = (H(i)u^{m_i})^x$ for each subblock m_i; The request information for the data is broadcast in the sensor network.

(2) Each network node searches for the stored subblock m_i' closest to the value according to $\{H_1(m_i)\}$, that is, $|H_1(m_i) - H_1(m_i')| \leq \text{dif}$. Then, the random number v_i will be selected for the subdata block i of the data block M, denoted as $Q = (i, v_i)$. The network node sends $\{H_1(m_i')\}$ and Q to the data publisher.

(3) The data issuer receives $\{H_1(m_i')\}$ from each network node and compares them with the $\{H_1(m_i)\}$ value, selecting the $H_1(m_i')$ value which is closest to each $H_1(m_i)$ value and adding the network node j that sent the $H_1(m_i')$ value to the node set J.

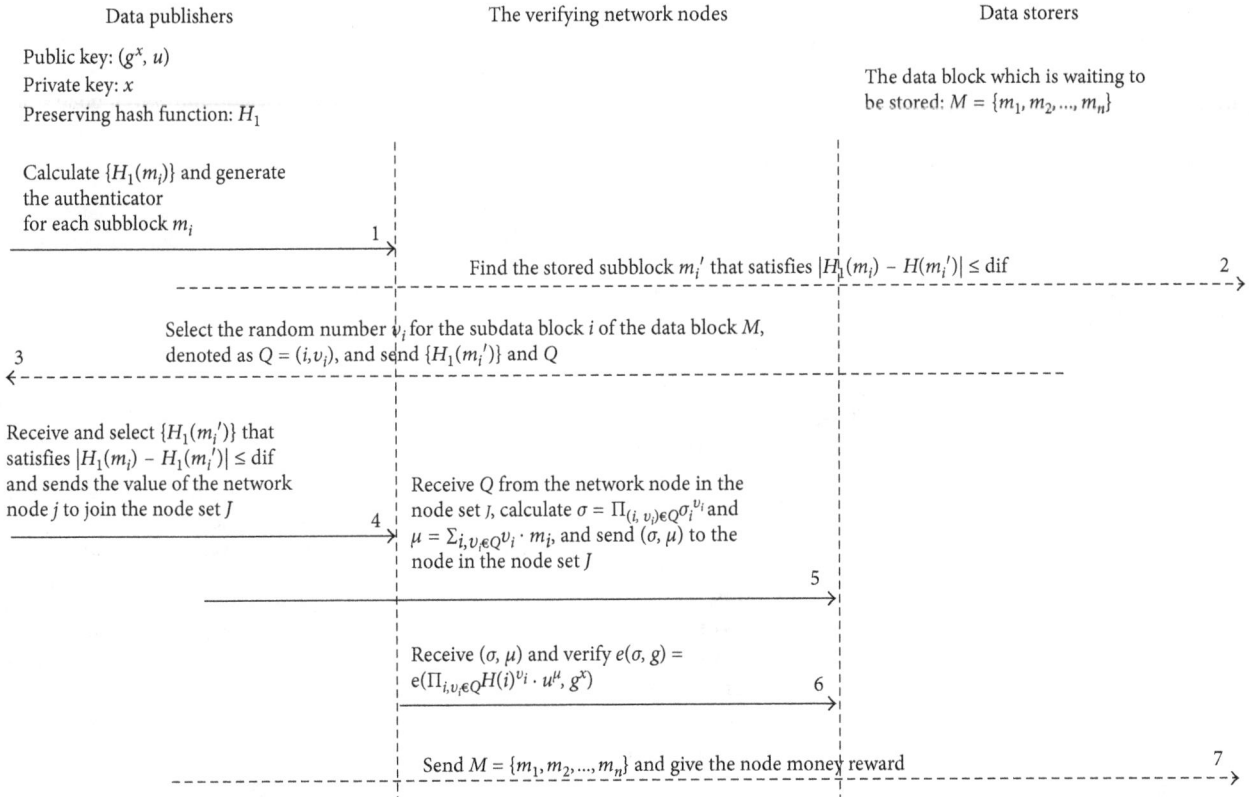

FIGURE 5: Incentive mechanism based on PDP.

Then, based on the Q received from the network node of the node set J, do the following calculation: $\sigma = \prod_{(i,v_i) \in Q} \sigma_i^{v_i}$ and $\mu = \sum_{(i,v_i) \in Q} v_i \cdot m_i$, and then send (σ, μ) to the network node of node set J.

(4) The network node of the node set J receives (σ, μ), verifying the following formula: $e(\sigma, g) = e(\prod_{(i,v_i) \in Q} H(i)^{v_i} \cdot u^\mu, g^x)$. If it is true, the data issuer will send the data block $M = \{m_1, m_2, \ldots, m_n\}$ to each network node of the set J for storage and give the node the digital currency reward.

(5) From the nature of the preserving hash function, it can be seen that the original data block of each network node in the set J contains data similar to the new data block $M = \{m_1, m_2, \ldots, m_n\}$. It only needs to store the part that is not the same as the original data. Therefore, through the strategy, it can greatly reduce the required storage space. The scheme is shown in Figure 5.

5.4.2. Parameters in Our Scheme. In our scheme, we take the pairing function $e : G_1 \times G_1 \rightarrow G_T$, where $|G_1| = |G_T|$ and g, u are generators of the group G_1. The practical constructions of pairings are done on hyperelliptic curves defined over a finite field. $E(F_q)$ is a set of points on an elliptic curve E defined over the finite field F_q. G_1 is taken as a subgroup of $E(F_q)$, and G_T is taken as a subgroup of $F^*_{q^{k'}}$, where k' is the embedding degree. The hash function H hashes a binary string of arbitrary length into G_1, and u is

a random element of G_1. σ also is an element of G_1, and μ belongs to Z_p. The Barreto–Naehrig (BN) curves are suitable for our scheme.

5.4.3. Efficient Storage. In our scheme, the network node stores l data segments $\{(m_j, \sigma_j)\}$, $j \in I$ and $|I| = l$. I is the set of indices of M corresponding to these l segments, and σ_j is the tag of the segment m_j. If SHA-256 is used to compute these hash values, then the size of each σ_j becomes 256 bits. This generates a small storage requirement for each of the segments, though the number of segments in the data M is huge in general. Instead of the Merkle proof, the nodes store a small tag and authenticator of size 256 bits along with each segment. Therefore, a network node in our scheme enjoys around 256 bits less storage overhead per segment.

6. Discussion

Compared with the PDP mechanism, the POR (proofs of retrievability) mechanism can effectively identify whether a file is damaged, and at the same time, it can recover the errors that have occurred in the data file through fault tolerance technology to ensure that the file is available. The POR mechanism can be further adapted in our scheme to improve the fault tolerance of the system.

The PDP mechanism can quickly determine whether the data on the remote node are damaged or not and pay more attention to efficiency. POR mechanism can not only identify whether the data are damaged but also recover the damaged data. POR mechanism can not only detect data integrity but also further ensure data integrity. The publicly authenticated POR mechanism allows any third-party alternative user to initiate the integrity detection of data on a remote node. When the damage of the data was found less than a certain threshold ε, the error is recovered through the fault tolerance mechanism; otherwise, the data returned to the user fail.

Before the POR performs the initialization phase, it is needed to increase the redundant coded data preprocessing process to make the data file fault-tolerant, that is, to divide M into n blocks and then group n blocks. Then, for each group of data blocks, the Reed–Solomon error correction code can be used for fault tolerance coding to form a new data file. The same verification technology as PDP mechanism is adopted. For the POR mechanism, the assumption is that it is within the allowed error range (an error occurs once in 1000000 but passes the verification of the POR mechanism). Define $Y\omega = 1/\#B + (\rho n)^c/(n - c + 1)$, if $\varepsilon - \omega X$ is a negligible value, through $O(n/(\varepsilon - \omega))$ interactions, POR can recover the data with a failure rate of ρ. Here, B is the selection space of the random number when challenging the request, ρ is the data encoding rate, and c is the number of randomly selected data blocks.

7. Conclusions

In this paper, the first incentive mechanisms of nodes for data storage are built based on the blockchain technology in WSN. The data stored by every node are treated as a block of blockchain in our system. The reward for digital money will be obtained by the node who stored the data, and the reward for the node implementation increases as the data it store increases. In addition, it constructs two blockchains. One is to store data for each node, and the other for controlling the access of the data. Moreover, the provable data possession in the proposed scheme is used to substitute the proof of work (PoW) in primary bitcoins, which executes the mining and storage of the new data block. Compared with the PoW mechanism, it cuts down the computing power extremely. Furthermore, due to making use of the preserving hash functions, the new data can be stored in node which is nearest to the currently existing data. And only the different subblocks are stored. Therefore, the storage space of nodes in WSN can be highly saved.

Authors' Contributions

All the authors wrote the paper.

Acknowledgments

This work was supported by the NSFC (61772280, 61772454, 61702236, and 6171101570), the PAPD fund from NUIST, and Changzhou Science and Technology Program (CJ20179027).

References

[1] J. Wang, Y. Cao, B. Li, H. Kim, and S. Lee, "Particle swarm optimization based clustering algorithm with mobile sink for WSNs," *Future Generation Computer Systems*, vol. 76, pp. 452–457, 2017.

[2] B. Wang, X. Gu, L. Ma, and S. Yan, "Temperature error correction based on BP neural network in meteorological WSN," *International Journal of Sensor Networks*, vol. 23, no. 4, pp. 265–278, 2017.

[3] L. Min, W. Fan, Z. Guo, and G. Fan, "Wireless sensor networks data storage strategy based on RCFile," *Computer Science*, vol. 42, pp. 76–80, 2015.

[4] S. C. Lindsey and S. P. Raghavendra, "Power efficient gathering in sensor information systems," in *Proceedings of IEEE Aerospace conference*, pp. 1125–1130, IEEE, Big Sky, MT, USA, March 2002.

[5] J. Wang, C. Ju, H. J. Kim, R. S. Sherratt, and S. Lee, "A mobile assisted coverage hole patching scheme based on particle swarm optimization for WSNs," *Cluster Computing*, vol. 3, pp. 1–9, 2017.

[6] N. Zaman, L. T. Jung, and M. M. Yasin, "Enhancing energy efficiency of wireless sensor network through the design of energy efficient routing protocol," *Journal of Sensors*, vol. 2016, Article ID 9278701, 16 pages, 2016.

[7] T. M. Meyfroyt, S. C. Borst, O. J. Boxma, and D. Denteneer, "Data dissemination performance in large-scale sensor networks," in *Proceedings of International Conference on Measurement and Modeling of Computer Systems*, pp. 395–406, Austin, Texas, USA, June 2014.

[8] X. Shen, W. Liu, I. W. Tsang, Q. S. Sun, and Y. S. Ong, "Multilabel prediction via cross-view search," *IEEE Transactions on Neural Networks and Learning Systems*, vol. 99, pp. 1–15, 2018.

[9] R. Huang, X. Chu, J. Zhang, and Y. H. Hu, "Scale-free topology optimization for software-defined wireless sensor networks: a cyber-physical system," *International Journal of Distributed Sensor Networks*, vol. 13, no. 6, pp. 1–12, 2017.

[10] J. Wang, J. Cao, S. Ji, and J. H. Park, "Energy-efficient cluster-based dynamic routes adjustment approach for wireless sensor networks with mobile sinks," *Journal of Supercomputing*, vol. 73, no. 7, pp. 3277–3290, 2017.

[11] Y. Liu, Q. Zhang, and L. Ni, "Opportunity-based topology control in wireless sensor networks," *IEEE Transactions on Parallel and distributed systems*, vol. 21, no. 3, pp. 405–416, 2010.

[12] X. Shen, F. Shen, Q. S. Sun, Y. Yang, Y. H. Yuan, and H. T. Shen, "Semi-paired discrete hashing: learning latent hash codes for semi-paired cross-view retrieval," *IEEE Transactions on Cybernetics*, vol. 47, no. 12, pp. 4275–4288, 2018.

[13] X. Shen, F. Shen, L. Liu, Y. H. Yuan, W. Liu, and Q. S. Sun, "Multiview discrete hashing for scalable multimedia search," *ACM Transactions on Intelligent Systems and Technology*, vol. 9, no. 5, pp. 1–21, 2018.

[14] S. V. A. Jeba and R. S. Kumar, "Reliable anonymous secure packet forwarding scheme for wireless sensor networks," *Computers and Electrical Engineering*, vol. 48, pp. 405–416, 2015.

[15] J. R. M. Dios, K. Lferd, A. D. S. Bernabe, G. Nunez, A. Torres-Gonzalez, and A. Ollero, "Cooperation between UAS and wireless sensor networks for efficient data collection in large environments," *Journal of Intelligent and Robotic Systems*, vol. 70, pp. 491–508, 2013.

[16] D. Zeng, Y. Dai, F. Li, R. S. Sherratt, and J. Wang, "Adversarial learning for distant supervised relation extraction," *Computers, Materials and Continua (CMC)*, vol. 55, no. 1, pp. 243–254, 2018.

[17] H. Yetgin, K. T. K. Cheung, M. Ei-Hajjar, and L. Hanzo, "Network-lifetime maximization of wireless sensor networks," *IEEE Access*, vol. 3, pp. 2191–2226, 2015.

[18] R. I. Ogie, "Adopting incentive mechanisms for large-scale participation in mobile crowdsensing: from literature review to a conceptual framework," *Human-Centric Computing and Information Sciences*, vol. 6, no. 1, pp. 1–31, 2016.

[19] Z. M. Nezhad and S. Khorsandi, "Cooperation enforcement based on dynamic pricing in multi-domain sensor network," in *Proceedings of Consumer communications and networking conference 2011 (CCNC 2011)*, pp. 1055–1060, IEEE, Las Vegas, Nevada, USA, January 2011.

[20] S. Maity and J. Park, "Powering IoT devices: a novel design and analysis technique," *Journal of Convergence*, vol. 7, 2016.

[21] D. Yasmine, K. Bouabdellah, and F. K. Mohammed, "Using mobile data collectors to enhance energy efficiency and reliability in delay tolerant wireless sensor networks," *Journal of Information Processing Systems*, vol. 12, pp. 275–294, 2016.

[22] D. Goyal and M. R. Tripathy, "Routing protocols in wireless sensor networks: a survey," in *Proceedings of second international conference on advanced computing and communication technologies (ACCT 2012)*, pp. 256–275, IEEE, Rohtak, Haryana, India, January 2012.

[23] M. Li, E. Kamioka, and S. Yamada, "Pricing to simulate node cooperation in wireless Ad hoc networks," *IEICE Transactions on Communications*, vol. E90-B, no. 7, pp. 1640–1650, 2007.

[24] Y. Yuan and F. Wang, "Blockchain: the state of the art and future trends," *Acta Automatica Sinica*, vol. 42, no. 4, pp. 481–494, 2016.

[25] Q. Shao, C. Jin, Z. Zhang, W. Qian, and A. Zhou, "Blockchain: architecture and research progress," *Chinese Journal of Computers*, 2017, http://cjc.ict.ac.cn/online/cre/10xsqfs-2017127145754.pdf.

[26] Y. Ren, J. Shen, D. Liu, J. Wang, and J. Kim, "Evidential quality preserving of electronic record in cloud storage," *Journal of Internet Technology*, vol. 17, no. 6, pp. 1125–1132, 2016.

[27] B. Christian, M. Ueli, T. Daniel, and Z. Vassilis, "Bitcoin as a transaction ledger: a composable treatment," in *Proceedings of 36th annual international cryptology conference—Advances in Cryptology (CRYPTO 2017)*, pp. 324–356, Santa Barbara, CA, USA, August 2017.

[28] Y. Ren, J. Shen, Y. Zheng, J. Wang, and H. Chao, "Efficient data integrity auditing for storage security in mobile health cloud," *Peer-to-Peer Networking and Applications*, vol. 9, no. 5, pp. 854–863, 2016.

[29] W. Qian, Q. Shao, Y. Zhu, C. Jin, and A. Zhou, "Research problems and methods in blockchain and trusted data management," *Journal of Software*, vol. 29, pp. 150–159, 2018.

[30] Y. Tu, Y. Lin, J. Wang, and J. U. K. Kim, "Semi-supervised learning with generative adversarial networks on digital signal modulation classification," *Computers, Materials and Continua (CMC)*, vol. 55, no. 2, pp. 243–254, 2018.

[31] L. Xiang, Y. Li, W. Hao, P. Yang, and X. Shen, "Reversible natural language watermarking using synonym substitution and arithmetic coding," *Computers, Materials and Continua (CMC)*, vol. 55, no. 3, pp. 541–559, 2018.

[32] R. Qiao, S. Dong, Q. Wei, and Q. Wang, "Blockchain based secure storage scheme of dynamic data," *Computer Science*, vol. 45, pp. 57–62, 2018.

[33] R. Meng, S. Rice, J. Wang, and X. Sun, "A fusion steganographic algorithm based on faster R-CNN," *Computers, Materials and Continua (CMC)*, vol. 55, no. 1, pp. 1–16, 2018.

[34] P. He, G. Yu, Y. Zhang, and Y. Bao, "Survey on blockchain technology and its application prospect," *Computer Science*, vol. 44, pp. 1–8, 2017.

[35] Y. Ren, J. Shen, J. Wang, J. Han, and S. Lee, "Mutual verifiable provable data auditing in public cloud Storage," *Journal of Internet Technology*, vol. 16, pp. 317–323, 2015.

[36] D. He, N. Kumar, S. Zeadally, and H. Wang, "Certificateless provable data possession scheme for cloud-based smart grid data management systems," *IEEE Transactions on Industrial Informatics*, vol. 14, no. 3, pp. 232–242, 2018.

[37] H. Wang, K. Li, K. Ota, and J. Shen, "Remote data integrity checking and sharing in cloud-based health internet of things," *IEICE Transactions on Information and Systems*, vol. E99.D, no. 8, pp. 1966–1973, 2016.

[38] Q. Jiang, F. Wei, S. Fu, J. Ma, G. Li, and A. Alelaiwi, "Robust extended chaotic maps-based three-factor authentication scheme preserving biometric template privacy," *Nonlinear Dynamics*, vol. 83, no. 4, pp. 2085–2101, 2016.

[39] Z. Faheem, A. Khan, S. U. R. Malik et al., "A survey of cloud computing data integrity schemes: design challenges, taxonomy and future trends," *Computers and Security*, vol. 65, pp. 29–49, 2017.

[40] N. Gargn and S. Bawa, "Comparative analysis of cloud data integrity auditing protocols," *Journal of Network and Computer Applications*, vol. 66, pp. 17–32, 2016.

[41] K. Gu, W. Jia, and J. Zhang, "Identity-based multi-proxy signature scheme in the standard model," *Fundamenta Informaticae*, vol. 150, no. 2, pp. 179–210, 2017.

[42] Y. Wang and Q. Wu, "A survey on cryptographic technologies for data integrity checking in clouds," *Journal of Cyber Security*, vol. 2, pp. 23–35, 2017.

7

Energy-Efficient UAV Communication with Multiple GTs Based on Trajectory Optimization

Yu Xu,[1] Lin Xiao [iD],[1] Dingcheng Yang,[1] Laurie Cuthbert,[2] and Yapeng Wang[3]

[1]Information Engineering School, Nanchang University, Nanchang 330031, China
[2]Information Systems Research Centre, Macao Polytechnic Institute, Rua de Luis Gonzaga Gomes, Macao SAR, China
[3]MPI-QMUL Information Systems Research Centre, Macao Polytechnic Institute, Macao SAR, China

Correspondence should be addressed to Lin Xiao; xiaolin@ncu.edu.cn

Academic Editor: Jose F. Monserrat

Wireless communications with unmanned aerial vehicles (UAVs) is a promising technology offering potential high mobility and low cost. This paper studies a UAV-enabled communication system, in which a fixed-wing UAV is deployed to collect information from a group of distributed ground terminals (GTs). Considering the requirements for quality of service (QoS) (i.e., the throughput of each GT is above a given threshold) and GT scheduling, we maximize the energy efficiency (EE) of the UAV in bits/Joule by optimizing the UAV's flight trajectory. In this paper, a mixed integer nonconvex optimization problem is formulated. As that is difficult to solve, we divide the formulated problem into two subproblems and apply standard linear programming (LP) and successive convex optimization techniques. We further propose an efficient iterative algorithm that jointly optimizes GT scheduling and the UAV's trajectory. Moreover, we set two special cases as benchmarks to measure the performance of the proposed design. The numerical results show that our proposed design achieves much better performance than the other two benchmark designs.

1. Introduction

Unmanned aerial vehicles (UAVs) recently are attracting significant attention in many fields: they can be applied in many different scenarios, including surveillance, monitoring, mobile relays, and data collection [1]. In general, UAVs can provide line-of-sight (LoS) links and thus offer good link capacity. Due to their potential mobility, flexible deployment, and low cost, UAVs are available for many operations and applications, such as precision agriculture [2], search and rescue [3], and timely environment monitoring and disaster warning [4–6]. Furthermore, UAVs can be used as mobile relays to extend the capacity and coverage of networks [7]. Meanwhile, with increasing popularity in the field of information technology (IT), both Facebook Aquila Drone [8] and Google Loon Project [9] aim to provide ubiquitous internet access for users in remote locations by using UAVs. In addition, UAVs can be deployed as aerial base stations (BSs) for ground terminals (GTs) as they are flexibly reconfigured

[10–12]. UAVs therefore provide aerial platforms that can be widely applied in wireless communication systems, as they can provide the terrestrial-aerial communication service for terrestrial users in regions lacking terrestrial infrastructures or under overload conditions [13, 14].

Nevertheless, UAV communication systems still face many critical challenges [1]. One of these is the limited battery capacity so that the UAVs have to land for recharging, which severely restricts the endurance of UAVs. As a result, the energy efficiency (EE) in bits/Joule is an important performance metric in UAV wireless communication [15]. Note that, unlike conventional terrestrial systems, the UAV needs to consume propulsion power to remain aloft in addition to the power required for the communications [16]. Therefore, having an energy-efficient trajectory design for UAV communication systems is of paramount importance. Hence, the requirement is to guarantee high-rate communication for the network, with low propulsion energy consumption.

Moreover, a UAV communication system with multiple GTs unavoidably leads to a higher performance requirement for the UAV and adds many key issues (e.g., user scheduling, user fairness, and access delay) that are challenging to solve. Note that energy-efficient designs for UAV communication systems are significantly different from traditional cellular settings:

(i) Saving power and cost by energy-efficient designs for UAV communication systems is more critical than that for terrestrial communication systems because of the limited on-board energy.

(ii) In addition to the communication-related power consumption, the propulsion power requirement for UAVs leads to much higher overall consumption.

We take both of these factors into account in this paper for energy-efficient designs for UAV communications.

In previous work [12, 17, 18], the maximum coverage for GTs is studied by optimizing the deployment of UAVs. However, this work does not focus on the issue of the UAV's energy efficiency. The work in [19] studies the energy-efficient 3D placement of a UAV-BS to achieve maximum coverage for users with the minimum required transmit power, but it does not consider the information bit requirement of all users. In [20], the UAV is deployed in a circular trajectory to assist communication and an energy-efficient design is considered, but it does not take multiple GTs into consideration. The authors in [15] studied energy-efficient communication by optimizing the UAV's trajectory: a mathematical model for the propulsion energy consumption of a fixed-wing UAV was developed, but they discuss neither the condition of multiple GTs nor QoS requirements for GTs.

This paper aims to study the EE maximization design for a UAV communication system with a group of GTs over a finite time period. In this system, a UAV is dispatched to collect the uploading data from GTs. We assume that these GTs upload data to the UAV using time-division multiple access (TDMA) that can efficiently avoid the cochannel interference enabling the frequency band to be shared. In fact, we discuss a new system model jointly considering QoS requirements and GT scheduling. Our objective is to maximize the UAV's energy efficiency by optimizing the trajectory. In addition, we set two special cases as benchmarks for comparison as well as to indicate the optimal performance of the objective problem. For the multiple GT system, it is important to note how the UAV chooses to communicate with which GT in each time slot. Hence, the GT scheduling is crucial to improve the UAV's EE. Intuitively, the UAV needs to be close to the associated GT that is transmitting information to the UAV for a better UAV-GT channel [21], but the required propulsion energy of the UAV is unfortunately likely to be larger. In general, the solution to the problem of the UAV's EE maximization can be derived at an optimal balance point between throughput maximization and energy consumption minimization.

The rest of this paper is organized as follows: first, Section 2 presents the system model and the problem formulation. Then, Section 3 proposes an effective alternative iteration algorithm. The numerical results are presented in Section 4 to

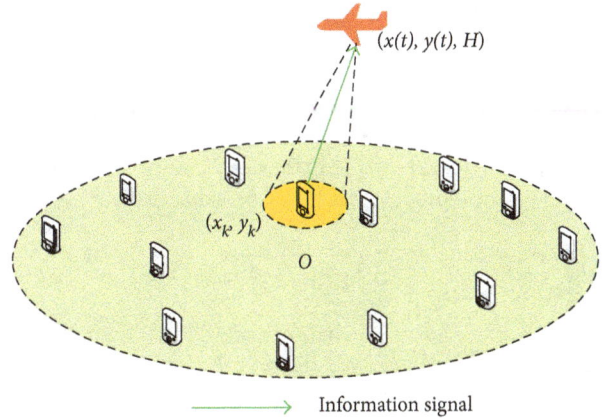

FIGURE 1: A UAV wireless communication system with multiple GTs.

demonstrate our proposed design. Finally, our main conclusions of the paper are summarized in Section 5.

2. System Model and Problem Formulation

2.1. System Model. As shown in Figure 1, we consider a UAV wireless communication system with N GTs denoted as the set $\mathcal{N} = \{1, 2, \ldots, N\}$. Without loss of generality, we consider a 3D Cartesian coordinate system. We assume that the group of GTs is distributed on a given circular geographical area with the geometrical center $[0, 0]^T$. The horizon coordinate of the GT k is predetermined and fixed at $\{\mathbf{w}_k = [x_k, y_k]^T\}_{k \in \mathcal{N}}$, $\mathbf{w}_k \in \mathbb{R}^{2 \times 1}$. In the system, the UAV is dispatched to collect the uplink data from GTs during the time horizon T. We assume that the altitude of the UAV is fixed at H meters that correspond to the minimum altitude to avoid collision. Consequently, the UAV trajectory can be denoted by $[x(t), y(t), H]^T, 0 \leq t \leq T$, with $x(t)$ and $y(t)$ denoting the time-varying x- and y-coordinates projected on the horizontal plane. Therefore, the UAV's projected position can be denoted by $\mathbf{q}(t) = [x(t), y(t)]^T$, where $0 \leq t \leq T$. For ease of formulation of the problem, we introduce a sufficiently small time step δ_t and then discretize the period time into M equal time slots with the step size δ_t, indexed by $n = 1, \ldots, M$. We assume that the trajectory of the UAV satisfies the following constraint:

$$\mathbf{q}[1] = \mathbf{q}[M], \tag{1}$$

$$\mathbf{v}[1] = \mathbf{v}[M], \tag{2}$$

where $\mathbf{v}[n]$ denotes the UAV's velocity in time slot n. From [15], we have the linear relationships as follows:

$$\mathbf{q}[n+1] = \mathbf{q}[n] + \mathbf{v}[n]\delta_t + \frac{1}{2}\mathbf{a}[n]\delta_t^2,$$
$$\mathbf{v}[n+1] = \mathbf{v}[n] + \mathbf{a}[n]\delta_t, \quad n = 1, 2, \ldots, M-1, \tag{3}$$

where $\mathbf{a}[n]$ denotes the UAV's acceleration during time slot n. Thus, in any time slot, we can express the time-varying distance between the UAV and the kth GT as

$$d_k[n] = \sqrt{\left\|\mathbf{q}[n] - w_k\right\|^2 + H^2}, \quad k \in \mathcal{N}. \tag{4}$$

Furthermore, we assume that the other GTs do not cause any interference on the current communication channel between the

UAV and the kth GT. Moreover, we consider that each UAV-GT channel follows a line-of-sight (LoS) link. Actually, practical UAV-GT channels can be well approximated by the LoS model. We also assume that the Doppler effect due to the UAV's mobility is perfectly compensated [22]. Therefore, the channel power gain from the kth GT to the UAV conforms to the free-space path loss model, which can be expressed as

$$h_k[n] = \beta_0 d_k^{-2}[n] = \frac{\beta_0}{\left\| \mathbf{q}[n] - w_k \right\|^2 + H^2}, \quad k \in \mathcal{N}, \quad (5)$$

where β_0 denotes the channel power at the reference distance $d_0 = 1$ meter. We define a binary variable $\alpha_k[n]$ that represents the communication scheduling factor for GTs; namely, it indicates whether or not the UAV collects data from the GT k during time slot n. If $\alpha_k[n] = 1$, it shows that the UAV communicates with the GT k in time slot n. Otherwise, we let $\alpha_k[n] = 0$. Denoted by P, the transmission power of each GT in per time slot is constant. The instantaneous channel capacity between the UAV and the GT k in bits/second/Hz (bps/Hz) can be written as

$$r_k[n] = \alpha_k[n] \log_2\left(1 + \frac{h_k[n]P}{\sigma^2} \right)$$

$$= \alpha_k[n] \log_2\left(1 + \frac{\gamma_0 P}{\left\| \mathbf{q}[n] - \mathbf{w}_k \right\|^2 + H^2} \right), \quad k \in \mathcal{N}, \quad (6)$$

where σ^2 is the white Gaussian noise power at each GT receiver and $\gamma_0 = \beta_0/\sigma^2$ denotes the reference-received signal-to-noise ratio (SNR) at $d_0 = 1$ meter. We assume that the communication bandwidth for the UAV is B, and each GT can share the whole frequency band using time-division multiple access (TDMA). Therefore, the throughput of the GT k in the period time T is given by

$$R_k = \delta_t B \sum_{n=1}^{M} r_k[n]$$

$$= \delta_t B \sum_{n=1}^{M} \alpha_k[n] \log_2\left(1 + \frac{P\gamma_0}{\left\| \mathbf{q}[n] - \mathbf{w}_k \right\|^2 + H^2} \right), \quad k \in \mathcal{N}. \quad (7)$$

Therefore, the total amount of information bits transmitted from all GTs to the UAV over the horizon time T can be expressed as

$$R(\{\mathbf{q}[n]\}, \{\alpha_k[n]\})$$

$$= \sum_{k=1}^{N} R_k$$

$$= \delta_t B \sum_{k=1}^{N} \sum_{n=1}^{M} \alpha_k[n] \log_2\left(1 + \frac{P\gamma_0}{\left\| \mathbf{q}[n] - \mathbf{w}_k \right\|^2 + H^2} \right), \quad k \in \mathcal{N}. \quad (8)$$

In our proposed system, the energy consumption of the UAV consists of two parts. The first is the communication-related energy consumption and the second is the propulsion energy consumption, which ensures the UAV remains in the air, supporting its mobility. In practice, the communication-related energy is small compared with the propulsion energy. Thus, we ignore the communication-related energy in this paper. From [15], the energy consumption of a fixed-wing UAV during T can be expressed as

$$E(\{\mathbf{q}[n]\}) = \delta_t \sum_{n=1}^{M} \left[c_1 \|\mathbf{v}[n]\|^3 + \frac{c_2}{\|\mathbf{v}[n]\|} \left(1 + \frac{\|\mathbf{a}[n]\|^2 - \left((\mathbf{a}^T[n]\mathbf{v}[n])^2 / \|\mathbf{v}[n]\|^2 \right)}{g^2} \right) \right] + \Delta_k, \quad (9)$$

where c_1 and c_2 are two constant parameters related to aerodynamics. The second term Δ_k denotes the UAV's kinetic energy, whose value only depends on the initial and final speed, that is, $\Delta_k = (1/2)m(\|\mathbf{v}[M]\|^2 - \|\mathbf{v}[1]\|^2)$. Obviously, under our hypothesis in (2), we can obtain $\Delta_k = 0$.

2.2. Problem Formulation. In this paper, our objective is to maximize the UAV's energy efficiency by jointly optimizing GT scheduling and the UAV's trajectory. With (8), (9), and some constrains like (1) and (2), the optimization problem can be formulated as

$$\max_{\{\mathbf{q}[n], \mathbf{v}[n], \mathbf{a}[n]\}, \{\alpha_k[n]\}} \frac{B \sum_{k=1}^{N} \sum_{n=1}^{M} \alpha_k[n] \log_2\left(1 + \left(P\gamma_0 / \left(\|\mathbf{q}[n] - \mathbf{w}_k\|^2 + H^2 \right) \right) \right)}{\sum_{n=1}^{M} \left(c_1 \|\mathbf{v}[n]\|^3 + (c_2/\|\mathbf{v}[n]\|) + \left(c_2 \|\mathbf{a}[n]\|^2 / g^2 \|\mathbf{v}[n]\| \right) \right)}, \quad (10)$$

$$\text{s.t. } \mathbf{q}[n+1] = \mathbf{q}[n] + \mathbf{v}[n] \cdot \delta_t + \frac{1}{2}\mathbf{a}[n] \cdot \delta_t^2, \quad n = 1, 2, \ldots, M-1, \quad (10a)$$

$$\mathbf{v}[n+1] = \mathbf{v}[n] + \mathbf{a}[n] \cdot \delta_t, \quad n = 1, 2, \dots, M-1, \quad \text{(10b)}$$

$$\mathbf{q}[1] = \mathbf{q}[M], \quad \text{(10c)}$$

$$\mathbf{v}[1] = \mathbf{v}[M], \quad \text{(10d)}$$

$$\|\mathbf{v}[n]\| \le V_{\max}, \quad \forall n, \quad \text{(10e)}$$

$$\|\mathbf{a}[n]\| \le a_{\max}, \quad \forall n, \quad \text{(10f)}$$

$$\sum_{k=1}^{N} \alpha_k[n] \le 1, \quad \forall n, \quad \text{(10g)}$$

$$\alpha_k[n] \in \{0, 1\}, \quad \forall k, n, \quad \text{(10h)}$$

$$B\delta_t \sum_{n=1}^{M} \alpha_k[n] \log_2\left(1 + \frac{P\gamma_0}{\|\mathbf{q}[n] - \mathbf{w}_k\|^2 + H^2}\right) \ge Q_m, \quad k \in \mathcal{N}, \quad \text{(10i)}$$

where the constraints (10c) and (10d) indicate that the UAV needs to get back to the initial location after the horizon time T with the same initial and final velocity, which is opportune for the UAV to take the next cycle flight via the same trajectory. In the constraints (10e) and (10f), $V_{\max}$ and $a_{\max}$ denote the maximum velocity and acceleration that the UAV can achieve. The constraints (10g) and (10h) represent that

the UAV can communicate with one GT at most during any time slot. The last constraint (10i) is the QoS requirement for all GTs, and Q_m indicates the minimum amount of uploading data for each GT in the period time T.

Note that problem (10) is difficult to solve due to two main reasons:

(i) Firstly, the constraints (10g) and (10h) involve integer constraints. Moreover, the constraint (10i) is nonconvex.

(ii) The objective function of (10) is nonconcave. As a result, problem (10) is a mixed-integer nonconvex problem that cannot be solved directly by using conventional convex optimization techniques in general.

Before discussing the solution to (10), we need to introduce two special cases, which can be used as benchmarks in this paper.

2.3. Special Case I: Average Time Allocation Tactics. In this special case, an approach using average time allocation tactics is presented: each GT always is allocated equivalent communication time during the horizon time T. In this case, we can use a vector $\alpha_k^{\text{Ave}}[n]$ to denote the GT scheduling. Similarly, $\alpha_k^{\text{Ave}}[n] = 1$ denotes that the UAV collects data from the kth GT in slot n. Otherwise, $\alpha_k^{\text{Ave}}[n] = 0$. There always is $\sum_{n=1}^{M} \alpha_k^{\text{Ave}}[n] = M/N, \forall k \in \mathcal{N}$, where M/N is assumed to be an integer. In this scenario, the problem of this case can be formulated as

$$\max_{\{\mathbf{q}[n],\mathbf{v}[n],\mathbf{a}[n],\alpha_k^{\text{Ave}}[n]\}} \frac{B\sum_{k=1}^{N}\sum_{n=1}^{M} \alpha_k^{\text{Ave}}[n] \log_2\left(1 + \left(P\gamma_0 / \left(\|\mathbf{q}[n] - \mathbf{w}_k\|^2 + H^2\right)\right)\right)}{\sum_{n=1}^{M}\left(c_1\|\mathbf{v}[n]\|^3 + (c_2/\|\mathbf{v}[n]\|) + \left(c_2\|\mathbf{a}[n]\|^2 / g^2\|\mathbf{v}[n]\|\right)\right)}, \quad \text{(11)}$$

$$\text{s.t. (10a)--(10f).}$$

$$B\delta_t \sum_{n=1}^{M} \alpha_k^{\text{Ave}}[n] \log_2\left(1 + \frac{P\gamma_0}{\|\mathbf{q}[n] - \mathbf{w}_k\|^2 + H^2}\right) \ge Q_m, \quad k \in \mathcal{N}. \quad \text{(11a)}$$

Compared to problem (10), except the constraint (11a), all constraints in problem (11) are convex. Nevertheless, the objective function of (11) is still nonconcave, and we need to convert it into a new form consisting of a concave numerator and a convex denominator that can be solved efficiently in Section 3.

2.4. Special Case II: Maximum EE without QoS. In this part, we discuss the other special case, in which we just consider

the optimal GT scheduling and UAV trajectory design rather than taking into account the QoS requirements, which means that this design has a higher probability of achieving a larger total system rate in each iteration but may result in the throughputs of some GTs being under the threshold. For this case, it can also be considered as a special case of problem (10), and this case can be formulated as

$$\max_{\{\mathbf{q}[n],\mathbf{v}[n],\mathbf{a}[n],\alpha_k[n]\}} \frac{B\sum_{k=1}^{N}\sum_{n=1}^{M} \alpha_k[n] \log_2\left(1 + \left(P\gamma_0 / \left(\|\mathbf{q}[n] - \mathbf{w}_k\|^2 + H^2\right)\right)\right)}{\sum_{n=1}^{M}\left(c_1\|\mathbf{v}[n]\|^3 + (c_2/\|\mathbf{v}[n]\|) + \left(c_2\|\mathbf{a}[n]\|^2 / g^2\|\mathbf{v}[n]\|\right)\right)}, \quad \text{(12)}$$

$$\text{s.t. (10a)--(10h).}$$

As problem (12) is a special case of problem (10), it can also be efficiently solved with the optimization technique proposed in Section 3.

3. Jointly Optimal Communication Scheduling and Trajectory Design

In this section, we propose a joint optimization method by applying the standard LP and successive convex optimization techniques to handle problem (10). We first split problem (10) into two subproblems. For the first subproblem, we can obtain $\{\alpha_k^{j+1}[n]\}$ by optimizing the GT scheduling factor $\{\alpha_k[n]\}$ with the given UAV's trajectory $\{\mathbf{q}_j[n]\}$ in the $(j+1)$th iteration. As to the second subproblem, we fix the GT scheduling $\{\alpha_k^{j+1}[n]\}$ and then optimize the trajectory of the UAV. Following this iterative method, the optimal energy-efficient trajectory of the UAV eventually can be obtained.

3.1. Optimal GT Scheduling with Fixed Trajectory. Considering that this subproblem is a 0-1 integer programming with the given UAV's trajectory $\{\mathbf{q}_j[n]\}$, we relax the binary variables $\{\alpha_k[n]\}$ in (10h) into continuous variables. Thus, we have the following optimization problem:

$$\max_{\{\alpha_k[n]\}} \sum_{k=1}^{N}\sum_{n=1}^{M} \alpha_k[n] \log_2\left(1 + \frac{P\gamma_0}{\left\|\mathbf{q}_j[n]-\mathbf{w}_k\right\|^2 + H^2}\right), \quad (13)$$

s.t. (10g),

$$0 \le \alpha_k[n] \le 1, \quad \forall n, k, \quad (13a)$$

$$B\delta_{\mathrm{t}} \sum_{n=1}^{M} \alpha_k[n] \log_2\left(1 + \frac{P\gamma_0}{\left\|\mathbf{q}_j[n]-\mathbf{w}_k\right\|^2 + H^2}\right) \ge Q_{\mathrm{m}}, \quad k \in \mathcal{N}. \quad (13b)$$

Obviously, problem (13) is a standard linear programming (LP) problem, so it can be solved easily and efficiently using optimization tools. By solving problem (13), we can achieve the optimal GT scheduling factor $\{\alpha_k^{j+1}[n]\}$, and then, $\{\alpha_k^{j+1}[n]\}$ serves as the input for problem (14) in the next subsection.

3.2. Optimal UAV's Trajectory Design with Fixed GT Scheduling. In this subsection, with the given $\{\alpha_k^{j+1}[n]\}$, we need to solve the following subproblem:

$$\max_{\{\mathbf{q}[n],\mathbf{v}[n],\mathbf{a}[n]\}} \frac{B\sum_{k=1}^{N}\sum_{n=1}^{M} \alpha_k^{j+1}[n] \log_2\left(1 + \left(P\gamma_0/\left(\left\|\mathbf{q}[n]-\mathbf{w}_k\right\|^2 + H^2\right)\right)\right)}{\sum_{n=1}^{M}\left(c_1\|\mathbf{v}[n]\|^3 + \left(c_2/\|\mathbf{v}[n]\|\right) + \left(c_2\|\mathbf{a}[n]\|^2/g^2\|\mathbf{v}[n]\|\right)\right)}, \quad (14)$$

s.t. (10a)–(10f),

$$B\delta_{\mathrm{t}} \sum_{n=1}^{M} \alpha_k^{j+1}[n] \log_2\left(1 + \frac{P\gamma_0}{\left\|\mathbf{q}[n]-\mathbf{w}_k\right\|^2 + H^2}\right) \ge Q_{\mathrm{m}},$$

$$k \in \mathcal{N}.$$

$$(14a)$$

Unfortunately, (14) is not a standard convex problem, but we can find the local optimal solution by applying the sequential convex optimization technique. Hence, we first introduce slack variables $\{\kappa_n\}$ to reform the denominator of the objective function in (14) as

$$P(\{\mathbf{q}[n]\}, \{\kappa_n\}) = \sum_{n=1}^{M}\left(c_1\|\mathbf{v}[n]\|^3 + \frac{c_2}{\kappa_n} + \frac{c_2\|\mathbf{a}[n]\|^2}{g^2\kappa_n}\right), \quad (15)$$

where $\|\mathbf{v}[n]\|^2 \ge \kappa_n^2$, $\forall n$. Note that this new constraint is nonconvex. We know that $\|\mathbf{v}[n]\|^2$ is convex; thus, its first-order Taylor expansion is the global underestimator. As a result, we adopt the Taylor approximation at the given local point $\{\mathbf{v}_j[n]\}$ and define the lower bound function as follows:

$$\rho_{lb}(\mathbf{v}[n]) \triangleq \left\|\mathbf{v}_j[n]\right\|^2 + 2\mathbf{v}_j^T[n]\left(\mathbf{v}[n]-\mathbf{v}_j[n]\right), \quad \forall n. \quad (16)$$

It is noteworthy that the numerator of the objective function in (14) is nonconcave with respect to $\{\mathbf{q}[n]\}$. However, if we consider $\|\mathbf{q}[n]-\mathbf{w}_k\|^2$ as a whole, the numerator is convex with respect to it. Therefore, with the given local point $\{\mathbf{q}_j[n]\}$ at the $(j+1)$th iteration, we can obtain the lower bound function of the numerator as follows:

$$R(\{\mathbf{q}[n]\}) = B \sum_{k=1}^{N}\sum_{n=1}^{M} \alpha_k^{j+1}[n] \log_2\left(1 + \frac{P\gamma_0}{\left\|\mathbf{q}[n]-\mathbf{w}_k\right\|^2 + H^2}\right)$$

$$\ge B \sum_{k=1}^{N}\sum_{n=1}^{M}\left[\varphi_j[n] - \psi_j[n]\left(\left\|\mathbf{q}[n]-\mathbf{w}_k\right\|^2 - \left\|\mathbf{q}_j[n]-\mathbf{w}_k\right\|^2\right)\right]$$

$$\triangleq R_{lb}(\{\mathbf{q}[n]\}), \quad \forall n,$$

$$(17)$$

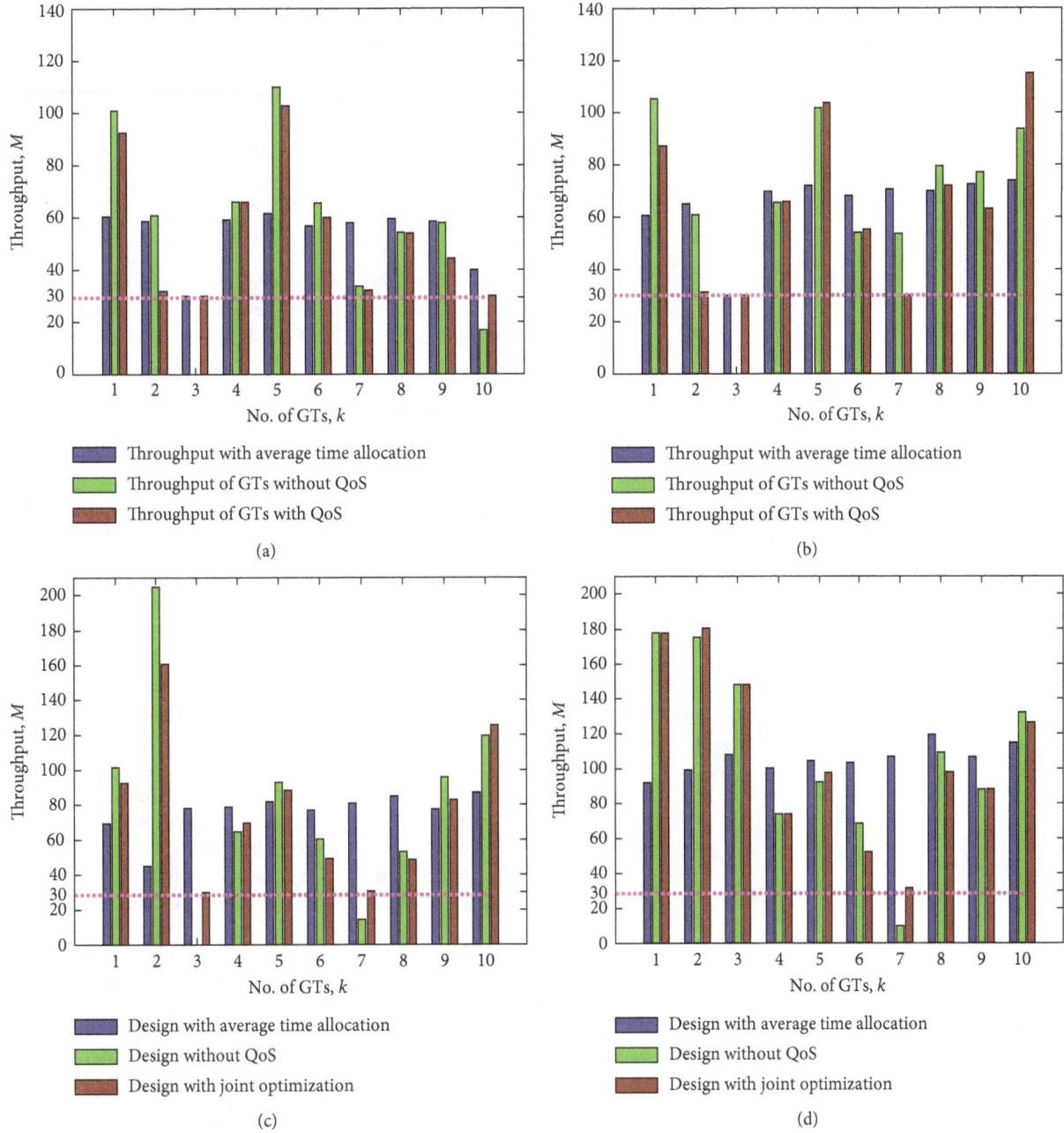

FIGURE 2: Comparison of the throughput of per GT with different period times T: (a) throughput of per GT, $T = 100$ s; (b) throughput of per GT, $T = 120$ s; (c) throughput of per GT, $T = 140$ s; (d) throughput of per GT, $T = 180$ s.

where

$$
\varphi_j[n] = \alpha_k^{j+1}[n] \log_2 \left(1 + \frac{P\gamma_0}{\left\| \mathbf{q}_j[n] - \mathbf{w}_k \right\|^2 + H^2} \right),
$$

$$
\psi_j[n] = \frac{1}{\ln 2} \cdot \frac{\alpha_k^{j+1}[n] P\gamma_0}{\left(\left\| \mathbf{q}_j[n] - \mathbf{w}_k \right\|^2 + H^2 \right) \left(\left\| \mathbf{q}_j[n] - \mathbf{w}_k \right\|^2 + H^2 + P\gamma_0 \right)}, \quad \forall n.
$$

(18)

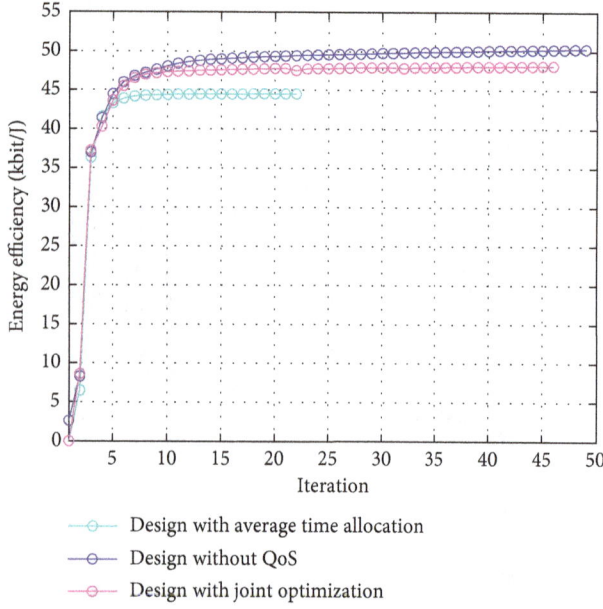

FIGURE 3: Convergence of the proposed Algorithm 1 for $T = 140$ s.

As a result, with any given local point $\{\mathbf{q}_j[n]\}$, we can reformulate the optimization problem (14) as the following problem:

$$\max_{\{\mathbf{q}[n],\mathbf{v}[n],\kappa_n,\mathbf{a}[n]\}} \frac{R_{lb}(\mathbf{q}[n])}{P(\{\mathbf{q}[n]\},\{\kappa_n\})}, \qquad (19)$$

s.t. (10a)–(10f),

$$\kappa_n \geq 0, \quad \forall n, \qquad (19a)$$

$$\rho_{lb}(\{\mathbf{v}[n]\}) \geq \kappa_n^2, \quad \forall n, \qquad (19b)$$

$$\delta_t \cdot R_{lb}(\{\mathbf{q}[n]\}) \geq Q_m, \quad k \in \mathcal{N}. \qquad (19c)$$

It is worth noting that the first-order Taylor expansions in (16) and (17) are tight at the given local point which means that problem (19) has the same objective value as that of problem (14). After problem (19), we can obtain the optimal trajectory $\{\mathbf{q}_{j+1}^*[n]\}$. In the next iteration, $\{\mathbf{q}_{j+1}^*[n]\}$ is used as the input for problem (13). In summary, the detail procedure is shown in Algorithm 1.

Note that problem (11) and problem (12) are special cases of problem (10). As a result, the proposed Algorithm 1 is also feasible to the special cases I and II.

4. Numerical Results

In this section, numerical results are presented to show the validation and effectiveness of the proposed designs. We assume that the altitude of the UAV is fixed at $H = 100$ m. The number of GTs is $N = 10$, and they are uniformly distributed within a 2D circular area with radius 0.8 km. The QoS requirement for each GT is $Q_m = 30$ Mbits. We set the communication bandwidth $B = 1$ MHz, and the noise power spectral density is $N_0 = -170$ dBm/Hz. The noise power is $\sigma^2 = N_0 B = -110$ dBm. We also assume that the constant

transmission power of each GT is $P = 10$ dBm, and the reference-received signal-to-noise ratio is $\gamma_0 = 80$ dB. Moreover, from [15], we set $c_1 = 9.26 \times 10^{-4}$ and $c_2 = 2250$. The threshold accuracy ε in Algorithm 1 is set as 10^{-6}. The initial and final locations of the UAV are assumed to be $\mathbf{q}[1] = \mathbf{q}[M] = [500, 0]^T$. The maximum velocity and acceleration of the UAV are set as $V_{max} = 100$ m/s and $a_{max} = 5$ m/s^2, respectively. The initial and final velocities are assumed as $\mathbf{v}[1] = \mathbf{v}[M] = [0, 30]^T$.

Figure 2 shows the throughput of each GT for different period times. We can see that the change of throughput per GT is closely related to the period time and location. For example, for the 10th GT, its throughput rapidly increases from $T = 100$ s to $T = 120$ s, which means that the UAV can achieve a better channel to communicate with the 10th GT in the later horizon time leading to a higher transmission rate. Note that, for the three designs, the throughput of each GT is not monotonically increasing with regard to the horizon time. In fact, the period time can affect the UAV's trajectory, and then, it has an effect on the channel conditions, which impacts the GT scheduling. On the other hand, if the UAV flies without considering QoS requirements (i.e., the special case II), it can be observed that the throughput of the 3rd and 7th GTs are below the QoS threshold, as shown in Figures 2(a)–2(c). What is worse, the throughput of the 3rd GT actually is zero in Figures 2(a)–2(c), which means that the 3rd GT is not allocated any slot to transmit data. Hence, in the special case II, the UAV cannot ensure the minimum communication requirement for each GT, which may result in a larger gap between maximum and minimum throughput. The special case II causes a severe limitation on the communication scenarios requiring users QoS. As for the design with average time allocation (i.e., the special case I), each GT is allocated an equal number of time slots for uploading data to the UAV. As a result, these GTs' throughputs are relatively regular and even. However, the optimal EE of this case actually is the lowest, as shown in Figure 3.

Figure 3 shows the convergence performance of the proposed Algorithm 1. In this result, we compare three design schemes for $T = 140$ s. It can be observed that the energy efficiency increases rapidly with the number of iterations at the beginning and then rises slowly until converges to the prescribed accuracy. In addition, it is observed that the UAV can achieve higher EE when it flies without considering QoS requirements, and the UAV's EE is minimum when it adopts the method of average time allocation, which is in line with our expectations.

Figure 4 shows the three cases of optimal energy-efficient trajectory obtained by Algorithm 1 for several different period times T. We compare the jointly optimal trajectory of the UAV with the special cases I and II. It can be observed that the disparity of the trajectory between jointly optimal design and special case II becomes smaller as the period time T increases. That is because there are more time slots for the UAV to utilize with the increasing period time. Thus, the ratio of the time satisfying QoS requirements to the horizon time T becomes increasingly small, which means that, in the great majority of time slots, the GT

(1) Initialize $\left\{\mathbf{q}_j[n], \mathbf{v}_j[n]\right\}$. Let $j = 0$, accuracy $\varepsilon > 0$.
(2) **repeat**
(3)   Solve problem (13) for the given $\left\{\mathbf{q}_j[n]\right\}$, and denote the optimal solution as $\left\{\alpha_k^{j+1*}[n]\right\}$.
(4)   Update $\alpha_k^{j+1}[n] = \alpha_k^{j+1*}[n]$.
(5)   Solve problem (19) for the given $\left\{\mathbf{q}_j[n], \mathbf{v}_j[n], \alpha_k^{j+1}[n]\right\}$, and denote the optimal solution as $\left\{\mathbf{q}_{j+1}^*[n], \mathbf{v}_{j+1}^*[n]\right\}$.
(6)   Update $\mathbf{q}_{j+1}[n] = \mathbf{q}_{j+1}^*[n]$, $\mathbf{v}_{j+1}[n] = \mathbf{v}_{j+1}^*[n]$.
(7)   Update $j = j + 1$.
(8)   **Until** converges to the prescribed accuracy ε.

ALGORITHM 1: Alternative iteration method for problem (10).

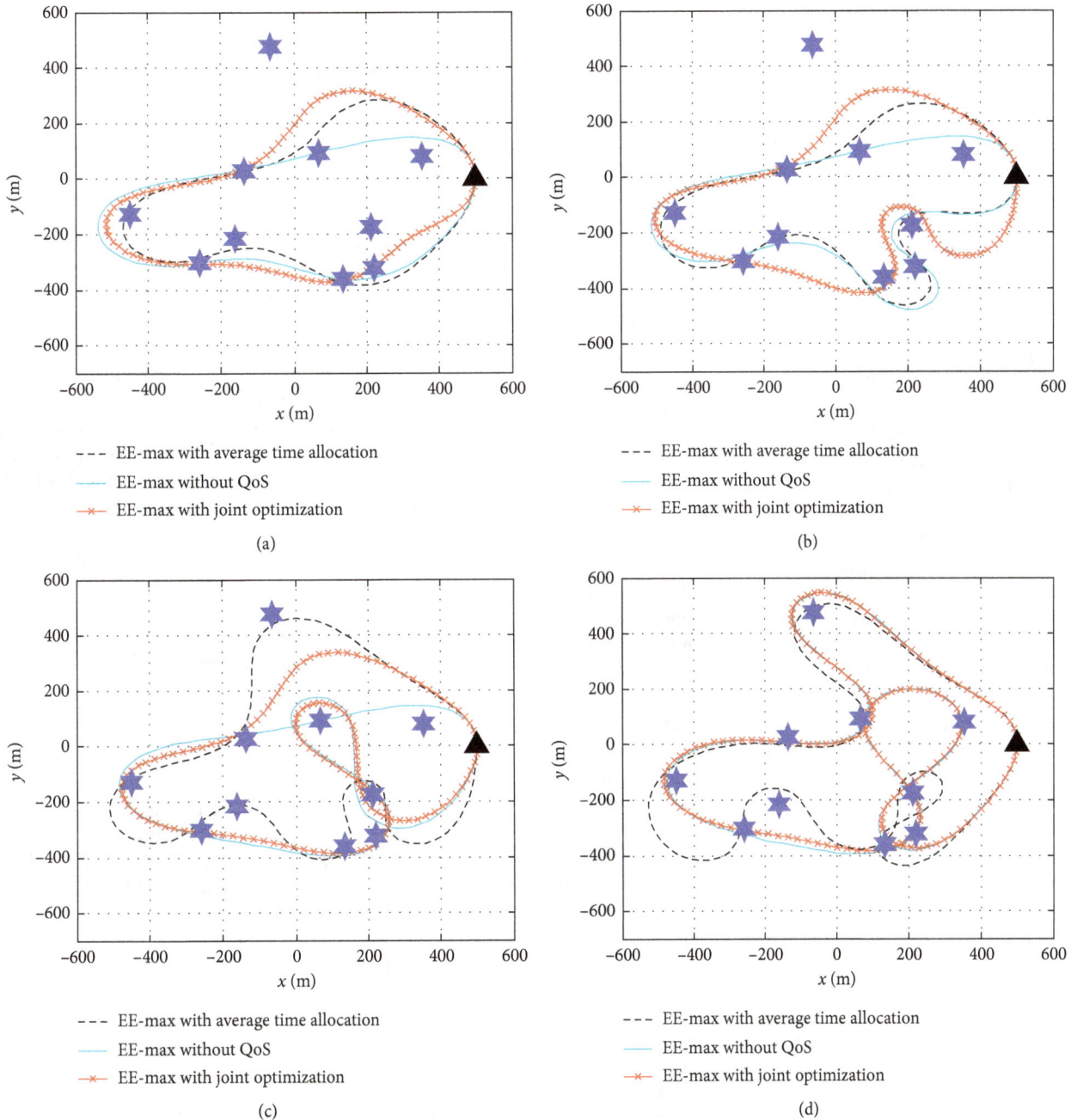

FIGURE 4: Comparison of optimal energy-efficient trajectory of the UAV with different period times T. The initial and final locations are marked by "▲," and the locations of GTs are marked by "★." (a) Optimal trajectory, $T = 100$ s. (b) Optimal trajectory, $T = 120$ s. (c) Optimal trajectory, $T = 40$ s. (d) Optimal trajectory, $T = 180$ s.

TABLE 1: Performance comparison for the two special cases with the proposed jointly optimal design.

Period time (s)	Energy efficiency (kbits/J)			Average rate (Mbps)		
	Average time allocation	Number of QoS requirements	Joint optimization	Average time allocation	Number of QoS requirements	Joint optimization
100	47.95	51.51	49.06	5.34	5.65	5.43
110	46.53	50.16	46.8	5.44	5.61	5.51
120	45.59	48.63	45.73	5.43	5.75	5.44
130	44.27	50.57	46	5.38	5.79	5.5
140	44.5	50.5	48.38	5.43	5.77	5.56
150	44.76	50.31	48.59	5.46	5.74	5.6
160	47.59	48.8	48.67	5.57	5.62	5.61
170	45.31	49.32	49.23	5.57	5.64	5.62
180	45.53	49.56	49.38	5.57	5.68	5.67

scheduling of these two designs is the same. As for the special case I, due to employing the average time allocation technique, the UAV is more likely to fly closer to each GT, especially when T is not large; this leads to the special case I having a better fairness compared with the other two designs, as shown in Figure 2.

Table 1, respectively, gives the comparison of the maximum energy efficiency and total average rate of the system achieved by the three energy-efficient trajectory designs in different periods T. It is observed that the EE of the design with average time allocation is invariably lowest as the average time allocation severely hinders the algorithm's optimization performance. For the design without QoS requirements, the UAV's EE is always largest in different period times T. The main reason is that no QoS requirements give the UAV more freedom to optimize its EE in each iteration. From this table, the performance of the joint optimization design falls in between, but it overcomes the main defects of the other two designs. It is interesting to note that, for a small period time (e.g., $T = 100$ s to 120 s), the energy-efficient performance of the joint optimization design is close to the design with average time allocation because there is little time freedom for the UAV to optimize the trajectory. Correspondingly, for a large period time (e.g., $T = 160$ s to 180 s), the UAV has sufficient time slots to optimize its trajectory, while the effect of the QoS constraint is inconspicuous. As a result, for a large period time, the design with joint optimization and the design without QoS requirements have the similar EE values and trajectory, as shown in Figure 4(d). For the period time within $120 \leq T \leq 160$, the jointly optimal design has more time freedom to maximize the EE values compared to the design with average time allocation, while the QoS constraint has a nonnegligible effect on the EE maximization compared to the design without QoS requirements. In this sense, the jointly optimal design actually strikes a trade-off between the special cases I and II.

In summary, the GT scheduling of the joint optimization design is more flexible compared to the special case I, and it leads to better performance. Although the EE of the joint optimization design is smaller compared to the case II, it can ensure the QoS requirements for all GTs, which is more in line with actual requirements and hence has more practical significance.

From Table 1, we also know that the EE of the UAV is not monotonically increasing with respect to the period time T; this gives us the freedom to find the optimal period time that can further improve the UAV's EE on the basis of this paper. This problem will be left to our future work.

5. Conclusions

In this paper, we investigate the energy efficiency of a UAV wireless communication system with multiple GTs. By deriving the propulsion energy consumption and communication rate models of the UAV, we establish the energy efficiency framework that jointly optimizes the scheduling of GTs and the UAV's trajectory. In addition, we establish two special cases as benchmarks to illustrate the optimal performance of our objective problem. To solve the formulated problem, we propose an iterative algorithm that jointly applies standard LP and successive convex optimization techniques. Using this algorithm, a locally optimal solution is obtained. Numerical results show that the energy efficiency of the UAV can be enhanced significantly with our design and that optimal performance is achieved compared with the other two benchmark designs, which is consistent with our expectations.

Acknowledgments

This work was supported in part by the National Natural Science Foundation of China (61703197, 61561032, and 61461029), China/Jiangxi Postdoctoral Science Foundation Funded Project (2014MT561879 and 2014KY046), Young Scientists Project Funding of Jiangxi Province (20162BCB23010 and 20153BCB23020), and the Natural Science Foundation of Jiangxi Province (20114ACE00200).

References

[1] Y. Zeng, R. Zhang, and T. J. Lim, "Wireless communications with unmanned aerial vehicles: opportunities and challenges," *IEEE Communications Magazine*, vol. 54, no. 5, pp. 36–42, 2016.

[2] J. Valente, D. Sanz, A. Barrientos, J. D. Cerro, A. Ribeiro, and

C. Rossi, "An air-ground wireless sensor network for crop monitoring," *Sensors*, vol. 11, no. 6, pp. 6088–6108, 2011.

[3] S. Waharte and N. Trigoni, "Supporting search and rescue operations with UAVs," in *Proceedings of the 2010 International Conference on Emerging Security Technologies (EST)*, pp. 142–147, Canterbury, UK, September 2010.

[4] M. Abdelkader, M. Shaqura, M. Ghommem, N. Collier, V. Calo, and C. Claudel, "Optimal multi-agent path planning for fast inverse modeling in UAV-based flood sensing applications," in *Proceedings of the 2014 International Conference on Unmanned Aircraft Systems (ICUAS)*, pp. 64–71, Orlando, FL, USA, May 2014.

[5] I. Maza, F. Caballero, J. Capitán, J. Martínez-de Dios, and A. Ollero, "Experimental results in multi-UAV coordination for disaster management and civil security applications," *Journal of Intelligent & Robotic Systems*, vol. 61, no. 1–4, pp. 563–585, 2011.

[6] C. Wu, X. Cao, R. Lin, and F. Wang, "Registration-based moving vehicle detection for low-altitude urban traffic surveillance," in *Proceedings of the 8th World Congress on Intelligent Control and Automation (WCICA 2010)*, pp. 373–378, Jinan, China, July 2010.

[7] Y. Zeng, R. Zhang, and T. J. Lim, "Throughput maximization for UAV-enabled mobile relaying systems," *IEEE Trans. Commun.*, vol. 64, no. 12, pp. 4983–4996, Dec. 2016.

[8] Facebook, *Connecting the World from the Sky*, Technical Report, Facebook, Inc., Menlo Park, CA, USA, 2014.

[9] S. Katikala, "Google project loon," *InSight: Rivier Academic Journal*, vol. 10, no. 2, 2014.

[10] M. Mozaffari, W. Saad, M. Bennis, and M. Debbah, "Drone small cells in the clouds: design, deployment and performance analysis," in *Proceedings of the IEEE Global Communications Conference (GLOBECOM 2015)*, pp. 1–6, San Diego, CA, USA, December 2015.

[11] M. Mozaffari, W. Saad, M. Bennis, and M. Debbah, "Unmanned aerial vehicle with underlaid device-to-device communications: performance and tradeoffs," *IEEE Transactions on Wireless Communications*, vol. 15, no. 6, pp. 3949–3963, 2016.

[12] J. Lyu, Y. Zeng, R. Zhang, and T. J. Lim, "Placement optimization of UAV-mounted mobile base stations," *IEEE Communications Letters*, vol. 21, no. 3, pp. 604–607, 2017.

[13] A. Merwaday and I. Guvenc, "UAV assisted heterogeneous networks for public safety communications," in *Proceedings of the IEEE Wireless Communications and Networking Conference*, pp. 329–334, New Orleans, LA, USA, March 2015.

[14] M. M. Azari, F. Rosas, and K.-C. Chen, "Optimal UAV positioning for terrestrial-aerial communication in presence of fading," in *Proceedings of the IEEE Global Communications Conference (GLOBECOM 2016)*, pp. 1–7, Washington, DC, USA, December 2016.

[15] Y. Zeng and R. Zhang, "Energy-efficient UAV communication with trajectory optimization," *IEEE Transactions on Wireless Communications*, vol. 16, no. 6, pp. 3747–3760, 2017.

[16] Y. Chen, S. Zhang, S. Xu, and G. Y. Li, "Fundamental tradeoffs on green wireless networks," *IEEE Communications Magazine*, vol. 49, no. 6, pp. 30–37, 2011.

[17] I. Bor-Yaliniz, A. El-Keyi, and H. Yanikomeroglu, "Efficient 3-D placement of an aerial base station in next generation cellular networks," in *Proceedings of the IEEE International Conference on Communications (ICC 2016)*, Kuala Lumpur, Malaysia, May 2016.

[18] M. Mozaffari, W. Saad, M. Bennis, and M. Debbah, "Efficient deployment of multiple unmanned aerial vehicles for optimal wireless coverage," *IEEE Communications Letters*, vol. 20,

no. 8, pp. 1647–1650, 2016.

[19] M. Alzenad, A. El-keyi, F. Lagum, and H. Yanikomeroglu, "3D placement of an unmanned aerial vehicle base station (UAV-BS) for energy-efficient maximal coverage," *IEEE Wireless Communications Letters*, vol. 6, no. 4, pp. 434–437, 2017.

[20] J. W. Zhang, Y. Zeng, and R. Zhang, "Spectrum and energy efficiency maximization in UAV-enabled mobile relaying," in *Proceedings of the IEEE International Conference on Communications (ICC 2017)*, Paris, France, May 2017.

[21] C. Zhan, Y. Zeng, and R. Zhang, "Energy-efficient data collection in UAV enabled wireless sensor network," *IEEE Wireless Communications Letters*, 2017, https://arxiv.org/abs/1708.00221.

[22] U. Mengali and A. N. D'Andrea, *Synchronization Techniques for Digital Receivers*, Springer, Berlin, Germany, 1997.

8

Outage Analysis and Power Allocation Optimization for Multiple Energy-Harvesting Relay System using SWIPT

Kang Liu [ID], Qi Zhu [ID], and Ying Wang

Department of Telecommunication and Information Engineering, Nanjing University of Posts and Telecommunications, Jiangsu 210003, China

Correspondence should be addressed to Qi Zhu; zhuqi@njupt.edu.cn

Academic Editor: Juan C. Cano

Energy harvesting (EH) combined with cooperative relying plays a promising role in future wireless communication systems. We consider a wireless multiple EH relay system. All relays are assumed to be EH nodes with simultaneous wireless and information transfer (SWIPT) capabilities, which means the relays are wirelessly powered by harvesting energy from the received signal. Each EH node separates the input RF signal into two parts which are, respectively, for EH and information transmission using the power splitting (PS) protocol. In this paper, a closed-form outage probability expression is derived for the cooperative relaying system based on the characteristic function of the system's probability density function (PDF) with only one relay. With the approximation of the outage probability expression, three optimization problems are built to minimize the outage probability under different constraints. We use the Lagrange method and Karush–Kuhn–Tucker (KKT) condition to solve the optimization problems to jointly optimize the relay's PS factors and the transmit power. Numerical results show that our derived expression of the outage probability is accuracy and gives insights into the effect of various system parameters on the performance of protocols. Meanwhile, compared with the no optimal condition, our proposed optimization algorithms can all offer superior performance under different system constraints.

1. Introduction

Recently, with the development of wireless communication networks, energy consumption has become a serious increasing problem. As a result, green communications have attracted many researchers' attention. Energy harvesting (EH), which can gain energy from the environment, appears to be an effective technique for green communications and has great potential to extend the lifetime of communication devices [1, 2]. On the contrary, cooperative relaying is an effective technique to improve the coverage of wireless networks, avoid the decrease of communication quantity caused by channel fading, and increase the channel gain. So, how to use EH in such networks with multiple relays is an interesting topic.

Plenty of studies have already been carried out about wireless systems which harvest energy from the environment. An optimal packet-scheduling strategy is proposed in

[3] in order to minimize the transmission time in the EH wireless communication system using additive white Gaussian noise (AWGN) channels. The traffic load and available energy are considered in order to adaptively change the transmission rate. Under the same AWGN channel scenario, an optimal energy allocation scheme which maximizes the throughput is obtained in [4] with the use of dynamic programming and convex optimization techniques. The AWGN channel capacity under energy constrains is studied in [5]. It is proved that given sufficient energy supply, the AWGN channel capacity of the EH model equals the capacity of the continuous energy supply model. A peer-to-peer communication system is analyzed in [3–5] without considering relays. In the literature [6], a communication system where the source and relays are EH nodes is considered. Its main work is to maximize the peer-to-peer system throughput using joint relay selection and power allocation schemes. For a slow-fading channel, Li et al. [7] analyze an EH

cooperative relaying network and derive the closed-form outage probability expression for the proposed protocol. Different from others, multiple source-destination pairs with an EH relay in a wireless cooperative network is considered in [8], and a power allocation scheme of the harvested energy is proposed. Meanwhile, the water-flooding method is used to achieve a better balance between the system performance and complexity.

Apart from the strategies that harvest energy from the environment, a new way is to use the radiofrequency (RF) signal [9]. The idea is first studied in [10], and a new energy transfer technology is proposed, that is, simultaneous wireless and information transfer (SWIPT). Two schemes are further proposed in [11] coordinating the information processing (IP) and EH, which are time switching (TS) and power splitting (PS). In [12–16], a relaying network where an energy-constrained relay node using SWIPT and TS or PS schemes is considered. In particular, the outage probability expressions and the ergodic capacity are derived in [12, 16], and effects of various system parameters are illustrated. The work in [12] is extended in [13] by assuming that the source also harvests energy from the relay. The outage performances of two schemes where one uses the EH relay and the other conveys data via direct links are analyzed in [14], and an optimization problem which minimizes the transmit power is also formulated. In two-way full-duplex (FD) relay networks, the capability is first investigated in [15], and two relay selection schemes are proposed in order to minimize the outage probability and maximize the sum capacity, respectively. Considering cooperative EH communications, a multiple-relay selection scheme using geometric programming (GP) is proposed in [17] in order to maximize data rate-based utilities over multiple coherent time slots. The work in [18] is extended to a cognitive two-way EH relay network. Specifically, closed-form expressions of the throughput are derived to maximize interference temperature apportioning parameter (ITAP) and the PS parameter. It is shown that they can be optimized separately.

Recent research work mainly focuses on the system performance of wireless networks with a single EH relay. However, few studies have been presented to analyze the system performance of multiple-relay EH systems using the PS protocol. In this paper, we consider an amplify-and-forward (AF) wireless network with multiple EH relays. We assume that all the relays use SWIPT to harvest energy from the RF signal and adopt the PS protocol for EH and IP. We analyze the system performance and derive closed-form expressions for the outage probability of the multiple-relay EH system. We also jointly optimize relays' PS factors and their transmit power levels to minimize the outage probability. The main contributions of this paper are summarized as follows:

(1) We use fitting algorithms to simplify the outage probability of the system with only one relay, and the result is approximated as a combination of several exponential functions. The closed-form expression for the outage probability of the multiple-relay EH system is obtained using the characteristic function of the probability density function (PDF).

(2) Based on the approximation of the outage probability expression, we propose three optimization algorithms to obtain power allocation policies which minimize the outage probability under different constraints. In particular, Algorithm 1 is to optimize relays' PS factors without considering the power constrains. Algorithm 2 is to optimize relay's transmit power levels with their total transmit power limited. Algorithm 3 is to jointly optimize the relay's PS factors and the transmit power of the source under the total power constrains of the whole system.

(3) For each algorithm, we use the Lagrange method and Karush–Kuhn–Tucker condition to solve the optimization problems. Our derived expression of the outage probability is consistent with simulation results and gives insights into the effect of various system parameters on the system performance. Meanwhile, our proposed optimization algorithms can offer superior performance, and the improvement is the same for different system conditions.

2. System Model

As shown in Figure 1, we consider a wireless communication system containing a source node, S, a destination node, D, and N relay nodes, R_i, $i = 1, 2, 3, \ldots, N$. The information is transferred from the source node to the destination node through relay nodes. Only relays can harvest energy from the RF signal. We assume that the channel state information (CSI) is only available at the source node, which means all relays are independent from each other. The source node broadcasts the information to all the relays, and the relays adopt the AF scheme to transfer the information to the destination node. Also, there is no direct link between the source node and the destination node.

At the relays, the received signals are split according to the PS protocol. As Figure 2 shows, the information transformation is conducted in two time slots. In the first slot, the relays receive signals from the source nodes. In the other slot, the relays send the signals to the destination node. In Figure 2, $y_r(t)$ is the signal received at the relay. P is the power of the received signal. The relay splits the signal into two parts according to a power ratio of $1 - \rho : \rho$. One part is for IP and the other is for EH. The harvested energy is used to power the transmission circuits.

We set the channel gain from the source node S to the relay node R_i as h_i and from the relay node R_i to the destination node D as g_i. The distances between the source node S and relay node R_i and between the relay node and the destination D are d_{1i} and d_{2i}, respectively. m is the path loss exponent. So that we define the equivalent channel gains as $\gamma_{sr_i} = |h_i|^2 / d_{1i}^m$ and $\gamma_{sr_i} = |g_i|^2 / d_{2i}^m$.

Assume that the information signal from the source is $s(k)$. In the first time slot, the signal at the relay used for information transmission can be given by

$$y_{r_i}(k) = \sqrt{(1 - \rho_i) P_s \gamma_{sr_i}} s(k) + \sqrt{1 - \rho_i} n_a^{r_i}(k) + n_c^{r_i}(k), \quad (1)$$

(1) Using the transmit power P_s and the energy conversion efficiency η_i in problem (37) to obtain power splitting factor ρ_i for each relay.
(2) Using ρ_i in $P_{r_i} = \eta_i \rho_i P_s \overline{\gamma_{r_i}}$ to obtain each relay's transmit power $P_{r_i\_temp}$.
(3) **if** $P_{r_i\_temp} \geq P^i_{r\,max}$,
(4) $P_{r_i} = P^i_{r\,max}$;
(5) **else** $P_{r_i} = P_{r_i\_temp}$
(6) **end**
(7) Using P_{r_i} to achieve the final outage probability.

ALGORITHM 1: The optimal power splitting factor design under each relay's power constraint.

(1) Initialization: Select a feasible μ_i^1. Choose the step size k and the error tolerance σ, and let $n = 1$
(2) **while** (1)
(3)   Using μ_i^n in $d(L(\mu))/d\rho_i = 0$ to solve the power splitting factor ρ_i^n for each relay.
(4)   Using ρ_i^n and ρ_i^{n-1} in $\log(P_{out})$ to obtain $\log(P_{out})_n$ and $\log(P_{out})_{n-1}$, respectively.
(5)   **if** $|\log(P_{out})_n - \log(P_{out})_{n-1}| < \sigma$
(6) $\rho_i^* = \rho_i^n$;
(7)   **break**;
(8)   **else**
(9)   Using ρ_i in formula (43) to update μ_i^{n+1}.
(10) $n = n + 1$;
(11)   **end**
(12)   if $n > 100$
(13)   **break**;
(14)   **end**
(15) **end**
(17) Using P_{r_i} to achieve the final outage probability.

ALGORITHM 2: The optimal power splitting factor design under all relays' total power constraint.

(1) Initialization: Let $\varphi = -(N/P_{max})$ and select a feasible P_s^1. Choose the step size k and the error tolerance σ, and let $n = 1$
(2) **while** (1)
(3)   Using P_s^n in $d(L(\varphi))/d\rho_i = 0$ to solve the power splitting factor ρ_i^n for each relay.
(4)   Using ρ_i^n and ρ_i^{n-1} in $\log(P_{out})$ to obtain $\log(P_{out})_n$ and $\log(P_{out})_{n-1}$, respectively.
(5)   **if** $|\log(P_{out})_n - \log(P_{out})_{n-1}| < \sigma$
(6) $P_s^* = P_s^n$
(7) $\rho_i^* = \rho_i^n$;
(8)   **break**;
(9)   **else**
(10)   Using ρ_i in formula (50) to update power P_s^{n+1}
(11) $n = n + 1$;
(12)   **end**
(13)   **if** $n > 100$
(14)   **break**;
(15)   **end**
(16) **end**
(17) Using P_s^*, ρ_i^* to achieve the final outage probability.

ALGORITHM 3: The optimal power splitting factor design under the system's power constraint.

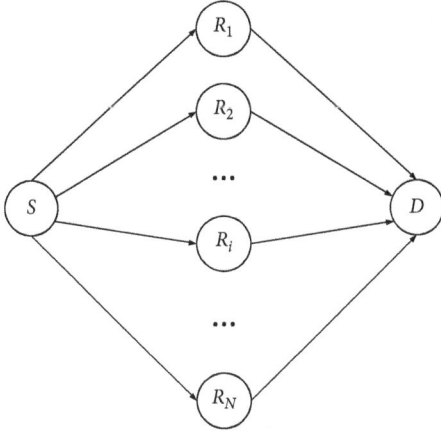

FIGURE 1: System diagram.

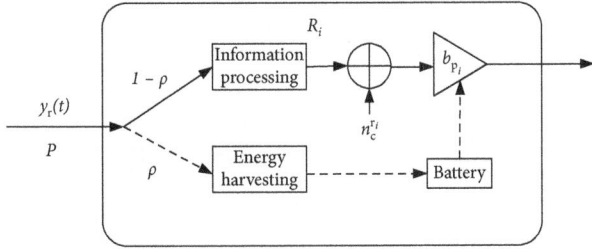

FIGURE 2: Relay node diagram [9].

where P_s is the transmit power at the source, $n_a^{r_i}(k)$ is the AWGN incurred at the radiofrequency, and $n_c^{r_i}(k)$ is the AWGN incurred in the RF-to-baseband conversion. Both of the noise signals are considered as a Gaussian random variable with mean zero and variance $\sigma^2_{n_a^{r_i}}$ and $\sigma^2_{n_c^{r_i}}$, respectively. ρ_i is the power factor that the received signal splits for EH. The energy that the relay harvests is given by $E_{hr_i} = \eta_i \rho_i P_s \gamma_{sr_i} T/2$, where $0 < \eta_i < 1$ is the energy conversion efficiency and $T/2$ is the duration of the time slot.

In the second time slot, the relay transfers the signal to the destination node. So the received signal at the destination node is given by

$$y_{d_i}(k) = \sqrt{\gamma_{rd_i}} \sqrt{P_{r_i}} b_{p_i} y_{r_i}(k) + n_a^{d_i}(k) + n_c^{d_i}(k), \quad (2)$$

where $P_{r_i} = E_{hr_i}/(T/2) = \eta_i \rho_i P_s \gamma_{sr_i}$ is the transmit power at the relay and b_{p_i} is the normalized transmit factor which is $b_{p_i} = 1/\sqrt{(1-\rho)P_s \gamma_{sr_i} + (1-\rho)\sigma^2_{n_a^{r_i}} + \sigma^2_{n_c^{r_i}}}$. $n_a^{d_i}(k)$ and $n_c^{d_i}(k)$ are the RF-front noise and the conversion noise with mean zero and variance $\sigma^2_{n_a^{d_i}}$ and $\sigma^2_{n_c^{d_i}}$, respectively.

In order to simplify the formula, we use $n^{r_i}(k) = \sqrt{1-\rho_i} n_a^{r_i}(k) + n_c^{r_i}(k)$ and $n_d^{r_i}(k) = n_a^{d_i}(k) + n_c^{d_i}(k)$ to represent the total AWGNs at the relay and destination node, respectively. Substituting (1) into (2),

$$y_{d_i}(k) = b_{p_i}\sqrt{(1-\rho_i)P_{r_i}P_s\gamma_{sr_i}\gamma_{rd_i}}s(k) + b_{p_i}\sqrt{P_{r_i}}n^{r_i}(k) + n_d^{r_i}(k). \quad (3)$$

Assume that the variances for the noise signal satisfy $\sigma^2_{n^{r_i}} = (1-\rho_i)\sigma^2_{n_a^{r_i}} + \sigma^2_{n_c^{r_i}}$, $\sigma^2_{n_d^{r_i}} = \sigma^2_{n_a^{d_i}} + \sigma^2_{n_c^{d_i}}$ and $\sigma^2_{n_a^{r_i}} = \sigma^2_{n_c^{r_i}} = \sigma^2$, $\sigma^2_{n_a^{d_i}} = \sigma^2_{n_c^{d_i}} = \sigma^2$. Substituting P_{r_i} and b_{p_i} in (3), the signal-to-noise ratio (SNR) at the destination is given by

$$\text{SNR}_i = \frac{P_s^2(1-\rho_i)\eta_i\rho_i\gamma^2_{sr_i}\gamma_{rd_i}}{P_s(1-\rho_i)\gamma_{sr_i}\sigma^2_{n_d^{r_i}} + P_s\eta_i\rho_i\gamma_{sr_i}\gamma_{rd_i}\sigma^2_{n^{r_i}} + \sigma^2_{n_d^{r_i}}\sigma^2_{n^{r_i}}}. \quad (4)$$

Let

$$A_i = P_s^2(1-\rho_i)\eta_i\rho_i,$$
$$B_i = P_s(1-\rho_i)\sigma^2_{n_d^{r_i}}, \quad (5)$$
$$C_i = P_s\eta_i\rho_i\sigma^2_{n^{r_i}}.$$

The final expression is given by

$$\text{SNR}_i = \frac{A_i\gamma^2_{sr_i}\gamma_{rd_i}}{B_i\gamma_{sr_i}\gamma_{rd_i} + C_i\gamma_{sr_i} + \sigma^2_{n_d^{r_i}}\sigma^2_{n^{r_i}}}. \quad (6)$$

2.1. Outage Probability Analysis. Assume that all the relays take part in the transmission. According to the maximal-ratio combining (MRC), the final SNR at the destination node is given by

$$\text{SNR} = \sum_{i=1}^{N} \text{SNR}_i. \quad (7)$$

Then, based on the Shannon capacity formula, the overall throughout is derived as

$$C = \frac{1}{2}\log_2(1 + \text{SNR}). \quad (8)$$

Let R be the minimized rate demand for users. So the outage probability is defined as $P_{out} = \Pr\{C < R\}$. Using (6)–(8), it becomes

$$P_{out} = \Pr\left\{\sum_{i=1}^{N} \text{SNR}_i < 2^{2R-1}\right\}$$
$$= \Pr\left\{\sum_{i=1}^{N} \frac{A_i\gamma^2_{sr_i}\gamma_{rd_i}}{B_i\gamma_{sr_i}\gamma_{rd_i} + C_i\gamma_{sr_i} + \sigma^2_{n_d^{r_i}}\sigma^2_{n^{r_i}}} < 2^{2R-1}\right\}. \quad (9)$$

Define the random variables $\omega_i = (A_i\gamma^2_{sr_i}\gamma_{rd_i})/(B_i\gamma_{sr_i}\gamma_{rd_i} + C_i\gamma_{sr_i} + \sigma^2_{n_d^{r_i}}\sigma^2_{n^{r_i}})$, and the cumulative distribution function (CDF) of ω_i is given in [3] as

$$P_{out}(x) = \Pr(\omega_i < x) = 1 - e^{-((C_i x)/(A_i\overline{\gamma_i}))}u_i K_1(u_i), \quad (10)$$

where $u_i = \sqrt{(4B_i x)/(A_i\overline{\gamma_{r_i}}\,\overline{\gamma_{d_i}})}$, $\overline{\gamma_{r_i}} = (E\{|h_i|^2\})/d_{1i}^m$, and $\overline{\gamma_{d_i}} = (E\{|g_i|^2\})/d_{2i}^m$. $K_1(\cdot)$ is the first-order modified Vessel function of the second kind [20].

Defining the random variable $W = \sum_{i=1}^{N} \omega_i$, our goal is to obtain the CDF of the W, $\Pr(W \leq x)$. According to [13], we will use the characteristic function to achieve the final expression.

In formula (10), the CDF of ω_i can be split into two parts. One part $e^{-((C_i x)/(A_i \overline{\gamma_{r_i}}))}$ is an exponential random variables, the other $u_i K_1(u_i)$ is too difficult to obtain the CDF directly. However, it is achievable if we approximate this part. Notice that $u_i K_1(u_i)$ is a function of an independent variable u_i, and $u_i = \sqrt{(4B_i x)/(A_i \overline{\gamma_{r_i}} \, \overline{\gamma_{d_i}})} = \sqrt{(4\sigma_{n^{r_i}}^2 x)/(P_s \eta_i \rho_i \overline{\gamma_{r_i}} \, \overline{\gamma_{d_i}})}$, which means the value of u_i is mainly dependent on $\sigma_{n^{r_i}}^2/P_s \overline{\gamma_{r_i}} \, \overline{\gamma_{d_i}}$. Due to the fact that $\overline{\gamma_{r_i}}$ and $\overline{\gamma_{d_i}}$ represent the equivalent channel gain for the source node to the relay and the relay to the destination node, respectively, $P_s \overline{\gamma_{r_i}} \, \overline{\gamma_{r_i}}$ can be recognized as the received power P_r. According to the basic demand of the system, the SNR at the receiver is always larger than 0 dB, which means $\sigma_{n^{r_i}}^2/P_s \overline{\gamma_{r_i}} \, \overline{\gamma_{d_i}} < 1$. So the independent variable u_i can be approximated as $u_i < 1$.

Let $f(u_i) = u_i K_1(u_i)$. Considering the character of Vessel function, if we let $u_i = 0$, then $f(u_i) = 1$ and if let $u_i \longrightarrow \infty$, then $f(u_i) = 0$. It is the same for the exponential random variables. Also, noticing that the former part of $P_{out}(x)$ is already an exponential random variable, it is not hard to consider using a combination of several exponential functions to approximate $f(u_i)$. $f(u_i)$ can be approximated as

$$f(u_i) = a_1 \cdot e^{-b_1 u_i^2} + a_2 \cdot e^{-b_2 u_i^2} + \cdots + a_L \cdot e^{-b_L u_i^2}, \quad (11)$$

where $a_1 + a_2 + \cdots + a_L = 1$.

We use fitting tools in MATLAB to obtain the coefficients of the approximate function of $f(u_i)$ and get

when $L = 1$, $a_1 = 1$, $b_1 = 0.6$

when $L = 2$, $a_1 = 0.12$, $a_2 = 0.88$, $b_1 = 6.2$, $b_2 = 0.38$

when $L = 3$, $a_1 = 0.9$, $a_2 = -2.3$, $a_3 = 2.4$, $b_1 = 0.4$, $b_2 = 7.8$, $b_3 = 7.8$

Figure 3 shows the approximate result of $f(u_i)$. For $L = 1$, the performance of fitting approach is not good, and the estimation errors between two functions are too large. But for $L = 2$ and $L = 3$, the results reflect the trend of the original function perfectly, and the degree of those two fitting function are nearly 0.99. Also, the complexity for $L = 2$ is lower than that for $L = 3$, so the final function for $P_{out}(x)$ can be approximated as

$$P_{out}(x) = 1 - e^{-\left((C_i x)/\left(A_i \overline{\gamma_{r_i}}\right)\right)} \cdot \left(0.12 e^{-6.2 u_i^2} + 0.88 e^{-0.38 u_i^2}\right). \quad (12)$$

Using u_i in formula (12), the CDF of ω_i is given by

$$\Pr\left(\omega_i < x\right) \approx 1 - e^{-\lambda_i^1 x} - e^{-\lambda_i^2 x}, \quad (13)$$

where $\lambda_i^1 = (C_i/(A_i \gamma'_{r_i})) + 0.62 u_i'^2$, $\lambda_i^2 = (C_i/A_i \gamma'_{r_i}) + 0.38 u_i'^2$, and $u_i' = \sqrt{((4B_i)/(A_i \overline{\gamma_{r_i}} \, \overline{\gamma_{d_i}}))}$. Using the characteristic function, the CDF of W is

$$\Pr(W \le x) = \sum_{A \in U, B \in U} \left\{ \sum_{i \in A} \left[\left(\prod_{k \in A, k \neq i} \frac{\lambda_k^1}{\lambda_k^1 - \lambda_i^1} \prod_{k \in B, k \neq i} \frac{\lambda_k^2}{\lambda_k^2 - \lambda_i^1}\right) \lambda_i^1 \left(1 - e^{-\lambda_i^1 w}\right)\right] \right.$$
$$\left. + \sum_{i \in B} \left[\left(\prod_{k \in A, k \neq i} \frac{\lambda_k^1}{\lambda_k^1 - \lambda_i^2} \prod_{k \in B, k \neq i} \frac{\lambda_k^2}{\lambda_k^2 - \lambda_i^2}\right) \lambda_i^2 \left(1 - e^{-\lambda_i^2 w}\right)\right] \right\}. \quad (14)$$

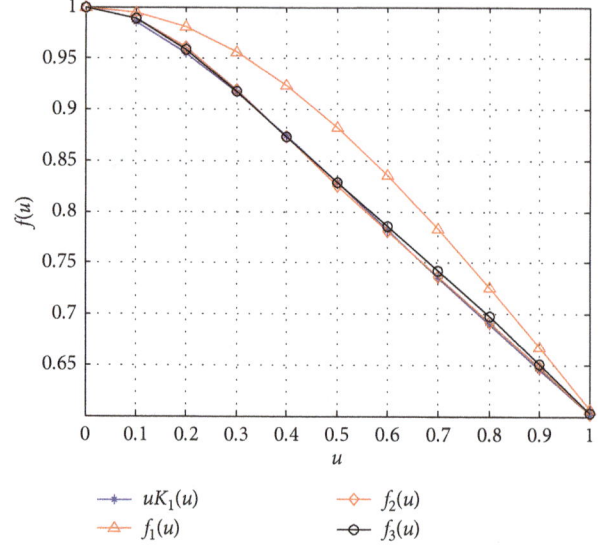

FIGURE 3: Different fitting function for $L = 1, 2, 3$.

Proof. The probability density function (PDF) of ω_i is

$$f(x) = \lambda_i^1 e^{-\lambda_i^1 x} + \lambda_i^2 e^{-\lambda_i^2 x}. \quad (15)$$

The characteristic function of ω_i is

$$g_{w_i}(t) = \left(1 - \frac{jt}{\lambda_i^1}\right)^{-1} + \left(1 - \frac{jt}{\lambda_i^2}\right)^{-1}. \quad (16)$$

According to the character of the characteristic function, the characteristic function of W is

$$g_W(t) = \prod_{i=1}^{N_1} \left[\left(1 - \frac{jt}{\lambda_i^1}\right)^{-1} + \left(1 - \frac{jt}{\lambda_i^2}\right)^{-1}\right]. \quad (17)$$

It can be rewritten as

$$g_W(t) = \sum_{A \in U, B \in U} \left[\prod_{i \in \{A\}} \left(1 - \frac{jt}{\lambda_i^1}\right)^{-1} \prod_{j \in \{B\}} \left(1 - \frac{jt}{\lambda_j^2}\right)^{-1}\right], \quad (18)$$

where $U = \{1, 2, 3, \ldots, N_1\}$ and A, B denote all the subsets of U that satisfy $A + B = U$. From formula (18), $g_W(t)$ is a combination of 2^{N_1} product terms. We only need to consider the PDF of one particular term and can obtain the others in the same way.

For $A = U$, $B = \varnothing$, the characteristic function is

$$g_{(A=U, B=\varnothing)}(t) = \prod_{i=1}^{N_1} \left(1 - \frac{jt}{\lambda_i^1}\right)^{-1}. \quad (19)$$

Based on the connection between the PDF and the characteristic function,

$$f_W(w) = \frac{1}{2\pi} \int_{-\infty}^{\infty} e^{-jtw} g_W(t) \, dt. \quad (20)$$

Using the method of partial fraction, $g_{(A=U, B=\varnothing)}(t)$ can be rewritten as

$$g_{(A=U,B=\varnothing)}(t) = \prod_{i=1}^{N_1}\left(1-\frac{jt}{\lambda_i^1}\right)^{-1} = \sum_{i=1}^{N_1}\frac{N_i}{\lambda_i^1-jt}, \quad (21)$$

where $\quad N_i = g_{(A=U,B=\varnothing)}(t)\cdot(\lambda_i^1-jt)|_{\lambda_i^1=jt} = \lambda_i^1\cdot$ $\prod_{k=1,k\neq i}^{N_1}(\lambda_i^1/(\lambda_k^1-\lambda_i^1))$.

Then,

$$f_{(A=U,B=\varnothing)}(\omega) = \sum_{i=1}^{N_1}\left(\prod_{k=1,k\neq i}^{N_1}\frac{\lambda_k^1}{\lambda_k^1-\lambda_i^1}\right)\lambda_i^1 e^{-\lambda_i^1 w}. \quad (22)$$

Formula (22) is the PDF of this particular case. For random A and B, we can obtain

$$f_{(A,B)}(\omega) = \sum_{i\in A}\left[\left(\prod_{k\in A,k\neq i}\frac{\lambda_k^1}{\lambda_k^1-\lambda_i^1}\prod_{k\in B,k\neq i}\frac{\lambda_k^2}{\lambda_k^2-\lambda_i^1}\right)\lambda_i^1 e^{-\lambda_i^1 w}\right]$$
$$+ \sum_{i\in B}\left[\left(\prod_{k\in A,k\neq i}\frac{\lambda_k^1}{\lambda_k^1-\lambda_i^2}\prod_{k\in B,k\neq i}\frac{\lambda_k^2}{\lambda_k^2-\lambda_i^2}\right)\lambda_i^2 e^{-\lambda_i^2 w}\right], \quad (23)$$

because

$$f_W(w) = \sum_{A\in U,B\in U} f_{(A,B)}(\omega). \quad (24)$$

The CDF of W is

$$\Pr(W\leq x) = \sum_{A\in U,B\in U}\left\{\sum_{i\in A}\left[\left(\prod_{k\in A,k\neq i}\frac{\lambda_k^1}{\lambda_k^1-\lambda_i^1}\prod_{k\in B,k\neq i}\frac{\lambda_k^2}{\lambda_k^2-\lambda_i^1}\right)\lambda_i^1\left(1-e^{-\lambda_i^1 w}\right)\right]\right.$$
$$\left.+\sum_{i\in B}\left[\left(\prod_{k\in A,k\neq i}\frac{\lambda_k^1}{\lambda_k^1-\lambda_i^2}\prod_{k\in B,k\neq i}\frac{\lambda_k^2}{\lambda_k^2-\lambda_i^2}\right)\lambda_i^2\left(1-e^{-\lambda_i^2 w}\right)\right]\right\}. \quad (25)$$

2.2. Power Allocation and Scheduling Design.

In this subsection, we propose different algorithms based on different power constrains by optimizing the transmit power of the source node and the power splitting factor of the relays, in order to minimize the outage probability. However, formula (14) is too complex for optimization. So, we choose a new approximate function for the outage probability which will give up some accuracy but is easy for further analysis.

The outage probability is now approximated as

$$P_{\text{out}}(x) = 1 - e^{-\left((C_i x)/\left(A_i\overline{\gamma_{r_i}}\right)\right)}\cdot e^{\left(-0.6u_i^2+1.4u_i-1.3\right)u_i^2}. \quad (26)$$

Proof. According to the formula in [20], $\int_0^\infty e^{-(\beta/4x)-\gamma x}\,dx = \sqrt{\beta/\gamma}K_1(\sqrt{\beta\gamma})$; let $u_i = \sqrt{\beta\gamma}$, we can get

$$u_i K_1(u_i) = \gamma\cdot\sqrt{\frac{\beta}{\gamma}}K_1(\sqrt{\beta\gamma}) = \gamma\cdot\int_0^\infty e^{-(\beta/4x)-\gamma x}\,dx. \quad (27)$$

Let $t = \gamma x$, the integration can be rewritten as

$$u_i K_1(u_i) = \gamma\cdot\int_0^\infty e^{-(\beta/4x)-\gamma x}\,dx = \int_0^\infty e^{-(\beta/4t)-t}\,dt$$
$$= \int_0^\infty e^{-\left(u_i^2/4t\right)}\cdot e^{-t}\,dt. \quad (28)$$

By using the mean value theorem of integrals, there must exit an ε between $(0,\ +\infty)$ and

$$u_i K_1(u_i) = e^{-\left(u_i^2/4\varepsilon\right)}\int_0^\infty e^{-t}\,dt = e^{-\left(u_i^2/4\varepsilon\right)}. \quad (29)$$

The value of ε changes with u_i. Assuming that $u_i K_1(u_i) = e^{f'(u_i)\cdot u_i^2}$ and using the fitting tools in MATLAB, the analytical expression is

$$f'(u_i) = -0.6u_i^2 + 1.4u_i - 1.3. \quad (30)$$

For optimization problems, it is not necessary to estimate the outage probability accurately. So we let $f(u_i) = 0.5$ for further simplification.

The CDF of ω_i is

$$\Pr(\omega_i < x) \approx 1 - e^{-\lambda_i x}, \quad (31)$$

where $\quad\lambda_i = (C_i/(A_i\overline{\gamma_{r_i}})) + 0.5\cdot u_i'^2\quad$ and $\quad u_i' = \sqrt{(4B_i)/(A_i\overline{\gamma_{r_i}}\,\overline{\gamma_{d_i}})}$. Using the characteristic function, the CDF of W is

$$\Pr(W\leq x)\approx\sum_{i=1}^{N_1}\left(\prod_{k=1,k\neq i}^{N_1}\frac{\lambda_k}{\lambda_k-\lambda_i}\right)\left(1-e^{-\lambda_i x}\right). \quad (32)$$

According to the conclusion in [19], the CDF of W can be written as

$$\Pr(W\leq x)\leq\frac{1}{(N_1+1)!}\cdot\prod_{i=1}^{N_1}\lambda_i\cdot\left(2^{(2+1)x}-1\right)^{N_1+1}. \quad (33)$$

Based on the formula (33), we propose three algorithms to obtain power allocation policies which minimize the outage probability under different constraints. In particular,

Algorithm 1: with the transmit power of the source node fixed, we consider each relay's transmit power is limited. The maximal power for each relay is $P_{\text{r max}}^i$.

Algorithm 2: with the transmit power of the source node fixed, we consider the total transmit power of all the relays is limited. The maximal power is $P_{\text{r max}}$.

Algorithm 3: the transmit power of the source node is not fixed and the total transmit power of the system is limited. The maximal power is P_{max}.

2.2.1. Algorithm 1: The Optimization Algorithm under Each Relay's Power Constraint.

The optimal power allocation scheme is based on minimizing the outage probability bound in (33). Considering $P_{\text{r}_i} = \eta_i\rho_i P_s\overline{\gamma_{r_i}}$, we can transfer the problem into finding the optimal power splitting factor ρ_i^* to minimizing the outage probability bound. Assuming that the power of the transmission signal from the source node is fixed, the optimization can be written as

$$P1 : \underset{\rho_i}{\text{argmin}} \quad P_{\text{out}} = \frac{1}{(N_1+1)!} \cdot \prod_{i=1}^{N_1} \lambda_i' \cdot \left(2^{(2+1)x}-1\right)^{N_1+1}$$

$$\text{s.t.} \qquad P_{r_i} \leq P_{r\,\text{max}}^i$$

$$0 < \rho_i < 1, \tag{34}$$

where $P_{r_i} = \eta_i \rho_i P_s \overline{\gamma_{r_i}}$. Using the monotonicity of the log function, our optimization problem can be written as

$$P1' : \underset{\rho_i}{\text{argmin}} \quad \log(P_{\text{out}}) = \sum_{i=1}^{N_1} \log(\lambda_i')$$

$$\text{s.t.} \qquad \eta_i \rho_i P_s \overline{\gamma_{r_i}} \leq P_{r\,\text{max}}^i \tag{35}$$

$$0 < \rho_i < 1,$$

where the fixed terms in the optimization bound is removed. It is easy to know that cost function in (35) is convex in ρ_i, which means the minimal value exists in the range $(0,1)$. Using λ_i' and P_{r_i} in the cost function, we can obtain the derivative of the function as

$$\frac{\partial(\log(P_{\text{out}}))}{\partial \rho_i} = \frac{\left(2^{(2+1)x}-1\right)^{N_1+1}}{(N_1+1)!}$$

$$\cdot \frac{\left(\eta_i \overline{\gamma_{d_i}} - 2\right) \cdot \rho_i^2 + 4\rho_i - 2}{\left(\eta_i \rho_i \overline{\gamma_{r_i}} + 2(1-\rho_i)\right)\left(\eta_i P_s (1-\rho_i)\rho_i \overline{\gamma_{r_i}}\, \overline{\gamma_{d_i}}\right)}. \tag{36}$$

Let $(\partial(\log(P_{\text{out}})))/\partial \rho_i = 0$, we can get

$$\rho_i = \frac{-4 + \sqrt{16 + 8\left(\eta_i \overline{\gamma_{d_i}} - 2\right)}}{2\left(\eta_i \overline{\gamma_{d_i}} - 2\right)} \quad \text{(negative reject)}. \tag{37}$$

We summerize it in Algorithm 1.

2.2.2. Algorithm 2: The Optimization Algorithm under All Relays' Total Power Constraint. According to [19, 21], it is necessary to consider the total power consumed by the system. In our condition, we first consider the total power of all the relays. Based on minimizing the outage probability bound in (33), we consider that the total power of all the relays are limited and formulate an optimization problem as

$$P2 : \underset{\rho_i}{\text{argmin}} \quad P_{\text{out}} = \frac{1}{(N_1+1)!} \cdot \prod_{i=1}^{N_1} \lambda_i' \cdot \left(2^{(N_1+1)x}-1\right)^{N_1+1}$$

$$\text{s.t.} \qquad \sum_i^{N_1} P_{r_i} \leq P_{r\,\text{max}}$$

$$0 < \rho_i < 1, \tag{38}$$

where $P_{r_i} = \eta_i \rho_i P_s \gamma_{sr_i}$. Similar to Algorithm 1, we can rewrite the optimization problem as

$$P2' : \underset{\rho_i}{\text{argmin}} \quad \log(P_{\text{out}}) = \frac{\left(2^{(N_1+1)x}-1\right)^{N_1+1}}{(N_1+1)!} \cdot \sum_{i=1}^{N_1} \log(\lambda_i')$$

$$\text{s.t.} \qquad \sum_{i=1}^{N_1} \eta_i \rho_i P_s \overline{\gamma_{r_i}} \leq P_{r\,\text{max}}^i$$

$$0 < \rho_i < 1. \tag{39}$$

This is a convex optimization problem in ρ_i. In order to solve the problem, we can use the Lagrangian method. We split the question into two steps. The first step is to use the Lagrangian of this optimization problem to solve the power splitting factor for each relay. It can be written as

$$L(\mu) = C \cdot \sum_{i=1}^{N_1} \log(\lambda_i') + \mu \left(P_{r\,\text{max}} - \sum_i^{N_1} P_{r_i} \right), \tag{40}$$

where $C = \left(\left(2^{(N_1+1)x} - 1\right)^{N_1+1}\right)/(N_1+1)!$ and μ represents the Lagrange multiplier associated with the power constraint in (39). The derivative of the function is

$$\frac{d(L(\mu))}{d\rho_i} = C \cdot \frac{\left(\eta_i \overline{\gamma_{d_i}} - 2\right) \cdot \rho_i^2 + 4\rho_i - 2}{\left(\eta_i \rho_i \overline{\gamma_{d_i}} + 2(1-\rho_i)\right)\left(\eta_i P_s (1-\rho_i)\rho_i \overline{\gamma_{r_i}}\, \overline{\gamma_{d_i}}\right)}$$

$$- \mu \eta_i P_s \overline{\gamma_{r_i}}. \tag{41}$$

Applying the KKT conditions, we let $d(L(\mu))/d\rho_i = 0$ and obtain the following necessary and sufficient conditions for a fixed μ as

$$\rho_i = \begin{cases} 0, & \frac{d(L(\mu))}{d\rho_i} = 0 \text{ has no answer within } (0,1), \\[2ex] \rho_i^*, & \frac{d(L(\mu))}{d\rho_i} = 0 \text{ has answer } \rho_i^* \text{ within } (0,1). \end{cases} \tag{42}$$

Step two is to update the Lagrange multiplier by the power constraint in problem (39) as follows at each iteration n.

$$\mu_i^{n+1} = \mu_i^n - k \left(P_{r\,\text{max}} - \sum_i^{N_1} P_{r_i} \right), \tag{43}$$

where k is the step size at the nth iteration.

We summerize it in Algorithm 2.

2.2.3. Algorithm 3: The Optimization Algorithm under the System's Power Constraint. Now, we consider the total power of the system. Based on minimizing the outage probability bound in (33), we consider that the total power of the whole system is limited and formulate an optimization problem as

$$P_3 : \underset{P_s, \rho_i}{\text{argmin}} \quad P_{\text{out}} = \frac{1}{(N_1 + 1)!} \cdot \prod_{i=1}^{N_1} \lambda_i' \cdot \left(2^{(N_1+1)x} - 1\right)^{N_1+1}$$

$$\text{s.t.} \quad P_s + \sum_{i}^{N_1} P_{r_i} \le P_{\max}$$

$$0 < \rho_i < 1,$$

$$(44)$$

where $P_{r_i} = \eta_i \rho_i P_s \gamma_{sr_i}$. Similar to Algorithm 1, we can rewrite the optimization problem as

$$P3' : \underset{\rho_i}{\text{argmin}} \quad \log(P_{\text{out}}) = \frac{\left(2^{(N_1+1)x} - 1\right)^{N_1+1}}{(N_1 + 1)!} \cdot \sum_{i=1}^{N_1} \log(\lambda_j')$$

$$\text{s.t.} \quad P_s + \sum_{i=1}^{N_1} \eta_i \rho_i P_s \overline{\gamma_{r_i}} \le P_{\text{r max}}^i$$

$$0 < \rho_i < 1.$$

$$(45)$$

This is also a convex optimization problem in both P_s and ρ_i and can be solved by the Lagrangian method. We can obtain the Lagrangian of this optimization problem:

$$L(\varphi) = C \cdot \sum_{i=1}^{N_1} \log(\lambda_i') + \varphi \left(P_{\max} - P_s - \sum_{i}^{N_1} P_{r_i} \right), \quad (46)$$

where φ represents the Lagrange multiplier. The derivative for P_s of the function is

$$\frac{dL(\varphi)}{dP_s} = -\frac{1}{P_s} - \varphi \left(1 + \sum_{i=1}^{N} \rho_i \eta_i \right). \quad (47)$$

Let $dL(\varphi)/dP_s = 0$ and we can obtain $\varphi = -(N/P_{\max})$. Also the derivative for ρ_i of the function is

$$\frac{d(L(\varphi))}{d\rho_i} = C \cdot \frac{\left(\eta_i \overline{\gamma_{d_i}} - 2\right) \cdot \rho_i^2 + 4\rho_i - 2}{\left(\eta_i \rho_i \overline{\gamma_{d_i}} + 2(1 - \rho_i)\right)\left(\eta_i P_s (1 - \rho_i) \rho_i \overline{\gamma_{r_i}} \, \overline{\gamma_{d_i}}\right)}$$

$$- \varphi \eta_i P_s \overline{\gamma_{r_i}}.$$

$$(48)$$

Use $\varphi = -(N/P_{\max})$ in formula (48) and obtain

$$\frac{d(L(\varphi))}{d\rho_i} = C \cdot \frac{\left(\eta_i \overline{\gamma_{d_i}} - 2\right) \cdot \rho_i^2 + 4\rho_i - 2}{\left(\eta_i \rho_i \overline{\gamma_{d_i}} + 2(1 - \rho_i)\right)\left(\eta_i P_s (1 - \rho_i) \rho_i \overline{\gamma_{r_i}} \, \overline{\gamma_{d_i}}\right)}$$

$$+ \frac{N}{P_{\max}} \eta_i P_s \overline{\gamma_{r_i}},$$

$$(49)$$

In order to solve the optimization problem, we split it into two steps.

(1) Optimal power splitting factor for each relay.

For a given P_s^n, let $d(L(\varphi))/d\rho_i = 0$, we can obtain the optimal power splitting factor ρ_i^n for each relay.

(2) Update the transmit power of the source node.

By the power constraint in problem (45), we can update P_s^n using the following formula as

$$P_s^{n+1} = \frac{P_{\max}}{1 + \sum_{i=1}^{N} \rho_i^n}. \quad (50)$$

We summerize it in Algorithm 3.

2.3. Numerical Results. We use MATLAB to conduct some simulations to prove the theoretical analysis presented in the former part. We assumed that the system contains a source node, a destination node, and multirelays. We set the distance between the source node and the destination node $d = 2.4$. All the relays are located in a circular manner whose the center is at the midpoint between the source node and the destination with radius $r = 1.2$. We adopt $P_s = 5w$ and set $\sigma^2 = 0.01$ [11]. The large scale fading exponent of the channel m is 2.7, and the small scale fading coefficients are generated as independent and identical (i.i.d.) Rayleigh random variables with mean zero and variance 1.

Figure 4 shows the outage probabilities for the theoretical analysis and the simulation versus normalized user transmission rate R bits/sec/Hz. From Figure 4, it is clear that the theoretical analysis is basically the same with the simulation results for different relay numbers. Also the outage probability increases as R increases. Meanwhile, comparing the outage probability for different relay numbers, we know that the outage probability decreases as relay number increases, which proves multiple relays can achieve the diversity gain of the channel.

Figure 5 shows the effect of the power splitting factor ρ. From the figure, we know that the outage probability is increasing when ρ is smaller but later, it starts decreasing, which means there is an optimal value for ρ. When the value of ρ is smaller than the optimal ρ, there is less power available for EH. Therefore, the transmit power of the relay node is not enough, which leads to the increase of the outage probability. On the contrary, when ρ is large than the optimal value, although the power for EH is enough, the power left for information transmission is low. Because of that, the signal strength at the destination node is not enough, and the outage probability increases too. Meanwhile, comparing different relay numbers $N = 4$ and $N = 5$ and different relay distribution $r = 1.2$ and $r = 1$, the trend of the outage probability is similar, but the value is much lower for $N = 5$ and $r = 1$. That is because when the relays are located near the midpoint between the source node and the destination, both the channel gain h_i and g_i are better.

Figure 6 plots the effect of the source transmit power P_s. As shown in Figure 6, the outage probability of the system decreases as P_s increases. This is because the large value of P_s leads to more received power at the relay nodes, which means more power for both EH and information transmission. Consequently, the outage probability for the system is small. However, the outage probability does not change much when P_s is large enough, which means there is no need to continually increase transmit power in order to decrease

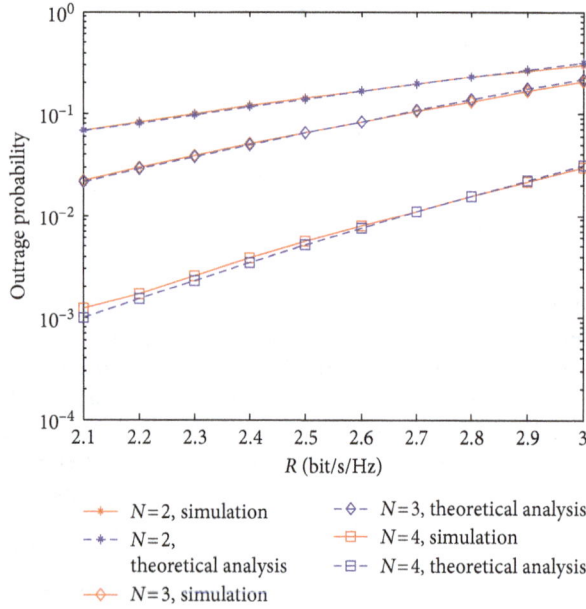

FIGURE 4: Outage probabilities for the theoretical analysis and the real scenario with different relay numbers.

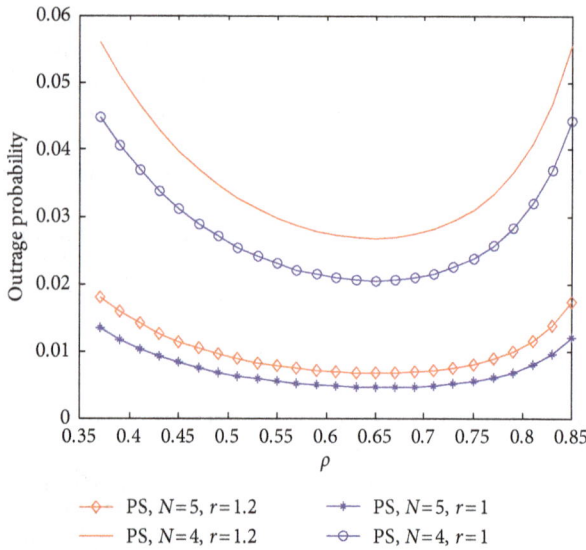

FIGURE 6: Outage probabilities vs. the source transmit power P_s for different relay numbers $N = 4$ and $N = 5$ and different source to destination distances $d = 1.2$ and $d = 1.4$.

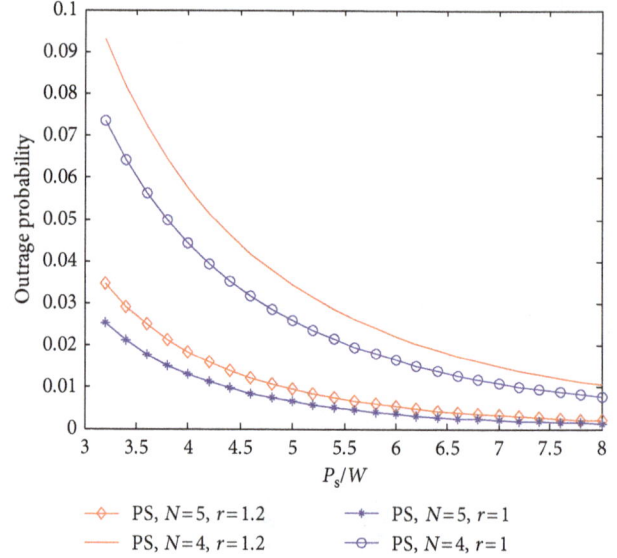

FIGURE 5: Outage probabilities vs. power splitting factor ρ for different relay numbers $N = 4$ and $N = 5$ and different source to destination distances $r = 1.2$ and $r = 1$.

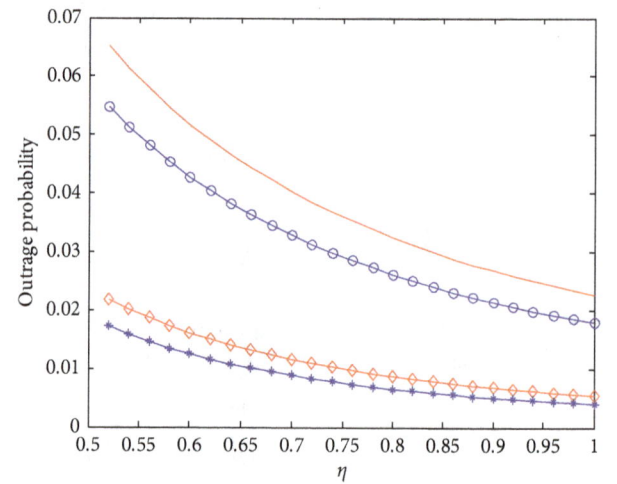

FIGURE 7: Outage probabilities vs. the effect of the energy conversion efficiency η for different relay numbers $N = 4$ and $N = 5$ and different source to destination distances $d = 1.2$ and $d = 1.4$.

the outage probability. Also, changing the relay number and the source to destination distance, the trend is the same, and the value for the large relay number and short source to destination distance is lower. Meanwhile, we notice that the variation of the tendency for the relay number $N = 5$ is not the obvious comparing with that for $N = 4$.

Figure 7 depicts the effect of the energy conversion efficiency. From this, we can observe that for smaller values of η, the outage probability is higher, and for larger values, the probability is lower, because the value of η decides the power from the relay node as $P_{r_i} \propto \eta_i$. Also, changing the relay

number and the source to destination distance, the trend is the same and the value for large relay number and short source to destination distance is lower. But from the figure, when the relay number $N = 5$, the value of the outage probability is almost unchanged. That means we should also limit the number of relays because the large number of relays may cause the restriction of the system performance.

Figures 8–10 show a comparison between no optimal power allocation and optimal power allocation schemes in Algorithm 2 and Algorithm 3, respectively. From Figure 8, we observe that the outage probability is lower by using the

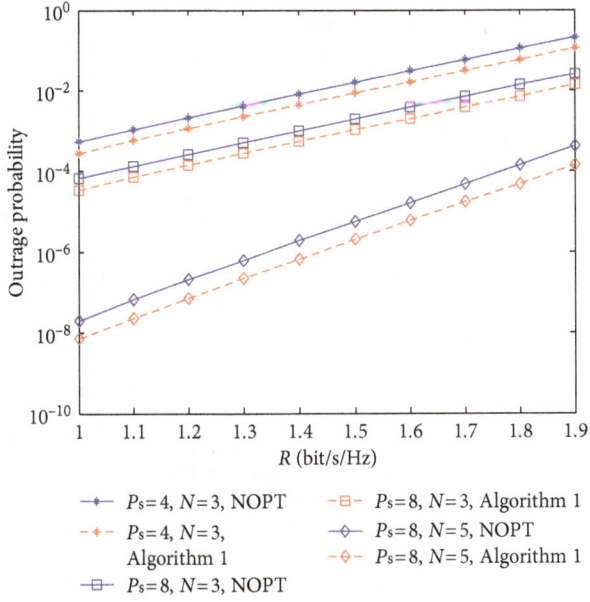

FIGURE 8: Comparison between no optimal power allocation and optimal power allocation scheme in Algorithm 1.

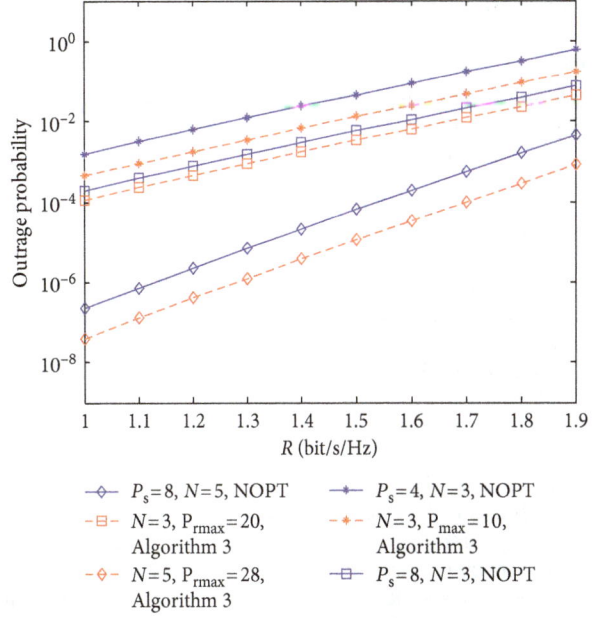

FIGURE 10: Comparison between no optimal power allocation and optimal power allocation scheme in Algorithm 3.

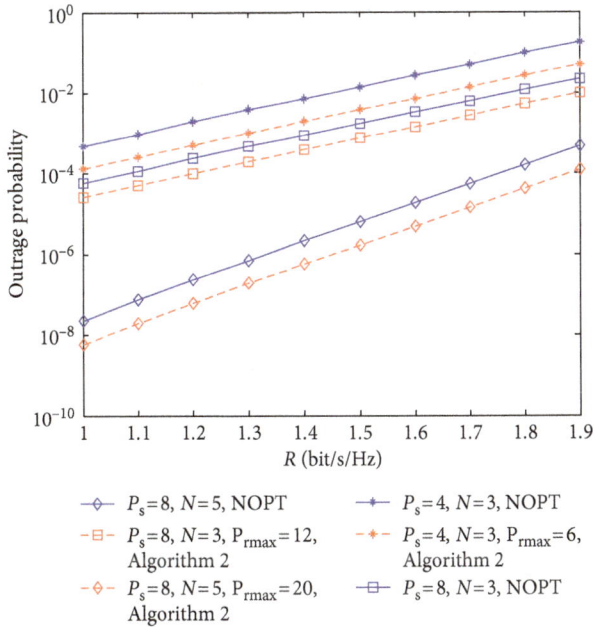

FIGURE 9: Comparison between no optimal power allocation and optimal power allocation scheme in Algorithm 2.

scheme in Algorithm 1. The improvement is the same for schemes in Algorithm 2 and Algorithm 3. Here, in order to make the numerical result more reliable, we set the power consumption of the two schemes equal. In particular, for Algorithm 2, we set the power splitting factor $\rho = 0.5$ for no optimal scheme, and the power constraint for all relays is $P_{r\max} = \rho \cdot P_s \cdot N$. For Algorithm 3, the power splitting factor is also $\rho = 0.5$, and the power constraint for the total system is $P_{r\max} = \rho \cdot P_s \cdot N$. Based on this condition, we can observe that the outage probability for optimal schemes is lower than that for no optimal situation. And it also has the improvement for different source power P_s and relay number N. From the figures, we can also see that the outage probability is growing with the increasing of the transmission rate R, and the variation between no optimal power allocation and optimal power allocation schemes is almost the same for a different source power P_s. But for the relay number $N = 5$, the growing rate and the variation between two algorithms are much higher than that for $N = 3$. This also proves that when we want to use multiple relays, the number should be under control to prevent the rapid deterioration of the system performance.

3. Conclusion

In this paper, we consider a wireless cooperative network with multiple relay nodes. All relays are considered as EH nodes with SWIPT capabilities. We derive a closed-form expression for the outage probability of the whole system. Based on the approximation of the final expression, we also propose three optimization algorithms to obtain power allocation policies which minimize the outage probability under different constraints. In the end, we show that the analytical expression of the outage probability is suitable for most communication scenarios and illustrate the effect of various system parameters on the performance of the protocols. We also show that optimal schemes in different algorithms can all result in a performance improvement compared to no optimal power allocation scheme.

Acknowledgments

This work was supported by the National Natural Science Foundation of China (61571234 and 61631020) and the Postgraduate Research & Practice Innovation Program of Jiangsu Province (KYZZ16 0257).

References

[1] S. Ulukus, A. Yener, E. Erkip et al., "Energy harvesting wireless communications: a review of recent advances," *IEEE Journal on Selected Areas in Communications*, vol. 33, no. 3, pp. 360–381, 2015.

[2] M. L. Ku, W. Li, Y. Chen, and K. J. Ray Liu, "Advances in energy harvesting communications: past, present, and future challenges," *IEEE Communications Surveys & Tutorials*, vol. 18, no. 2, pp. 1384–1412, 2016.

[3] Y. Jing and S. Ulukus, "Optimal packet scheduling in an energy harvesting communication system," *IEEE Transactions on Communications*, vol. 60, no. 1, pp. 220–230, 2012.

[4] C. K. Ho and R. Zhang, "Optimal energy allocation for wireless communications with energy harvesting constraints," *IEEE Transactions on Signal Processing*, vol. 60, no. 9, pp. 4808–4818, 2011.

[5] O. Ozel and S. Ulukus, "Information-theoretic analysis of an energy harvesting communication system," in *Proceedings of IEEE International Symposium on Personal, Indoor and Mobile Radio Communications Workshops*, pp. 330–335, IEEE, Istanbul, Turkey, September 2010.

[6] I. Ahmed, A. Ikhlef, R. Schober, and R. K. Mallik, "Joint power allocation and relay selection in energy harvesting AF relay systems," *IEEE Wireless Communications Letters*, vol. 2, no. 2, pp. 239–242, 2013.

[7] T. Li, P. Fan, and K. B. Letaief, "Outage probability of energy harvesting relay-aided cooperative networks over rayleigh fading channel," *IEEE Transactions on Vehicular Technology*, vol. 65, no. 2, pp. 972–978, 2016.

[8] Z. Ding, S. M. Perlaza, I. Esnaola, and H. Vincent Poor, "Power allocation strategies in energy harvesting wireless cooperative networks," *IEEE Transactions on Wireless Communications*, vol. 13, no. 2, pp. 846–860, 2014.

[9] L. R. Varshney, "Transporting information and energy simultaneously," in *Proceedings of IEEE International Symposium on Information Theory*, pp. 1612–1616, IEEE, Toronto, ON, Canada, July 2008.

[10] P. Grover and A. Sahai, "Shannon meets Tesla: wireless information and power transfer," in *Proceedings of IEEE International Symposium on Information Theory*, pp. 2363–2367, IEEE, Austin, TX, USA, July 2010.

[11] A. A. Nasir, X. Zhou, S. Durrani, and R. A. Kennedy, "Relaying protocols for wireless energy harvesting and information processing," *IEEE Transactions on Wireless Communications*, vol. 12, no. 7, pp. 3622–3636, 2013.

[12] X. Zhou, R. Zhang, and C. K. Ho, "Wireless information and power transfer: architecture design and rate-energy tradeoff," *IEEE Transactions on Communications*, vol. 61, no. 11, pp. 4754–4767, 2013.

[13] Y. Chen, R. Shi, W. Feng, and N. Ge, "AF relaying with energy harvesting source and relay," *IEEE Transactions on Vehicular Technology*, vol. 66, no. 1, p. 1, 2017.

[14] I. Krikidis, "Simultaneous information and energy transfer in large-scale networks with/without relaying," *IEEE Transactions on Communications*, vol. 62, no. 3, pp. 900–912, 2014.

[15] D. Wang, R. Zhang, X. Cheng et al., "Relay selection in two-way full-duplex energy-harvesting relay networks," in *Proceedings of 2016 IEEE Global Communications Conference (GLOBECOM 2016)*, pp. 1–6, IEEE, Washington, DC, USA, December 2016.

[16] Z. Mheich and V. Savin, "Cooperative communication protocols with energy harvesting relays," in *Proceedings of Wireless Days*, pp. 60–65, IEEE, Porto, Portugal, March 2017.

[17] A. Alsharoa, H. Ghazzai, A. E. Kamal, and A. Kadri, "Optimization of a power splitting protocol for two-way multiple energy harvesting relay system," *IEEE Transactions on Green Communications and Networking*, vol. 1, no. 4, pp. 444–457, 2017.

[18] S. Singh, S. Modem, and S. Prakriya, "Optimization of cognitive two-way networks with energy harvesting relays," *IEEE Communications Letters*, vol. 21, no. 6, pp. 1381–1384, 2017.

[19] K. G. Seddik, A. K. Sadek, W. Su, and K. J. R. Liu, "Outage analysis and optimal power allocation for multinode relay networks," *IEEE Signal Processing Letters*, vol. 14, no. 6, pp. 377–380, 2007.

[20] I. S. Gradshteyn and I. M. Ryzhik, "14-determinants," in *Table of Integrals, Series, and Products*, I. S. Gradsheyn, I. M. Ryzhik, D. A. Jeffrey, and D. Zwillinger, Eds., Academic Press, Cambridge, MA, USA, 1980.

[21] A. Doosti-Aref and A. Ebrahimzadeh, "Adaptive relay selection and power allocation for OFDM cooperative underwater acoustic systems," *IEEE Transactions on Mobile Computing*, vol. 1, no. 1, pp. 1–15, 2017.

9

Resource Allocation Schemes based on Intelligent Optimization Algorithms for D2D Communications Underlaying Cellular Networks

Xujie Li [ID],[1] **Lingjie Zhou** [ID],[1] **Xing Chen** [ID],[1] **Ailin Qi** [ID],[1] **Chenming Li** [ID],[1] and **Yanli Xu**[2]

[1]*College of Computer and Information, Hohai University, Nanjing 211100, China*
[2]*College of Information Engineering, Shanghai Maritime University, Shanghai 201306, China*

Correspondence should be addressed to Xujie Li; lixujie@hhu.edu.cn

Academic Editor: Michael Vassilakopoulos

In this paper, the resource allocation problem for device-to-device (D2D) communications underlaying cellular networks is formulated and analyzed. In our scenario, we consider that the number of D2D user equipment (DUE) pairs is far larger than that of cellular user equipments (CUEs). Meanwhile, the resource blocks are divided into two types: resource blocks for CUEs and the ones for CUEs and DUEs. Firstly, the system model is presented, and the resource allocation problem is formulated. Then, a resource allocation scheme based on the genetic algorithm is proposed. To overcome the problem that the dedicated resource is not fully shared in the genetic algorithm, an improved harmony search algorithm is proposed for resource allocation. Finally, the analysis and simulation results show that the performances of the proposed genetic algorithm and the improved harmony search algorithm outperform than that of the random algorithm and are very close to that of the exhaustive algorithm. This result can provide an effective optimization for resource allocation of D2D communications.

1. Introduction

As the demand for mobile network data continues to grow, the scarcity of radio spectrum resources impels us to develop new technology so as to improve the spectral efficiency. D2D communication is proposed as the key technology of 5G to overcome this issue. D2D communication can implement the direct communication of device to device under the control of base station, which provides high-quality experience for users [1].

In the D2D communication system, when D2D pairs simultaneously share the same spectrum resource with the CUE, adjacent D2D pairs and the CUE will interfere with each other inevitably. Therefore, interference and resource allocation are the crucial issues which attract many researchers all around the world. In [2], the CUEs can share the spectrum resources with socially close D2D users. Based on the resource allocation framework, the authors proposed a two-step joint game. First, users with close relationships formed the same community. Then, the optimal resource

allocation issue was solved for D2D pairs. In [3], in order to maximize the weighted D2D sum rate under cellular rate constraints, the authors proposed a two-step approach by introducing energy-splitting variable such that mixed-mode allocation and resource allocation can be decoupled and optimized independently. The two-step method had little signalling overhead and low computational complexity. In [4], in order to minimize the transmission power, a distributed strategy based on Q-learning and softmax decision-making was proposed. Considering the interference of users under the same frequency, the candidate D2D user set was firstly reduced according to the outage probability constraint. Then, a graph was constructed based on this. Next, the Hungarian algorithm was used to search for the optimal spectral matching scheme to maximize the sum rate [5]. In [6], a stable matching approach was adopted to handle the energy efficiency optimization issue. Firstly, game theory was used to analyze the interactions and correlations among user equipments (UEs), and an iterative power allocation algorithm was applied to establish mutual preferences. Then the Gale–Shapley algorithm was applied to match D2D users with cellular users. In [7], the analytical and exact symbol error rate (SER) expression of the D2D system by using multicarrier band limited quadrature amplitude modulation (QAM) signalling was derived. In this model, the system resources were centrally configured, and a new resource allocation algorithm for multicarrier D2D video transmission was proposed, which improved the performance of the communication system effectively. In [8], the authors analyzed the D2D communication under the full-duplex cellular network. D2D users can reuse the uplink or downlink resources of the cellular network to improve the spectral efficiency of the system. A joint resource allocation and power control scheme based on a new graph coloring was proposed, which improved the performance of the system effectively. In [9], in order to improve the utility of all D2D pairs, a learning framework based on the Markov chain with specific problems was firstly proposed. Based on the design of the local equilibrium equation of the Markov chain, the transition probability of distributed implementation was derived. Then, a novel two-stage algorithm was proposed to perform mode selection and resource allocation at each stage. In order to reduce the computational complexity of the learning framework, two kinds of resource allocation algorithms based on matching theory were proposed for different scenarios. In [10], a low-complexity heuristic algorithm had also been proposed to realize the high logarithmic sum of the average user data rates. In [11], a new adaptive subcarrier allocation scheme was designed and a novel power allocation scheme was proposed, which can provide an optimal solution with low computation complexity. In [12], a greedy heuristic algorithm and two approximate algorithms were proposed to solve maximizing the revenue of telecom operators and efficient use of frequency resource. Besides, the greedy heuristic algorithm was proposed to maximize the number of satisfied users. In [13], the authors analyzed the deployment of an unmanned aerial vehicle (UAV) as a base station to provide the wireless communications. The coverage and rate are mainly analyzed in two scenarios: a static UAV and a mobile UAV. Simulation results showed that the optimal values for the UAV altitude can lead to maximum sum rate and coverage probability. In [14], an approach and a self-organizing algorithm were proposed to solve the problem of the wireless resource allocation for wireless small cell networks. Simulation results showed that gathering socially connected users can lead to the reduction of amount of traffic. In [15], the authors presented a framework for optimizing the caching of popular content in D2D communications. The problem of the collaborative effect of a set of UEs on content offloading was formulated as a cooperative game. The simulation results showed that the proposed framework can efficiently improve system performance.

Meanwhile, some researchers adopt intelligent optimization algorithms to solve the resource allocation issue. In [16], a joint resource allocation and user matching scheme based on the genetic algorithm was proposed to minimize the intracell interference. This algorithm was used to globally search for optimal user matching solutions to maximize

system throughput. In [17], a user machine scheme based on the genetic algorithm was proposed and discussed, which had optimal power allocation to achieve the multidimensional optimization. And the genetic algorithm was applied to obtain the near-optimal user matching in the whole network. In [18], because of the energy efficiency optimization target based on the system throughput insurance, rather than the traditional system throughput, the scheme based on the improved genetic algorithm was applied to solve the problem fusion and the lower complexity of the nondeterministic polynomial time. In [19], a genetic algorithm was proposed to maximize the minimum data rate achieved by D2D communication. In [20], the authors formulated the network model with stochastic geometry. Then, they combined the spectrum resource distribution with power selection and proposed a heuristic chaotic genetic algorithm, which was associated with four-color theorem to maximize the total capacity. The system capacity was improved and the complexity of the algorithm was reduced. In [21], a fully autonomous multiagent Q-learning algorithm which showed relatively fast convergence was proposed to implement the selection of the autonomous channel and power level. In [22], a simple particle swarm optimization (SPSO) algorithm was proposed for resource allocation to enhance the system capacity.

However, these papers mainly focus on the case that one CUE and one DUE pair sharing one subchannel. To take full advantage of spectrum resources, the resource allocation problem under the condition that the number of DUE pairs is far larger than that of CUEs is analyzed in this paper. In our scenario, the resource blocks are divided into two types: resource blocks for CUEs and the ones for CUEs and DUEs.

The main contributions of our work are as follows:

(1) We propose a resource allocation scheme based on the genetic algorithm for D2D communications underlaying cellular networks

(2) To overcome the problem that the dedicated resource is not fully shared in the genetic algorithm, an improved harmony search algorithm is proposed for resource allocation

(3) We evaluate system performances in terms of capacity, the average transmission power of CUEs, and energy efficiency.

The rest of this paper is organized as follows. In Section 2, we describe our system model. In Section 3, a resource allocation scheme based on the genetic algorithm for D2D communications is proposed and analyzed. Next, an improved harmony search algorithm is proposed for resource allocation in Section 4. In Section 5, system performances in terms of capacity, the average transmission power of CUEs, and energy efficiency are simulated and discussed. Finally, the conclusion is drawn out in Section 6.

2. System Model and Problem Formulation

There are two types of user equipment (UE) in cellular networks: cellular user equipment (CUE) in the conventional cellular network and D2D user equipment (DUE).

DUEs are in pairs, and a pair of DUEs includes a D2D transmitting user equipment (DTUE) and a D2D receiving user equipment (DRUE). Let us consider a single cellular network, where N CUEs and M DUE pairs coexist. We assume that all CUEs and DUEs are uniformly distributed in a cell. Because it is advantageous to use uplink resources for the D2D link, we only focus the case that the D2D links use uplink cellular resources in this paper.

2.1. Channel Allocation Mode. To guarantee the QoS of DUEs and accommodate more DUE pairs, we allocate dedicated resource blocks for DUE pairs, as shown in Figure 1. There are two kinds of resource blocks: blue colored resource blocks for CUEs and DUEs, yellow colored resource blocks for DUEs.

For example, there are four subchannels where one subchannel is dedicated for DUE pairs and other three subchannels are shared by the CUEs and DUE pairs. Meanwhile, we consider that there are 8 DUE pairs and 3 CUEs in this example. Then, we can get an initial resource allocation matrix as follows:

$$\text{group initial} = \begin{bmatrix} 3 & 2 & 1 & 1 & 3 & 4 & 2 & 1 \\ 4 & 2 & 2 & 2 & 3 & 1 & 2 & 4 \\ 3 & 2 & 4 & 2 & 1 & 3 & 4 & 1 \\ 3 & 2 & 1 & 2 & 2 & 2 & 3 & 1 \end{bmatrix}. \tag{1}$$

In this matrix, each row denotes a resource allocation scheme of 8 D2D users, and the element of every row denotes the index of the channel which is allocated to the DUE pair. For example, the first row indicates that DUE 3, DUE 4, and DUE 8 share subchannel 1, DUE 2 and DUE 7 share subchannel 2, and DUE 1 and DUE 5 share subchannel 3. At the same time, subchannel 4 is dedicated for DUE 6.

2.2. Definition of the Fitness Functions. In D2D communications, we need to get the maximum capacity. To guaranty the QoS of UEs, the signal to interference plus noise ratio (SINR) value should be greater than SINR threshold. Therefore, during the uplink period for the CUE i, the SINR can be written as

$$\beta_i = \frac{P_i / r_i^\alpha}{\sum_{k \in \Re_i} \left(P_T / d_k^\alpha \right) + N_0}, \tag{2}$$

where P_i is the transmitting power of CUE i, r_i is the distance between CUE i and the BS, P_T is the transmitting power of the DTUE k, d_k is the distance between DTUE k and BS, α is the path loss exponent, and N_0 is the noise power.

At the same time, during the uplink period for the DRUE j, the SINR can be written as

$$\gamma_j = \frac{\left(P_T / l_j^\alpha \right)}{\left(P_m / d_{m,j}^\alpha \right) + \sum_{k \in \Re_m k \notin m} \left(P_T / d_{k,j}^\alpha \right) + N_0}, \tag{3}$$

where l_j is the distance between DTUE j and DRUE j, $d_{m,j}$ is the distance between CUE m and DRUE j, $d_{k,j}$ is the distance between DTUE k and DRUE j, and $\Re_m$ denotes the set of the equipments which share mth subchannel.

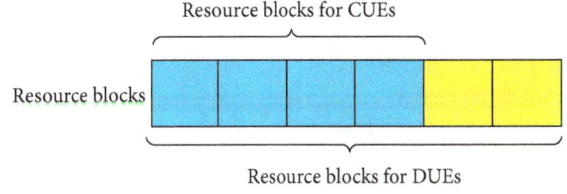

FIGURE 1: Resource blocks for CUEs and DUEs.

Obviously, the total capacity consists of two parts: CUEs and DUE pairs. For CUE i, the capacity is written as

$$Cc_i = B \log_2 \left(1 + \beta_i \right), \tag{4}$$

where B is the bandwidth of one subchannel.

Similarly, for DUE pairs, we consider that DUE pairs j belong to package m, i.e., $j \in \Re_m$. Then, we have

$$Cd_j = B \log_2 \left(1 + \gamma_j \right). \tag{5}$$

Therefore, the capacity function is denoted as

$$C(U_x) = \sum_{i=1}^{N} Cc_i + \sum_{j=1}^{M} Cd_j = B \sum_{i=1}^{N} \log_2 \left(1 + \beta_i \right)$$
$$+ B \sum_{j=1}^{M} \log_2 \left(1 + \gamma_j \right), \tag{6}$$

where U_x denotes some chromosome.

What is more, the average transmission power of CUEs P_{average} can be written as

$$P_{\text{average}} = \frac{\sum_{i=1}^{N} \left(\sum_{k \in \Re_i} \left(P_T / d_k^\alpha \right) + N_0 \right) \cdot \tau \cdot r_i}{N}, \tag{7}$$

where τ represents the threshold of SINR.

Then, the energy efficiency EE can be written as

$$\text{EE} = \frac{C(U_x)}{\sum_{i=1}^{N} P_i + \sum_{j=1}^{M} P_T}. \tag{8}$$

3. Resource Allocation Scheme Based on Genetic Algorithm

In this section, a resource allocation scheme based on the genetic algorithm for D2D communications underlaying cellular networks is proposed. The genetic algorithm can be divided into five main steps as follows [23].

3.1. Coding. Without loss of generality, we consider that ith subchannel is allocated to CUE i. The DUEs share the channel resource with the CUEs. In this case, the chromosome can be coded as a dimensional row vector like $G = (g_1, \cdots, g_j, \cdots, g_M)$, $g_j \in (1, 2, \cdots, N)$. Here, j is the index of the DUE pairs. The value of g_j is the index of the subchannel which is shared by jth DUE pair.

3.2. Population Initialization. In general, the number of the populations is set as 10. In every chromosome, element g_j is a discrete random variable which is a number between 1 and

N. The probability mass function of g_j can be written as $p(a) = P\{g_j = a\}$. g_j usually has equivalent probability for each value. And there is

$$p(1) = p(2) = \cdots = p(M) = \frac{1}{M}. \qquad (9)$$

3.3. Set Fitness Function. We select the formulae of the capacity, the average transmission power of CUEs, and energy efficiency as the fitness function, respectively.

3.4. Breeding Process. The population of the genetic algorithm is similar to the operation of breeding process, and we can obtain an optimal solution after multiple iterations. The breeding process can be divided into 4 steps as follows:

(a) Selection

The classical roulette wheel selection scheme is adopted in our scheme. At this point, the probability that individual I_k is selected can be written as $p(I_k)$. Then, we have

$$p(I_k) = \frac{C(I_k)}{\sum_{x=1}^{10*N} C(I_x)}. \qquad (10)$$

(b) Crossover

Every crossover action can generate two new individuals by swapping some elements between two selected parent chromosomes. A uniform crossover scheme that every element has same crossover probability P_0 is adopted in our algorithm. Therefore, it is possible that several elements are simultaneously selected to implement crossover in every crossover action. The uniform crossover can actually belong to the multipoint crossover, and its specific operation is to determine how each element of a new individual is provided by which parent, by setting a mask word. The main operation process of uniform crossing is as follows:

(1) Randomly generate a mask word with the same length as the individual code string $\omega = \omega_1 \omega_2 \cdots \omega_i \cdots \omega_M$, where M is the length of the individual code string

(2) Two new progeny individuals A' and B' are generated from two parent individuals A and B by the following rules:

(i) If $\omega_i = 0$, the ith element of A' will inherit the corresponding element of A, and the ith element of B' will inherit the corresponding element of B.

(ii) If $\omega_i = 1$, the ith element of A' will inherit the corresponding element of B, and the ith element of B' will inherit the corresponding element of A

(c) Mutation

The uniform mutation operation is that, with a small probability P_1, to replace the original element of the chromosome with a random number from 1 to N. The main operation process of uniform mutation is as follows:

(1) Each element of the chromosome is specified as a mutation point with the probability P_1 in sequence

(2) For each mutation point, a random number is taken from 1 to N to replace the original element

(d) Amendment

In each iteration, we select only the chromosomes in the population that satisfy the constraints of the objective function for each operation. Therefore, when generating a new population, we need to remove the chromosomes that do not satisfy the constraint conditions to form the offspring. For the new offspring, we rank the fitness function value of them and parent population. Then, we select the first $10 * N$ individuals as the new population.

3.5. Stopping Criteria. After *Num* iterations, the population will eventually evolve to a convergence. Then, we can calculate the optimal result based on the best chromosome.

Remark 1. If the strategy that D2D users share dedicated channel resources do not exist in the initialized solution space, D2D users fail to fully share the dedicated channel resources. For example, element "4" does not exist in the last line of the matrix of Equation (1). If the optimization strategy based on the genetic algorithm is adopted, the crossover step in the evolution process is realized through the crossover between the row vector and the other row vector; that is, the individuals in the population crossover each other. If D2D users want to share the dedicated channel resources, the genetic algorithm relies mainly on the process of mutation. But the probability of variation is small, and the dedicated channel resource is not fully shared.

To overcome this problem, the harmony search algorithm uses the probability evaluated from the memory bank, and the feasible solution in the memory bank can get out of the constraint of the initialized solution space by the probabilistic optimization strategy so as to find more feasible solutions. Therefore, we propose an improved harmony search algorithm based on the conventional harmony search algorithm to enhance the system performance in terms of capacity, the average transmission power of CUEs, and energy efficiency.

4. Harmony Search Algorithm

In 2001, Geem Z. W was inspired by musical performance [24]. A novel intelligent optimization algorithm was proposed. The harmony search algorithm shows better performance in global search. The harmony search algorithm can realize beautiful harmony by constantly modifying the tone of each instrument.

4.1. Description of the Conventional Harmony Search Algorithm. The steps of the conventional harmony search algorithm are as follows: firstly, randomly generate the harmony memory (HM) space that meets the constraint condition and discriminate whether a new variable is generated from the initialized solution space according to the probability evaluated from the harmony memory bank HMCR. If the HMCR value condition is satisfied, adjust the new value slightly according to the probability of pitch tuning *PAR*; if the HMCR value condition is not satisfied, a new solution is generated from the solution space outside the harmony memory bank. The objective function value in the problem model is used to measure the advantages and disadvantages of the solution. The new solution with a better objective function value is used to replace the worst solution in the initialized solution space. When the continuous iteration fails to significantly optimize the objective function value, the iteration is stopped. Then, decode the coded solution space so as to obtain an optimal resource allocation strategy.

4.2. Improved Harmony Search Algorithm.

4.2. Improved Harmony Search Algorithm. Due to the organic combination of the interior and exterior of the initialized solution space of the harmony search algorithm, the diversity of feasible solutions is effectively increased, but the convergence speed is affected. In order to improve the convergence speed of the algorithm, we propose an improved harmony search algorithm for resources allocation of D2D communications. The detailed steps are as follows.

4.2.1. Initialization. Firstly, initialize each system parameter and a harmony memory bank whose size is HMS. HMS denotes the number of individuals. HMCR_{max} denotes the maximum value of the probability evaluated from the harmony memory bank. HMCR_{min} denotes the minimum value of the probability evaluated from the harmony memory bank. g_n denotes the number of iterations, and MAXI denotes the maximum number of iterations. PAR denotes the slight adjusting probability. PAR_{min} denotes the minimum value of PAR. PAR_{max} denotes the maximum value of PAR. U_{max} and U_{min} are the maximum and minimum values of the tone, respectively. a denotes the adjustment factor, and d denotes the direction adjustment factor. The initial value of these parameters are as follows: $\text{PAR}_{\text{min}} = 0.01$, $\text{PAR}_{\text{max}} = 0.99$, $\text{HMCR}_{\text{max}} = 0.95$, $\text{HMCR}_{\text{min}} = 0.6$, $\text{HMS} = 10$, $a = 0.24$, $d = 0.5$, $\text{MAXI} = 50$, $U_{\text{min}} = 1$, and $U_{\text{max}} = N$.

4.2.2. The Generation of New Solutions. Compare a value which is randomly generated from 0 to 1 with HMCR. If it is less than HMCR, we use the selection mechanism which is similar to the roulette, according to the ratio of the fitness of the current solution to the total fitness of the memory bank, to select a pair of solutions. Then in this pair of solutions, the one with higher fitness is selected as the new harmony variable. According to the slight adjusting probability PAR, the tune of the corresponding position is then adjusted slightly through the slight adjusting bandwidth which is denoted as bw. If the adjusted tune is out of the range, the initial value is retained. If it is larger than HMCR, a random variable is generated from the solution space that is outside of the harmony memory bank.

4.2.3. The Update of Harmony Memory Bank. Repeat Step 2 to generate K new solutions so as to make full use of the accumulated information in the harmony memory bank, where K is equal to the number of rows of matrix 1. In every iteration, the values of HMCR, PAR, and bw will be updated. The related formulae are as follows:

$$\text{HMCR} = \frac{\text{HMCR}_{\text{max}} - (\text{HMCR}_{\text{max}} - \text{HMCR}_{\text{min}}) \cdot g_n}{\text{MAXI}},$$

$$\text{PAR} = \frac{\text{PAR}_{\text{min}} + (\text{PAR}_{\text{max}} - \text{PAR}_{\text{min}}) \cdot g_n}{\text{MAXI}},$$

$$\text{bw} = \frac{(U_{\text{max}} - U_{\text{min}})}{g_n^a}.$$

(11)

Then, we rank these new solutions and the initial solutions in the harmony memory bank according to their fitness functions and select the optimal HMS solutions as the ones in the new harmony memory bank.

4.2.4. Stopping Criteria. Usually, by iterating for MAXI generations, the harmony memory bank will eventually evolve to a convergence. Finally, we get the best solution and calculate the optimal result.

In conclusion, the proposed detailed improved harmony search algorithm (Algorithm 1) is as follows:

The flowchart of the improved harmony search algorithm is shown as Figure 2.

Remark 2. In our problem, we need to traverse all possible combination. Actually, this is a NP-hard problem which is similar to the travelling salesman problem. So it is very hard to get the global optimal solution using the traditional mathematic method. This problem can be solved using the intelligent optimization algorithms such as the genetic algorithm and the improved harmony search algorithm. But in fact, the intelligent algorithms can commonly get the suboptimal solution of the original problem. Although the global optimal solution may not be obtained, the convergence speed of the intelligent algorithms is very fast. Therefore, the intelligent optimization algorithms are very suitable to our scenario where massive UEs exist and the speed of change of channel state information is fast. Compared with the genetic algorithm, the improved harmony search algorithm can effectively balance the global searching and local searching, which can enhance the robustness and improve the performance. Based on the analysis above, the improved harmony search algorithm is adopted to solve our problem in this paper. In the genetic algorithm and the improved harmony search algorithm, the reason of ranking the solutions is to select and keep the good solutions for the next iteration. Although the global optimal solution may not be obtained by the method of ranking solutions in the intelligent algorithms, the obtained suboptimal solution is usually very close to the global optimal solution whilst keeping fast convergence speed.

5. Simulations and Discussions

In this section, we discuss some important conclusions obtained from the simulation and analyze the algorithmic

Input: PAR_{min}, PAR_{max}, $HMCR_{min}$, HMS, g_n, MAXI, d, U_{min}, U_{max}, L, α
Output: we use $C(U_x)$, $P_{average}$ or EE as the results
(1)   Step 1: initialization and coding
(2)   Step 2: generate new solutions and update the harmony memory bank
(3)   While g_n <= MAXI
(4)   For $j = 1:K$
(5)   For $i = 1:M$
(6)   If rand < HMCR then
(7)   index1 = roultte(fitness function)
(8)   index2 = roultte(fitness function)
(9)   If index1 < index2
(10)   index = index2
(11)   Else
(12)   index = index1
(13)   End
(14)   If rand < PAR
(15)   If rand < d
(16)   Newharmony(i) = HM(index, i) + bw
(17)   Else
(18)   Newharmony (i) = HM(index, i) – bw
(19)   End
(20)   Else
(21)   Newharmony (i) = HM(index, i);
(22)   End
(23)   Else
(24)   Newharmony (i) = rand(U_{min}, U_{max})
(25)   End
(26)   End
(27)   End
(28)   rank these new solutions that satisfied the QoS of UEs
(29)   update the harmony memory bank
(30) End
(31) Step3: return the best fitness value
(32)
(33) Function index = roultte (fitness function)
(34)   len = length(fitness function)
(35)   Total = sum(fitness function)
(36) P = fitness function/Total
(37)   randnumber = rand
(38)   While randnumber == 0
(39)   randnumber = rand
(40)   End
(41)   For $j = 1:$ len
(42)   randnumber = randnumber-P(j)
(43)   If randnumber < 0
(44)   index = j
(45)   return index
(46)   Break
(47)   End
(48)   End
(49) End function

ALGORITHM 1: Improved harmony search algorithm.

complexity and overhead. In our simulation, we assume that CUEs and DTUEs follow a uniform distribution in the cell. The radius of the cell is denoted as R, and the DRUEs uniformly locate in the circle of radius L centered on the corresponding DTUEs. The simulation parameters are summarized in Table 1.

5.1. Numerical Results. We evaluate the capacity of the communication system under different resource allocation algorithms. From Figure 3, it can be seen that the performance of the random algorithm (RA) is the worst, and the performance of the exhaustive algorithm (EA) is the best. From Figures 3 and 4, compared with the genetic

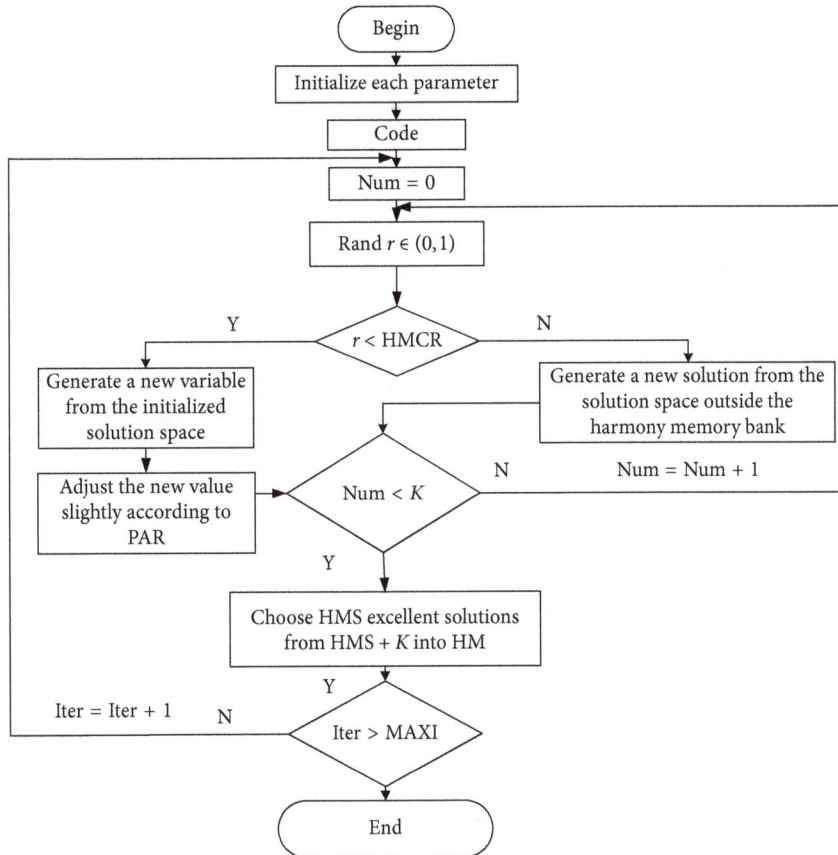

FIGURE 2: The flowchart of the improved harmony search algorithm.

TABLE 1: Simulation parameters.

Parameter	Value
Cell radius, R	600 m
L	20 m
Path loss factor, α	4
SINR threshold, β	4.6 dB
N_0	−90 dBm
K	10
P_0	0.1
The number of DUE pairs	20
The number of CUEs	3
The maximum transmission power of CUEs	2 W
The transmission power of DTUEs	0.01 W
The bandwidth of the subchannel B	0.15 MHz
P_1	0.01

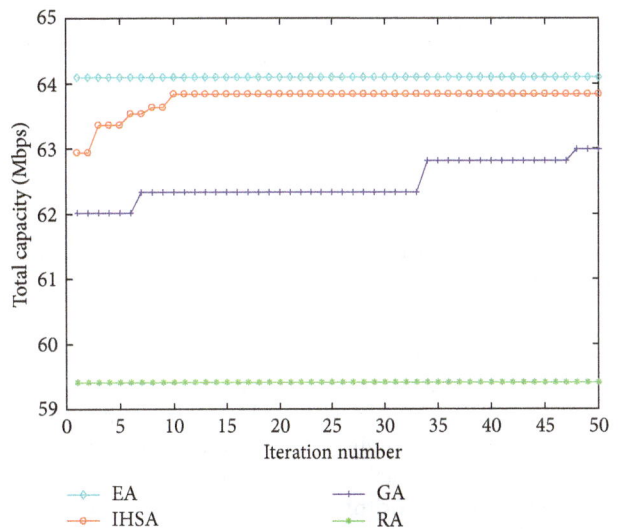

FIGURE 3: The capacity of D2D communications.

algorithm (GA) which has good performance in the field of D2D resource allocation, the improved harmony search algorithm (IHSA) has superior global search performance by dynamically adjusting algorithm parameters. Therefore, the improved harmony search algorithm proposed in this paper is closer to the result of the exhaustive algorithm.

Figure 5 and 6 show the average transmission power of CUEs under different algorithms and the cumulative distribution function of the average transmission power of CUEs. We can see from the figures that the improved

harmony search algorithm is better than the genetic algorithm. This improved harmony search algorithm can find the better resource allocation scheme and reduce the transmission power of the cellular users effectively.

Figure 7 and 8, respectively, shows the energy efficiency of the terminals under different algorithms and the cumulative distribution function of energy efficiency. Usually,

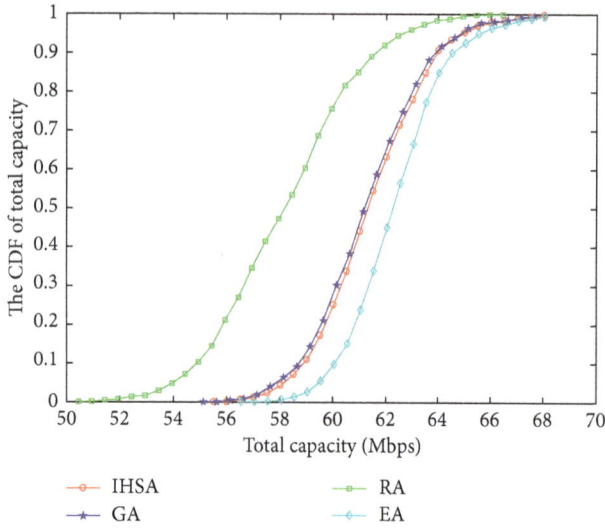

FIGURE 4: The CDF of total capacity.

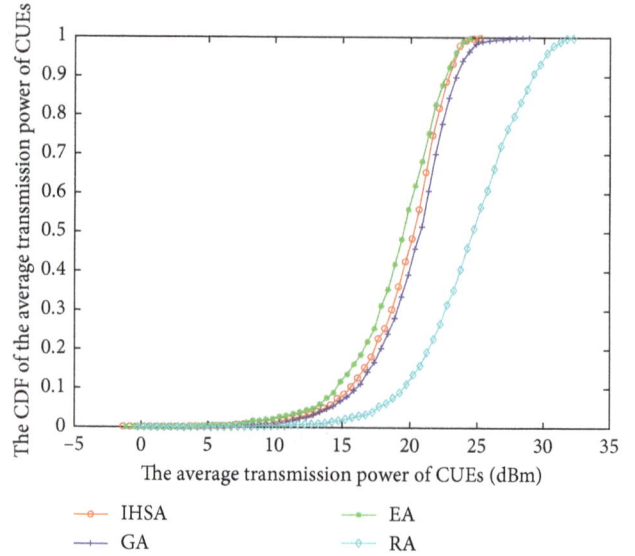

FIGURE 6: The CDF of the average transmission power of CUEs.

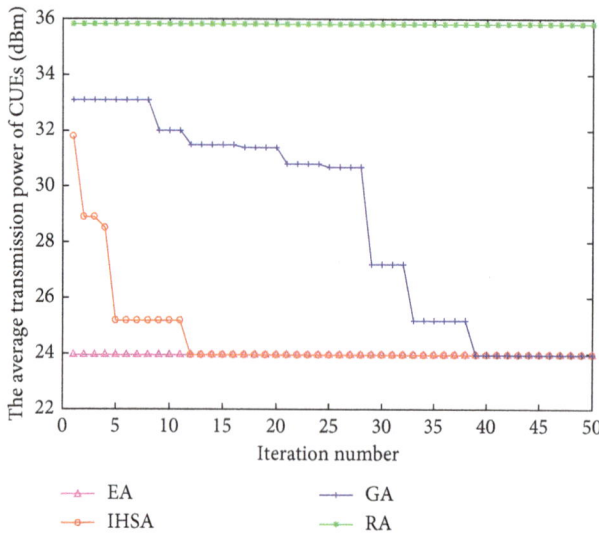

FIGURE 5: The average transmission power of CUEs under different algorithms.

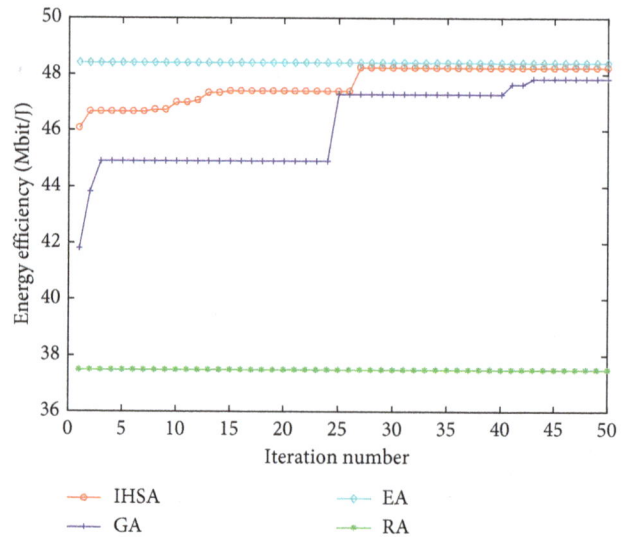

FIGURE 7: The energy efficiency of terminals.

when the CUEs use the traditional power control mode, the power consumption of CUEs will be decreased as the interference decreased. Then the channel capacity will be increased, and the energy efficiency of the terminal will be improved. We can obtain from the graph that the improved harmony search algorithm has better performance of resource allocation. It can reduce the interference among the terminals of same frequency through reasonable optimization so as to improve the energy efficiency.

5.2. Algorithmic Complexity Analysis. In these algorithms, the main computational complexities are from the fitness function calculation and random function calculation. So we mainly analyze the times of running the random function and fitness function in these algorithms. For the genetic algorithm, the number of iteration is set to *MAXI*. For each

crossover operation of the genetic algorithm, we need to run random function 1 time and fitness function $P_c * 2$ times. For each mutation operation of the genetic algorithm, we need to run random function 1 time and fitness function P_m times. Then in each iteration, we need to run crossover operation $K/2$ times and mutation operation $K * M$ times. Here, $K/2$ means that K individuals in one population are divided to $K/2$ pairs. On the whole, we need to run random function $K/2 + K * M$ times and fitness function $K/2 * P_c * 2 + K * M * P_m$ times in each iteration. In the end, we need to run random function $MAXI * (K/2 + K * M)$ times and fitness function $MAXI * (K/2 * P_c * 2 + K * M * P_m)$ times in each algorithm process.

At the same time, we can also get the complexities of the improved harmony search algorithm, random algorithm, and exhaustive algorithm, as shown in Table 2.

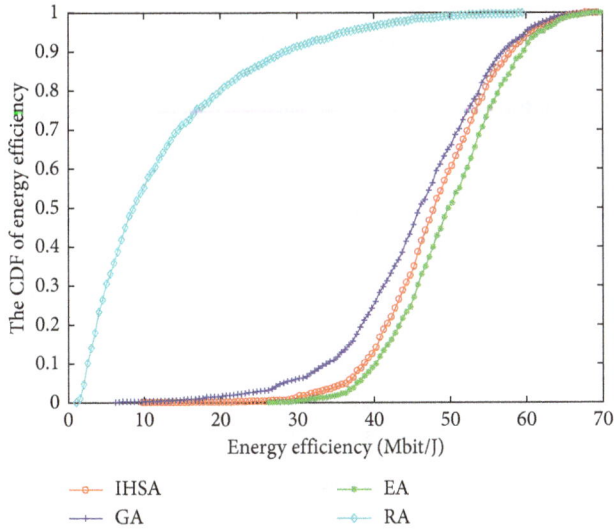

FIGURE 8: The CDF of the energy efficiency.

TABLE 2: The complexities of the algorithms.

	The times of running fitness function	The times of running random function
GA	$MAXI * (K/2 * P_c * 2 + K * M * P_m)$	$MAXI * (K/2 + K * M)$
EA	N^M	0
IHSA	$MAXI * (K * HMCR * PAR + K * (1 - HMCR))$	$MAXI * (M * K + M * K * HMCR + M * K * HMCR * PAR + M * K * (1 - HMCR))$
RA	1	M

5.3. Overhead Analysis. In D2D communications, the additional signalling overhead and required message passing can be measured by the number of the required pilot signals which are used to detect the channel state information. From the received SINR formulae of the CUEs and DRUEs, it can be seen that complete channel state information should be perfectly known at the receivers. During the uplink transmission, in order to calculate the received SINR of the CUEs, the state information of the channels from the CUEs to the BS and from DTUEs to the BS should be obtained. Similarly, in order to calculate the received SINR of the DRUEs, the state information of the channels from the CUEs to DRUEs and from the DTUEs of other pairs to the DRUEs should also be obtained.

For the genetic algorithm and the improved harmony search algorithm, any possible combinations may occur in each iteration. Hence, we need to get all the possible combinations of the CUEs and DUE pairs. Then in this case, the overhead can be calculated as $M + 2 \times C_M^2$. Here, M represents the number of the channels between the DTUEs and the BS. $2 \times C_M^2$ is the number of channels between any DRUEs and the DTUEs of other pairs.

For the exhaustive algorithm, we also need to traverse all the possible solutions. The state information of the channels between any transmitters and any receivers should be obtained. So the overhead is also equal to $M + 2 \times C_M^2$.

For the random algorithm, we only need one combination which corresponds to one row of Equation (1). Without loss of generality, we can consider that the M DUE pairs are equally allocated to N package which corresponds to N CUEs. Then, for the DRUEs in one package, the state information of $2 \times C_{M/N}^2$ channels should be known. Meanwhile, the state information of M channels should be obtained for all CUEs. Therefore, the overhead can be written as $M + 2 \times N \times C_{M/N}^2$ in the random algorithm.

6. Conclusions

In this paper, the resource allocation problem for D2D communication is formulated and discussed. In our scenario, we consider that the number of D2D user equipment (DUE) pairs is far larger than that of cellular user equipments (CUEs) and the resource blocks are divided into two types. Then a resource allocation scheme based on the genetic algorithm is proposed. To improve system performance, an improved harmony search algorithm is proposed for resource allocation. Finally, the analysis and simulation results show that the performances of the proposed genetic algorithm and the improved harmony search algorithm outperform that of the random algorithm and are very close to that of the exhaustive algorithm. This result can be applied for design and optimization of D2D communications.

Acknowledgments

This work was supported in part by the Fundamental Research Funds for the Central Universities (No. 2017B14214), Six Talent Peaks Project in Jiangsu (DZXX-008), and National Natural Science Foundation of China (Nos. 51669014, 61601283, and 61301110).

References

[1] D. Xu, Y. Li, J. Li, M. Ahmed, and P. Hui, "Joint topology control and resource allocation for network coding enabled D2D traffic offloading," *IEEE Access*, vol. 5, pp. 22916–22926, 2017.

[2] F. Wang, Y. Li, Z. Wang, and Z. Yang, "Social-community-aware resource allocation for D2D communications underlaying cellular networks," *IEEE Transactions on Vehicular Technology*, vol. 65, no. 5, pp. 3628–3640, 2016.

[3] H. Tang and Z. Ding, "Mixed mode transmission and resource allocation for D2D communication," *IEEE Transactions on Wireless Communications*, vol. 15, no. 1, pp. 162–175, 2016.

[4] J. Pérez-Romero, J. Sánchez-González, R. Agustí, B. Lorenzo, and S. Glisic, "Power-efficient resource allocation in a heterogeneous network with cellular and D2D capabilities," *IEEE Transactions on Vehicular Technology*, vol. 65, no. 11, pp. 9272–9286, 2016.

[5] L. Wang, H. Tang, H. Wu, and G. L. Stüber, "Resource allocation for D2D communications underlay in Rayleigh fading channels," *IEEE Transactions on Vehicular Technology*, vol. 66, no. 2, pp. 1159–1170, 2017.

[6] Z. Zhou, K. Ota, M. Dong, and C. Xu, "Energy-efficient matching for resource allocation in D2D enabled cellular networks," *IEEE Transactions on Vehicular Technology*, vol. 66, no. 6, pp. 5256–5268, 2017.

[7] P. Wu, P. C. Cosman, and L. B. Milstein, "Resource allocation for multicarrier device-to-device video transmission: symbol error rate analysis and algorithm design," *IEEE Transactions on Communications*, vol. 65, no. 10, pp. 4446–4462, 2017.

[8] T. Yang, R. Zhang, X. Cheng, and L. Yang, "Graph coloring based resource sharing (GCRS) scheme for D2D communications underlaying full-duplex cellular networks," *IEEE Transactions on Vehicular Technology*, vol. 66, no. 8, pp. 7506–7517, 2017.

[9] S. M. A. Kazmi, N. H. Tran, W. Saad et al., "Mode selection and resource allocation in device-to-device communications: a matching game approach," *IEEE Transactions on Mobile Computing*, vol. 16, no. 11, pp. 3126–3141, 2017.

[10] J. Gu, S. J. Bae, S. F. Hasan, and M. Y. Chung, "Heuristic algorithm for proportional fair scheduling in D2D-cellular systems," *IEEE Transactions on Wireless Communications*, vol. 15, no. 1, pp. 769–780, 2016.

[11] A. Sultana, L. Zhao, and X. Fernando, "Efficient resource allocation in device-to-device communication using cognitive radio technology," *IEEE Transactions on Vehicular Technology*, vol. 66, no. 11, pp. 10024–10034, 2017.

[12] J. R. Bhat, J. P. Sheu, and W. K. Hon, "Resource allocation schemes for revenue maximization in multicast D2D networks," *IEEE Access*, vol. 5, pp. 26340–26353, 2017.

[13] M. Mozaffari, W. Saad, M. Bennis, and M. Debbah, "Unmanned aerial vehicle with underlaid device-to-device communications: performance and tradeoffs," *IEEE Transactions on Wireless Communications*, vol. 15, no. 6, pp. 3949–3963, 2016.

[14] O. Semiari, W. Saad, S. Valentin, M. Bennis, and H. V. Poor, "Context-aware small cell networks: how social metrics improve wireless resource allocation," *IEEE Transactions on Wireless Communications*, vol. 14, no. 11, pp. 5927–5940, 2015.

[15] M. Naderi Soorki, W. Saad, M. H. Manshaei, and H. Saidi, "Social community-aware content placement in wireless device-to-device communication networks," *IEEE Transactions on Mobile Computing*, vol. 99, p. 1, 2018.

[16] C. Yang, X. Xu, J. Han, W. U. Rehman, and X. Tao, "GA based optimal resource allocation and user matching in device to device underlaying network," in *Proceedings of 2014 IEEE Wireless Communications and Networking Conference Workshops*, pp. 242–247, Istanbul, Turkey, April 2014.

[17] C. Yang, X. Xu, J. Han, and X. Tao, "GA based user matching with optimal power allocation in D2D underlaying network," in *Proceedings of IEEE Vehicular Technology Conference*, pp. 1–5, Vancouver, Canada, September 2014.

[18] C. Yang, X. Xu, J. Han, and X. Tao, "Energy efficiency-based device-to-device uplink resource allocation with multiple resource reusing," *Electronics Letters*, vol. 51, no. 3, pp. 293–294, 2015.

[19] M. Hamdi, D. Yuan, and M. Zaied, "GA-based scheme for fair joint channel allocation and power control for underlaying D2D multicast communications," in *Proceedings of International Wireless Communications and Mobile Computing Conference*, pp. 446–451, Valencia, Spain, June 2017.

[20] M. Ru, S. Yin, and Z. Qu, "Power and spectrum allocation in D2D networks based on coloring and chaos genetic algorithm," *Procedia Computer Science*, vol. 107, pp. 183–189, 2017.

[21] A. Asheralieva and Y. Miyanaga, "An autonomous learning-based algorithm for joint channel and power level selection by D2D pairs in heterogeneous cellular networks," *IEEE Transactions on Communications*, vol. 64, no. 9, pp. 3996–4012, 2016.

[22] Y. F. Huang, T. H. Tan, B. A. Chen, S. H. Liu, and Y. F. Chen, "Performance of resource allocation in device-to-device communication systems based on particle swarm optimization," in *Proceedings of International Conference on Systems*, pp. 400–404, Banff, Canada, October 2017.

[23] Y. Sun, X. Yan, X. Li, Y. Gu, and C. Li, "A resource allocation scheme based on genetic algorithm for D2D communications underlaying multi-channel cellular networks," in *Proceedings of 2017 International conference on 5G for Future Wireless Networks*, pp. 675–684, Beijing, China, April 2017.

[24] W. G. Zong, J. H. Kim, and G. V. Loganathan, "A new heuristic optimization algorithm: harmony search," *Simulation*, vol. 2, no. 2, pp. 60–68, 2001.

10

Bringing Geospatial Data Closer to Mobile Users: A Caching Approach based on Vector Tiles for Wireless Multihop Scenarios

Chao Li,[1] Huimei Lu,[1] Yong Xiang (iD),[2] Zhuoqun Liu,[3] Wanli Yang,[4] and Ruilin Liu[5]

[1]*School of Computer Science and Technology, Beijing Institute of Technology, Beijing, China*
[2]*Department of Computer Science and Technology, Tsinghua University, Beijing, China*
[3]*Neoclub Information Technology Company Limited, Shanghai, China*
[4]*The First Research Institute of the Ministry of Public Security, Beijing, China*
[5]*Department of Computer Science, Rutgers University, Piscataway, NJ, USA*

Correspondence should be addressed to Yong Xiang; xyong@csnet4.cs.tsinghua.edu.cn

Academic Editor: Joaquín Torres-Sospedra

Mobile applications based on geospatial data are nowadays extensively used to support people's daily activities. Despite the potential overlap among nearby users' geospatial data demands, it has not been feasible to share geospatial data with peer wireless devices directly. To address this issue, we designed a scheme based on vector tiles to organize spatial data and proposed a system named GeoTile for geospatial data caching and sharing. In GeoTile, a tile request from the mobile client relies on multihop communication over intermediate nodes to reach the server. Since GeoTile enables all network nodes to cache and process geospatial data tiles, requests may be handled before they actually reach the server. We implement the GeoTile prototype system and conduct comprehensive real-world experiments to evaluate the performance. The result shows that the GeoTile system can serve vector tiles for users conveniently and friendly. In addition, the caching mechanism based on vector tiles can substantially reduce the response time and network throughput under the wireless multihop scenarios.

1. Introduction

With the ubiquitous penetration of mobile devices, especially smartphones, mobile applications based on geospatial data are nowadays considered as one of the necessities of people's daily life. About half of mobile users around the world have map applications installed in their own devices [1]. People consume geospatial data for local services (e.g., point of interest recommendation and navigation) and also generate local data to share among different applications (e.g., [2]) through crowdsourcing [3, 4].

Despite the benefits brought by the map-related application, accessing geospatial data incurs nontrivial data consumption and sometimes can even be annoying. For instance, free public hot spots may be too sparse to provide real-time data updates, especially in developing countries. Mobile data usage over the cellular network, on the other hand, can be expensive, leading to a high cell phone bill and

consequently user's cost concern [5]. Offline data packages, that is, map data downloaded beforehand, are considered as a naive solution to this problem. Nevertheless, while the offline data packages are usually bulky (up to tens of megabytes, e.g., [6, 7]), they are now treated as atomic and exclusive: (i) offline map data can be used only when the entire data package is downloaded and does not support partial updates and (ii) such data packages are formatted as per application, and users may find it difficult to share among peers [8, 9].

More recently, researchers explore to tackle the challenge of geospatial data access by exploiting the wireless communication and distributed storage among multiple mobile devices. Nowadays, the storage space of mobile devices is usually adequate to carry a considerable amount of data, which makes every mobile device a potential content provider. Considering the prevalence of the mobile device and the fact that similar geospatial data are frequently requested from hot regions, such as urban centers or transportation hubs [10], the geospatial data requested by a user can be very likely available in other nearby users' devices. This idea is known as "Device-to-Device (D2D)" [11], and if implemented properly, it can result in an instant and cost-effective solution for users' geospatial data needs.

In this paper, with the idea of D2D for geospatial data sharing in wireless multihop scenarios, we present GeoTile, a novel system that enables geospatial data caching and access in a distributed and collaborative manner. GeoTile adopts the GeoJSON format with optional compression to partition the large-scale geospatial data into small units, that is, vector tiles, for efficient data sharing. Moreover, GeoTile also explores the possibility to use the peer cache and amalgamation technique to reuse the data cached locally in order to reduce both the response time and network data usage.

Specifically, the paper makes three main contributions:

(i) Based on the state-of-art geospatial data formats, we propose a new geospatial exchange standard, which constructs vector tiles with optional compression to efficiently index, exchange, and organize geospatial data with various granularities and level of details (LoDs).

(ii) We implement the GeoTile system for geospatial data caching and sharing. GeoTile not only contains the vector tile system in the data-provider server but also enables network intermediate nodes to cache vector tiles and to make responses to the geospatial data request that overlaps with the local cache.

(iii) We evaluate the GeoTile system in real-world settings and show that the GeoTile system provides efficient vector tile service in a convenient and user-friendly way. Meanwhile, the evaluation result also demonstrates that the caching mechanism based on vector tiles can substantially reduce the response time, wireless network traffic, and the dependency on root data sources in distributed scenarios.

The rest of this paper is organized as follows. First, we revisit the related works in Section 2. Then, in Section 3, we describe the main idea of the vector tile system. After that, Section 4 elaborates on the design of the GeoTile system by introducing tile service component and mobile prototype component, respectively. Performance evaluations and analysis of GeoTile under realistic environments are presented in Section 5. Finally, we conclude the paper together with a short future work discussion in Section 6.

2. Related Works

In this section, we take a brief review of literatures on multiple aspects concerning our work in this paper.

2.1. Tile-Based Mapping and Geospatial Information Service.
Map tiling has been proved as the most effective way of delivering spatial data over the recent decade [12]. By creating tiny map fragments at numerous predefined scales, map tiling enables users to just retrieve the pieces of data needed for the current view to reduce the data transfer volume and response time. Meanwhile, tile caching at the server side reduces the computational resource requirement and increases the concurrent request rate a server can handle.

Lately, the mobile geographical information system (GIS) [13] stretches users' ability of geospatial information access to mobile devices. However, the mobile GIS follows the client/server architecture as in traditional Internet GIS [14]: an active connection to the GIS server is required for a user to fetch the desired spatial data. In our work, by utilizing cached geospatial data at intermediary nodes, users can get required data from locations as near as possible instead of accessing server directly.

2.2. The Emergence of Vector Tiles.
Raster tiles dominated the earlier years of map tiling (especially for mobile clients), mainly because of its simplicity and the limited computation capacities on the mobile client [15]. However, as mobile users' demand for a more interactive and informative mapping service growth [16], vector tiles have come to light recently: vector tiles provide readable, descriptive, and extensible content, which facilitates the provision of data supplementation, manipulation, and customization. For instance, by using vector tiles, users can scale the map smoothly without the problem of pixelation [17]. Expressive features, such as highlight, labeling, styling, and animation, are also well supported by popular libraries and mainstream browsers.

To bring the idea of the vector tile into reality, one scheme of the vector tile is proposed in [18]. Those objects crossing multiple tiles are clipped to the boundary they belong to. Although it makes little impact on visualization, the integrity of original data is damaged and extra points are brought in. Mobile users will find it challenging to collect full information of objects and restore them. Another shortage of splitting tiles according to fixed areas is that the size of tiles varies greatly. In order to improve the efficiency of transmission, in [19] tiles are generated according to data density, so the tiles are of irregular shapes. But this method is inimical to sharing as it is hard to define global rules that guarantee each tile can be identified and reused by everyone.

2.3. Quadtree Indexing Structure.
The pyramid-structured quadtree indexing structure has become the de facto standard of online map and has been adopted by most major web map service providers. The most representative model is the XYZ Map Tiling System (a.k.a. Slippy Map) [20]. As a matter of fact, such structure, owing to its characteristics of top-down hierarchy and multiscale [21], has wide applications in a variety of areas of science and technology (e.g., image processing [22], indexing [23], artificial intelligence [24], movement prediction [25], privacy protection [26], and just to name a few). In Slippy Map, multiple zoom levels are defined to represent different scales. Tiles follow a recursive subdivision strategy. The region of one tile on zoom level N is equally divided into four portions, namely, four tiles on zoom level $N + 1$. Each tile is indexed by a unique group of values (x, y, z). With the naming scheme, each node can identify demanded tiles unambiguously. Users or intermediate

nodes can clearly know if they have contents of certain locations. It is the foundation of tile sharing among mobile counterparts. In our work, we follow this indexing structure and naming scheme to identify and arrange vector tiles.

2.4. The Encoding of Vector Data. Various formats can be the container of vector data [27, 28]. ESRI Shapefile and OSM XML are mainly used for online data release. Shapefile [29] stores data in three separate parts (geometry of features, positional index of features, and attributes of features) and professional tools are necessary to process it. OSM XML [30] stores complex relation information in addition to the geometry data and metadata. The user has to traverse the file along the relations to get complete object geometry data. GML (Geography Markup Language) [31] and KML (Keyhole Markup Language) [32] are two formats based on XML. KML has a more concise schema, while GML provides advanced features to describe complex maps. However, being subject to the heavyweight nature of XML syntax, KML and GML are usually bigger in size [33].

Comparatively speaking, lightweight formats are more appropriate for data-interchange among mobile devices. GeoJSON [34] is an open standard format of widespread use for vector tiles. It is dedicated for encoding geographic data along with various nonspatial attributes. All structures and columns are human readable and straightforward to users. We here note that the significance of the readability is that it enables users to find information or make modifications easily without any specialized software. Another JSON-based format TopoJSON [35] can achieve smaller size, and reduce data redundancy by storing the shared boundary only once. As the expense, it impairs the independence and completeness of original objects. Thus, it is not proper for further data editing and sharing. Moreover, the coordinates have been transformed to integers so that it could not be edited or resolved before decoding. Protocol buffers [36], developed by Google, is a binary serialization method. Although it may have better performance and much smaller size [37], it is not human readable and specialized tools, or libraries to resolve them are not widespread among mobile users. Additionally, the schema (proto file) which contains a description of the data structure and data types should be made aware by users beforehand to parse received data.

2.5. Compression Techniques Applied in GIS. Compression is usually deemed as an effective way to preserve and distribute geospatial data in GIS systems, as it reduces the amount of data on transmission and storage. General-purpose methods can also be applied directly to geodata units, such as ZIP used in KMZ [38], LZMA and DEFLATE used in [39, 40], gzip used in [41], and LZW used in [42]. The description of the above mainstream compression methods can be found in [43]. Apart from these, some dedicated-purpose methods are proposed to make optimizations for certain formats. For example, methods are summarized in [44] for GML. In our work, compression is also of importance as the bandwidth and storage resources are quite precious in wireless scenario. Generally speaking, every

compression method has its characteristics. We should take factors such as data type, platform, processing speed, and ·application scenario into consideration to make the most suitable decision. We apply a new promising compression method, Brotli [45], to our GeoTile system to see whether expected effect can be achieved. The reasoning of selecting Brotli is explained in Section 3.3.

2.6. Simplification and Quantization of Vector Geospatial Objects. Another way to address the oversize problem of geodata units is the simplification and quantization. Both of them can be categorized into lossy compression. The former one reduces the visually insignificant vertice of objects, while the latter one rounds and shortens the coordinates. Simplification (a.k.a. generalization) [46, 47] is often used in progressive transmission [48, 49]. By dividing the original vector object into multiresolution representations, a subset of the data returns to user upon request at first, and then subsequently it will be refined incrementally. The main goal of progressive transmission is avoiding transmitting too much detailed data at one time to manage near real-time response at client side, that is, no longer than 1 or 2 seconds [50].

However, to our best knowledge, most progressive transmission methods require stable Internet connections as the prerequisite to perform frequent and timely data transferring. But under mobile circumstances, multiple rounds of data transmitting may not finish due to intermittent links. When it is impossible to establish connection again, the client has to resort to other nodes for missing parts of the object, which will result in longer delay. In aspect of correctness, as simplification techniques are not mature and still under research, features with distortions or topological errors may appear [51, 52]. Moreover, as object information is divided, much differential data should be maintained, which incurs higher complexity on data identification, exchanging and caching for mobile devices. In order to better suit the mobile environment, we will use units that are relatively complete and independent to accomplish the goal of facilitating the sharing of partial data among users and providing the demanded data for requesters cooperatively within a moderate delay.

The quantization method shrinks the data size by reduction of the digits and precision of the real-valued coordinates [53]. Compared with simplification, it can maintain the topological relationship of objects. If data are not being excessively quantized, the difference on visualization is inconspicuous. Even so, we want to emphasize that the visualization is not the only use of geospatial data units. Driven by the enthusiasm of crowdsourcing and rapid development of related technologies, the amount of volunteered geographic information (VGI) grows tremendously [54, 55]. Correspondingly, researches have shown that geodata from OSM in some parts of the world is locationally and semantically more accurate, complete, and timely than proprietary commercial datasets or information provided by government agencies [56, 57]. With such data, users can perform editing and updating, distance measurement, pathfinding [58], surrounding POI lookup [59], and so on. Our future work is also towards this orientation. Undoubtedly, the loss in data precision will undermine the value of data which users have collected and contributed for future sharing and potential use.

2.7. Caching for Data Dissemination. Caching has been regarded as a highly effective solution for data dissemination in mobile wireless networks. With replicas residing closer to the requesting client, we can reduce the communication traffic over the limited wireless channel and hence shorten the overall latencies. Caching is especially crucial to improve data accessibility in multihop mobile networks [60, 61] because intermittent links often impede users from contacting the root data sources. Recently, as D2D communications gain research attentions, caching on terminal devices also starts to show its significance [62]. For that reason, researchers are making efforts to answer questions around caching such as where to cache (placement) [63, 64], what to cache (content type, popularity, and mobility awareness) [65], and how to cache (replacement policies) [66, 67]. In this work, we adopt the idea of caching for vector tile sharing, and we thus introduce the definition of vector tile layers and naming scheme for data caching and exchanging.

3. Vector Tile System

The foundation of data caching and exchange is the format of the data. In this section, we elaborate on the definition of our vector tiles, that is, the basic unit for data caching and exchange in the GeoTile system. Specifically, we cover the partition, naming, and format of vector tiles, together with the optional compression approach and the client-side amalgamation to reduce the data communication volume, and finally, the vector tile's support to custom extensions.

3.1. Generation of Geospatial Data Tiles. Tiles are the fundamental units for data caching and exchanging in the GeoTile system. When breaking bulky geospatial data into smaller tiles, each of them can be sent separately to effectively speed up the map delivery. We use the naming scheme of the Slippy Map system, the most popular way of organizing map tiles. In this way, we can utilize the data provided by other online resources for the layers that our server cannot provide or to supplement our data to add additional customization.

In our vector tile system, each tile is associated with a coordinate in the form of (x, y, z), where x, y, and z represent the longitude index, latitude index, and zooming level of the tile, respectively. We use the following steps to build the vector representation for the tile.

3.1.1. Geographic Boundary of a Tile. We create vector tiles according to the rules defined in XYZ Quadtree Indexing System [68] to determine the location and the scope of the tile. Specifically, given the coordinate (x, y, z) of a tile t, we can get the west longitude line and the north latitude line of t using (1) and (2). Likewise, the east longitude line and the south latitude line of t can also be determined by reusing (1) and (2) on the tile t': $(x + 1, y + 1, z)$, which is diagonally adjacent to t at the bottom right direction. This is because that

TABLE 1: Several rules from zoom level reference.

Category	Type	Zoom level				
		13	14	15	16	17
Highway	Primary	✓	✓	✓	✓	✓
Highway	Tertiary	✓	✓	✓	✓	✓
Highway	Pedestrian			✓	✓	✓
Railway	Station	✓	✓	✓	✓	✓
Amenity	Hospital				✓	✓
Amenity	Restaurant					✓
Shop	Supermarket				✓	✓

the east longitude line and south latitude line of t overlap with the west longitude and the south latitude lines of t':

$$\text{lon} = \frac{x}{2^z} \cdot 360 - 180, \tag{1}$$

$$\text{lat} = \arctan\left(\sinh\left(\pi - \frac{y}{2^z} \cdot 2\pi\right)\right) \cdot \frac{180}{\pi}. \tag{2}$$

3.1.2. Level of Detail. As demanded by most online map services, we apply the idea of LoD in our tile system. At zooming levels of low values, only limited objects with coarse-grained appearance are included into the tile, whereas more objects will present on the map when a user zooms into a more granular level. To achieve this, we extract a set of rules (some instances are shown in Table 1) from the well-known render engine Mapnik [69]. Based on these rules and the zoom level of t, we can tell which objects should be included in t.

Based on the LoD rules, we include in the tile t those objects which have an intersection with t and are qualified to show at zoom level z. This implies that if an object crosses multiple tiles, all these tiles will contain it. This approach will introduce some redundancy, whereas the integrity of objects is not broken. Users can trust that what they receive is the description of the entire object without worrying about seeking for missing parts when they need to render full objects or to perform calculations.

3.1.3. Formatting and Postprocessing. We select GeoJSON as the format of vector tiles for the following reasons. First, the accuracy of raw geometries, such as position coordinates, will be kept in GeoJSON. It is meaningful as we expect that many calculations and navigations could be performed locally to reduce the reliance on infrastructures and to obtain speedups. Second, it is supported natively by HTML5 and mapping tools such as Leaflet and OpenLayers, which eases the utilization at the terminal [70]. Third, it is human readable. Users can easily make modifications and contributions to original data with common editors. Finally, it is more compact compared to XML-based formats.

After the tile t is published using the GeoJSON format, we make some optimizations. First, we use the compact mode of GeoJSON to remove unnecessary spaces and line breaks. Then, we delete fields whose value is null. Lastly, we remove some fields which are meaningless to users, such as the information of database or server.

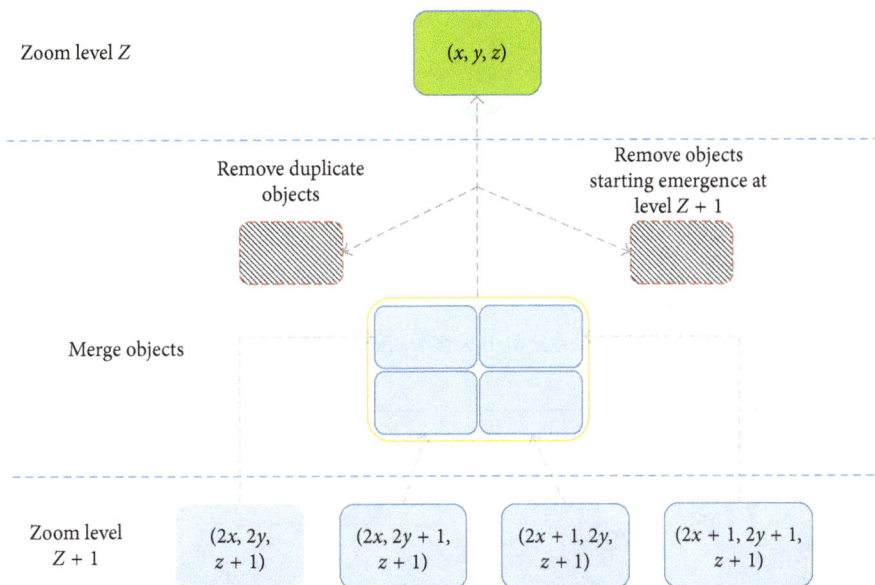

FIGURE 1: The procedure of amalgamating local tiles.

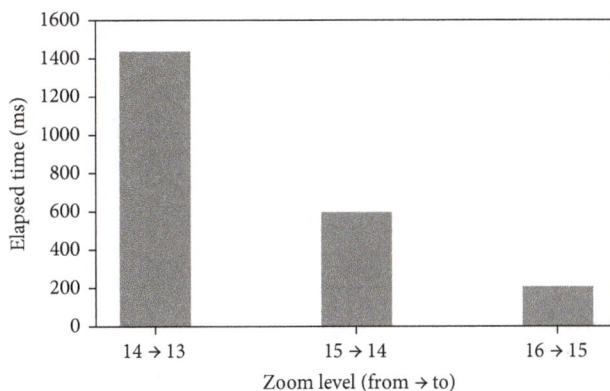

FIGURE 2: Time cost of tile amalgamation on Nexus 7. The result in each setting is based on the average processing time of 30 random tile amalgamations.

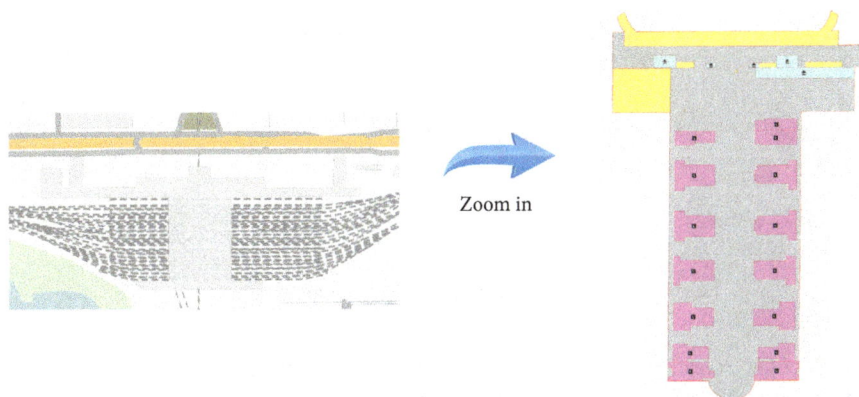

FIGURE 3: An example of interior map of Beijing west railway station.

3.2. Tile Compression. Given the tile definition and formatting presented in Section 3.1, one can expect the size of tile grows substantially as the zoom level goes up. The main reason is that some large objects, such as trunk roads, railways, or district boundaries, are more likely to be included by more tiles with the scope of each tile enlarged. To handle this issue, we apply compressions on the tiles using Brotli [45], a generic-purpose lossless compression

FIGURE 4: Map with traffic condition.

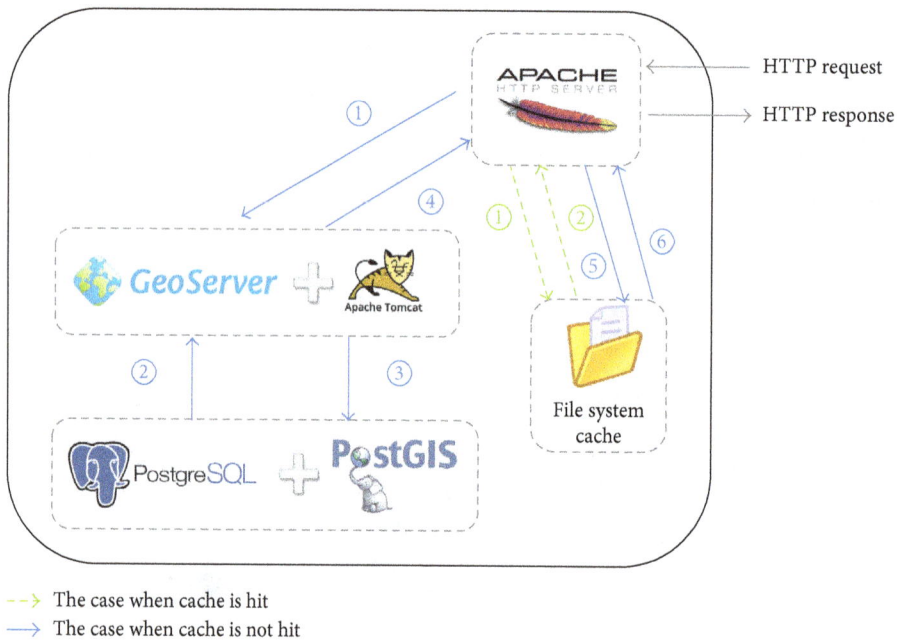

FIGURE 5: The structure and working flow on server.

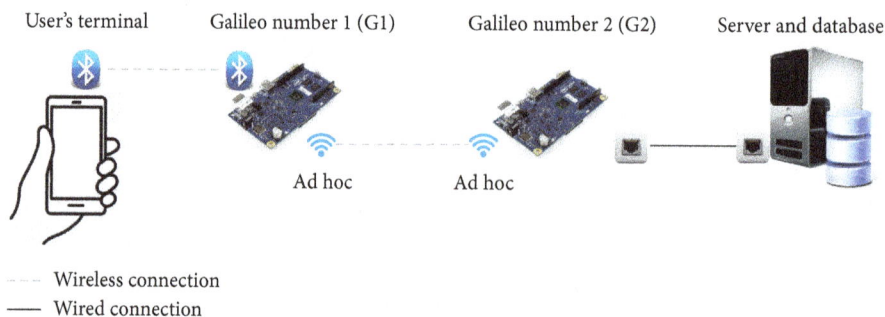

FIGURE 6: The structure of mobile prototype component.

algorithm presented by Google recently. Brotli achieves better compression ratio, especially for small-sized text file [71, 72], and provides fast decompression speed at the client mobile device, regardless of the compression quality levels. Although Brotli is relatively slow on compression, considering most compressions are performed at the powerful server side and only happen infrequently, we choose Brotli as our default tile compression method and decide to use compression level 10 to trade-off the compression time and the compression ratio. Based on both the benchmark test [73] and the real test on our server, Brotli high-compression levels, for example, level 10 or level 11, achieve about 20%~30% savings in the file size compared to the middle-quality levels (level 7~9). However, level 11, which provides the best compression ratio, reduces the file size by 3%~5% than level 10 but incurs 30% longer compression time.

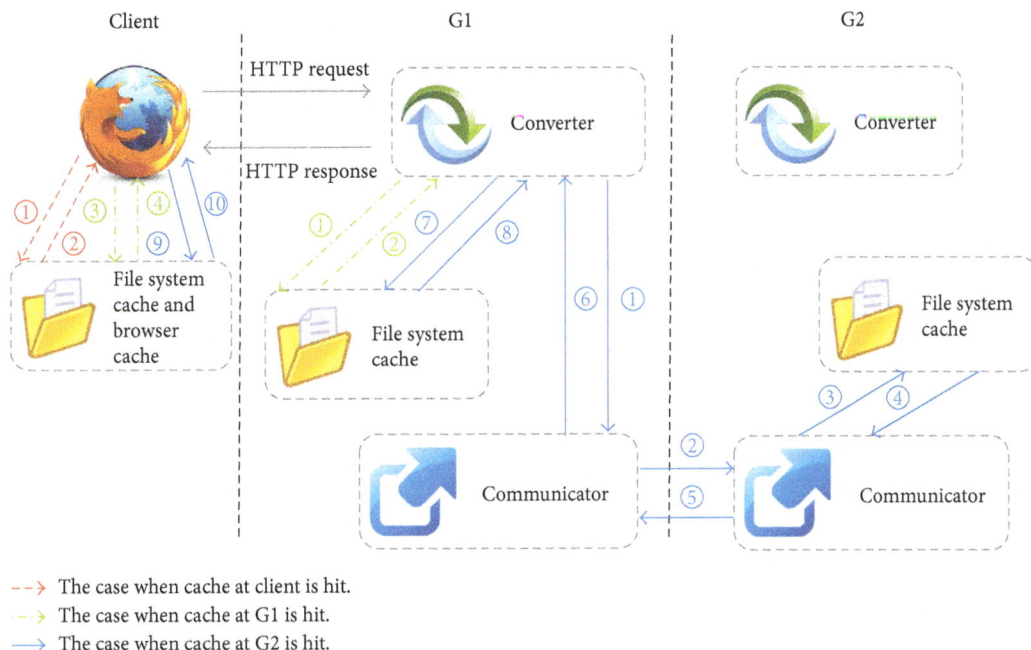

--→ The case when cache at client is hit.
-·→ The case when cache at G1 is hit.
——→ The case when cache at G2 is hit.

FIGURE 7: The working flow of the mobile prototype when the cache is available at one of the locations.

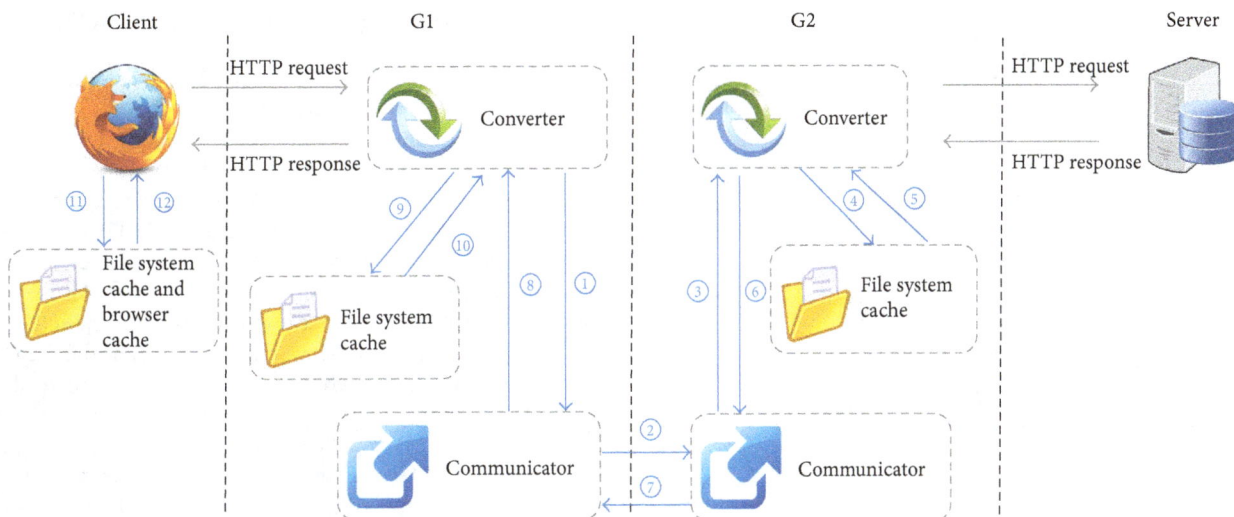

FIGURE 8: The working flow of the mobile prototype when the cache is not available.

3.3. Client-Side Tile Amalgamation. When users are under extremely poor network conditions, even the compressed tiles may take too much time to fetch. Tiles cached at the terminal will help shorten the waiting period and lessen the data traffic. Based on the rules for organizing the tiles above, we can use local tiles to generate corresponding upper layer tiles directly without requests. As illustrated in Figure 1, given the (x, y, z) coordinate of target tile t, we check the existence of the four lower-layer tiles in the local cache, that is, tile $(2x, 2y, z + 1)$, tile $(2x + 1, 2y, z + 1)$, tile $(2x, 2y + 1, z + 1)$, and tile $(2x + 1, 2y + 1, z + 1)$. If all the four tiles exist, we will merge all objects into a new tile and remove duplicate objects or those objects whose emerging zoom level is $z + 1$.

As shown in Figure 2, the computation overhead for amalgamation is trivial for smaller tiles, for example, hundreds of milliseconds for tiles in level 14 or 15. For tiles of larger size, for example, level 13 and above, even though a long amalgamation processing period may be needed, it is still useful when a user has very limited network data budget or when the network connection is so poor that fetching the tile instead of generating it using amalgamation takes even longer time.

3.4. Support for Extended Information. Besides the basic geometric information, the flexibility of GeoJSON brings numerous new possibilities to the tile unit. Users can append

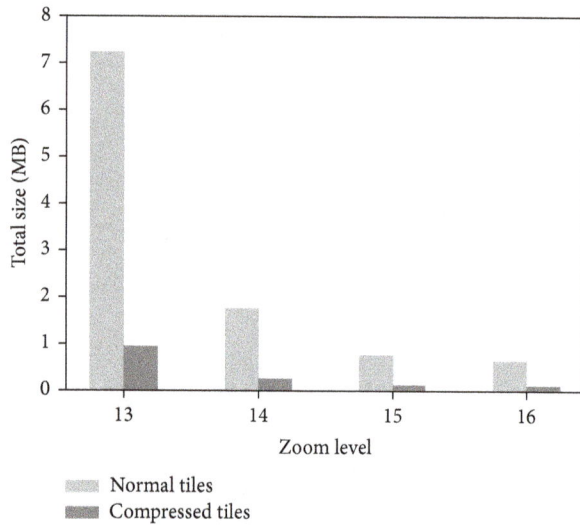

FIGURE 9: Transferred data size of the initial map of each experiment group.

TABLE 2: The cache availability of experiments of normal tiles.

Group	Device			
	Terminal	G1	G2	Server
None	🚫	🚫	🚫	🚫
Server	🚫	🚫	🚫	GeoJSON
G2	🚫	🚫	GeoJSON	GeoJSON
G1	🚫	GeoJSON	GeoJSON	GeoJSON

TABLE 3: The cache availability of experiments of compressed tiles.

Group	Device			
	Terminal	G1	G2	Server
None	🚫	🚫	🚫	🚫
Server	🚫	🚫	🚫	Brotli
G2	🚫	🚫	Brotli	Brotli
G1	🚫	Brotli	Brotli	Brotli

extra key-value pairs to make original objects more informative. To exemplify, we implement an extension of building interior structure and an extension of road traffic conditions on our vector map. Figure 3 shows the zoom in effect for the interior structure of Beijing West Railway Station. To implement this effect, we add a "room" attribute, which indicates the type of room, and a "floor" attribute, which indicates which floor the room locates, to the geometry records of objects. Likewise, we add a "speed" attribute to road objects in the basic map to include traffic conditions calculated in a travel time estimation system and visualize it in Figure 4.

With the support for the extended information, it can be expected that maps with additional specific functions could be easily transformed from the basic vector map. For instance, Beijing Government has released 12 kinds of specialized map recently focusing the convenience for civilians, including air quality, housing, and tourism. The customization potential of vector maps enables the integration and presentation of diversified information. A user can just switch the content layer defined by the customized data to browse different information on the same basic vector map.

4. System Structure and Design of GeoTile

4.1. GeoTile Server. Figure 5 demonstrates the architecture of our GeoTile server, which provides the vector tile service for mobile applications. The implementation of the GeoTile server is built upon Ubuntu OS 14.04 and a series of open source software packages [74]. The GeoTile server consists of four main modules, that is, the Apache HTTP server, the caching system, the web server based on GeoServer, and the data storage module based on PostgreSQL [75]. All the raw data from OpenStreetMap are imported into PostgreSQL with PostGIS [76] support. When a request arrives at the GeoTile server, Apache will first check whether the demanded tile is available at the local cache. If the cache hits, the tile is returned as an HTTP response right away. Otherwise, Apache will pass the request to the GeoServer. The GeoServer deployed on top of Tomcat queries the PostgreSQL database and publishes the data using GeoJSON format through a web feature service (WFS), where our vector tile logic presented in Sections 3.1, 3.2, and 3.4 is implemented. Finally, the tile is returned through Apache with also the result updated in the cache.

As stated above, we use an on-the-fly way to generate tiles gradually rather than preprocessing. The design decision is based on the following considerations. (i) Not all tiles are equally demanded by users. For the sake of resource efficiency, we should keep those tiles needed by most users. (ii) Generating all vector tiles in advance may be a waste when the map data get updated periodically, for example, for the road condition use case, and may exhaust the resource of the server.

4.2. Mobile Prototype Component. In order to test the feasibility and performance of our design under the realistic environment, we implement a mobile multihop prototype as another part of GeoTile. The topology of the prototype is

(a)

(b)

(c)

(d)

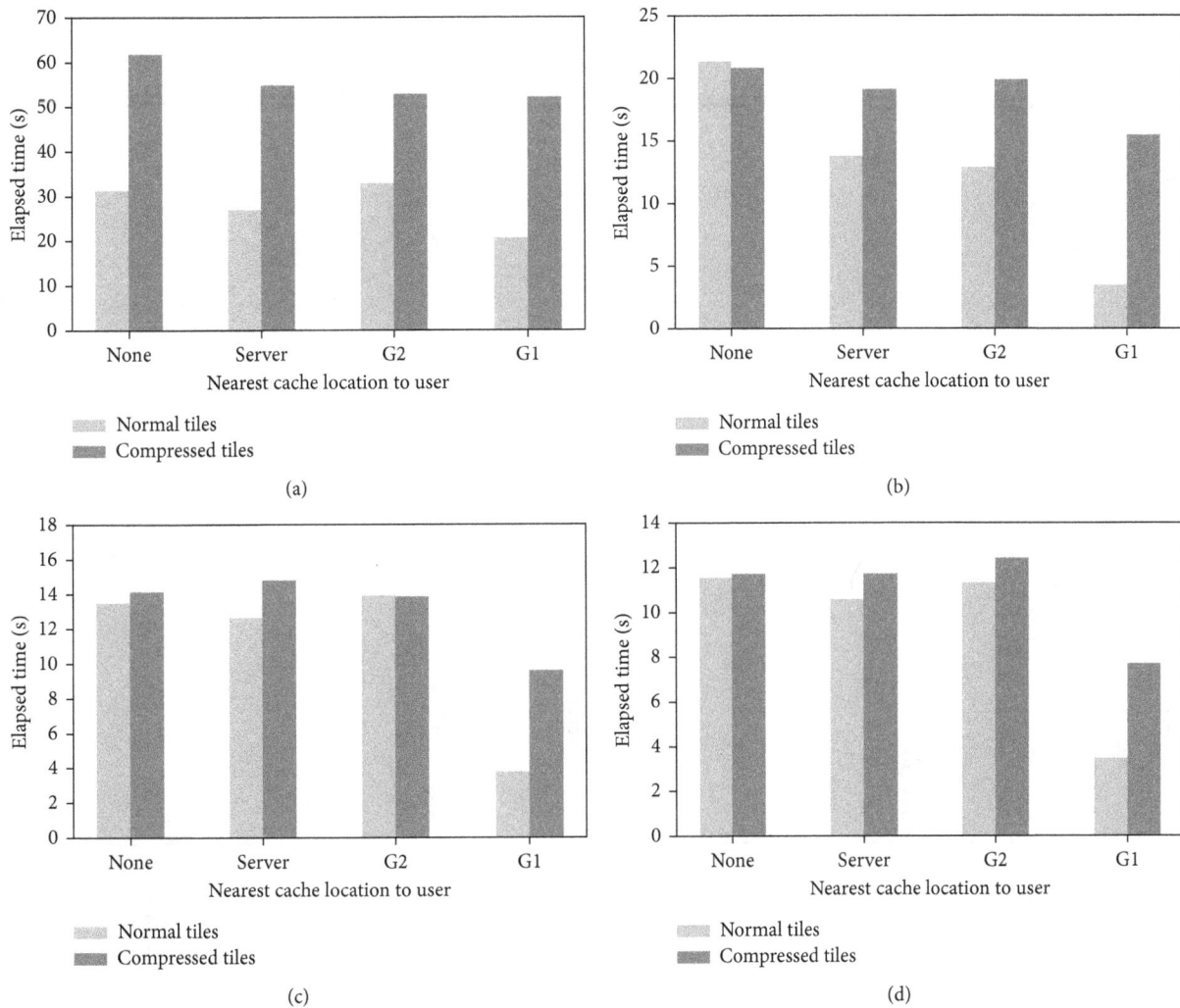

FIGURE 10: Time cost from terminal's view on showing the initial map (cache location varies at different zoom levels). Zoom level: (a) 13; (b) 14; (c) 15; (d) 16.

demonstrated in Figure 6. The requesting client (also called terminal) device is Nexus 7 (Qualcomm Snapdragon S4 Pro 8064 Quad-Core 1.5 GHz, 2 GB memory). We use a cross-platform Firefox browser and a webpage based on HTML5 and Leaflet for vector map request and presentation. The two intermediate wireless nodes are Intel Galileo Gen 2 board (Intel Quark SoC X1000 400 MHz, 256 M DRAM, and 8 GB Class-4 MicroSD card as storage). The terminal connects to the first intermediate node (called G1 hereafter) via Bluetooth. The reason why we choose Bluetooth is that Bluetooth 4.0 is energy-efficient compared to Wifi-based device-to-device methods to provide satisfied transmission for small data traffic [77]. The connection between G1 and the second intermediate node (called G2 hereafter) is an ad hoc network to support mobility. G2 downloads data from the GeoTile server through a wired link. The converter module inside intermediary nodes is implemented with Anyproxy [78].

Both G1 and G2 have the ability to cache received tiles. The demanded tile available at the nearest location to the user is used to respond to a user request. Note that, when we say a tile is "available" locally, it can either be cached locally

or can be generated from other tiles by using the method described in Section 3.3. The procedure of how the mobile prototype works is illustrated in Figures 7 and 8.

5. Experiment and Evaluation

In this section, we conduct practical experiments on GeoTile from multiple angles. We start by investigating the performance of the implementation from user's standpoint. More specifically, we quantify how much time and data transfer are needed to load the initial map at the client side and the impact of available cache location, together with the improvement brought by using tile compression. After that, we take a closer look into the mobile prototype component. Through studying the detailed time cost of each phase, we will gain a deeper understanding of whether the system works properly and accordingly find the room for further improvements.

5.1. Performance from User's Perspective: Response Time. In this part, we evaluate the performance of GeoTile in terms of the response time from a user's standpoint. In particular, we

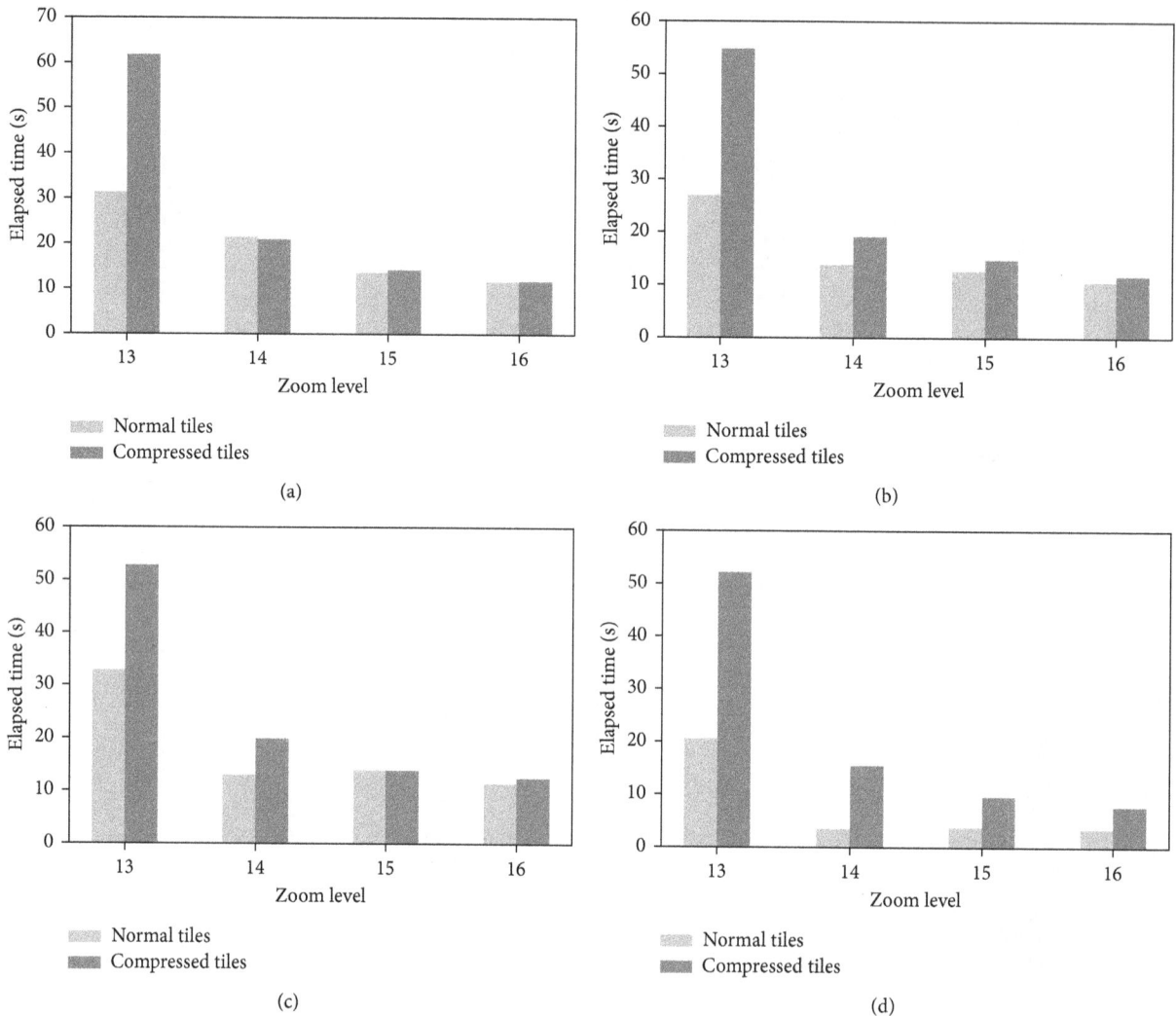

FIGURE 11: Time cost from terminal's view on showing the initial map (zoom level varies at different cache locations): (a) cache is unavailable at all the locations; (b) cache is available at server; (c) cache is available at G2; (d) cache is available at G1.

measure the time needed to load a complete initial map in the browser. The response time is measured from the moment when the request is issued from the terminal until when all data are successfully returned to the terminal and loaded into the browser. The total volume of the transferred data for presenting the initial map in each group is given in Figure 9. Each experiment group is repeated for three times, and the average number is presented as the result. The available location(s) of cache in each experiment setting is given in Tables 2 and 3, for normal tiles and compressed tiles, respectively.

Figure 10 demonstrates the response time when the cache-hit location varies at different zoom levels. Generally speaking, it takes less time to show the map completely when the cache hit is closer to the requesting client.

For the normal tile groups, cache hit at the GeoTile server saves about 10%~20% overall time compared to that without any cache hits. The cache hit at G1 can significantly save the response time, resulting in approximately 25%~30% of the response time when the cache hit only happens at the GeoTile server. This means when the cache hits just 1 hop away from the user, the user could get them quite instantly

without reaching the server. The response time when the cache hits at G2 is sometimes higher than others, and we will explain the reason in the following parts. For the compressed tile groups, as the required data volume of layer 16~14 is very low, not much time is used for tiles generating, transferring, and processing, so the effect brought by the cache at server or G2 is subtle. However, cache hit at G1 still achieves the most response time save.

Figure 11 presents the response time when cache hits at the same place, but the zoom level varies. We can find that the response time for both normal tile and compressed tile decreases as the zoom level number increases. For the normal tile, the response time differences between level 16 and level 14 are not very significant, as they all have the similar size of data transfer. At level 13, since the data transfer size is 3~4 times higher, much more time is spent on transmission among nodes, rendering at the terminal.

Figures 10 and 11 also show that the response time of the compressed tiles is generally longer than that of normal tile groups under the same conditions. This means that although the compressed tiles have less data amount compared to

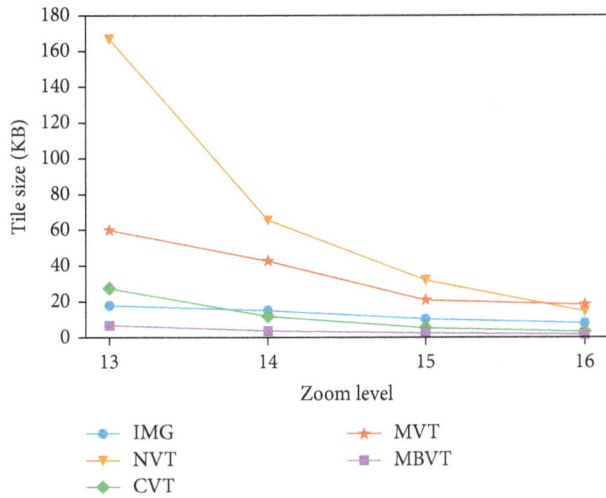

Figure 12: Size of different tile types.

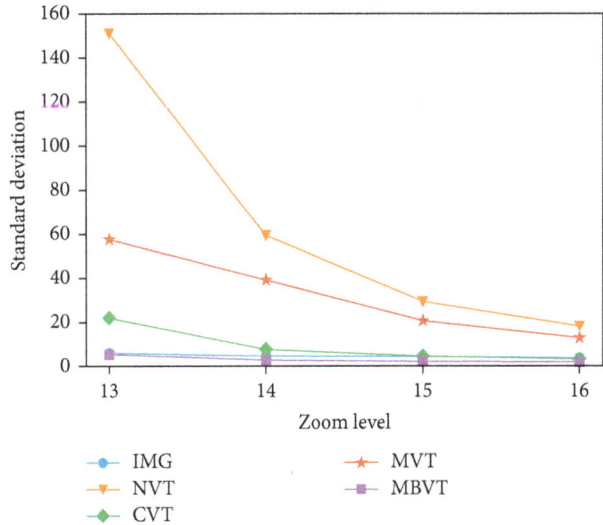

Figure 13: The standard deviation of size of different tile types.

normal tiles, the compression and decompression undermine the time saved within the transmission procedure. Therefore, we conclude that when the quality of the network is poor, the compressed tiles are more recommended as they have smaller size and better tolerate the connectivity failure. When the network quality is good, or when the computation capability of the user's device is not so powerful, normal vector tiles without compression should be preferred to achieve shorter response time.

5.2. Performance from User's Perspective: Data Usage. In addition to the response time, the data usage, that is, the volume of data transfer, is a metric that a user may also be interested in. To evaluate the performance of GeoTile in data transfer efficiency, we conducted two experiments to evaluate (i) the size comparison of different tile types and (ii) the data transfer volume when a user zooms out. The first dimension clarifies the characteristics of our tile scheme in the aspect of size and shows the effect of our tile compression method, while the second additionally shows the effect of client-side tile amalgamation.

5.2.1. Tile Size Comparison. In this part, we compare the size of five different tile types, that is, OSM PNG (IMG), Mapzen Vector Tiles (MVT) [79], Mapbox Vector Tiles (MBVT) [80], our vector tile format (called as NVT), and our compressed tile format (called as CVT). Figures 12 and 13 present the mean and the standard deviation of the tile data usage at different zoom levels. IMG, MVT, and MBVT, which are the tile types available publicly, are included as the comparison baselines. The key difference between MVT and NVT is that MVT clips vector objects along tile boundaries. We sampled 100 tiles of each kind of tiles from each zoom level.

From Figure 12, we can clearly see that the size of CVT is comparable with that of image tiles and is significantly lower than MVT and NVT, which proves that the compression provides a great advantage on reducing the size of vector tiles. The tile size of NVT and MVT increases as zoom level goes up because the scope of each tile becomes larger resulting in

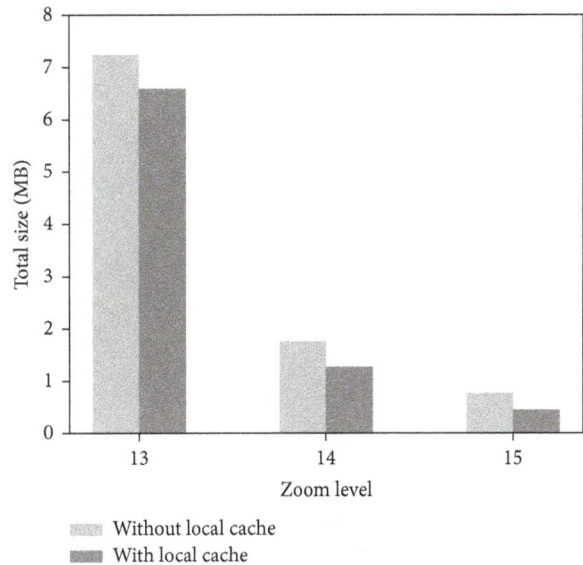

Figure 14: The comparison on transferred data size with and without utilizing the local cache.

more objects being included. At lower zoom levels, NVT and MVT have similar average tile size. However, at upper zoom levels, tile size of NVT becomes much higher than that of MVT due to the redundancy discussed in Section 3.1.

From Figure 13, the standard deviation and average size of the MBVT are both the minimum among all these types. This can be ascribed to the following: (1) this format is based on protocol buffers, and its binary serialization shows good effect on reducing size; (2) it also clips objects around the tile boundary; and (3) the objects are simplified and the coordinates are rounded. Likewise, the two metrics of image tiles are also stable and keep in very low level because they have fixed resolution and do not have a direct linear correlation with how much geospatial information they contain. The standard deviation of NVT is very high at upper zoom levels, indicating that the size of NVT distributes over a wide range due to the fact

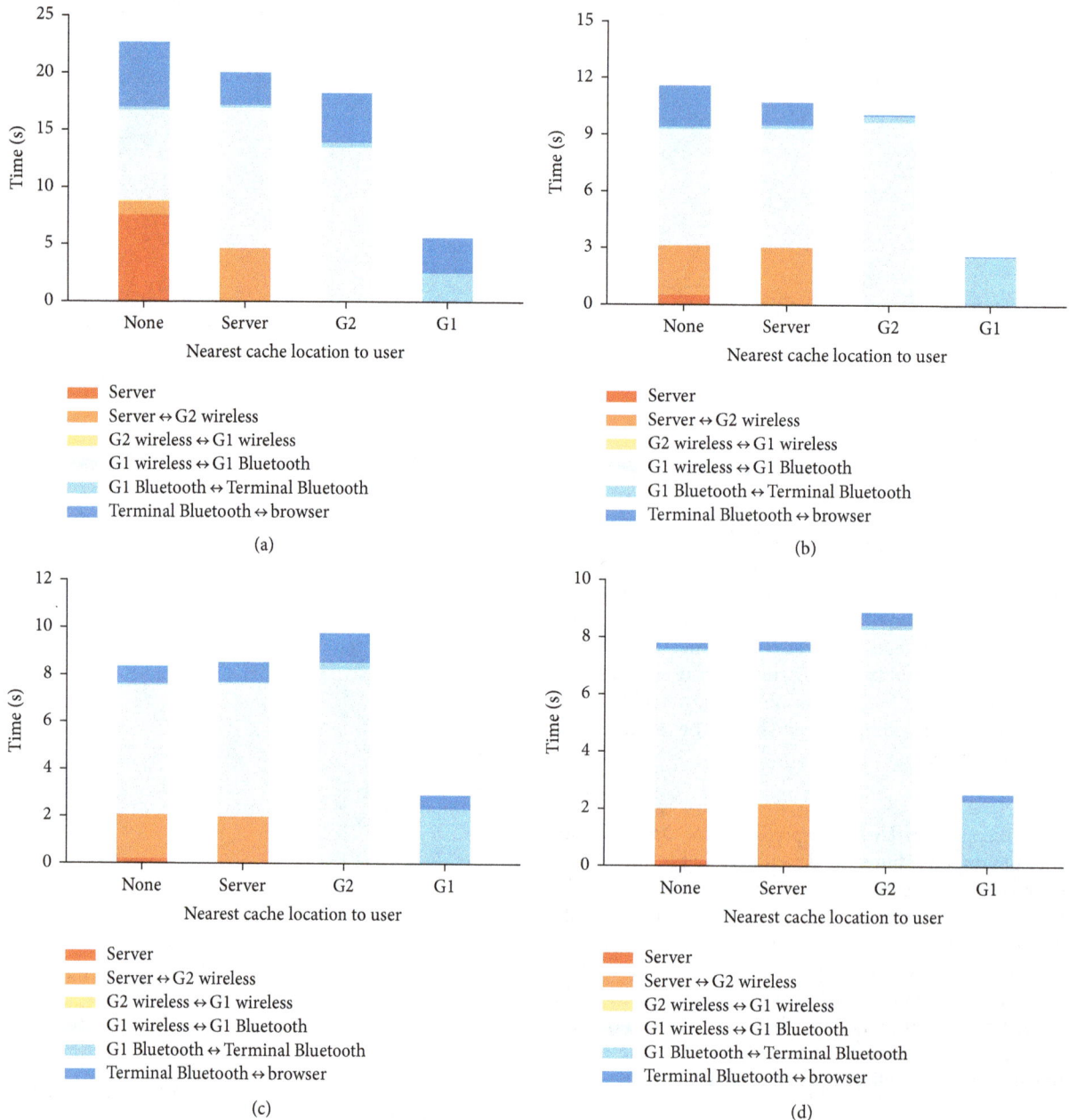

Figure 15: Time cost by each phase of transmission procedure of normal tiles. Zoom level: (a) 13; (b) 14; (c) 15; (d) 16.

that some tiles incorporate large objects. After compression, the size differences among tiles get diminished. So when GeoJSON and compression are used together, all tiles returned to users are guaranteed to be relatively small. Although some of NVT tiles of upper layers may be oversized, such zoom levels are probably too macroscopic for daily use of pedestrians or drivers. According to statistical results, for commonly used zoom levels, the majority of NVT does not contain an exceeding amount of data and is acceptable for GeoTile.

5.2.2. Effect of Client-Side Tile Amalgamation.
During this experiment, we use normal vector tiles and enable the local cache at the terminal. After loading a complete initial map in the browser, we make a zoom out operation to an upper zoom-level view. As presented in Section 3.3, GeoTile will

check the local cache and generate as many usable tiles as possible and request only the missing ones. Theoretically, if the map screen includes 20 tiles, amalgamation can save up to 25% data usage. From Figure 14, we can see that in both experiments when zooming from level 16 to 15 and when zooming from level 15 to 14, 4 tiles are generated from the local cache and about 20% of transferred data are saved. In the experiment, when zooming from level 14 to level 13, the number of tiles grows up in the level 13 view, resulting in more tiles to be requested. Additionally, because of the size limit of the local cache, many level-13 tiles cannot find the corresponding level-14 tiles in the cache for amalgamation. These two factors cause the saving proportion to become small. If there are enough tiles in the cache, more tiles could be produced locally.

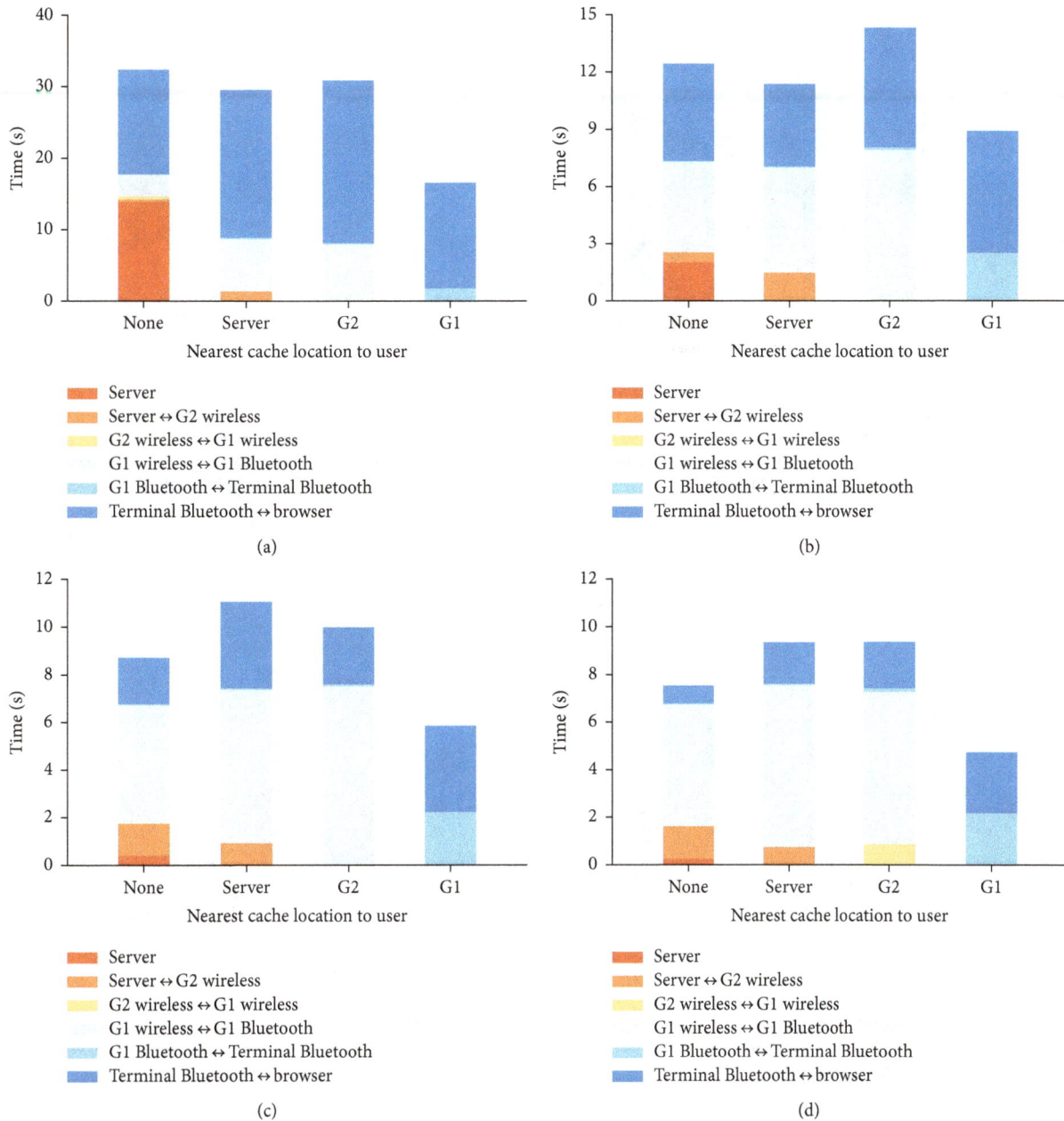

FIGURE 16: Time cost by each phase of transmission procedure of compressed tiles. Zoom level: (a) 13; (b) 14; (c) 15; (d) 16.

5.3. Performance from Prototype's Perspective: Initial Map. In this part, we will investigate the detail of the mobile prototype to find out how much time each phase costs. This will help us realize the advantages and deficiencies of our design or implementation for future improvements.

Figures 15 and 16 illustrate the total time cost for a tile from request to finally load in the browser. The total time represented by a bar in the figures is partitioned into various colors to represent the time taken in different phases. Each phase contains the time cost in two directions, that is, both request and response, and reports the average value of all tiles in one experiment.

Figure 15 shows the result of normal tile groups. It can be observed that the server has to generate demanded tiles when there is no cache. However, this phase only costs trivial

time unless the tile size is large (e.g., level 13). When the cache is available on the server, no generation time is needed. But the conversion time between request and data at G2 takes longer, indicating that the requests and data are queued at this point. The time cost of transmission between G2 and G1 is nearly invisible. It is mainly because of the following reasons: First, there is no conversion operation at this phase. Second, the transmission speed of ad hoc port is fast enough and is not the bottleneck. Third, we use a RAMDisk for caching to accelerate file I/O performance. These reasons can be also applied to explain the short time between the Bluetooth port of G1 and the terminal.

Another thing to be addressed is the long time spent between the G1 wireless port and the Bluetooth port. As we

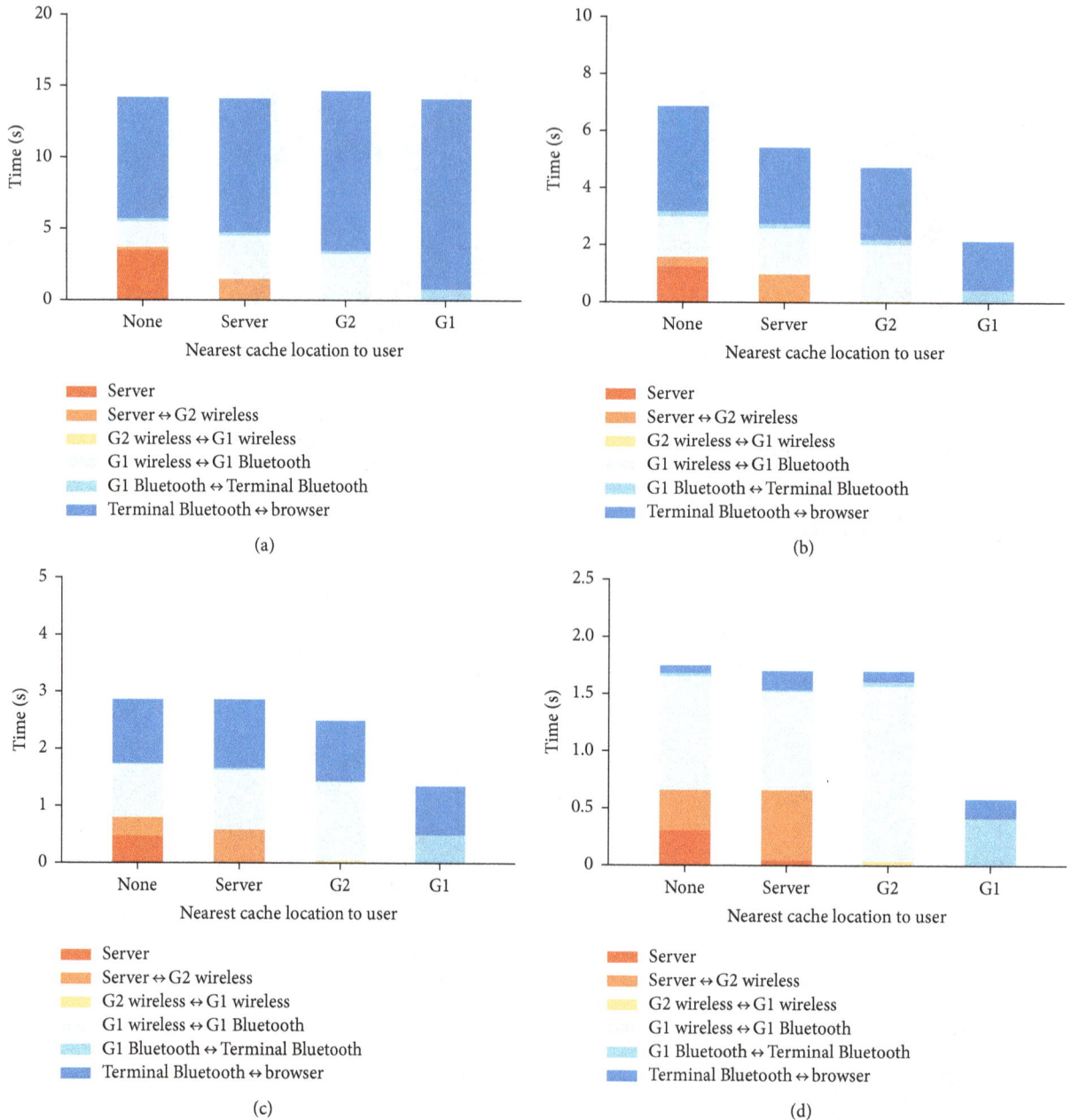

FIGURE 17: Time cost by each phase of transmission procedure of normal tiles by move operation. Zoom level: (a) 13; (b) 14; (c) 15; (d) 16.

have verified that the link capacity is not fully utilized and the speed of accessing local cache is fast, we can deduce the possible reasons as follows: (i) the support for parallel processing of Anyproxy (implemented in NodeJS) is limited, and we do not find any better alternatives yet; (ii) the computational ability of Galileo is restricted so that the processing speed is relatively slow. These will result in the phenomenon that tiles wait for processing or transfer at the converter. This problem is more obvious when cache hits at G2. In this situation, the request can be turned into data by G2 immediately and send back to G1, which will make the queuing phenomenon at G1 much worse. This will also explain why sometimes G2 group takes longer overall time than other groups. When cache hits at G1, data is returned from G1's

Bluetooth port quickly. Note that the time spent at the terminal shows considerable randomness across the groups. We consider that it may have something to do with the internal working mechanisms of the browser which we cannot control so far.

Figure 16 shows the result of compressed tile groups. Compared to the normal tile groups, the time cost at server side and terminal side rises up notably due to compression and decompression. This is also the main reason that leads to longer overall response time. On the other hand, because the size of tiles is reduced, the time needed by transferring is shorter and queuing problem is mitigated to some extent.

5.4. *Performance from Prototype's Perspective: Move Operation.* In this part, we conduct experiments on move and drag

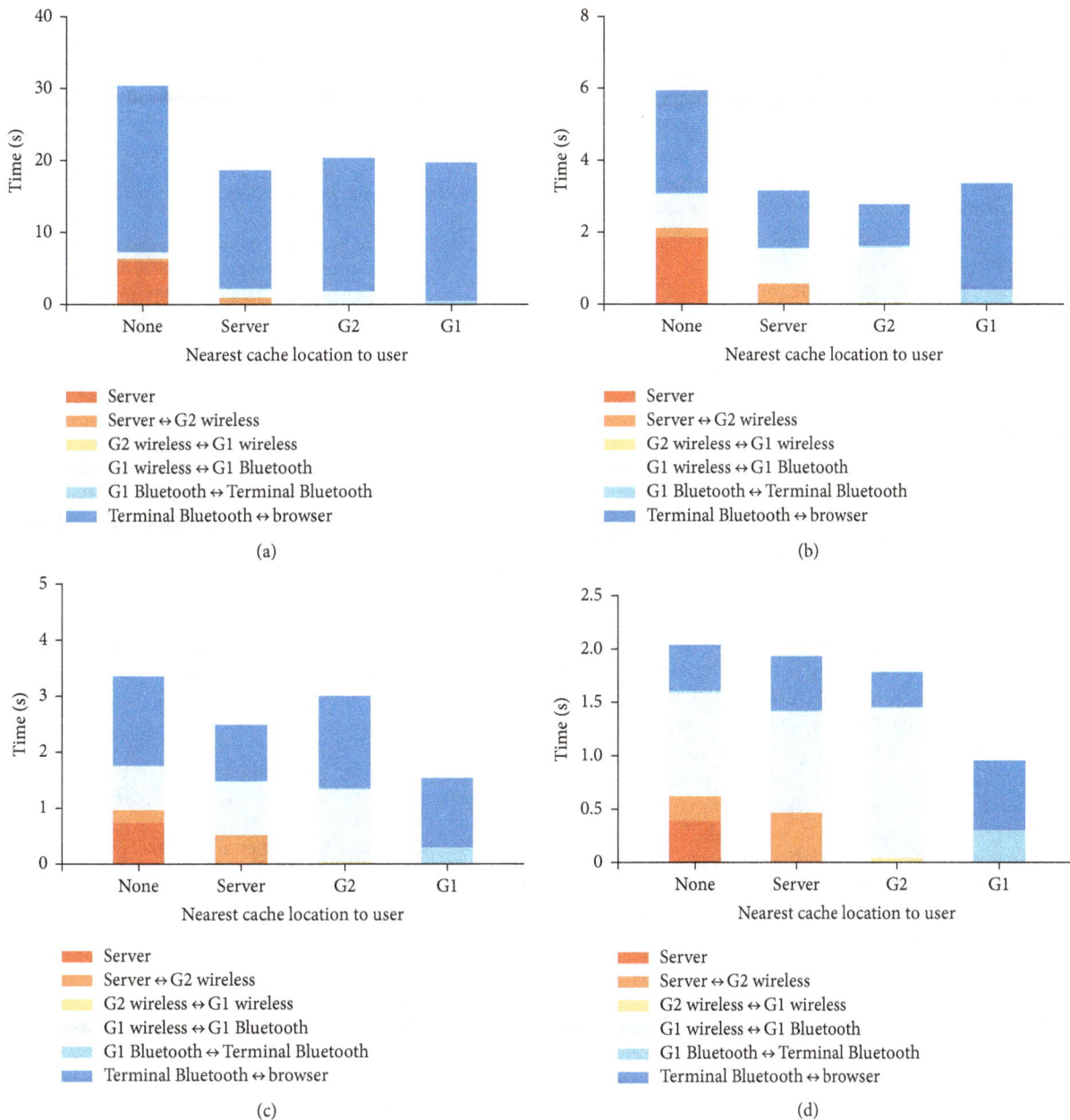

(a)

(b)

(c)

(d)

FIGURE 18: Time cost by each phase of transmission procedure of compressed tiles by move operation. Zoom level: (a) 13; (b) 14; (c) 15; (d) 16.

operations. The action of loading the initial map is taken as the case in which many tiles are requested simultaneously, while the drag and move behavior represents the case where only a few supplementary tiles are requested. We intend to test how our prototype performs under the situation of light pressure.

In the experiment, we execute two move operations, and each time 4 new tiles are requested upon each movement. Figure 17 is the result of normal tile groups. Figure 18 is the result of compressed tile groups. It can be noticed that the queuing phenomenon at the server side and the G2 side is clearly mitigated, proving that too many requests at one time will overload the system's capability. Moreover, the average overall time of each tile goes down, indicating that GeoTile is more suitable for exchanging a small amount of data. Another

problem that needs explanation is that, in the compressed tile groups, it takes much longer time at the browser. It is mainly because the move operation will incur reprocessing and rerendering data on the whole screen, so the overall time is not less than that in initial map experiments.

5.5. *Evaluation Summary.* Based on the experiment results, our evaluation reveals the following findings. (i) Loading the initial map made of smaller tiles (i.e., tiles each of which covers a smaller area) is less time-consuming compared with larger tiles. The waiting time to load a same initial map becomes shorter as the location of the cache hit draws nearer to the user. However, the problem of queuing caused by the converter module undermines the advantage of caching to some extent. (ii) The compressed tiles can reduce the transmission time and queuing time, but the extra compression and decompression time are required. They are suitable for those users who would like to get tiles more quickly and use a device with good computational abilities. They will also benefit scenarios with poor communication quality or intermittent links. (iii) A small number of tiles provoked by user's move or drag operation is well supported by GeoTile. Users can incrementally obtain vector tiles on demand efficiently. (iv) The tile amalgamation approach can save around 20% of user's traffic given the tiles of the initial map of lower zoom level in the cache. If there is plenty of tile resource in the cache, users can generate more new tiles without requesting.

6. Conclusion

In this paper, we present GeoTile, a novel system to share and exchange vector geospatial data in the distributed and collaborative way. GeoTile adopts the GeoJSON format with optional compression to partition the large-scale geospatial data into small units, that is, vector tiles, to enable efficient data sharing. In addition, GeoTile also explores the possibility to use peer cache and amalgamation technique to reuse the data cached locally or nearby in order to reduce both the response time and network data usage.

We implement the GeoTile system, including the tile service component and mobile prototype component. The experiments show the following findings: (i) Caching at locations closer to the user will shorten the overall waiting period. (ii) The tile amalgamation based on the local cache can save time for some requests. (iii) Users can get tiles through GeoTile in an incremental way effectively. (iv) Using compressed tiles can relieve the problem of oversizing and overloading and is suitable for the networks with low quality.

In our future work, we plan to explore a more flexible and progressive way for client-side amalgamation so that if there is a part of needed tiles available, by complementing small missing tiles, the amalgamation can also be conducted. In addition, we also plan to introduce the disruption tolerant techniques to better support the mobile environment with intermittent connectivity and to make better use of the distributed cache resources.

Acknowledgments

The authors would like to express their gratitude to Dr. Jingbang Wu for his support on wireless prototype system and suggestions on experiments and to Yu Zhang for his support on road traffic information.

References

[1] *The Mobile Consumer-A Global Snapshot*, http://www.nielsen.com/content/dam/corporate/us/en/reports-downloads/2013%20Reports/Mobile-Consumer-Report-2013.pdf.

[2] M. D. Dunlop, M. Roper, M. Elliot, R. McCartan, and B. McGregor, "Using smartphones in cities to crowdsource dangerous road sections and give effective in-car warnings," in *Proceedings of SEACHI 2016 on Smart Cities for Better Living with HCI and UX*, pp. 14–18, San Jose, CA, USA, November 2016.

[3] C. Heipke, "Crowdsourcing geospatial data," *ISPRS Journal of Photogrammetry and Remote Sensing*, vol. 65, no. 6, pp. 550–557, 2010.

[4] M. Goetz and A. Zipf, "The evolution of geo-crowdsourcing: bringing volunteered geographic information to the third dimension," in *Crowdsourcing Geographic Knowledge: Volunteered Geographic Information (VGI) in Theory and Practice*, D. Sui, S. Elwood, and M. Goodchild, Eds., pp. 79–95, Springer, Berlin, Germany, 2014.

[5] *2016 Global Mobile Consumer Survey: US Edition*, https://www2.deloitte.com/content/dam/Deloitte/us/Documents/technology-media-telecommunications/us-global-mobile-consumer-survey-2016-executive-summary.pdf.

[6] *Comparing Google Maps and HERE Maps Offline*, https://360.here.com/2015/11/17/comparing-google-maps-and-here-maps-offline/.

[7] *Osmand Offline Map Data Package Indexes List*, http://osmand.net/srtm-countries/.

[8] *Google Maps Help-Download Areas and Navigate Offline*, https://support.google.com/maps/answer/6291838.

[9] TIME, *How to Use Google Maps Offline*, TIME, New York, NY, USA, 2016, http://time.com/4437599/how-to-use-google-maps-offline/.

[10] O. Cats, Q. Wang, and Y. Zhao, "Identification and classification of public transport activity centres in Stockholm using passenger flows data," *Journal of Transport Geography*, vol. 48, pp. 10–22, 2015.

[11] Y. Wu, S. Yao, Y. Yang et al., "Challenges of mobile social device caching," *IEEE Access*, vol. 4, pp. 8938–8947, 2017.

[12] A. Clouston and M. P. Peterson, "Tile-based mapping with opacity," in *Developments in the Theory and Practice of Cybercartography Applications and Indigenous Mapping*, D. Taylor and T. P. Lauriault, Eds., pp. 79–95, Elsevier, Amsterdam, Netherlands, 2014.

[13] R. Abdalla, "Mobile GIS and location-based services (LBS)," in *Introduction to Geospatial Information and Communication Technology (GeoICT)*, pp. 83–103, Springer, Berlin, Germany, 2016.

[14] W. Alsabhan and S. Love, "Platforms and viability of mobile GIS in real-time hydrological models: a review and proposed model," *Journal of Systems and Information Technology*, vol. 13, no. 4, pp. 425–444, 2011.

[15] J. Gaffuri, "Toward web mapping with vector data," in *Proceedings of International Conference on Geographic Information Science, GIScience 2012*, pp. 87–101, Columbus, OH, USA, September 2012.

[16] A. Poplin, "How user-friendly are online interactive maps? Survey based on experiments with heterogeneous users," *Cartography and Geographic Information Science*, vol. 42, no. 4, pp. 358–376, 2015.

[17] S. Torbert, *Applied Computer Science*, CRC Press, Boca Raton, FL, USA, 2nd edition, 2016.

[18] V. Antoniou, J. Morley, and M. Haklay, "Tiled vectors: a method for vector transmission over the web," in *Proceedings of the 9th International Symposium on Web and Wireless Geographical Information Systems*, pp. 56–71, Maynooth, Ireland, December 2009.

[19] A. Dufilie and G. Grinstein, "Feathered tiles with uniform payload size for progressive transmission of vector data," in

Proceedings of the 13th International Symposium on Web and Wireless Geographical Information Systems, pp. 19–35, Seoul, South Korea, May 2014.

[20] *Google Map—Tile and Coordinate Specification*, https://developers.google.com/maps/documentation/javascript/maptypes.

[21] H. Samet, "The quadtree and related hierarchical data structures," *ACM Computing Surveys*, vol. 16, no. 2, pp. 187–260, 1984.

[22] T. Markas and J. Reif, "Quad tree structures for image compression applications," *Information Processing and Management*, vol. 28, no. 6, pp. 707–721, 1992.

[23] E. El-Qawasmeh, "A quadtree-based representation technique for indexing and retrieval of image databases," *Journal of Visual Communication and Image Representation*, vol. 14, no. 3, pp. 340–357, 2003.

[24] H. Noborio, T. Naniwa, and S. Arimoto, "A quadtree-based path-planning algorithm for a mobile robot," *Journal of Robotic Systems*, vol. 7, no. 4, pp. 555–574, 1990.

[25] M. Xu, D. Wang, and J. Li, "DESTPRE: a data-driven approach to destination prediction for taxi rides," in *Proceedings of the 2016 ACM International Joint Conference on Pervasive and Ubiquitous Computing, UbiComp'16*, pp. 729–739, Heidelberg, Germany, September 2016.

[26] X. Ju and K. G. Shin, "Location privacy protection for smartphone users using quadtree entropy maps," *Journal of Information Privacy and Security*, vol. 11, no. 2, pp. 62–79, 2015.

[27] C. Portele, *Encoding of Geographic Information*, Springer, Berlin, Germany, 2011.

[28] F.-J. Behr, K. Holschuh, D. Wagner, and R. Zlotnikova, "Vector data formats in Internet based geoservices," in *Advances in Web-based GIS, Mapping Services and Applications*, pp. 349–367, CRC Press, Florida, USA, 2011.

[29] *ESRI Shapefile Technical Description*, https://www.esri.com/library/whitepapers/pdfs/shapefile.pdf.

[30] *OSM XML Format*, http://wiki.openstreetmap.org/wiki/OSM XML.

[31] R. Lake, "The application of geography markup language (GML) to the geological sciences," *Computers and Geosciences*, vol. 31, no. 9, pp. 1081–1094, 2005.

[32] *KML Tutorial*, https://developers.google.com/kml/documentation/kml_tut.

[33] S. Steiniger and A. J. Hunter, "Data structure: spatial data on the web," in *International Encyclopedia of Geography: People, the Earth, Environment and Technology*, Wiley, Hoboken, NJ, USA, 2016.

[34] *GeoJSON*, http://geojson.org/geojson-spec.html.

[35] *TopoJSON*, https://github.com/topojson/topojson-specification.

[36] *Protocol Buffers Overview*, https://developers.google.com/protocol-buffers/docs/overview.

[37] A. Sumaray and S. K. Makki, "A comparison of data serialization formats for optimal efficiency on a mobile platform," in *Proceedings of the 6th International Conference on Ubiquitous Information Management and Communication*, p. 48, Kuala Lumpur, Malaysia, 2012.

[38] *Compressed KMZ Files*, https://developers.google.com/kml/documentation/kmzarchives.

[39] W. Li, M. Song, B. Zhou, K. Cao, and S. Gao, "Performance improvement techniques for geospatial web services in a cyberinfrastructure environment—a case study with a disaster management portal," *Computers, Environment and Urban Systems*, vol. 54, pp. 314–325, 2015.

[40] H. Shao and W. Li, "A comprehensive optimization strategy for real-time spatial feature sharing and visual analytics in cyberinfrastructure," *International Journal of Digital Earth*, pp. 1–20, 2018.

[41] S. Sikdar, K. Teymourian, and C. Jermaine, "An experimental comparison of complex object implementations for big data systems," in *Proceedings of the 2017 Symposium on Cloud Computing*, pp. 432–444, Santa Clara, California, September 2017.

[42] Y. Zhao, Z. Cheng, H. Dong, and J. Fang, "One efficient compression method used on WebGIS," in *Proceedings of the 19th International Conference on Geoinformatics*, pp. 1–5, Shanghai, China, June 2011.

[43] D. Salomon and G. Motta, *Handbook of Data Compression*, Springer, Berlin, Germany, 5th edition, 2010.

[44] Q. Wei, J. Guan, S. Zhou, and X. Wang, "A new and effective approach to GML documents compression," *Computer Journal*, vol. 57, no. 11, pp. 1723–1740, 2014.

[45] *Brotli Compressed Data Format*, https://tools.ietf.org/html/rfc7932.

[46] B. P. Buttenfield, "Transmitting vector geospatial data across the Internet," in *Proceedings of the International Conference on Geographic Information Science*, pp. 51–64, Boulder, CO, USA, September 2002.

[47] L. Zhang, L. Zhang, Y. Ren, and Z. Guo, "Transmission and visualization of large geographical maps," *ISPRS Journal of Photogrammetry and Remote Sensing*, vol. 66, no. 1, pp. 73–80, 2011.

[48] B. Yang, R. Purves, and R. Weibel, "Efficient transmission of vector data over the Internet," *International Journal of Geographical Information Science*, vol. 21, no. 2, pp. 215–237, 2007.

[49] R. Miao, J. Song, and M. Feng, "A feature selection approach towards progressive vector transmission over the Internet," *Computers and Geosciences*, vol. 106, pp. 150–163, 2017.

[50] R. E. Roth, "Interactive maps: what we know and what we need to know," *Journal of Spatial Information Science*, no. 6, pp. 59–115, 2013.

[51] M. Guo, Y. Huang, Q. Guan, Z. Xie, and L. Wu, "An efficient data organization and scheduling strategy for accelerating large vector data rendering," *Transactions in GIS*, vol. 21, no. 6, pp. 1217–1236, 2017.

[52] W. Shi and C. Cheung, "Performance evaluation of line simplification algorithms for vector generalization," *Cartographic Journal*, vol. 43, no. 1, pp. 27–44, 2006.

[53] *OpenLayers 3 Vector Rendering with Topology Simplification*, https://boundlessgeo.com/2014/03/openlayers-vector-rendering/.

[54] J. J. Arsanjani, P. Mooney, M. Helbich, and A. Zipf, "An exploration of future patterns of the contributions to OpenStreetMap and development of a contribution index," *Transactions in GIS*, vol. 19, no. 6, pp. 896–914, 2015.

[55] J. J. Arsanjani, M. Helbich, M. Bakillah, and L. Loos, "The emergence and evolution of OpenStreetMap: a cellular automata approach," *International Journal of Digital Earth*, vol. 8, no. 1, pp. 76–90, 2015.

[56] S. Maguire and M. Tomko, "Ripe for the picking? Dataset maturity assessment based on temporal dynamics of feature definitions," *International Journal of Geographical Information Science*, vol. 31, no. 7, pp. 1334–1358, 2017.

[57] D. Zielstra, H. H. Hochmair, P. Neis, and F. Tonini, "Areal delineation of home regions from contribution and editing patterns in OpenStreetMap," *ISPRS International Journal of Geo-Information*, vol. 3, no. 4, pp. 1211–1233, 2014.

[58] J. Itoi, M. Sasabe, J. Kawahara, and S. Kasahara, "An offline mobile application for automatic evacuation guiding in outdoor environments," *Scientific Phone Apps and Mobile Devices*, vol. 3, no. 1, 2017.

[59] D. Bahrdt, "OSMFIND: fast textual search on OSM data—on smartphones and servers," in *Proceedings of the 2nd ACM SIGSPATIAL International Workshop on Mobile Geographic Information Systems*, pp. 35–42, Orlando, FL, USA, November 2013.

[60] S. Glass, I. Mahgoub, and M. Rathod, "Leveraging MANET based cooperative cache discovery techniques in VANETs: a survey and analysis," *IEEE Transactions on Wireless Communications*, vol. 19, no. 4, pp. 2640–2661, 2017.

[61] W. Gao, G. Cao, A. Iyengar, and M. Srivatsa, "Cooperative caching for efficient data access in disruption tolerant networks," *IEEE Transactions on Mobile Computing*, vol. 13, no. 3, pp. 611–625, 2014.

[62] D. Liu, B. Chen, C. Yang, and A. F. Molisch, "Caching at the wireless edge: design aspects, challenges, and future directions," *IEEE Communications Magazine*, vol. 54, no. 9, pp. 22–28, 2016.

[63] J. Rao, H. Feng, C. Yang, Z. Chen, and B. Xia, "Optimal caching placement for D2D assisted wireless caching networks," in *Proceedings of the 2016 IEEE International Conference on Communications (ICC)*, pp. 1–6, Jiangsu, China, May 2016.

[64] G. Zheng, H. A. Suraweera, and I. Krikidis, "Optimization of hybrid cache placement for collaborative relaying," *IEEE Communications Letters*, vol. 21, no. 2, pp. 442–445, 2017.

[65] S. Wang, X. Zhang, Y. Zhang, L. Wang, J. Yang, and W. Wang, "A survey on mobile edge networks: convergence of computing, caching and communications," *IEEE Access*, vol. 5, pp. 6757–6779, 2017.

[66] M. Akon, M. T. Islam, X. Shen, and A. Singh, "OUR: optimal update-based replacement policy for cache in wireless data access networks with optimal effective hits and bandwidth requirements," *Wireless Communications and Mobile Computing*, vol. 13, no. 15, pp. 1337–1352, 2013.

[67] S. Drakatos, N. Pissinou, K. Makki, and C. Douligeris, "A future location-aware replacement policy for the cache management at the mobile terminal," *Wireless Communications and Mobile Computing*, vol. 9, no. 5, pp. 607–629, 2009.

[68] *Slippy Map Tilenames*, http://wiki.openstreetmap.org/wiki/Slippy map tilenames.

[69] *Mapnik: An Open Source Toolkit for Developing Mapping Applications*, http://www.mapnik.org.

[70] G. Farkas, "Applicability of open-source web mapping libraries for building massive web GIS clients," *Journal of Geographical Systems*, vol. 19, no. 3, pp. 273–295, 2017.

[71] *Text Compression in R: brotli, gzip, xz and bz2*, https://cran.r-project.org/web/packages/brotli/vignettes/benchmarks.html.

[72] *Compression Benchmarks: brotli, gzip, xz, bz2*, https://www.opencpu.org/posts/brotli-benchmarks/.

[73] *Squash Compression Benchmark*, https://quixdb.github.io/squash-benchmark/.

[74] S. Steiniger and A. J. Hunter, "The 2012 free and open source GIS software map—a guide to facilitate research, development, and adoption," *Computers, Environment and Urban Systems*, vol. 39, pp. 136–150, 2013.

[75] *PostgreSQL: The World's Most Advanced Open Source Database*, https://www.postgresql.org/.

[76] *PostGIS–Spatial and Geographic objects for PostgreSQL*, http://postgis.net/.

[77] R. Friedman, A. Kogan, and Y. Krivolapov, "On power and throughput tradeoffs of WiFi and Bluetooth in smartphones," *IEEE Transactions on Mobile Computing*, vol. 12, no. 7, pp. 1363–1376, 2013.

[78] *Anyproxy*, https://github.com/alibaba/anyproxy.

[79] *Mapzen Vector Tile Service*, https://mapzen.com/projects/vector-tiles/.

[80] *Mapbox Vector Tile Specification*, https://www.mapbox.com/vector-tiles/specification/.

11

Flexible Wi-Fi Communication among Mobile Robots in Indoor Industrial Environments

Jetmir Haxhibeqiri ⓘ, Elnaz Alizadeh Jarchlo, Ingrid Moerman ⓘ, and Jeroen Hoebeke ⓘ

Imec, IDLab, Department of Information Technology, Ghent University, iGent Tower, Technologiepark-Zwijnaarde, B-9052 Ghent, Belgium

Correspondence should be addressed to Jetmir Haxhibeqiri; jetmir.haxhibeqiri@ugent.be

Academic Editor: Alessandro Bazzi

In order to speed up industrial processes and to improve logistics, mobile robots are getting important in industry. In this paper, we propose a flexible and configurable architecture for the mobile node that is able to operate in different network topology scenarios. The proposed solution is able to operate in presence of network infrastructure, in ad hoc mode only, or to use both possibilities. In case of mixed architecture, mesh capabilities will enable coverage problem detection and overcoming. The solution is based on real requirements from an automated guided vehicle producer. First, we evaluate the overhead introduced by our solution. Since the mobile robot communication relies in broadcast traffic, the broadcast scalability in mesh network is evaluated too. Finally, through experiments on a wireless testbed for a variety of scenarios, we analyze the impact of roaming, mobility and traffic separation, and demonstrate the advantage of our approach in handling coverage problems.

1. Introduction

A never-ending process in industry is the automation process. With increasing automation, industries try to improve efficiency, reduce energy consumption, increase benefits, and working conditions for workers. The increasing number of mobile nodes, such as automated guided vehicles (AGVs) [1], and other field devices increased the need for new ways of communication inside the factory floor, apart from wired communication. Hereby, the wireless technology is seen as an important enabling technology in such cases. By adopting wireless technology, industries enable the device and personnel mobility, reduce cable costs, and connect hard-to-reach areas. This trend of connecting everything to a network inside of factory floor is referred as Industry 4.0 [2].

Mobile robots, such as AGVs, are included in different tasks, from delivering raw materials to production lines, moving materials during the production process, moving finished goods, and up to trailer loading. In order to increase the flexibility of mobile robot usage, recent AGVs are incorporating wireless solution to communicate with each other as well as with the controller.

In industrial environments, deploying wireless communication solutions is challenging. Different radio propagation effects such as reflections, multipaths, shielding, and so on will result in network coverage problems, packet losses, and communication outage for mobile robots. Since factory environments are large areas, multiaccess point (AP) systems will be used to cover the whole area. However, as mobile robots may move around with fast speeds (up to 2 m/s [3–5]), communication paths will change frequently due to handovers from one AP to the other or due to link breakage in case of ad hoc networks. In addition to this, communication between robots and the controller requires low latencies in order to have real-time robot coordination.

Within this challenging environment, robust and reliable wireless communication solutions must be realized. Different problems arise when robots rely only on infrastructure network, like increased latency during handover time, uncovered areas, and so on. In this paper, we will discuss such potential problems. We based our solution on real requirements from an AGV producer. This paper will consider a wide range of scenarios that can be solved with a flexible and modular solution. We propose a mixed architecture for the mobile node that exploits the possibilities of multiple interface usage.

The main key contributions of the paper are

(1) the full implementation of the mobile node architecture,

(2) the evaluation of the architecture in a real testbed with real nodes,

(3) a solution based on real requirements from producers of industrial AGVs.

This paper is an extended version of the paper published in [6], which provided an initial version of the mixed architecture and preliminary results. In this paper, we provide a more in depth description of the different architectural elements and discusses in detail newly added system functionality for dealing with coverage problems. This is further complemented with novel measurement results.

The outline of the paper is as follows. Section 2 further details the potential problems that arise when solely relying on the presence of a wireless infrastructure network and motivates our decision of adding meshing capabilities. Section 3 discusses related work in this domain, whereas Section 4 presents our resulting node and network architecture. In Section 5, we give a low-level description of the functionalities enabling our communication system architecture for the mobile nodes. In Section 6, we discuss our solution to handle the coverage problems by using mesh capabilities with the proposed mixed architecture. In Section 7, we illustrate our solution overhead in packet latency and throughput. Additionally, we evaluate the broadcast scalability for a group of nodes in mesh network. Further, we illustrate the potential performance issues in infrastructure networks through real-life experiments in a wireless testbed and show how our combined solution can deliver improved performance and flexibility. The achieved latency values of the proposed mixed architecture are benchmarked with the values from infrastructure-only based solution and ad hoc-only based solution. Also, the values are benchmarked with the list of requirements from Section 2. We end Section 7 with results of experiments to assess the behavior of our solution for tackling coverage problems. Finally, Section 8 concludes the work and provides an outlook to future work.

2. Problem Statement

Since wireless communication systems are widely used and deployed even in industrial environments, they can be used also to provide connection for AGVs in the network. However, the challenges that arise when using only infrastructure network in industrial environments are the increased latency during handover time and uncovered areas due to shielding effects.

Different scenarios need to be taken into account. The provided mobile node system architecture needs to function in absence of the infrastructure network. This situation can happen in two cases: where there is no infrastructure network or where it is not allowed to be used due to the interference with other processes that use it.

In scenarios where there is infrastructure in place, the mobile node system architecture needs to handle the possibilities of connectivity problems. Connectivity problems arise due to uncovered areas from APs. The mobile node needs to ensure permanent connectivity to the network and continuous outage time needs to be under certain time duration.

For a network consisting of a multitude of access points (APs), fast movement speeds of AGVs (0–2 m/s [5]) will result in frequent handovers. Such handovers greatly increase the communication latency. For the particular real-life use case we consider here, frequent time-critical broadcast exchanges between mobile robots are required for their distributed coordination, in addition to less time-critical but reliable unicast traffic to and from controllers. More specifically, the latency of broadcast packets has a strict upper bound (20 ms) in order to reach targeted mobile robots in time. The upper bound latency is calculated based on the path accuracy of the mobile robot. In [3], for 100 mm path accuracy, they ask for an overall latency of 50 ms, including also the processing latency. For 20 ms communication latency, the path accuracy will be 40 mm for highest speed of 2 m/s, without taking into account the processing latency. Every handover involves a series of packet exchanges, which consumes valuable time. Hence, frequent handovers may have a detrimental impact on the required performance, as we will show in Section 7. Moreover, robots are not allowed to travel more than two meters without communication. Considering the maximal speed of robots, the maximal continuous outage time should be lower than one second. As many small to medium enterprises, where these communication systems will be deployed, are searching for solutions without high operational costs, the use of unlicensed spectrum has been put forward as another requirement. Finally, as requirements to the mobile robot system may change over time, for example, when scaling up the network, it should support dynamic adaptation of the communication behavior.

The above observations and performance requirements lead to a challenging set of functional requirements for our mobile robot system, which we have summarized in Table 1 together with KPIs. All the summarized requirements are based on requirements from an AGV manufacturer company. Based on the above requirements, it is clear that we need to target a design that is capable of connecting either to existing enterprise networks (RQ2) to create its own mesh network (RQ1) or to do both (RQ3).

These requirements have led to a modular and configurable communication system for mobile robots, consisting of two wireless interfaces that can operate either in ad hoc or infrastructure mode and offering the possibility to control in a fine-grained way how traffic is being handled. As such, the system can support a variety of different networking architectures, potentially combining both infrastructure communication and mesh communication and supporting the separation or duplication of different traffic streams according to configuration settings. The design of the system and the supported network architectures are discussed in more detail in Section 4, whereas the advantages of our

TABLE 1: Requirements and KPIs.

Requirements	Description	KPI
RQ1	Function in the absence of fixed wireless infrastructure (network of APs)	
RQ2	Exploit the presence of available fixed infrastructure	
RQ3	Deal with occasional/sudden coverage holes in wireless infrastructure	Relay node role
RQ4	Reliably deliver unicast traffic	98%
RQ5	Timely deliver frequent broadcast traffic	<20 ms
RQ6	Deal with mobility	0–2 m/s
RQ7	Limit continuous outage times	<1 sec
RQ8	Function in unlicensed spectrum	2.4 or 5 GHz
RQ9	Adapt to future needs	Modular architecture

architecture for our particular use case at hand are experimentally evaluated in Section 7 (Subsections 7.4–7.7).

Security issues are outside the scope of this paper. There are plenty of possibilities to tackle security issues, such as dual authentication scheme [7], randomized authentication schemes [8], shared key encryption, and so on. Of course, depending on the complexity of the security method chosen, it may come with a performance penalty.

3. Related Work

Until now, there were several studies on mixed wireless system networks where ad hoc communication is supported by infrastructure. These studies mostly focused on the capacity improvement when an infrastructure network is used next to mesh capabilities [9–11]. More recent ones also include the delay performance and delay-throughput trade-off for such networks [12, 13]. In [9], they prove theoretically that hybrid wireless networks have greater throughput capacity and smaller average packet delay than pure ad hoc networks. In [10], they studied the effect of network dimensions on the capacity of hybrid networks. Apart from throughput, communication delays are important too in industrial applications. In [12], authors propose a multichannel ad hoc network with infrastructure support that offers a lower average delay compared to ad hoc network or infrastructure only. In [13], authors propose an analytical framework to characterize the communication delay distribution of the network.

Recently, device-to-device communication is being discussed as part of the 5G research. Current research mainly focuses on peer discovery [14, 15], resource allocation [16, 17], and power control [18]. Even though some of the concepts of D2D communication in 5G networks can be applied to use cases currently considering network technologies such as IEEE 802.11, there is still work to be done with respect to dynamic switching between operational modes, IP-based routing in such mixed networks, and multihop relaying support. Apart from technical challenges, pricing is often another problem in case of device-to-device communication using cellular network support [19]. As we described in Section 2, many small to medium size enterprises are looking for communication solutions that do not imply operational costs, like payments for small cellular cell

installation and other network operator costs. As such, this paper considers IEEE 802.11-based infrastructure network with mesh capabilities of the mobile nodes by offering dynamic switching between ad hoc and infrastructure network. The network can operate in unlicensed spectrum, and cheap chipsets are available. Moreover, the whole mobile node architecture is based on open solutions (click modular router [20]) and works on top of COTS chipsets.

In a recently published patent [21], authors describe a solution to increase the number of mobile nodes served by an ad hoc network by introducing infrastructure support. As distributing the network control information through the ad hoc network is bandwidth costly, the authors propose to use the infrastructure network for signaling traffic. In our solution, we go one step further by offering the possibility to separate any traffic type between fixed and ad hoc network, for example, based on broadcast or unicast data traffic.

Apart from hybrid communication possibilities, communication systems of multiple mobile robots form an interesting research domain that is gaining importance in manufacturing in order to improve performance and increase automation. In [22–24], literature reviews regarding mobile robot systems, communication, and heterogeneous network are given together with open research issues and architecture used. In [22], authors highlight localization problems, coverage problems, robust communication needs, and environment hardships in manufacturing environments as important open research issues. In [23], a survey regarding the coordination in multirobot systems is presented, including the communication technologies. The authors highlight the importance of explicit communication, that is, direct message exchanges between robots, to ensure accuracy of the information. In our solution, we offer direct communication between robots through broadcast messages using the ad hoc network.

In [25], authors give a model for integrating three different areas, namely, wireless sensor networks (WSNs), mobile robotics, and teleoperation for use in different fields from medicine to military. Also, they give a survey of the literature regarding the research challenges, including routing and connectivity maintenance on ad hoc mobile robot networks. One of the solutions to maintain connectivity is to give robots a certain fixed role in order to route the traffic [25]. In our solution, we take a different approach, designing the mobile node to take up multiple roles at the same time (e.g., can be end node and relay node).

During recent years, mobile robot communication experienced an evolution in their application as well as the protocols being used. Many works put forward ad hoc or mesh communication as a promising solution for realizing interrobot communication. For instance, [26] illustrates how an infrastructure network can be extended with multihop relaying functionality. In [27, 28], authors propose a model of a cooperative robot's system, where relaying robots (follower robots) will assist in establishing connectivity between the operating center and the robot that does the task (tank robot). Researchers in [29] construct an ad hoc network for communication between robots by classifying each robot as either a search robot or relay robot. This way they increase the area of communication (coverage zone). However, the robot roles are strictly defined based on their position in the network topology. We also consider multihop communication capabilities as one of the key requirements for our communication solution, but we also consider direct ad hoc or mesh communication between all mobile robots. Moreover, we think that there is no need to classify robot roles into relaying ones and main robots (like in [25, 27–29]) as their role can change depending on their position and the network topology, requiring support for multiple roles at the same time. In our solution, one of the goals of using the multihop functionalities is to extend the coverage zone of the APs of the infrastructure network. So far, most research into interrobot communication has focused on pure ad hoc networking. For instance, in [30], a review for routing protocols that can be used in robot networks is given. They show that the AODV routing protocol can be used in scenarios where robots have speeds up to 6 km/h, which is similar to the robot speeds that we consider in this paper. An architecture for mobile nodes using multiple interfaces is presented in [31]. However, they implemented their solution only in a network simulator, whereas our solution has been implemented on real nodes and has been tested in a testbed. Moreover, our proposed solution is capable of combining both mesh and infrastructure communication in a variety of ways and offering flexibility to distribute traffic.

In ad hoc networks, link management and neighbor discovery mechanisms are crucial in the performance of the routing protocol. In [32], link break detection is done within the routing protocol by using hello messages. We use the same approach by sending beacons in the ad hoc network to announce the presence of a node and to detect link breaks in the absence of beacons. In [33], an analytical model for the neighboring mechanism of OLSR is given while in [34], a hybrid asynchronous algorithm for neighbor discovery that leads to 24% shorter time for discovery is presented.

In industrial settings, it is also important to be able to meet the performance and latency requirements as we have indicated in Section 2. In [35], a routing algorithm for mesh networks is presented for use in industrial applications. They use a QoS manager which, after a calibration phase, manages QoS flows based on the requests from stations on specific QoS flow requirements, packet data unit (PDU) size, and destination. The calibration phase makes the solution more difficult to be deployed in highly dynamic environments. Finally, [36] describes a solution for wireless mesh network infrastructure with extended mechanisms to foster QoS support for industrial applications. Like in [35], they propose a mesh network with a central admission unit to decide for the communication flows requested by different applications. In [36] with their solution they could offer streams with RTT less than 100 ms. The mechanisms are only applied to a mesh case, whereas we believe that a mixed solution such as the one we propose can offer additional benefits, especially when further extended with more advanced QoS mechanisms.

4. Communication System and Network Architecture

In the following subsections, we will describe the designed mobile robot communication system and potential network architectures that can be realized.

4.1. Mobile Robot Communication System Architecture. In Section 2, we motivated our decision to design a communication solution that makes use of two wireless communication interfaces. Each of these interfaces can either operate in ad hoc mode for establishing mesh communication or in infrastructure mode in order to connect to an existing enterprise network. From an application point of view, it should not matter which interface is being used for transmitting packets or how this interface has been configured. Similarly, external components, such as a controller, that want to communicate with a particular mobile robot, should also not be bothered with underlying communication details.

Figure 1 gives an overview of the high-level architecture we designed for the communication system of the mobile robot. We provide an abstraction layer that transparently manages and dynamically configures the underlying network interfaces. Application layer will communicate with single virtual interface using single IP address. This one is regardless of which physical interface will be used for actual communication. Additional logic for routing and traffic management is designed that is able to consider the specifics of the underlying physical interfaces. A basic implementation of the dynamic mobile ad hoc networks (MANETs) on demand (DYMO) routing protocol [37] is being used for unicast mesh routing. Regarding traffic management, the node design foresees a number of traffic classification components that can be dynamically configured. According to their configuration, unicast and broadcast traffic streams can be separated and directed to different interfaces or traffic can be even duplicated for redundancy purpose. Also, a neighbor discovery mechanism in mesh network based on beacon generation is designed.

To tackle coverage problems, a mechanism is needed to detect when a node goes outside of coverage zone of access points. Based on the coverage problem detection mechanism, the node will start looking for other communication possibilities within the network through mesh links it can establish with its neighbors.

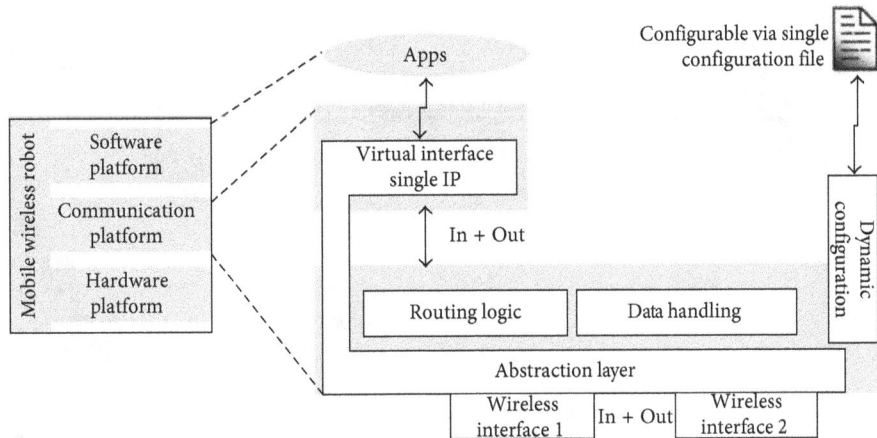

FIGURE 1: High-level mobile robot communication architecture.

4.2. Network Architecture. As motivated in Section 2, the system should be able to function in different use case scenarios. In this subsection, we discuss a number of potential network architectures that are derived from the real use cases in industrial environments which are shown in Figure 2. The proposed mobile robot system solution is able to support each of the network architectures by simply reconfiguring a set of parameters.

Many small to medium warehouses, where mobile robot systems are needed, are not willing to make additional investment in network infrastructure. In other factory environments where infrastructure already exists, it might be in use by other production processes. Figure 2(a) shows the first architecture that can be realized in such cases (RQ1). Both wireless interfaces can then operate in ad hoc mode, forming a mesh network with parallel links that operate on different frequencies increasing the capacity. In case wireless infrastructure is present and can be used, a mixed network can be established as shown in Figure 2(b) (RQ2). One of the interfaces is used to connect to the existing network, whereas the other interface is used to form a mesh network. Depending on additional configuration settings, it can be further decided how traffic is distributed over the different interfaces by exploiting different techniques for traffic separation. This is shown in Figure 2(c) for the case of a multimesh configuration, where one of the interfaces is used for unicast traffic and the other interface is used for broadcast traffic. Finally, Figure 2(d) shows how the communication can be configured in order to tackle coverage problems by making use of mesh functionality in the specific area that experiences these coverage problems (RQ3).

5. Low-Level Description

The modular Click Router framework [20, 38] was used for the communication system. Click Router is software architecture for building flexible and configurable routers. A router in Click is created by a chain of the packet processing modules. Each individual module implements simple router functions that chained with all other needed modules providing the router's functionalities. Different basic

functions are already implemented by Click elements such as packet queuing, scheduling, interfacing with network interfaces, and so on.

As one of the requirements (RQ8) was that the system should be adaptable to future needs, Click Router is a good choice. It has a modular structure, and it is easy to support future extensions or the replacement of existing elements with more advanced element versions. We have extended the Click framework with additional features for event handling, flexible configuration, and dynamic interface management in order to fulfill the requirements for our mobile robot communication system. Elements can subscribe for events of interest and will take specific actions when the event is announced by any other element. This facilitates information sharing between Click elements that are not neighbors in the Click chain.

Further, all configurable parameters are specified in a text file. Such parameters include network interface roles, interface classifiers, timing parameters for the neighbor discovery, link breaking and mesh routing, IP address for virtual interface, and so on. This way the system can be configured dynamically, enabling administrators to define the system behavior in a single configuration file without changing the Click chain. Finally, the dynamic interface management makes the Click chain independent of the number of used interfaces.

Figure 3 gives an overview of the different functionalities and building blocks that implement them, as well their interactions. In the following subsections, we will give a short description of the main functionalities.

5.1. Convergence Layer and Interface Management. Each mobile robot will have one unique IP address, regardless of the fact that there are two network interfaces. A virtual interface with single IP on top of multiple physical interfaces is created. It enables the application layer to communicate through only one interface irrespective of which underlying physical interface is used for communication. The actual selection of which physical interface should be used can be done based on the interface role and/or traffic type (Section 5.2). The interface role determines whether the interface is

(a)

(b)

(c)

(d)

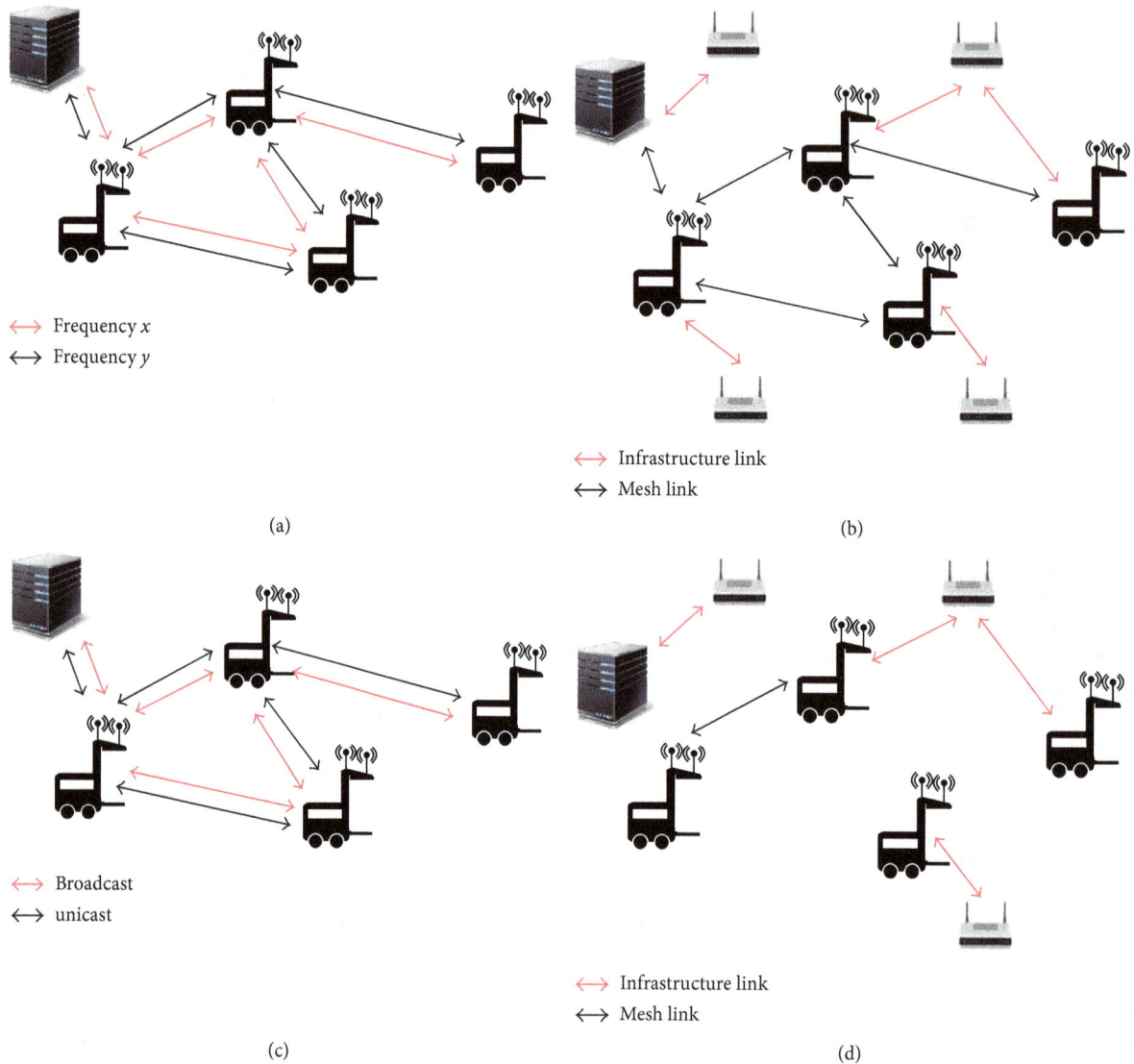

FIGURE 2: Potential network architectures that can be realized by reconfiguring the designed mobile robot communication system. (a) Multiinterface mesh network in the absence of fixed infrastructure or in case fixed infrastructure cannot be used. (b) Mixed mesh (single interface) and infrastructure network. (c) Multiinterface mesh with traffic separation. (d) Infrastructure network with mesh support for handling coverage problems.

used in ad hoc mode or connects to an access point. The traffic type specifies for which traffic (unicast, broadcast, or both) the interface can be used. Once a decision is made on which interface will be used for sending the packet, the packet is tagged with the MAC address of the physical interface. The convergence inspects the tag and takes care of the actual transmission.

Interfaces can be added or removed dynamically, and their role can be changed dynamically. An interface manager monitors the changes on interfaces such as connectivity, whether the interface is up and running, and so on.

5.2. Traffic Classifiers. Different levels of traffic classifications are foreseen in order to properly route the packets through the Click chain. Firstly, packets are classified based on their type: IP or ARP packets. ARP packets will be handled by Click

functions for reply/request of ARP packets. IP packets are further classified into routing control packets (i.e., DYMO in our case) and other data packets. Routing control packets are handed over to the Click element that implements routing protocol logic. The other packets go through a next level of classification, namely, into broadcast and unicast traffic. Since there is a beaconing system for neighbor discovery, each neighbor beacon packets are filtered out from the broadcast packets and given to neighbor discovery functions. On the contrary, the remaining broadcast traffic and unicast traffic will be delivered to the virtual interface, routed based on the main routing table (unicast) or tagged according to the interface configuration for traffic separation.

5.3. Neighbor Discovery and Link Break Detection. Neighbor detection in the mesh network is done based on keep-alive

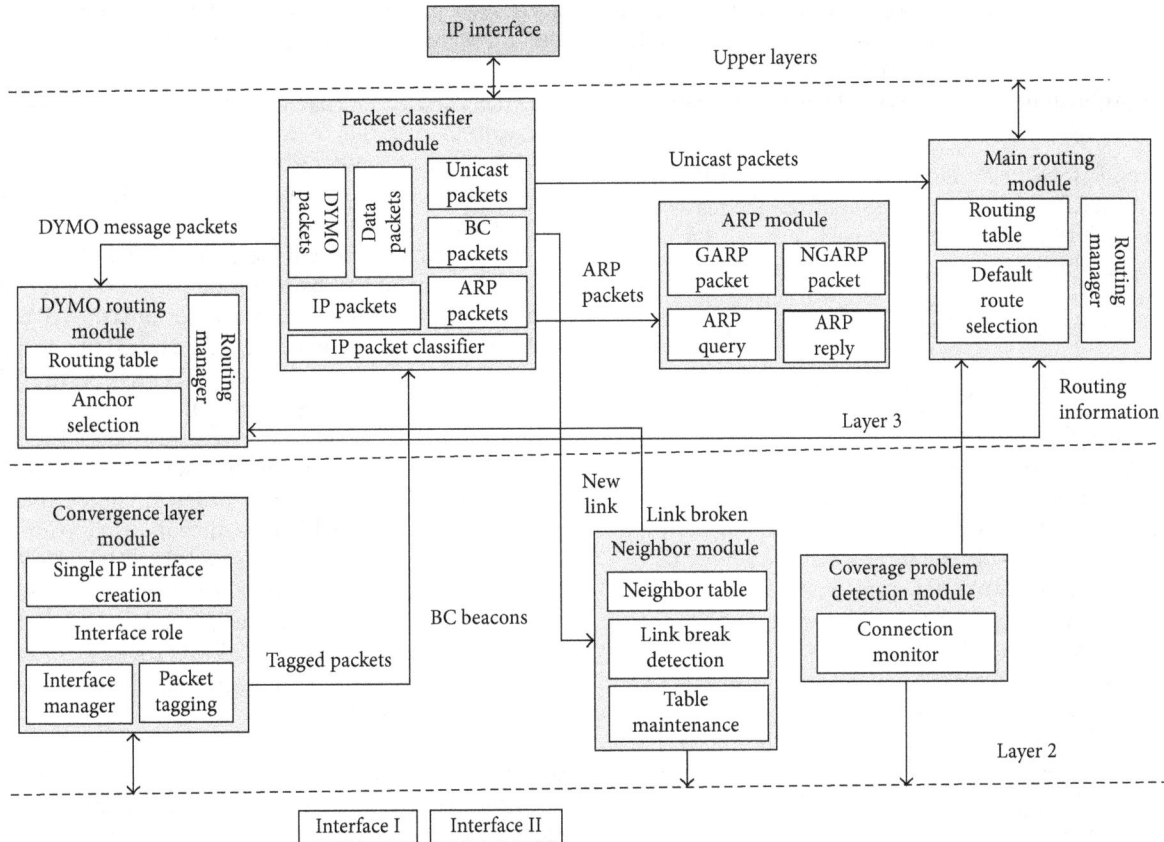

FIGURE 3: Detailed node architecture of the mobile robot communication system, implemented in Click Router.

beacons. Each mobile robot will broadcast beacons in the mesh network every Nms seconds. All other mobile nodes will update their neighbor's table based on the reception of these beacons. For each entry in the neighbor's table, a timer is kept. As soon as the timeout is passed and no more beacons are heard from that peer, the entry will be removed from the table. Parameters such as keep-alive beacons interval and neighbor's table entry timeout are dynamically configurable.

Given the fact that in our use case, the broadcast traffic is used frequently to communicate between mobile robots, real broadcast traffic can be used in addition to beacons for neighbor discovery. Hereby, we enable the suppression of the real beaconing traffic in order to reduce the network load.

The neighbor discovery module will maintain the neighbor table and will use the eventing mechanism to inform routing module about neighbor changes.

5.4. Routing.
Each mobile robot can make use of two interfaces: for example, one to connect to an access point and another one to establish a mesh network. In this case, routing is done in two different networks implying the need to have a routing table that incorporates routes from both networks.

For routing in the mesh network, the DYMO routing protocol is used. A DYMO routing table for unicast routing

in the mesh network is maintained. DYMO message packets (Route Request (RREQ), Route Reply (RREP), and Route Error (RERR)) are generated to find and maintain needed routes.

The main routing table will contain the default route towards the access point as well as other routes in the mesh network. Hereby, the main routing table is subordered to the DYMO routing table, meaning that it will be updated based on the DYMO routing table updates.

Contrary to unicast routing, broadcast routing does not require a routing table. The only thing needed for broadcast packets is the tag that defines over which interface the broadcast packet has to be transmitted. The assignment of this tag is again based on the interface configuration for traffic separation.

5.5. Coverage Problems Detection.
The first step to handle the coverage problems is to detect when they happen. To detect the disconnection from an access point, we use the wpa_supplicant [39] control interface. The connection status is probed periodically, and the specific actions are taken when the node is detected to have entered the uncovered zone. After the disconnection detection, the mobile robot will only use the ad hoc interface for communication. Changing from two interfaces to one interface has impact on the address resolution protocol (ARP), which will be handled by issuing Gratuitous ARP (GARP) packets and Non-

Gratuitous ARP (GARP) packets. The AP monitoring interval parameter can be configured directly in the configuration file. For further understanding, our solution for coverage problem handling we refer the reader to Section 6.

5.6. Address Resolution Protocol (ARP) Module. ARP module handles the ARP requests and response from the network. It contains ARP responder table where it saves all the MAC addresses for which it can reply. At the same time, it maintains the ARP reply table by the replies it receives. During the anchor selection process, it issues GARP and NGARP packets.

6. Handling Coverage Problems

As we discussed in Section 4, one of the network topologies that can be supported by our solution is an infrastructure network with mesh support to overcome coverage problems. This section will discuss in detail how we exploit the mesh capabilities in order to tackle those coverage problems.

In order to minimize the problems of "dead zones" where robots will become unreachable, other mobile robots, which do have connectivity to an AP, are used as AP coverage extenders. Each node which has a direct link to an AP and helps other nodes to communicate with the rest of network will be further referred to as an anchor node. Each node that does not have a direct AP connection, but has a connection through an anchor, will be further referred to as anchored node. Each node that does not have a direct AP connection, but has a multihop connection to an anchor, will be referred to as a multihop anchored node. Assuming that the traffic separation configuration mandates unicast traffic to be sent over the infrastructure network, a mechanism that allows the unicast traffic to be transmitted through mesh links to the anchor node has to be incorporated.

The proposed solution considers the problem of link breakage and link breakage healing in mesh networks, as well as the detection of the coverage problems from APs. As soon as one mobile robot detects the disconnection from the infrastructure network, it starts looking for another mobile robot which is able to act as an anchor for the unicast traffic. For this, the anchor node will use its second network interface, whereas its first interface is connected to an AP. In addition, the anchor node should be able to inform the network that it has been selected as anchor for certain mobile nodes. Thereby, it should be able to reroute all unicast traffic destined for the anchored mobile nodes as well as to reply on incoming ARP requests on behalf of anchored nodes, as the other nodes have to use the MAC address of the anchor interface (i.e., interface of the anchor node that is connected to the infrastructure network) to reach the anchored node. Further, existing ARP entries for a newly anchored node must be updated based on which interface is being used by the mobile node.

Finally, also mechanisms to detect unidirectional links between anchor and anchored mobile nodes and vice versa are provided. The combined mechanisms enable multihop routing between anchored and anchor mobile nodes and make sure that the shortest route (in terms of hops) in the mesh network is always used for unicast traffic. The anchorage process is triggered by the node which is in the uncovered zone and maintained by it.

6.1. Anchor Selection Process. The two Click functional modules that are responsible for facilitating the anchor selection process are the Coverage Problem Detection Module and Neighbor Module.

The Coverage Problem Detection module uses the wpa-cli [39] library functions to probe periodically for the AP connectivity. When the connection towards the AP becomes inactive, a disconnection event will be raised that will trigger the anchor selection process in the DYMO routing module. Simultaneously, the same event will clear the default route towards the AP in the main routing table. The routing module will wait for the anchor selection process to be finished by DYMO routing in order to choose one of the paths towards its neighbor as default route. Conversely, when the node is already anchored the Coverage Problem Detection module will monitor whether the node re-enters a covered zone. In that case it will issue a connection event, which will trigger a process in DYMO routing to release the anchor and to inform the network that the node is directly connected through an AP.

The neighbor module will facilitate the welfare of the communication between the anchor and anchored node during the time of anchorage. Beacon mechanism is used for recognizing neighbors. In case neighbor beacons are not being received for a certain period, called the link breakage timeout, the node will remove the link from its neighbor list. If the link between the anchor and anchored node is broken, then the anchored node will initiate again the anchor selection process.

Two packets that are used to inform the network that a node is anchored or released are the GARP reply and the NGARP reply [40, 41]. Both of these packets are broadcasted by the anchor through the ARP module. GARP and NGARP replies are ARP reply packets that are sent without any request. Both the targeted MAC and IP addresses are set to broadcast addresses. In our case, the sender IP will be the IP of the anchored node while the sender MAC will be the MAC of the interface that provides connectivity to an AP. In case of the GARP, the ARP entry will be updated, while in case of the NGARP, the ARP module will clear the ARP entry match from the table. This way the ARP tables can be updated across the entire network for nodes that are being or no longer being anchored. If there is no entry in the ARP table for a specific destination and no reply is being received upon ARP requests, then the sender is enforced to buffer all the outgoing traffic.

For routing in this mixed network, we use the DYMO routing protocol with an extension for finding a default route through the mesh network in case the mobile node enters an uncovered zone. These mesh routes are stored in the DYMO routing table and trigger updates in the main routing table. As such, when in a covered zone, the main routing will contain a default route entry through an AP, whereas in an uncovered zone, the main routing table will contain a default route through the mesh network.

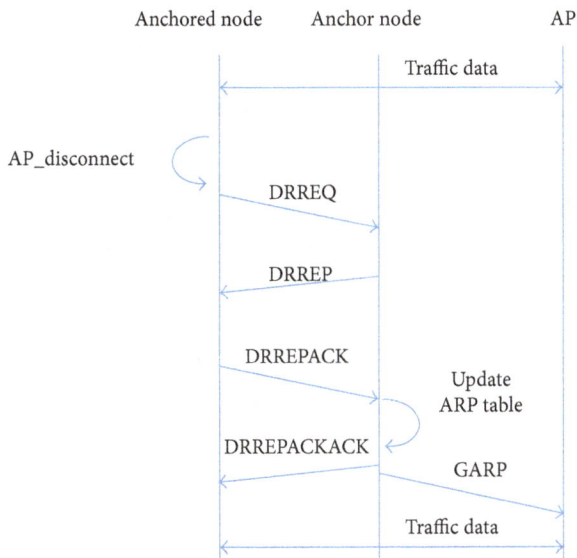

FIGURE 4: Time diagram of packets for anchor selection process.

The time flow of the packets between an anchor and an anchored node during the anchorage process is given in Figure 4. When the AP disconnection event is raised, the mobile node will broadcast a Default Route Request (DRREQ) packet for anchor discovery. We use a four-way handshake to select the anchor node, avoiding the possibility of selecting two anchors at the same time and creating confusion in the network. After issuing the DRREQ, other nodes that receive the DRREQ and that have a direct default route towards an AP in their main routing table will send back a Default Route Reply (DRREP) packet. The mobile node might receive multiple DRREPs from different nodes but will choose the first received DRREP (which is typically the one with the lowest number of hops) and ignore the others. In reply to the DRREP, the mobile node will send a unicast Default Route Reply Acknowledgement (DRREPACK) packet to the node that issued the DRREP. Upon reception of the DRREPACK, the anchor node will issue a GARP in the infrastructure network. This way the rest of network will be informed that the IP of anchored node is now reachable through the anchor, enabling the anchor to intercept all packets destined for the anchored node. At the same time, the anchor updates the ARP response table in the ARP module with the IP address of the anchored node to be able to respond to future ARP requests from the network for the anchored node IP. Finally, the anchor will inform the anchored node with the Default Route Reply Acknowledgement of Acknowledgement (DRREPACKACK) packet. After receiving the DRREPACKACK from the anchor node, bidirectional communication is possible between the anchored node and any other node in the network.

In order to ensure that the anchor is selected correctly and the network is informed on time, there are timers for each of the packets issued by the mobile nodes during the anchorage process. If any of the packets is lost, the timer expires and the mobile node has to restart the anchorage process.

6.2. Handling Unidirectional Link Problem between Anchor and Anchored Node. The link between the anchored node and its anchor should always be bidirectional as it will be used for bidirectional traffic. However, the beaconing process to detect neighboring nodes does not ensure that links are bidirectional. It might happen that the anchored node is hearing beacons from the anchor but not the other way around, resulting in a unidirectional link. In order to eliminate such unidirectional links, Route Error (RERR) packets in the mesh network are used to inform about the existence of the unidirectional links.

The time flow of the packets for the case of a unidirectional link from the anchor to the anchored node is given in Figure 5(a). In this case, the link break event is raised at the anchor because of the absence of beacon packets from the anchored node. The anchor will broadcast a RERR with default IP address 0.0.0.0 in its payload, indicating that the anchored node should no longer use the default route through the anchored node. At the same time, it will inform the network with an NGARP broadcast packet that the anchored node is no longer reachable on its MAC address. The IP address of the anchored node will be removed from the ARP responder table in the ARP module in the anchor as well. As the link in the direction from the anchor to the anchored node is active, the anchored node will receive the RERR packet. It will remove its default route towards the anchor from its main routing table, which in turn will raise the "remove default route" event. This event will trigger the start of a new anchor selection process in the DYMO routing module.

The time flow of the packets for the case of a unidirectional link from the anchored node to the anchor is given in Figure 5(b). In this case, the link break event is raised at the anchored node. The only difference from previous case is that the RERR packet is generated anchored node.

It is obvious that in case when the link is broken in both directions, the link break is detected at both sides, and the anchorage data will be cleared at both sides.

This way we ensure that at any time, the link between the anchored and the anchor node is bidirectional. Moreover, the use of the NGARP packet will ensure that any time the anchor loses its connection towards the anchored node (either as a result of the presence of a unidirectional link or of a total link break), the network is informed and should wait until the anchored node is reanchored again on another node. This will minimize the number of lost packets due to the wrong or outdated ARP and routing table information in the network.

6.3. Multihop Anchor Selection Process. If the node density is high, then most probably the anchored node will find its way towards an AP in just one hop. However, this is not necessarily true in case the node density is relatively low or the uncovered regions are large. In those cases, a multihop connection and packet forwarding mechanism is needed between the anchored node and the anchor. On top of the normal forwarding procedure already offered by the default DYMO routing protocol, each node needs to take certain actions when forwarding special packets for handling the

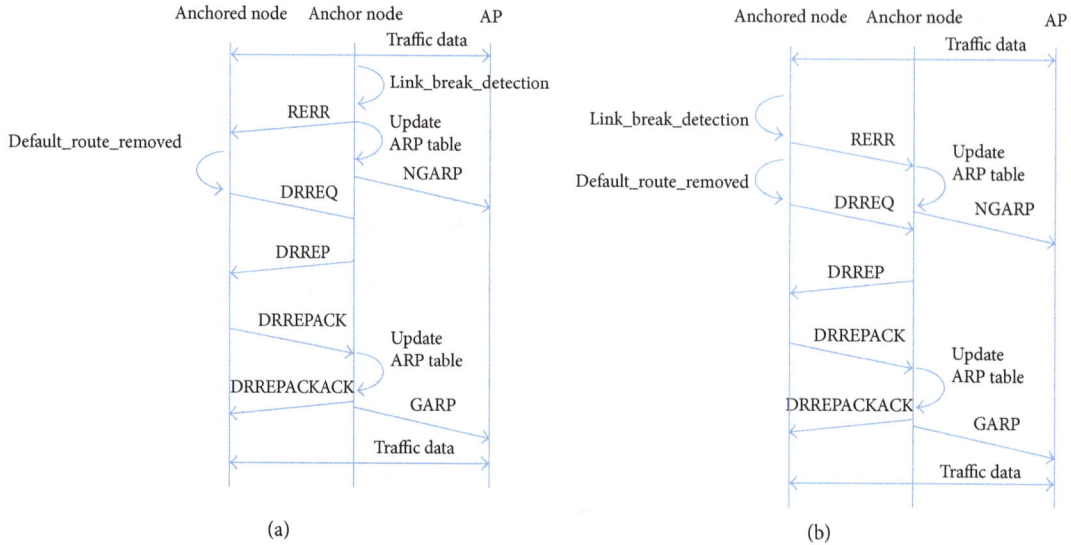

FIGURE 5: Unidirectional link detection. (a) From anchor to anchored node. (b) From anchored node to anchor.

network coverage problems like DRREQ, DRREP, DRRE-PACK, or DRREPACKACK. The time flow of the packets in case of a multihop anchor selection process is given in Figure 6.

When a node forwards the DRREPACK, it has to update its ARP response table with the source IP address of the DRREPACK packet and the MAC of its AP interface. This is due to the fact that in case the forwarding node reconnects to an AP, it has to directly notify the network about the shorter path to the multihop anchored node by sending a GARP containing the IP of that node. In addition, the forwarding node will broadcast a RERR packet with the IP address of the multihop anchored node in its payload on the mesh network. When the old anchor receives such a RERR packet, it will remove the multihop anchored node from its anchored node table.

Every time the mesh link between the anchored node and the anchor expires due to the absence of traffic, the anchorage process has to start all over. Otherwise, if traffic is going on, the link will not expire as the traffic is used for updating the mesh link expiration timer.

7. Performance Analysis

It is clear that the proposed communication system enables several networking topologies. Combined with the flexibility on how to distribute the traffic, it is interesting to investigate how this flexibility can be exploited in order to deal with the other requirements that are specific for our targeted use case (RQ4-7). For this, we conducted a set of experiments on the w-iLab.t wireless testbed [42], which are now discussed in the following subsections. We tested the Click packet processing overhead, broadcast scalability, and the different network topologies, which we already showed in Figure 2.

7.1. Tools for Running Experiments. Hostapd [43] and wpa-supplicant [39] are used as user space daemon to realize

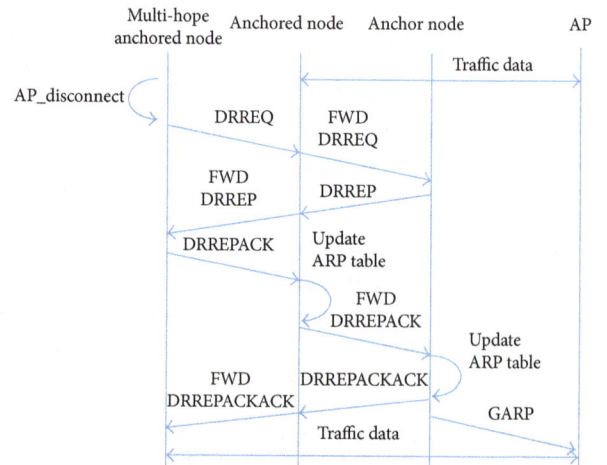

FIGURE 6: Multihop anchor selection process.

access point and client functionality, respectively. The mobile robots consist of embedded PC nodes which are running Linux and our Click Router implementation presented in Section 5. The access points are static embedded PC nodes running Linux. The Wi-Fi cards of all devices have Atheros AR93 chips.

To perform proper diagnostics and performance analysis, we have created on each node a database with three tables: an EVENT table, TOPOLOGY table, and ROUTES table. The EVENT table captures all internal Click Router events. The TOPOLOGY table contains link information, consisting of the time, neighbor MAC address, and event type ("add new link" or "link is broken"). This enables us to derive at each point in time the topology of the mesh network. Finally, the ROUTES table collects all routing information, including time, destination IP, and next hop IP.

7.2. Click Packet Processing Overhead. As already mentioned, all networking modules have been developed in Click

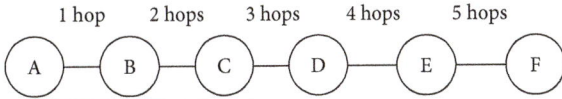

FIGURE 7: Setup to assess the impact of Click packet processing overhead.

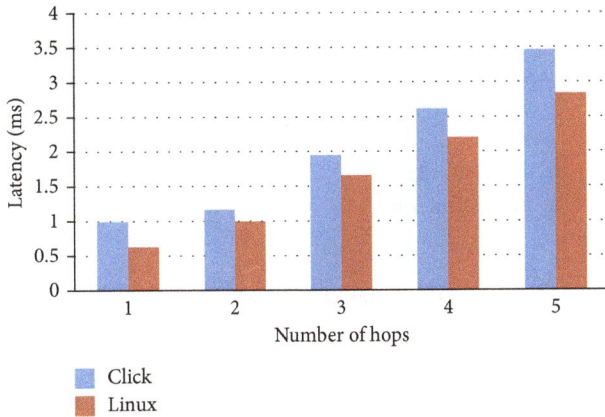

FIGURE 8: Latency versus numbers of hops.

Router, running in user level. Consequently, Click Router packet processing introduces additional overhead compared to kernel-level packet processing. Therefore, it is important to assess the introduced performance penalty. To this end, we measure the latency and UDP throughput for both cases using the setup shown in Figure 7. We use wired connections in order to isolate Click packet processing and avoid performance losses introduced by wireless communication links.

The keep-alive beacon interval is set to 50 ms and packets are sent every 100 ms. We measure the latency for different hops, running every experiment for 50 seconds, and averaging the latency over the measurement time. In Figure 8, the latency for the different number of hops is given. Due to the packet processing in Click Router, the packet latency is about 20% higher compared to normal Linux stack processing, but sufficiently low to be able to meet our latency requirements. The only exception is for one-hop connection, where the latency is approximately 48% higher.

Regarding the UDP throughput, we did measurements with data rates up to 100 Mbps, not noticing any losses. The performance started to decrease when the data rate became higher than 500 Mbps, observing packet losses from 1% up to 40% for the highest sending rate of 1 Gbps over a one-hop direct link. So considering the typical data rates of wireless cards, it is clear that the packet processing overhead in Click is not a bottleneck as it can handle rates up to 500 Mbps without losses.

7.3. Broadcast Traffic Scalability.
Our use case heavily relies on broadcast communication between mobile robots. One of the interesting things to analyze is the broadcast scalability, for example, how frequently broadcast data packets can be transmitted within a group of nearby mobile nodes without

having high packet losses. By quantifying the packet loss ratio of directly broadcasted and rebroadcasted packets, we can assess which percentage of traffic was received through direct links and which was received through a multihop path.

We use two different packets with payload size of 50 and 20 bytes, respectively, and two different group sizes, with 8 and 12 nodes, respectively. From each node, we send 10000 broadcast packets at the same frequency.

If the number of nodes in the group is N and the broadcast frequency is f_{BC}, then we will have $N*f_{BC}$ generated and $(N-1)*f_{BC}$ received directly broadcasted packets per time, while we will have $N*(N-1)*f_{BC}$ generated and $N*(N-1)*2*f_{BC}$ received rebroadcasted packets per time. At each node, we classify the received packets based on source IP and source MAC address enabling us to calculate packet losses of directly broadcasted packets and rebroadcasted packets on each node.

In Figure 9(a), average packet losses are given for the case with ping packets having a payload of 50 bytes and for two different group sizes, 12 nodes and 8 nodes, respectively. It can be seen that by increasing the broadcast frequency, the directly broadcasted packet losses increase drastically: from 1% for 3.33 Hz and group of 8 nodes to 48% for 33.33 Hz. For the group with 12 nodes, these values are even higher, starting at 9% for 3.33 Hz up to 66% for a frequency of 33.33 Hz. The rebroadcasted packets exhibit even higher losses because they are more prone to collisions as they will be retransmitted by all nodes nearly at the same time. However, all the nodes received all the packets at least once with just small loses (~2%) in case of a frequency of 33.33 Hz. From this, we can conclude that by increasing the broadcast frequency and number of nodes that are in the range of each other, packets will still be able to go through but will more likely take a path with more than one hop. For example, in case of a frequency of 20 Hz and packet payload of 50 bytes, 57% of packets arrived through a path with more than one hop in case of group with 12 nodes, while in case of a group with 8 nodes, this was 40%.

Also the packet size has an impact on the packet losses. In Figure 9(b), the average packet loss for ping packets with a payload size of 20 bytes is shown. It can be noticed that all values are lower than those in the previous case. In our use case, the packets payload will be low since the mobile nodes will just transmit their positions and other data that are related to their battery life.

7.4. Wireless Infrastructure Network Only.
In this scenario, we assume the presence of fixed access points and do not make use of any meshing capabilities. Every mobile robot is connected to an access point, and selection of the most suitable access point is based on signal strength. Mobile robots move around in the environment covered by access points and get attached and detached to/from access points. As mobile robots can drive at relatively high speeds, such handovers may take place frequently and will affect the communication performance. To quantify this effect on the performance of unicast and broadcast traffic, we set up an

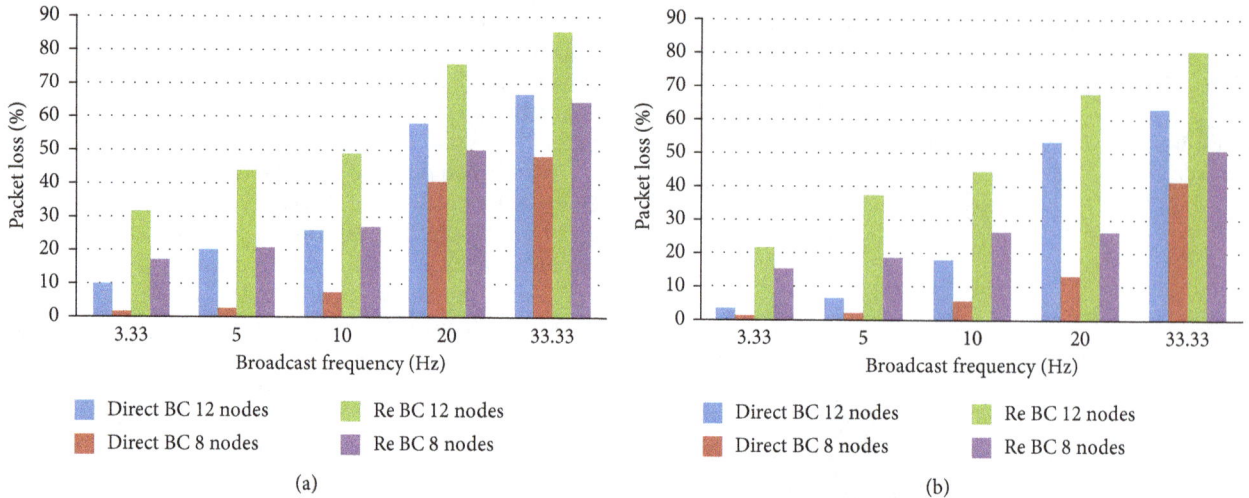

FIGURE 9: Average packet losses for broadcast scalability tests. (a) Packet payload size 50 bytes. (b) Packet payload size 20 bytes.

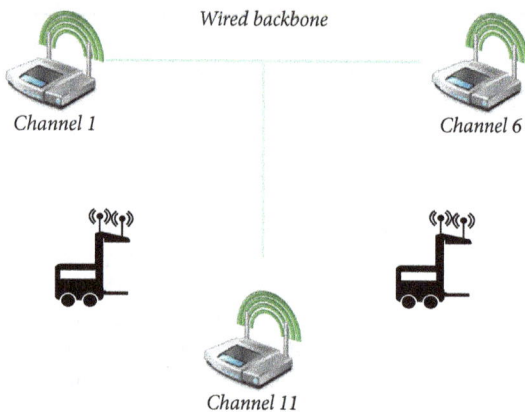

FIGURE 10: Experimental setup to assess the impact of handovers on the communication performance.

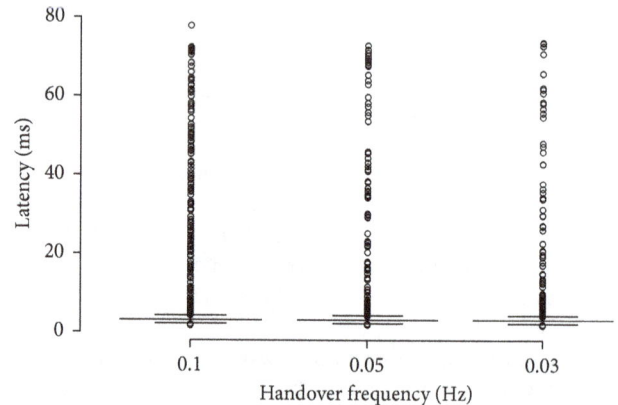

FIGURE 11: Latency of unicast traffic for different roaming frequencies.

experiment in the w-iLab.t testbed [42] as shown in Figure 10. Three APs and two mobile robots are used.

Three nonoverlapping channels (1, 6, and 11) in 2.4 GHz frequency band have been used. To trigger handovers of mobile clients between APs in a small area (limited by the physical space of the testbed), the transmit powers of the APs are configured during the experiment. The mobile robots are limited to scan only over the mentioned channels to prevent time and energy consuming procedure for scanning all available channels. During the experiment, both mobile robots are communicating with each other through the infrastructure wireless network. Figure 11 shows the latency distribution of 10000 unicast packets during a measurement period of 200 seconds. Unicast packets are exchanged every 20 ms and the roaming among access points is configured to be once every 10, 20, and 30 seconds. As can be seen, in most cases, the latency is lower than 4 ms, which is close to the average amount. However, it can become as high as 78 ms during the roaming procedure. Further, the more frequently roaming happens among the access points, the higher the packet latency can become. The reason behind this is that every time a client performs a handover between access

points, it gets dissociated, has to look for stronger signal strength, and needs to associate to a new access point. Table 2 shows the latency statistics, presenting the first and third quartile of the results shown in Figure 11.

Figure 12 presents the latency of 10000 broadcast packet transmissions within the same 200 seconds time period. Again, the roaming procedure happens every 10, 20, and 30 seconds. As shown in Table 3, in contrast to the unicast latency, the broadcast latency is now not around the average value but around the third quartile value. The results also show a much more profound negative impact of handovers on the broadcast latencies, due to the way broadcasts are disseminated through the network. Every broadcast from a mobile robot needs to be rebroadcasted to other devices connected to the same access point as well as to all other devices connected to the other access points. This is visible in Figure 13 where every time the mobile robots were connected to the same access point, the latency was around 5 ms while when roaming took place the latency increased up to 100 ms. It is clear that even in this simple setup, our mobile robot solution will never be able to meet the envisioned latency requirements (<20 ms) of broadcast traffic.

TABLE 2: Unicast latency statistics in ms.

Statistics	10 s roaming frequency	20 s roaming frequency	30 s roaming frequency
Average	3.67	3.57	3.52
1st quartile	3.21	3.18	3.13
Minimum	2.13	2.12	2.09
Maximum	78	73	73.8
3rd quartile	3.75	3.66	3.66

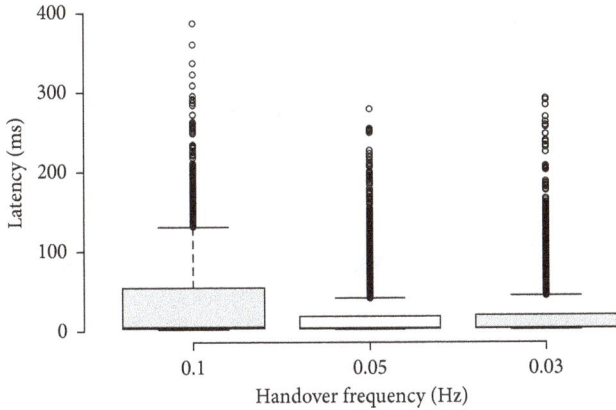

FIGURE 12: Latency of broadcast traffic for different roaming frequencies.

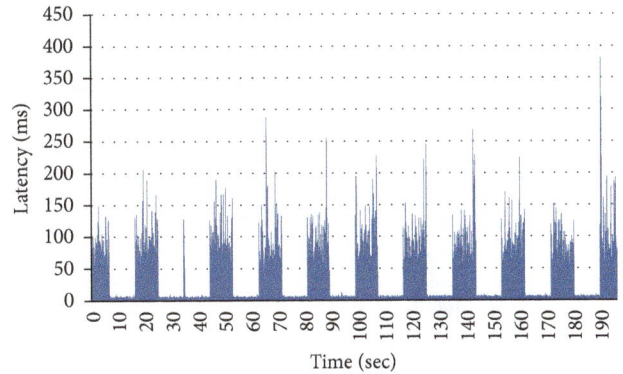

FIGURE 13: Latency of broadcast traffic for a roaming frequency of 0.1 Hz.

7.5. Mesh Network Only. In this scenario, only a mesh network is being used as shown in Figure 2(a). As mentioned, unicast traffic uses a simple reactive routing protocol, whereas broadcast traffic uses blind flooding with duplicate detection. Using this setup, we again measure the impact of mobility of mobile robots on the latency of packet transmissions. In order to be able to mimic a variety of speeds and thus link breaks, we used a forced mobility approach, where MAC filtering is being used to artificially change the mesh topology as shown in Figure 14. While nodes c1 and c5 are communicating, c1 establishes a new link with nodes c2, c3, c4, and c5, respectively, breaking the old link and gradually changing the number of hops over which the packets need to travel.

Figure 15 presents the impact of link breaks and the resulting change in topology and hop count on unicast and broadcast packet transmissions with transmissions being generated every second. In this experiment, latency for unicast and broadcast traffic varies between 17.2 ms and 2.62 ms and 19.9 ms and 3.04 ms, respectively. It is also visible that the latency decreases with the hop count between the sender and the receiver.

In the scenario shown in Figure 15, the beacon interval was set to a very small value (20 ms), making it possible to very quickly react to link breaks in this small topology. In addition, with traffic only being generated every second, no significant unicast packet losses occurred, illustrating only the impact of hop count on latency in a mesh setting. In other settings, the performance of unicast traffic, however, is also strongly affected by the link break detection and routing mechanism.

In reality, the protocol might react slower, traffic generation can happen more frequently or the topology is more complex. These first two aspects are shown in Figure 16, where unicast traffic is being generated every 120 ms. Keep-alive beacons are sent less frequently, that is, every 500 ms, with the detection of a link break in the absence of beacons after 2500 ms. Further, upon the detection of a link break, all traffic for a destination that has become unreachable is being buffered until the route has been established. This has two consequences. First of all, unicast traffic in the presence of link breaks in the mesh network exhibits much higher packet losses than in an infrastructure network, with the amount of lost packets directly related to the efficiency of the underlying link break detection mechanism as shown in Figure 16. Secondly, route recovery takes some time, resulting in higher latencies of the packets that were buffered between the detection of the link break and the moment the route has been recovered. Broadcast traffic does not experience these drawbacks as it can make use of any available link and does not depend on route establishment.

7.6. Combined Network. The third scenario being considered is a mixed setup, where every mobile robot uses one interface to connect to the infrastructure network and one interface to set up a mesh network as shown in Figure 2(b). In order not to overload the wired network with broadcast traffic, the communication system is configured to send broadcast traffic over the mesh interfaces. To avoid frequent routing inside the mesh network, unicast traffic is configured to run over the other wireless interface. Again, we measure the latency of unicast and broadcast traffic in order to investigate the advantages and feasibility of a hybrid configuration with traffic separation. In this scenario, we use three interconnected access points (as in Figure 10) and four mobile robots. Two of them are communicating using unicast traffic

TABLE 3: Broadcast latency statistics in ms.

Statistics	10 s roaming frequency	20 s roaming frequency	30 s roaming frequency
Average	6.1	5.02	4.99
1st quartile	4.44	4.18	4.46
Minimum	3.06	3.17	3.26
Maximum	381	275	288
3rd quartile	54.5	19.2	20.6

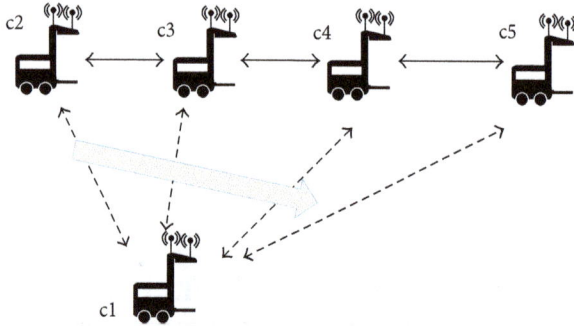

FIGURE 14: Fully mesh network among mobile robots.

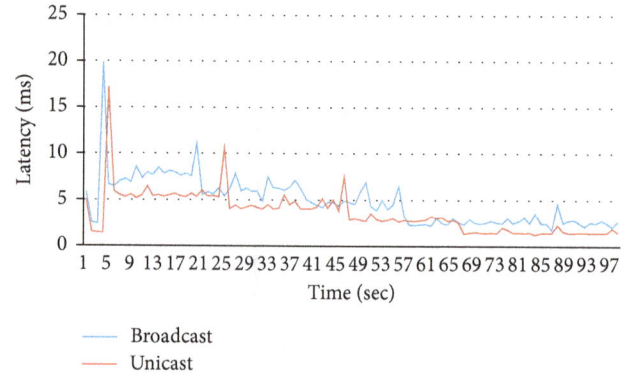

FIGURE 15: Latency of broadcast and unicast traffic with link breaks occurring every 20 s.

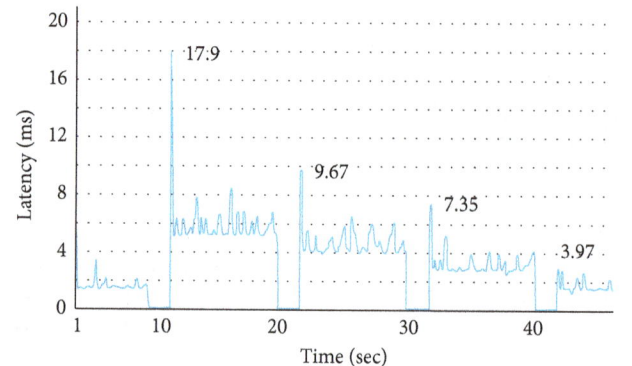

FIGURE 16: Unicast packet transmission latency in the presence of link breaks every 10 s.

through access points while two others are generating broadcast traffic. All of them are connected to one of the APs. One mobile robot is configured to reply to the broadcast packets. Channel 6 is used for communication within the mesh network while channel 1 is used for communication with APs. The handover and link break frequency in this case are both 0.1 Hz.

Figure 17(a) shows the latency of 10000 unicast transmissions during 200 seconds, whereas Figure 17(b) shows the latency of 10000 simultaneous broadcast transmissions. As it is shown in Table 4, the mixed scenario that exploits the possibility to separate different traffic streams combines the best of both worlds. Broadcast traffic can meet the strict latency requirements by using the mesh network, whereas unicast traffic achieves low latency by avoiding the complexity of ad hoc routing. Compared to Table 2, it can be seen that the maximal values for unicast traffic latencies have now dropped from ~70 to ~11 ms. The same conclusion can be drawn for broadcast traffic where maximal latencies have dropped from ~300 to ~10 ms.

Compared to solutions proposed in [32, 36], our solution achieves lower latencies. In [36], when using the full QoS feature set of the system with two high priority flows, they reach an average latency of 7.1 ms for unicast traffic compared to an average latency of 3.8 ms with our system. Moreover, their maximal latencies are in the order of 170 ms compared to 11.3 ms in our case.

7.7. Mesh Capabilities for Handling Coverage Problems. In this scenario, we consider the mesh capabilities for handling coverage problems by using the mobile robot as range extenders of the APs, as shown in the network topology in Figure 2(d). All mobile robots are using two network interface cards, one for connecting to the AP and the other one for mesh communication towards other mobile robots.

We motivate the traffic separation based on the results from subsection F. In order to assess the solution for handling the coverage problems, we use the setup as shown in Figure 18(a). To emulate the situation in uncovered zones, the node is disconnected from the AP using the wpa-cli [39] disconnection method. Ping packets are sent from node E to node A. Every 10 seconds, we disconnect sequentially the nodes from the AP. This way, we increase the number of hops between the anchored node and the anchor, testing thus also the multihop anchorage process. During the first 10 seconds, node A is connected directly to the AP. After the disconnection from the AP, it is forced to search for an anchor and will select node B as its anchor. After 20 seconds, node B is disconnected from the AP too. This will enforce both nodes A and B to start searching for a new anchor. After the anchor selection process, node C will become the direct anchor for node B and one-hop anchor for node A. After disconnection

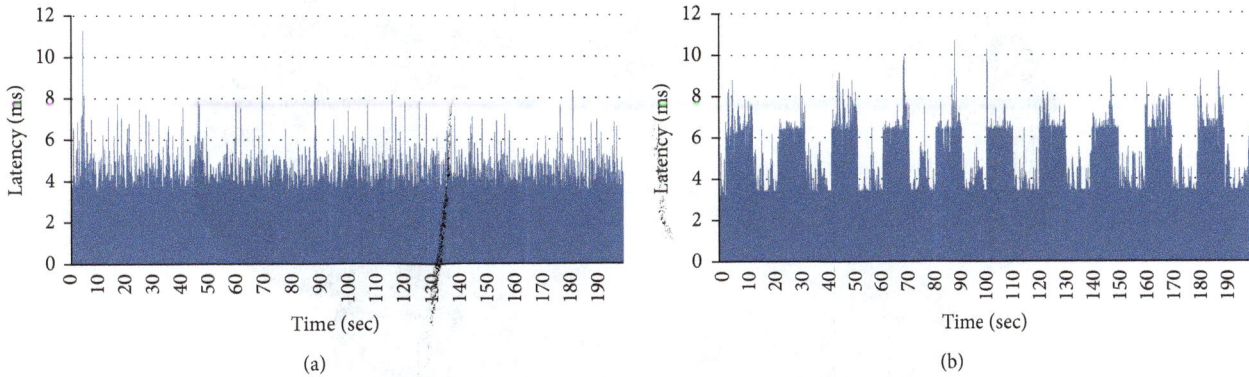

FIGURE 17: Unicast latency and broadcast latency in a mixed scenario with traffic separation. (a) Packet payload size 50 bytes. (b) Broadcast latency.

of node C, there will be three hops between the multihop anchored node A and its main anchor node D. After 40 seconds, we reestablish the AP connections for nodes C, B, and A sequentially every 10 s.

In Section 6, the key parameters of this scenario were introduced: the AP monitoring interval is set to 1s, link break detection time is set to 150 ms, and beacon interval is set to 50 ms. Measurements are performed for two scenarios, one with ping packets sent every 100 ms, and the other one every 200 ms. The expected maximum number of packets to be lost is calculated based on the number of AP disconnections multiplied by the number of packets sent during one AP connection monitor time. We expect packet losses to occur when the AP disconnection happens until the disconnection is detected by the system. In our case, we should have at most 10 or 5 lost packets per AP disconnection for ping frequencies of 100 and 200 ms, respectively. On the other hand, when the AP connection is reestablished, we should see a higher latency for some packets at the beginning of the connection due to other information packets (GARP and NGARP) that needs to be processed.

In Figure 19, the latency for both packet rates is given. We see that by increasing the hop count between the anchored node and the anchor, the latency increases from 2 ms for one hop up to 6 ms for three hops. Also it can be noticed that every 10 seconds, when the AP disconnection happens, some packets are lost until the disconnection is detected. In the first case (ping packets every 100 ms), in total 18 packets are lost, or ~2% of packets. We have three AP disconnections during the measurement time, so the maximum number of lost packets we expect is 30. For the other case where we send a packet every 200 ms, in total 8 packets are lost, or ~2% of packets, while the maximum number to be expected 15. Moreover, when the AP connection is reestablished for the mobile nodes, we see that the first packet exhibits a higher latency due to the fact that the communication link moves to new anchor. However, in this case, no packets are lost since there is no communication outage time while switching from the old anchor to the new anchor.

To check the communication outage time during the tests, we parsed the ROUTE table from the databases of all nodes involved in the communication. The total communication outage time was 1.72 s for the first case and 1.67 s for the

TABLE 4: Latency statistics of unicast and broadcast packet transmission in ms.

Statistics	Unicast traffic	Broadcast traffic
Average	3.8	4.55
1st quartile	3.2	3.4
Minimum	2.26	2.73
Maximum	11.3	10.7
3rd quartile	3.49	5.3

second case, with the largest continuous outage being 0.63 s and 0.55 s, respectively. This communication outage time fulfills one of the requirements for our system too, namely, RQ7. Based on the time relation between the EVENT table and ROUTE table, we observed that all of the communication outage time happened during the time until the AP disconnection was detected and the anchor was selected. There was no communication outage time due to other reasons. So the communication outage time is related to the configuration parameters: AP monitoring interval, link break detection time, and beaconing time interval.

Another key issue for the proper functioning of the anchor selection process is the existence of a bidirectional link between the anchored node and the anchor itself. In order to test our solution, we use the setup shown in Figure 7. We use a MAC filtering Click element to create unidirectional links between the anchor and the anchored node. This MAC filtering element filters out all the incoming packets with a certain MAC address, emulating thus the absence of beacons from a specific neighbor. We send pings from node D to node A every 100 ms and 200 ms, respectively. The AP monitoring interval is 1 s, link break detection time is set to 150 ms, and beaconing time interval is set to 50 ms. During the first 10 s, node A is connected directly to the AP. After 10 s, we break the link between node A and AP. Since MAC filtering for node C is enabled in node A, node A will choose node B as its anchor. After 10 s of communication, we enable MAC filtering for node B and disable MAC filtering for node C in node A. This way we create a unidirectional link from node A to node B. As such, the anchored node A should initiate a new anchor selection process. In this case, it will select node C as its anchor. After 10 s, we do it the other way around. This way, the anchored node A will alternate

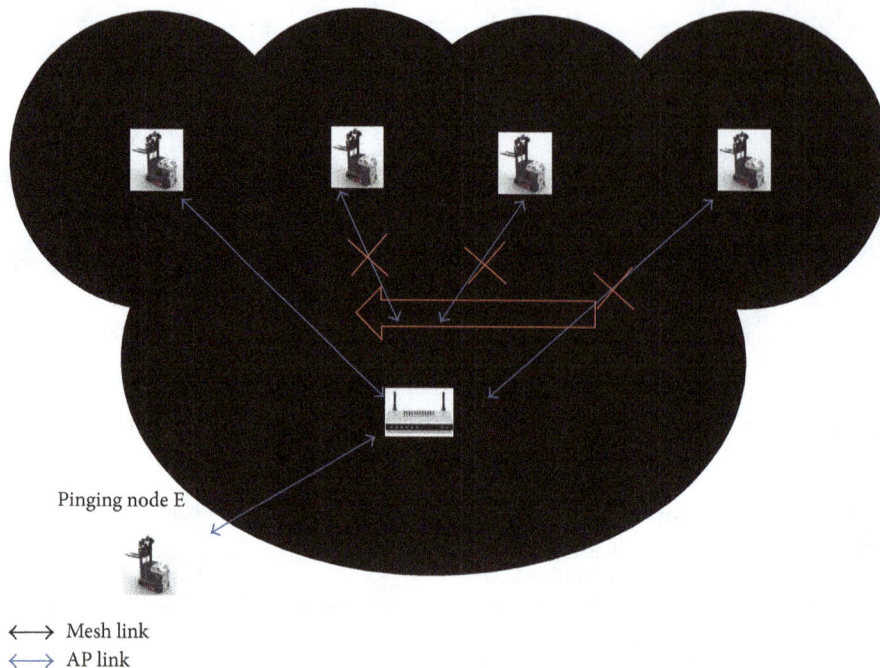

(a)

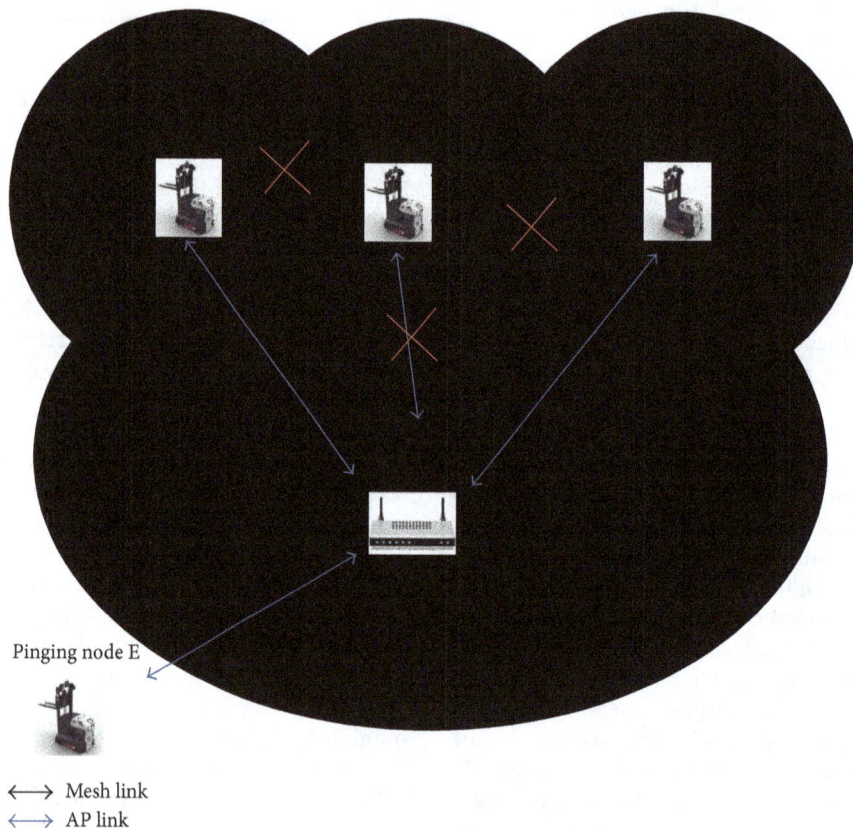

(b)

FIGURE 18: Mesh capabilities to handle coverage problems. (a) Setup for testing the mutihop anchor selection process. (b) Setup for testing the anchor selection process in case of unidirectional links.

anchors every 10 seconds due to unidirectional link breakage. After 40 seconds, we reestablish the link towards the AP and stop the measurements.

In Figure 20, the latency of unicast packets for this setup is given. Initially the latency is low for direct communication through AP, around 3 ms. When the communication is

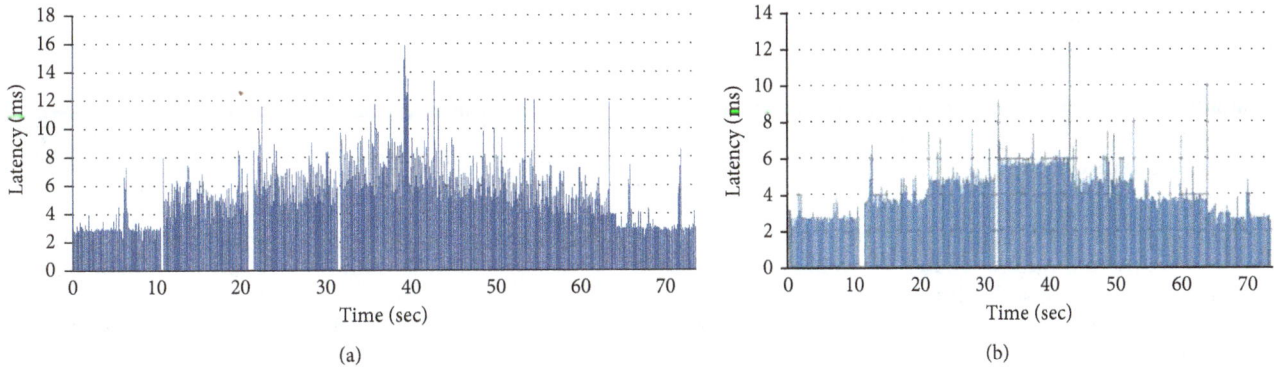

FIGURE 19: Latency for unicast traffic from an anchored node. (a) Packet rate 10 Hz. (b) Packet rate 5 Hz.

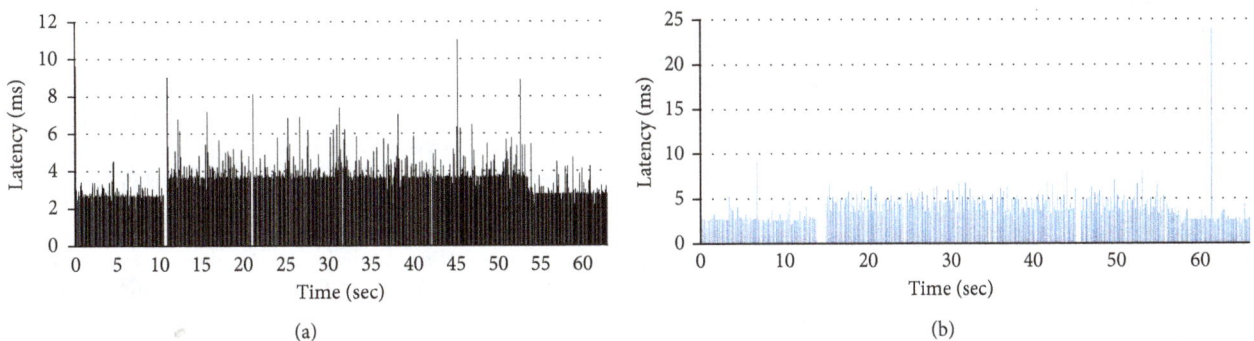

FIGURE 20: Latency for unicast traffic from an anchored node regarding the unidirectional link. (a) Packet rate 10 Hz. (b) Packet rate 5 Hz.

going through an anchor, the latency increases up to 4 ms due to the increase in number of hops. It can be noticed that some packets are lost until the node detects the AP disconnection. Afterwards, every 10 s, when the node will have to switch anchor due to the presence of a unidirectional link, again some packets are lost due to the link break detection mechanism. Since we send packets every 100 ms and the link break detection time is 150 ms, at most 2 packets can be lost in the first case. In the second case (ping packets every 200 ms), we can lose 1 packet at most.

We checked the ROUTE table with the OTM tool, and we found out that the total communication outage time was 0.75 s and 1.11 s respectively. Since we had one AP disconnection, and three link break detections, the maximal total communication outage time to be expected was 1.45 s. In total, 8 packets were lost (~1% of packets) in the first case (ping packets every 100 ms), while in the second case (ping packets every 200 ms) 9 packets (~2% of packets) in total were lost. Correlating the timing between the ROUTE table and EVENT table, the largest communication outage time was during the AP disconnection detection. It was 0.45 s and 0.6 s, respectively, while the rest was due to the link break detection mechanism.

This test shows that the packet loss and communication outage time is a function of the parameters of the mobile communication system we designed. Based on the requirements of the system, these parameters can be easily changed in the configuration file by the administrators.

Regarding packet loss rates, the proposed mobile communication system shows similarities with the system proposed in [36]. Here we have losses up to 2% where in [36], losses are between 1.4 and 3.2%. However, in our communication system, losses depend on configuration parameters.

8. Conclusions

Many existing solutions in industrial settings that make use of mobile robots utilize the existing enterprise network. In this paper, we discussed the potential drawbacks of such an approach. For our particular use case at hand, a key requirement was the ability to deliver broadcast traffic with very low latencies, a requirement that could not be fulfilled in an enterprise network where handovers take place frequently, as shown on our testbed. We proposed flexible and modular system architecture for the mobile node that makes use of two physical interfaces. The mobile node is able to function in different network topologies. It can make use of infrastructure network only; it can use only ad hoc capabilities or use both of them at the same time.

The proposed architecture is able to exploit both the advantages of the presence of an infrastructure network and the advantages of a mesh network. In this paper, we showed the feasibility of implementing such architecture and the advantage of the mixed architecture with traffic separation.

We show that the mixed architecture was able to deal with the occurrence of coverage holes. The mesh capabilities were used to extend the AP coverage zones by using intermediate robots to enable the communication for robots outside of coverage zones. Moreover, the communication outage time was related solely to the configuration of different parameters, being the AP connection monitor time and link break detection time.

Since the use case under consideration relies on broadcast traffic communication between mobile robots, we evaluated the broadcast scalability in the mesh network. We showed that by increasing the broadcast frequency and number of nodes per group, every node will be able to receive all packets at least once; however, most of the packets will take a route that spans more than one hop.

The solution was validated in a testbed, and the outcome figures were benchmarked according to the initial requirements. As future work, other tests in larger scale setup need to be done to prove the feasibility further.

References

[1] S. Arumugam, R. K. Kalle, and A. R. Prasad, "Wireless robotics: opportunities and challenges," *Wireless Personal Communications*, vol. 70, no. 3, pp. 1033–1058, 2013.

[2] R. Drath and A. Horch, "Industrie 4.0: hit or hype? [industry forum]," *IEEE Industrial Electronics Magazine*, vol. 8, no. 2, pp. 56–58, 2014.

[3] M. Essers, *Design of a Novel, Hybrid Decentralized, Distributed, Modular Architecture for Manufacturing Systems*, Universiteit Twente, Enschede, Netherlands, 2016.

[4] M. S. Essers and T. H. J. Vaneker, "Design of a decentralized modular architecture for flexible and extensible production systems," *Mechatronics*, vol. 34, pp. 160–169, 2016.

[5] "Manual of AGVs from motum producer," 2017, http://www.motum.be/wp-content/uploads/2015/10/Br 2Move16 en.pdf.

[6] E. A. Jarchlo, J. Haxhibeqiri, I. Moerman, and J. Hoebeke, "To mesh or not to mesh: flexible wireless indoor communication among mobile robots in industrial environments," in *Proceedings of the International Conference on Ad-Hoc Networks and Wireless*, pp. 325–338, Lille, France, 2016.

[7] P. Vijayakumar, M. Azees, A. Kannan, and L. J. Deborah, "Dual authentication and key management techniques for secure data transmission in vehicular ad hoc networks," *IEEE Transactions on Intelligent Transportation Systems*, vol. 17, no. 4, pp. 1015–1028, 2016.

[8] W. Jiang, F. Li, D. Lin, and E. Bertino, "No one can track you: randomized authentication in vehicular ad-hoc networks," in *Proceedings of the IEEE International Conference on Pervasive Computing and Communications (PerCom)*, pp. 197–206, Kona, HI, USA, 2017.

[9] P. Li, C. Zhang, and Y. Fang, "Capacity and delay of hybrid wireless broadband access networks," *IEEE Journal on Selected Areas in Communications*, vol. 27, no. 2, pp. 117–125, 2009.

[10] B. Liu, P. Thiran, and D. Towsley, "Capacity of a wireless ad hoc network with infrastructure," in *Proceedings of the 8th ACM International Symposium on Mobile Ad Hoc Networking and Computing*, pp. 239–246, Montreal, QC, Canada, September 2007.

[11] A. Zemlianov and G. De Veciana, "Capacity of ad hoc wireless networks with infrastructure support," *IEEE Journal on Selected Areas in Communications*, vol. 23, no. 3, pp. 657–667, 2005.

[12] H.-N. Dai, R. C.-W. Wong, and H. Wang, "On capacity and delay of multichannel wireless networks with infrastructure support," *IEEE Transactions on Vehicular Technology*, vol. 66, no. 2, pp. 1589–1604, 2017.

[13] L. Chen, W. Luo, C. Liu, X. Hong, and J. Shi, "Capacity-delay trade-off in collaborative hybrid ad-hoc networks with coverage sensing," *Sensors*, vol. 17, no. 2, p. 232, 2017.

[14] W. Lee, J. Kim, and S.-W. Choi, "New D2D peer discovery scheme based on spatial correlation of wireless channel," *IEEE Transactions on Vehicular Technology*, vol. 65, no. 12, pp. 10120–10125, 2016.

[15] C. F. M. e Silva, T. F. Maciel, R. L. Batista, L. Elias, A. Robson, and F. R. P. Cavalcanti, "Network-assisted neighbor discovery based on power vectors for D2D communications," in *Proceedings of the IEEE 81st Vehicular Technology Conference (VTC Spring)*, pp. 1–5, Glasgow, UK, May 2015.

[16] Q. Sun, L. Tian, Y. Zhou, J. Shi, and X. Wang, "Energy efficient incentive resource allocation in D2D cooperative communications," in *Proceedings of the IEEE International Conference on Communications (ICC)*, pp. 2632–2637, London, UK, June 2015.

[17] J. Zheng, B. Chen, and Y. Zhang, "An adaptive time division scheduling based resource allocation algorithm for D2D communication underlaying cellular networks," in *Proceedings of the IEEE Global Communications Conference (GLOBE-COM)*, pp. 1–7, San Diego, CA, USA, December 2015.

[18] S. Xiao, X. Zhou, D. Feng, Y. Yuan-Wu, G. Y. Li, and W. Guo, "Energy-efficient mobile association in heterogeneous networks with device-to-device communications," *IEEE Transactions on Wireless Communications*, vol. 15, no. 8, pp. 5260–5271, 2016.

[19] M. N. Tehrani, M. Uysal, and H. Yanikomeroglu, "Device-to-device communication in 5G cellular networks: challenges, solutions, and future directions," *IEEE Communications Magazine*, vol. 52, no. 5, pp. 86–92, 2014.

[20] http://read.cs.ucla.edu/click/click, 2017.

[21] V. Firoiu, B. T. Decleene, M. M. Leung, S. Nanda, and C. Tao, "Mobile infrastructure assisted ad-hoc network," US Patent 9,596,619, 2017.

[22] M. Shneier and R. Bostelman, *Literature Review of Mobile Robots for Manufacturing*, National Institute of Standards and Technology, US Department of Commerce, Gaithersburg, MD, USA, 2015.

[23] Z. Yan, N. Jouandeau, and A. A. Cherif, "A survey and analysis of multi-robot coordination," *International Journal of Advanced Robotic Systems*, vol. 10, no. 12, p. 399, 2013.

[24] T. Qiu, N. Chen, K. Li, D. Qiao, and Z. Fu, "Heterogeneous ad hoc networks: architectures, advances and challenges," *Ad Hoc Networks*, vol. 55, pp. 143–152, 2017.

[25] A. Wichmann, B. D. Okkalioglu, and T. Korkmaz, "The integration of mobile (tele) robotics and wireless sensor networks: a survey," *Computer Communications*, vol. 51, pp. 21–35, 2014.

[26] F. B. Saghezchi, A. Radwan, and J. Rodriguez, "Energy efficiency performance of wifi/wimedia relaying in hybrid ad-hoc networks," in *Proceedings of the Third International Conference on IEEE Communications and Information Technology (ICCIT)*, pp. 285–289, Beirut, Lebanon, June 2013.

[27] H.-J. Im, C.-E. Lee, M. Jang, Y.-J. Cho, and S. Kim, "Implementation of cooperative routing robot system for long

distance teleoperation," in *Proceedings of the Autonomous Mobile Systems*, pp. 129–136, Stuttgart, Germany, 2012.

[28] H.-J. Im, C.-E. Lee, and Y.-J. Cho, "Implementation of a seamless communication scheme using cooperative routing robots for teleoperation," in *Proceedings of the IEEE International Symposium on Robot and Human Interactive Communication (RO-MAN)*, pp. 310-311, Gyeongju, South Korea, August 2013.

[29] H. Sugiyama, T. Tsujioka, and M. Murata, "Integrated operations of multi-robot rescue system with ad hoc networking," in *Proceedings of the IEEE 1st International Conference on Wireless Communication, Vehicular Technology, Information Theory and Aerospace & Electronic Systems Technology (Wireless VITAE)*, pp. 535–539, Aalborg, Denmark, May 2009.

[30] W. Vandenberghe, I. Moerman, and P. Demeester, "Adoption of vehicular ad hoc networking protocols by networked robots," *Wireless Personal Communications*, vol. 64, no. 3, pp. 489–522, 2012.

[31] R. Tahar, A. Belghith, and R. Braham, "A generic mobile node architecture for multi-interface heterogeneous wireless link layer," in *Proceedings of the IEEE Third International Conference on the Network of the Future (NOF)*, pp. 1–5, Tunis, Tunisia, November 2012.

[32] T. Lindhorst and E. Nett, "Dependable communication for mobile robots in industrial wireless mesh networks," in *Cooperative Robots and Sensor Networks*, pp. 207–227, Springer, Berlin, Germany, 2015.

[33] E. Khorov, A. Kiryanov, A. Lyakhov, and D. Ostrovsky, "Analytical study of neighborhood discovery and link management in OLSR," in *Proceedings of the IEEE IFIP Wireless Days (WD)*, pp. 1–6, Dublin, Ireland, November 2012.

[34] F. Tian, B. Liu, H. Cai, H. Zhou, and L. Gui, "Practical asynchronous neighbor discovery in ad hoc networks with directional antennas," *IEEE Transactions on Vehicular Technology*, vol. 65, no. 5, pp. 3614–3627, 2016.

[35] A. Herms, E. Nett, and S. Schemmer, "Real-time mesh networks for industrial applications," *IFAC Proceedings Volumes*, vol. 41, no. 2, pp. 9314–9319, 2008.

[36] T. Lindhorst, G. Lukas, and E. Nett, "Wireless mesh network infrastructure for industrial applications: a case study of teleoperated mobile robots," in *IEEE 18th Conference on Emerging Technologies & Factory Automation (ETFA)*, pp. 1–8, Cagliari, Italy, September 2013.

[37] C. Perkins, S. Ratliff, J. Dowdell, and Internet Draft, "Dynamic MANET on-demand (dymo) routing," Technical Report draft-ietf-manet-dymo-26, 2006.

[38] E. Kohler, R. Morris, B. Chen, J. Jannotti, and M. F. Kaashoek, "The click modular router," *ACM Transactions on Computer Systems (TOCS)*, vol. 18, no. 3, pp. 263–297, 2000.

[39] https://wiki.archlinux.org/index.php/WPA supplicant, 2017.

[40] S. Cheshire, B. Aboba, and E. Guttman, "Dynamic configuration of IPv4 link-local addresses," Technical Report, 2005.

[41] S. Cheshire, "IPv4 address conflict detection," 2008.

[42] http://wilab2.ilabt.iminds.be, 2017.

[43] https://w1.fi/hostapd/, 2016.

12

EpSoc: Social-Based Epidemic-Based Routing Protocol in Opportunistic Mobile Social Network

Halikul Lenando [ID][1] **and Mohamad Alrfaay**[2]

[1]*Faculty of Computer Science and Information Technology, Universiti Malaysia Sarawak, Kota Samarahan, Malaysia*
[2]*Faculty of Computer and Information Sciences, Jouf University, Jouf, Saudi Arabia*

Correspondence should be addressed to Halikul Lenando; cool@unimas.my

Academic Editor: Fabio Gasparetti

In opportunistic networks, the nature of intermittent and disruptive connections degrades the efficiency of routing. Epidemic routing protocol is used as a benchmark for most of routing protocols in opportunistic mobile social networks (OMSNs) due to its high message delivery and latency. However, Epidemic incurs high cost in terms of overhead and hop count. In this paper, we propose a hybrid routing protocol called EpSoc which utilizes the Epidemic routing forwarding strategy and exploits an important social feature, that is, degree centrality. Two techniques are used in EpSoc. Messages' TTL is adjusted based on the degree centrality of nodes, and the message blocking mechanism is used to control replication. Simulation results show that EpSoc increases the delivery ratio and decreases the overhead ratio, the average latency, and the hop counts as compared to Epidemic and Bubble Rap.

1. Introduction

Opportunistic mobile social network (OMSN) [1–4] is a promising networking model for data dissemination. In the OMSN, mobile nodes grab the opportunity of encountering the peer (they are in the communication range of each other) to forward the data. The OMSN incurs intermittent and disruptive connectivity due to node mobility. To tackle with this complex environment, the store-carry-forward scheme is applied in the OMSN. If no connection is available at a particular time, a mobile node stores data in its buffer and carries them until it encounters other mobile nodes to forward the data [5–8]. Various approaches have been proposed to address the information delivery problem in the OMSN such as in [9–12]. The main concerns of OMSN routing approaches are yielding high delivery ratio, low delay, and low overhead or cost on networks and nodes.

Flooding is one of the dominant schemes to disseminate data in the OMSN [13]. Each message will be flooded to every node in the network. Multiple copies of each message are generated and spread in the network. Epidemic [9] routing protocol is the flooding-based routing protocol. When two nodes encounter, they exchange all of their messages. This results in messages spread over the whole network by pairwise contacts between two nodes. If no buffer constraints are applied, Epidemic represents the upper bound in message delivery and latency. Epidemic routing is used as a benchmark and a reference for the most of other routing protocols in the opportunistic network. The main drawback of the Epidemic scheme is its high overhead. Many schemes are proposed to decrease the overhead in Epidemic-based approaches by limiting the number of message replicas [14–16]. An effective scheme to control replication spread is the vaccine [17]. It applies the antipacket mechanism to control replica distribution in Epidemic-based routing. In [18], a new scheme is proposed to control the replication of epidemically distributed information. Based on the vaccine scheme, signal distribution is early controlled by the fully immunized vaccine. In addition, a partially immunized vaccine is initiated when there is a local-forwarding opportunity to vaccine more packets.

In the OMSN, mobile devices are portable by humans so that social features of people can be exploited for networking purposes [19–21]. Social-based protocols utilize social properties of mobile users such as similarity, centrality, and friendship to improve routing efficiency in the opportunistic mobile social network. This is because social features are more stable and less changeable than other features like mobility patterns. HiBOp [22] and CiPRO [23] exploit the similarity social feature and the user's context information to forward data. LASS [24] takes into account the difference of members' activity within the node's community for data dissemination. In ML-SOR [25], node centrality (different types of centralities), the similarity between communities, and social ties are all exploited to effectively select the forwarding node. MCAR [26] exploits the preferred communities of people during their daily lives for effective information delivery. Direct (inside one community) and indirect (via different communities) contacts are considered in MCAR. In SPRINT-SELF [27], network and node overhead is decreased by exploiting social information of mobile users. The authors consider the social community of nodes and utilize it to predict future behavior based on contact history. In addition, they proposed a new mechanism to avoid selfish nodes for more improvements.

Social features are utilized widely for buffer management. Liu et al. [28] utilized social features and the congestion level to develop the forwarding strategy that drops the message with the minimum social link rather than random dropping. In SRAMSW [29], the buffer management mechanism is combined with social features to enhance the spray-based routing. Expired messages are deleted, successfully delivered messages are acknowledged, and messages are prioritized according to their spray times and residence time. In addition, three social features: centrality, similarity, and friendship, are adopted for better forwarding decision and to avoid the dead-end problem.

We hypothesize that combining social features with the Epidemic-based forwarding scheme improves the efficiency of routing in the OMSN. In this paper, we present an Epidemic-based Social-based routing protocol (EpSoc) that combines the advantages of the forwarding strategy used in the Epidemic routing protocol with the positive impact of exploiting social features. EpSoc exploits the degree centrality social feature to adapt the time to live (TTL) of the routed message. If a message is forwarded to a node which has higher degree centrality (socially active node), its TTL value will be decreased. If these messages are dropped in the active node when TTL is zero, the blocking mechanism is used to reject receiving replications of these messages.

The rest of this paper is organized as follows: the next section reviews the related work. We describe in detail our proposed algorithm EpSoc in Section 3. In Section 4, we introduce the performance evaluation and the discussion results. Finally, Section 5 concludes the paper.

2. Related Works

Delivering data to the destination at the right time with minimum resources is an optimal condition for any given routing protocols. Epidemic has optimal performance in terms of delivery ratio and latency. However, it suffers from high overhead cost. One of the solutions is to exploit social information to improve Epidemic-based routing protocols in the OMSN. Degree centrality is a social feature that is exploited widely in the literature to improve routing in the mobile social networks. For example, Bubble Rap [11] utilizes the mobile node's degree centrality to provide cost-effective routing as compared to Epidemic routing. Bubble Rap exploits two social and structural metrics, namely, centrality and community. It selects high centrality nodes and community members of destination as relays. In the Bubble Rap algorithm, nodes belong to different sizes of communities and have different levels of popularity (i.e., rank). Each node is assumed to have two rankings: global denotes the popularity (i.e., connectivity) of the node in the entire society and local denotes the popularity within its community. Messages are forwarded to nodes that have higher global ranking until a node in the destination's community is found. Then, the messages are forwarded to nodes having a higher local ranking within the destination's community.

Similar to Bubble Rap, CAOR [30] exploits degree centrality and similarity social features to improve routing. However, CAOR constructed autonomous communities based on common interest locations between mobile nodes. Members of a community with high centralities act as the home of this community. CAOR also has a mechanism to turn the routing between lots of nodes to the routing between a few community homes. It also applies the reverse Dijkstra algorithm to determine the optimal relays and compute the minimum expected delivery delay. In [31], cultural algorithm (CA), ant colony optimization (ACO), and social connectivity between users are combined to address the routing problem. Social metrics of nodes including degree and betweenness centralities are analyzed to support forwarding decision in the opportunistic network environment.

Besides exploited in the routing problem, social features were also applied to solve different type of problems. For example, works in [32] exploit social features (degree and betweenness centralities) to solve the throwbox placement problem based on the given graph. A user's degree is equal to the total number of its neighbors, and betweenness is the total number of shortest paths passing through the node. The work also introduced the concept of location degree to measure how many mobile users have the location as one of their top visited locations and the location betweenness to show how important the location is for the entire social graph. Social metrics such as degree centrality, social activeness, and community acquaintance are also applied to enhance data delivery in VSNs in [33]. The social interactions between nodes where nodes of similar interests or nodes belonging to the same community have greater probability to encounter each other. Social centrality is also utilized for congestion control in the postdisaster environment [34].

Unlike the aforementioned works, instead of considering community or homing structuring, our proposed protocol is based on the flooding-based forwarding strategy of Epidemic. This is because we intended to design our forwarding protocol to have high delivery ratio and low delay as the Epidemic-based forwarding strategy. Moreover, we also aimed to avoid extra time required to form and maintain the community structure. Thus, to decrease the overhead, we utilized degree centrality and adapted the message's TTL (Time to live) to control the forwarding nodes' activity. The message's TTL is considered in literature when developing efficient routing protocols. Miao et al. [33] proposed the adaptive multistep routing protocol for mobile delay-tolerant networks (MDTNs). Their aim is to get the balance between the delay and cost of message delivery. The time to live of the message is used in order to allocate the minimum number of copies necessary to achieve a given delivery probability. In [34], the authors considered the contact information of nodes and the time to live message's property to make route decision and improve performance. They built a replica distribution criterion between two encountered nodes based on residual messages' TTL. In our proposed solution, we do not limit the number of replicas but allow messages to be spread in the network and then we utilize degree centrality to decrease the TTL and consequently decrease the overhead. Also, the blocking mechanism is adopted in the active node to cancel receiving replications of the same message which is seen previously.

3. EpSoc Routing Protocol

EpSoc is the routing algorithm designed to decrease overhead in the Epidemic protocol by embedding social features in routing the message in the opportunistic network. To achieve this objective, we propose two mechanisms. First, the message's TTL is adapted based on degree centrality of the nodes in the OMSN. The second one is the message blocking mechanism that is used to prevent receiving the replications of runout TTL messages in active nodes.

3.1. Degree Centrality. Node centrality indicates the popularity of the node in the network, whereby node centrality in a social network is the reflection of its social relative importance [35]. A higher node degree centrality means the node connects with many numbers of nodes in the network. A degree centrality for a given node i can be calculated as follows:

$$DC_i = \sum_{k=1}^{N} a(i, k), \tag{1}$$

where N is the number of nodes in the network and $a(i, k) = 1$ if a direct link exists between node i and node k and $i \neq k$.

We adopt the CWindow [11] calculation algorithm to calculate degree centrality. CWindow divides the day into time windows and calculates the average of nodes' degree over these windows to estimate the node's centrality. To decrease the processing overhead results from processing any changes of degree centrality, CWindow calculates the node's degree centrality at regular intervals instead of every time the centrality changes.

We pick the CWindow algorithm to calculate degree centrality because it takes into account the changes in node centrality over time and averages the node's centralities for few window intervals which is suitable for people's behavior in the OMSN. It is also used by other social-based protocols that consider the node's degree centrality such as Bubble Rap [11], Dlife [36], and SCORP [37].

3.2. EpSoc Forwarding Strategy. Figure 1 depicts the forwarding process in EpSoc, where two mechanisms are applied.

In Figure 1(a), node $N1$ encountered three nodes: $N2$, $N3$, and $N4$. $N2$ has higher degree centrality (DC) than $N1$. The TTL value of the forwarded messages to $N2$ is decreased by dividing it by the DC value of $N2$. We call these messages socially infected messages. Node $N2$ registers the ID of these messages as seen messages in its blocking register. Node $N2$ will drop the socially infected message when it expires. In Figure 1(b), the Epidemic forwarding strategy which we adopted in our algorithm causes replications of socially infected messages to be sent to node $N2$ from other nodes such as $N4$. In this case, node $N2$ will reject receiving the message with the reason that it is a seen message (its ID is stored in the block register).

Based on (1), the TTL value is adapted using the following equation:

$$TTL_{new} = \frac{TTL_{old}}{DC_k} \quad \{if\ DC_k > DC_i\}. \tag{2}$$

The two combined mechanisms adopted in EpSoc enhance routing performance. If a node is socially active, it meets more nodes in the network and has a higher potential to deliver more messages. Decreasing messages' TTL value in active nodes results in releasing space in their buffers and increases the ability to deliver more messages. This will increase the delivery ratio. The blocking mechanism controls the replications by preventing decreased TTL messages from hitting again the previously traversed actives nodes. The result is decreasing network overhead. Regarding average latency, the messages that are delivered by active nodes have a shorter end-to-end delay compared to other messages delivered by lower activity nodes, so average latency will be decreased too. We conclude that decreasing message life socially by combining the message blocking scheme affects positively the performance of the Epidemic routing scheme in the OMSN.

3.3. Pseudocode of the EpSoc Message Forwarding. We used CWindow to calculate the centrality of the node. Each node N_i records the encountered nodes in the network. When node N_i encounters N_j, the degree centrality values are calculated using CWindow. Then, centrality values are exchanged between both nodes. N_i compares its centrality value DC_i with the N_j centrality value DC_j. If DC_j is greater than DC_i, which means that N_j is socially more active than N_i, then for each message m_i in the N_i buffer Buf_{N_i}, the TTL value $m_{i_{TTL}}$ is decreased by dividing it by the node N_j centrality value DC_j. The ID of each socially infected message is registered in the N_j blocking register $Block_{N_j}$. The time complexity of the EpSoc forwarding algorithm is $O(M)$, where M is the number of messages carried in the node's buffer and needed to be sent or forwarded.

Algorithm 1 shows the complete pseudocode of EpSoc.

4. Performance Evaluations

4.1. Data Set. We adopt the Cambridge experimental data set of haggle [38]. This data set includes traces of Bluetooth

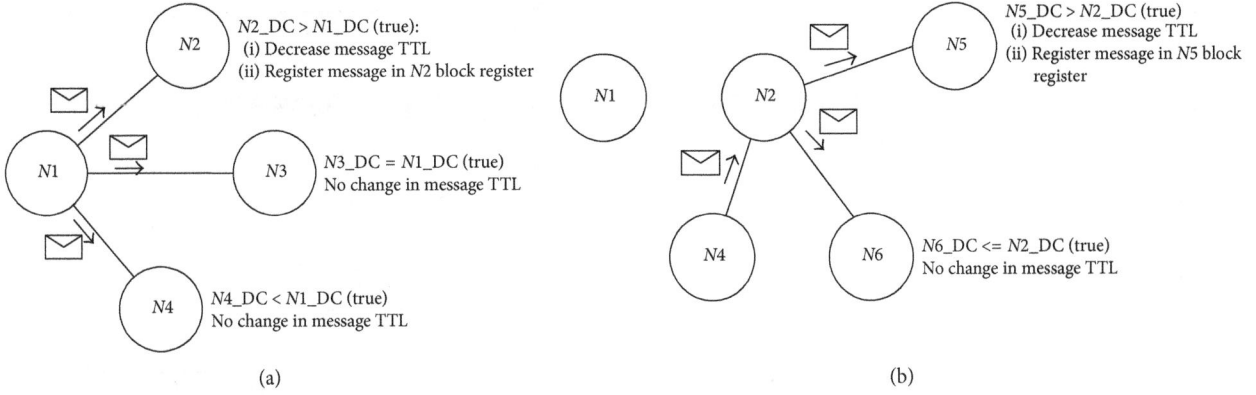

FIGURE 1: EpSoc forwarding scheme.

```
(1)   For all N_i ∈ N
(2)     if N_i encounter N_j
(3)       Calculate DC_i, DC_j
(4)       DC_i ⇆ DC_j
(5)       For all m_i ∈ Buf_{N_i}
(6)         if m_{i_{ID}} ∉ Block_{N_j}
(7)           if DC_j > DC_i
(8)             m_{i_{TTL}} ← m_{i_{TTL}}/DC_j
(9)             Buf_{N_j} ← m_i forward message to N_j
(10)            Block_{N_j} ← m_{i_{ID}}
(11)          End if
(12)        End if
(13)      End for
(14)    End if
(15) End for
```

ALGORITHM 1: Pseudocode of the forwarding strategy in EpSoc.

TABLE 1: Simulation settings.

Simulation time	987529 seconds
Interface	Bluetooth interface
No. of nodes	36
Transmit speed	250k (2 Mbps)
Mobility	Real trace data (Cambridge)
Buffer size	1, 5, 15, 25, 35, 45, and 55 MB
Routing protocols	Epidemic, EpSoc, and Bubble Rap
Message size	128k
Event interval	30 to 40 seconds
Initial message TTL	10 m, 30 m, 1 h, 3 h, 5 h, 12 h, 1 d, 1.5 d, 2.5 d, 3 d, 4 d, and 1 w

sightings by groups of users carrying small devices (iMotes) for a number of days in campus environments. The experiments are conducted in the computer laboratory that includes the undergraduate first-year and second-year students and also some Ph.D. and postgraduate students which lasted for 11 days.

4.2. Simulation Setup. We use the opportunistic network environment (ONE) [39] simulator to evaluate our algorithm. Also, comparison with Epidemic and Bubble Rap routing protocols is included to measure the performance of our proposed algorithm EpSoc. We want to justify that the Epidemic algorithm performs better when considering social features in forwarding messages. The simulator settings are tabulated in Table 1.

In each experiment, we compare the performance of the protocols EpSoc, Epidemic, and Bubble Rap based on the following metrics.

4.2.1. Successful Delivery Ratio. It is the ratio between the number of delivered messages and the total number of created messages. The ideal value of the successful delivery ratio is 1.0 when all created messages are delivered to their destinations.

4.2.2. Overhead Ratio. It is the additional bytes that are sent for successfully delivering a message to a destination.

4.2.3. Average Latency. It is the average of the time elapsed between message creation and delivery.

4.2.4. Average Hop Count. It is the average of the number of hops that messages must take in order to reach the destination.

4.3. Experiments and Discussion. To evaluate our work, we will consider two features: buffer size and message TTL value. These two features have a high impact on routing performance in the OMSN. In our work, we affect these two features. We adjust the TTL value socially and use the blocking mechanism to manage buffer storing. Consequently, our experiments are to measure the performance of EpSoc when varying the TTL and buffer size. The comparison will be with Epidemic and Bubble Rap protocols.

4.3.1. Varying Buffer Size. For varying the buffer size, we fixed the value of TTL to 2.5 d. Figures 2–5 show the performance comparisons between EpSoc, Epidemic, and Bubble Rap in terms of delivery ratio, overhead ratio, average latency, and average hop count, respectively.

Figure 2 shows the delivery ratio with buffer size. Generally, for Epidemic, Bubble Rap, and EpSoc, increasing the

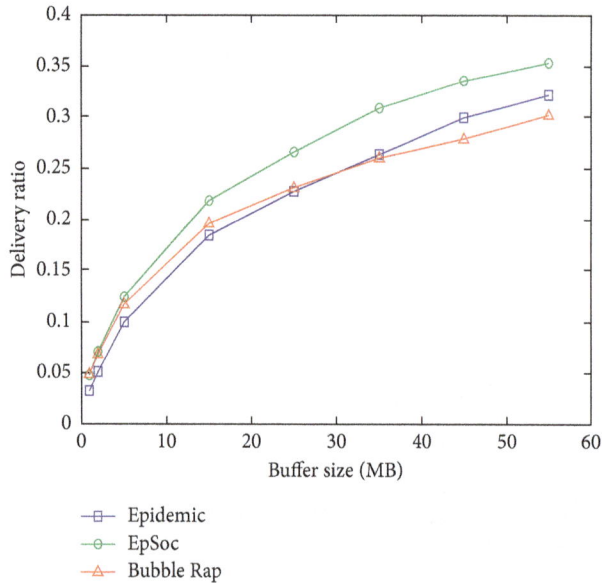

FIGURE 2: Delivery ratio versus buffer size.

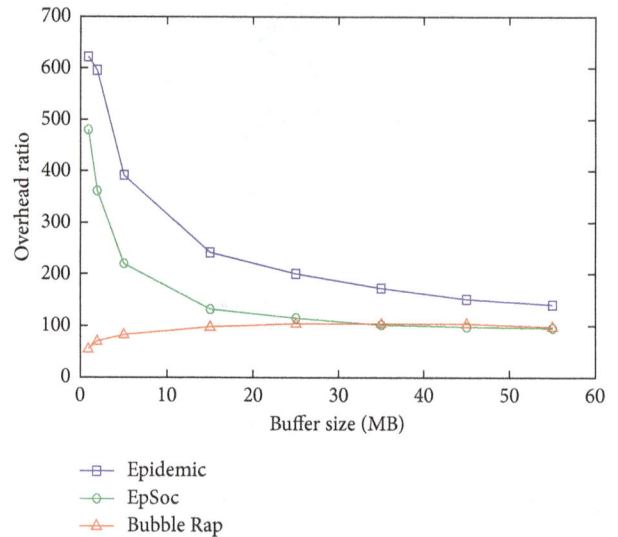

FIGURE 3: Overhead ratio versus buffer size.

buffer size will increase the delivery ratio. This is because more messages can be carried by intermediate nodes which consequently deliver more messages to destinations. Changing the number of delivered messages affects the delivery ratio, overhead ratio, average latency, and average hop count. Regarding the delivery ratio, it is clear that delivering more messages results in a higher value, while the decrease in delivered messages results in a lower value. In the lower buffer size scenario (1–25 MB), the delivery ratio of Epidemic is the lowest because of the redundancy. Bubble Rap and EpSoc outperform Epidemic due to utilizing social features. In higher buffer size scenarios (35–55 MB), Epidemic achieves higher delivery ratio than Bubble Rap. Larger buffer size alleviates the negative impact of dropping messages because of replication and enables Epidemic to deliver more messages. Our protocol EpSoc outperforms both Epidemic and Bubble Rap. Blocking runout TTL messages from being received by active nodes results in more space in their buffer and therefore can carry more different messages when encountering other nodes and later deliver them to destinations. In addition, the decrease of TTL of the messages that are forwarded to the more active nodes results in better utilization of the buffer's space. Decreased TTL message copies are dropped earlier enabling carrying more other messages. Active nodes deliver the message quickly. Therefore, the number of delivered messages is increased, and consequently, the delivery ratio is increased.

The relation of overhead ratio with buffer size is shown in Figure 3. When increasing the buffer size, the overhead is decreased for all algorithms. Regarding Bubble Rap, overhead is decreased with buffer increase because no replication exists. We observe from Figure 3 that both protocols Bubble Rap and EpSoc outperform Epidemic significantly and EpSoc outperforms Epidemic. We mentioned formerly that applying the strategy of blocking messages and decreasing TTL increase the number of delivered messages. In addition, the blocking message to be resent to active nodes decreases

the number of replications in the network and hence decreases the forwardings. As a result, decreasing the forwardings and increasing the delivered messages decrease the overhead ratio in the network. Bubble Rap achieves lower overhead than EpSoc for lower buffer size scenarios (1, 2, 5, and 15 MB). This is because of two reasons: first, the replication and Epidemic strategy adopted in EpSoc, and second, low buffer size causes dropping relayed messages early due to buffer overflow which in turn decreases the efficiency of our proposed algorithm compared to Bubble Rap in terms of overhead. However, for larger buffer size scenarios (25, 35, 45, and 55 MB), EpSoc is more efficient and its overhead ratio is very close to that of Bubble Rap (for 35, 45, and 55 MB, it is slightly better than Bubble Rap).

Figure 4 shows that average latency for all the three protocols, which increases when the buffer size is increased. A low-sized buffer only relays low-latency messages which will reach their destinations quickly. On the other hand, a higher-sized buffer allows messages to be carried for a longer time which contribute to a higher average latency. EpSoc decreases the average latency significantly. EpSoc optimizes the buffer usage where messages forwarded to active nodes have low TTL. With the low TTL, a relay node has more free space for new messages in its buffer as the buffer quickly dropped the old messages.

In Figure 5, the average hop count is recorded. The average hop count increases when the buffer size is increased. Epidemic has more hop count which indicates more nodes experiencing the duplicated messages, whereas Bubble Rap has lower hop count because it prevents replication of messages. For EpSoc, the number of hop count is contributed by allowing replication as in Epidemic.

The worst achievement is because of redundancy. Bubble Rap does not apply replication so that it outperforms EpSoc. EpSoc achieves better average hop count than Epidemic. This is because of the message blocking strategy which decreases the number of replications in the network and hence decreases the forwardings.

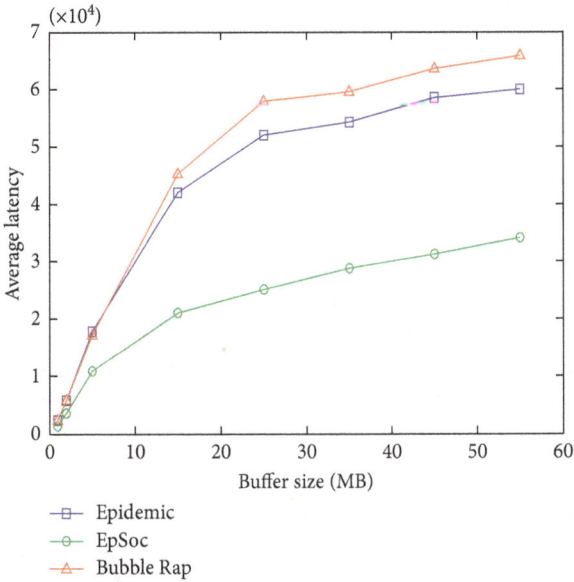

FIGURE 4: Average latency versus buffer size.

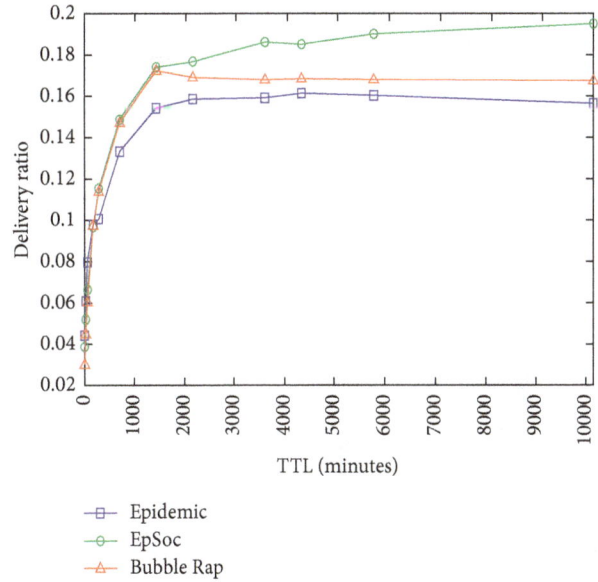

FIGURE 6: Delivery ratio versus TTL.

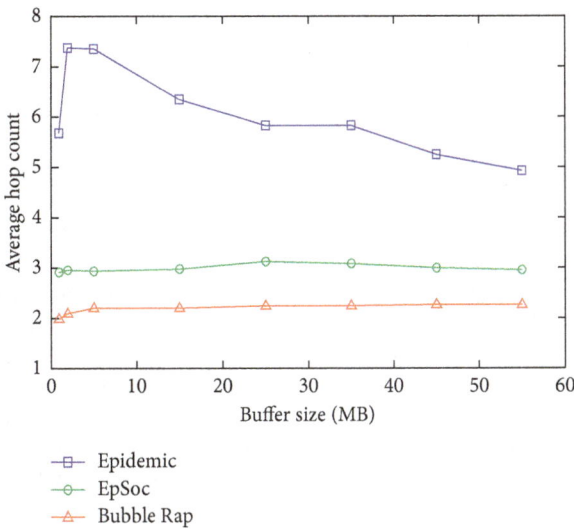

FIGURE 5: Average hop count versus buffer size.

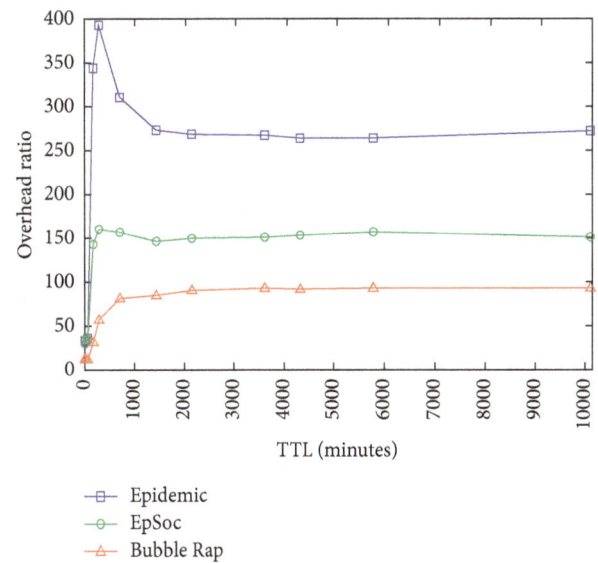

FIGURE 7: Overhead ratio versus TTL.

4.3.2. Varying Initial TTL. TTL determines the life of the message in the network. A high value of TTL gives high chances of the message delivered to the target destination and vice versa.

In Figure 6, the relation of delivery ratio with TTL is depicted. For lower TTL (10 m–3 h), the delivery ratio of Epidemic is slightly higher than that of Bubble Rap and EpSoc. The reason is that, for short TTL messages, exploiting social features will not be very effective due to quickly messages dropping. In addition, the higher replication occurred in Epidemic increases the number of delivered messages. When TTL is increased, Bubble Rap and EpSoc outperform Epidemic due to utilizing the social features. In high TTL scenarios (1.5 d–1 w), EpSoc outperforms Bubble Rap. This indicates an efficiency of EpSoc in its social feature selection.

Shortening the life of messages in the active nodes and blocking runout TTL messages from hitting again the active nodes cause an increment in the delivered messages and decrement in forwardings. Therefore, delivery ratio grows up.

Figure 7 compares the performance of the protocols with different values of TTL in terms of overhead ratio. When TTL is very low (10 m–1 h), Epidemic, EpSoc, and Bubble Rap achieve low overhead. The reason is that messages are dropped quickly. For high TTL values, Bubble Rap achieves the best and Epidemic the worst.

Our protocol EpSoc manages to decrease the overhead better than Epidemic. This is because EpSoc is a combination of the flooding-based forward strategy with social features.

From Figure 8, in terms of average latency, EpSoc appears to be outperforming others especially with high TTL

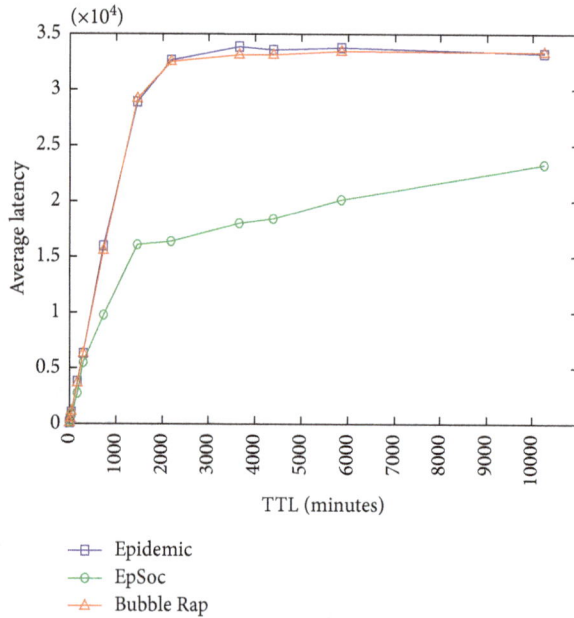

FIGURE 8: Average latency versus TTL.

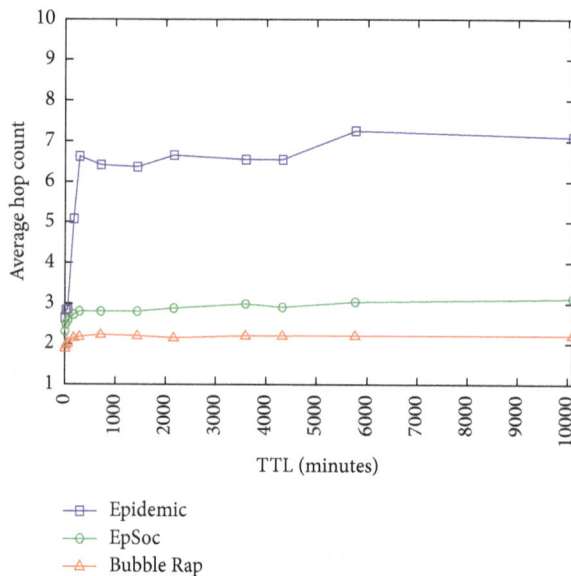

FIGURE 9: Average hop count versus TTL.

based forwarding strategy. For EpSoc, the average hop count is decreased significantly compared to Epidemic because of the social effect of active nodes.

5. Conclusion

In this paper, we investigate the flooding-based forwarding strategy, that is, Epidemic with social features to improve the routing performance in the opportunistic mobile social network (OMSN). Inspired by the advantages of Epidemic routing protocols in terms of delivery ratio and delivery delay and by exploiting the social activity of nodes, we formulated a flooding-based social-based routing protocol named as EpSoc. Simulation experiments using real data sets are conducted to evaluate our protocol performance. From the presented results, our approach increases the delivery ratio and decreases the delivery overhead, average latency, and average hop count as compared to the Epidemic protocol. As for benchmark social-based routing protocol performance (Bubble Rap), our protocol decreases the average latency significantly with a better delivery ratio in high buffer size and low TTL scenarios. Generally, we manage to exploit the advantage of Epidemic and Bubble Rap to improve the data dissemination in the OMSN.

Additional Points

Significance. Social features of a node can be utilized to have an effective role in the routing protocol in the OMSN network. This paper presents the routing protocol that exploits degree centrality to increase the delivery ratio and decrease the overhead and latency. Moreover, exploiting other social features such as similarity and community may lead to being more efficient routing protocol. Therefore, combining social features with other forwarding techniques such as the Epidemic-based strategy is significant to have an efficient forwarding strategy in the OMSN to support green technologies.

References

[1] A. Chaintreau, A. Mtibaa, L. Massoulie, and C. Diot, "The diameter of opportunistic mobile networks," in *Proceedings of the 2007 ACM CoNEXT Conference*, p. 12, ACM, New York, NY, USA, December 2007.

[2] M. Conti, S. Giordano, M. May, and A. Passarella, "From opportunistic networks to opportunistic computing," *IEEE Communications Magazine*, vol. 48, no. 9, pp. 126–139, 2010.

[3] K. Zhu, W. Li, X. Fu, and L. Zhang, "Data routing strategies in opportunistic mobile social networks: taxonomy and open challenges," *Computer Networks*, vol. 93, pp. 183–198, 2015.

[4] Y. Liu, Z. Yang, T. Ning, and H. Wu, "Efficient quality-of-service (QoS) support in mobile opportunistic networks," *IEEE Transactions on Vehicular Technology*, vol. 63, no. 9, pp. 4574–4584, 2014.

[5] L. Pelusi, A. Passarella, and M. Conti, "Opportunistic networking: data forwarding in disconnected mobile ad hoc

values (12 h–1 w). This is because our algorithm always enables the active node to carry messages with lower TTL.

From Figure 9, we observed that if TTL is very low (10 m–1 h), all routing protocols in the experiment have low average hop count. This is due to the low number of forwardings between nodes as message life exhausted quickly. Bubble Rap and EpSoc have almost stable performance when TTL is increased. Exploited social features in Bubble Rap and EpSoc and applying the blocking mechanism in EpSoc decrease the effect of changing the value of TTL on the number of traversed nodes to the destination. On the contrary, when the TTL is increased, the average hop count of Epidemic increases significantly because of the flooding-

networks," *IEEE Communications Magazine*, vol. 44, no. 11, pp. 134–141, 2006.

[6] T. Le and M. Gerla, "Social-distance based anycast routing in delay tolerant networks," in *Proceedings of the 2016 Mediterranean Ad Hoc Networking Workshop (Med-Hoc-Net)*, Vilanova i la Geltru, Spain, June 2016.

[7] K. Fall, "A delay-tolerant network architecture for challenged internets," in *Proceedings of the 2003 Conference on Applications, Technologies, Architectures, and Protocols for Computer Communications*, New York, NY, USA, August 2003.

[8] M. Alajeely, R. Doss, and A. Ahmad, "Routing protocols in opportunistic networks–a survey," *IETE Technical Review*, pp. 1–19, 2017.

[9] A. Vahdat and D. Becker, "Epidemic routing for partially-connected ad hoc networks," Tech. Rep. CS-200006, Duke University, Durham, NC, USA, 2000.

[10] A. Lindgren, A. Doria, and O. Schelen, *Probabilistic Routing in Intermittently Connected Networks*, Springer, Berlin, Germany, 2004.

[11] P. Hui, J. Crowcroft, and E. Yoneki, "BUBBLE Rap: social-based forwarding in delay-tolerant networks," *IEEE Transactions on Mobile Computing*, vol. 10, no. 11, pp. 1576–1589, 2011.

[12] P. Pholpabu and L. L. Yang, "Routing protocols for mobile social networks achieving trade-off among energy consumption, delivery ratio and delay," in *Proceedings of the 2015 IEEE/CIC International Conference on Communications in China (ICCC)*, Shenzhen, China, November 2015.

[13] M. Nikoo, F. Ramezani, M. Hadzima-Nyarko, E. K. Nyarko, and M. Nikoo, "Flood-routing modeling with neural network optimized by social-based algorithm," *Natural Hazards*, vol. 82, no. 1, pp. 1–24, 2016.

[14] E. Bulut, Z. Wang, and B. K. Szymanski, "Cost-effective multiperiod spraying for routing in delay-tolerant networks," *IEEE/ACM Transactions on Networking (TON)*, vol. 18, no. 5, pp. 1530–1543, 2010.

[15] T. Matsuda and T. Takine, "(p,q)-epidemic routing for sparsely populated mobile ad hoc networks," *IEEE Journal on Selected Areas in Communications*, vol. 26, no. 5, pp. 783–793, 2008.

[16] P. Mundur, M. Seligman, and J. N. Lee, "Immunity-based epidemic routing in intermittent networks," in *Proceedings of the 2008 5th Annual IEEE Communications Society Conference on Sensor, Mesh and Ad Hoc Communications and Networks*, San Francisco, CA, USA, June 2008.

[17] P. Y. Chen, S. M. Cheng, and K. C. Chen, "Optimal control of epidemic information dissemination over networks," *IEEE Transactions on Cybernetics*, vol. 44, no. 12, pp. 2316–2328, 2014.

[18] C. Y. Aung, I. W. H. Ho, and P. H. J. Chong, "Store-carry-cooperative forward routing with information epidemics control for data delivery in opportunistic networks," *IEEE Access*, vol. 5, pp. 6608–6625, 2017.

[19] Y. Zhu, B. Xu, X. Shi, and Y. Wang, "A survey of social-based routing in delay tolerant networks: positive and negative social effects," *IEEE Communications Surveys & Tutorials*, vol. 15, no. 1, pp. 387–401, 2013.

[20] C. C. Sobin, V. Raychoudhury, G. Marfia, and A. Singla, "A survey of routing and data dissemination in delay tolerant networks," *Journal of Network and Computer Applications*, vol. 67, pp. 128–146, 2016.

[21] P. Pholpabu and L.-L. Yang, "Social contact probability assisted routing protocol for mobile social networks," in *Proceedings of the IEEE 79th Vehicular Technology Conference (VTC Spring)*, Seoul, South Korea, May 2014.

[22] C. Boldrini, M. Conti, and A. Passarella, "Exploiting users' social relations to forward data in opportunistic networks: the HiBOp solution," *Pervasive and Mobile Computing*, vol. 4, no. 5, pp. 633–657, 2008.

[23] H. Nguyen and S. Giordano, "Context information prediction for social-based routing in opportunistic networks," *Ad Hoc Networks*, vol. 10, no. 8, pp. 1557–1569, 2012.

[24] Z. Li, C. Wang, S. Yang, C. Jiang, and X. Li, "LASS: local-activity and social-similarity based data forwarding in mobile social networks," *IEEE Transactions on Parallel and Distributed Systems*, vol. 26, pp. 174–184, 2015.

[25] A. Socievole, E. Yoneki, F. De Rango, and J. Crowcroft, "ML-SOR: message routing using multi-layer social networks in opportunistic communications," *Computer Networks*, vol. 81, pp. 201–219, 2015.

[26] P. Pholpabu and L.-L. Yang, "A mutual-community-aware routing protocol for mobile social networks," in *Proceedings of the Global Communications Conference (GLOBECOM)*, Austin, TX, USA, December 2014.

[27] R. Ciobanu, C. Dobre, V. Cristea, F. Pop, and F. Xhafa, "SPRINT-SELF: social-based routing and selfish node detection in opportunistic networks," *Mobile Information Systems*, vol. 2015, Article ID 596204, 12 pages, 2015.

[28] Y. Liu, K. Wang, H. Guo, Q. Lu, and Y. Sun, "Social-aware computing based congestion control in delay tolerant networks," *Mobile Networks and Applications*, vol. 22, no. 2, pp. 1–12, 2016.

[29] J. Guan, Q. Chu, and I. You, "The social relationship based adaptive multi-spray-and-wait routing algorithm for disruption tolerant network," *Mobile Information Systems*, vol. 2017, Article ID 1819495, 13 pages, 2017.

[30] M. Xiao, J. Wu, and L. Huang, "Community-aware opportunistic routing in mobile social networks," *IEEE Transactions on Computers*, vol. 63, pp. 1682–1695, 2014.

[31] A. C. K. Vendramin, A. Munaretto, M. R. Delgado, M. Fonseca, and A. C. Viana, "A social-aware routing protocol for opportunistic networks," *Expert Systems with Applications*, vol. 54, pp. 351–363, 2016.

[32] Y. Zhu, C. Zhang, X. Mao, and Y. Wang, "Social based throwbox placement schemes for large-scale mobile social delay tolerant networks," *Computer Communications*, vol. 65, pp. 10–26, 2015.

[33] J. Miao, O. Hasan, S. B. Mokhtar, L. Brunie, and G. Gianini, "A delay and cost balancing protocol for message routing in mobile delay tolerant networks," *Ad Hoc Networks*, vol. 25, pp. 430–443, 2015.

[34] H. Chen and W. Lou, "Contact expectation based routing for delay tolerant networks," *Ad Hoc Networks*, vol. 36, pp. 244–257, 2016.

[35] F. Xia, L. Liu, J. Li, J. Ma, and A. V. Vasilakos, "Socially aware networking: a survey," *IEEE Systems Journal*, vol. 9, pp. 904–921, 2015.

[36] W. Moreira, P. Mendes, and S. Sargento, "Opportunistic routing based on daily routines," in *Proceedings of the World of wireless, mobile and multimedia networks (WoWMoM)*, San Francisco, CA, USA, June 2012.

[37] W. Moreira, P. Mendes, and S. Sargento, "Social-aware opportunistic routing protocol based on user's interactions and interests," in *Proceedings of the International Conference on Ad Hoc Networks*, vol. 129, pp. 100–115, Springer International Publishing, Barcelona, Spain, August 2014.

[38] Website of CRAWDAD project, http://crawdad.org/cambridge/haggle/20090529/.

[39] A. Keränen, J. Ott, and T. Kärkkäinen, "The ONE simulator for DTN protocol evaluation," in *Proceedings of the 2nd International Conference on Simulation Tools and Techniques*, Brussels, Belgium, March 2009.

13

Robust System Design using BILP for Wireless Indoor Positioning Systems

Kriangkrai Maneerat and **Kamol Kaemarungsi**

National Electronics and Computer Technology Center, NSTDA, Pathumthani, Thailand

Correspondence should be addressed to Kamol Kaemarungsi; kamol.kaemarungsi@nectec.or.th

Academic Editor: Joaquín Torres-Sospedra

In wireless indoor positioning system designs, reference node (RN) failures during the online phase cause received signal strength values to be unavailable. This leads to accuracy performance degradation and a lack of system reliability in smart office systems. Moreover, the major design concern in the reliability of indoor positioning systems under the faulty RNs during the online phase has not been yet investigated in previous works. To address these gaps, we propose a novel mathematical formulation using a Binary Integer Linear Programming (BILP) approach that employs the Simulated Annealing (SA) solution technique. The proposed robust system design aims to put in place a suitable number of RNs and to determine their optimum locations, which may be located on a single floor or on multiple floors. In particular, the proposed system design provisions to support robust operation both during a normal situation and when there are some RN failures. Experimental results and comparative performance evaluation revealed that the proposed robust system design outperformed other system designs and was able to achieve the highest location accuracy performance in both fault-free and RN-failure scenarios. Specifically, when nine of the RNs in a three-story building failed, the proposed system design achieved 84.6%, 54.7%, and 32.9% more accurate performance than the Uniform, the MSMR, and the PhI-Uni, respectively.

1. Introduction

The rapid growth of wireless communication technologies is the major driving force behind the integration of the various embedded devices and systems in the vision of smart cities. The systems in a smart city deploy a combination of data collection, processing, and allocation technologies in conjunction with network and computer technologies. These provide several advantages, such as an effective resource allocation as well as higher data security and privacy, facilitating the development of smart cities and enhancing the quality of life for their citizens [1]. One of the applications currently deployed for smart cities is location-based services. For example, Li et al.[2] presented a wireless indoor positioning system that can monitor and track the location of the visitors in a smart office building. Nazari et al. [3] proposed the development of a hospital solution framework for smart health systems, in which the staff can quickly check the location and availability of the needed equipment in an emergency situation. Wang et al. [4] presented indoor location services for smart shopping systems, which enable customers to ask the system to find the shortest path for them to obtain the desired product in a supermarket. Therefore, this article focuses on indoor positioning systems inside smart buildings and smart offices that can be considered as important parts of smart cities.

A wireless indoor positioning system is one of the key capabilities of context-aware computing, allowing the system to determine the location of a mobile station inside a building. The physical information of a mobile station with respect to a set of reference positions within a service area is analyzed and used to estimate its location by using a localization algorithm. Thus, the challenges of indoor positioning systems in terms of location estimation are not only to achieve performance accuracy in finding the required location, but also to overcome completely various indoor positioning issues, such as the problems arising from a dense indoor environment (e.g., multipath effect) [5], technology capabilities (e.g., hardware limitations) [6], and the types of service areas (e.g., inside multistory buildings) [7]. Generally, the operation of a wireless indoor positioning system is divided into two processes: the system design phase and the localization phase.

First, in the system design phase, various wireless network configurations are considered and determined by the system designer. These are provisioned to support the indoor positioning system before implementing and deploying the actual localization system. This is a necessary and indeed essential process for an indoor positioning system to achieve any required performance goal. An example of wireless network configurations that are considered in the system design phase include the type of applications, the number and installed locations of reference devices, size of service area, and wireless device specifications (e.g., wireless technologies, radio frequency channels, and transmit power) [5, 8].

In the existing literature, on the system design phase, the research focused on various issues related to meeting required indoor positioning performance. In [9–11], the authors investigated the impact of reference node (RN) placement on location accuracy performance. They compared the positioning performance obtained from different RN placements, such as comparing the placement of various basic geometric layouts [9], symmetrical/unsymmetrical layouts [10], or overlapping and hierarchical clustering [11]. Other system design researches aimed at developing the RN-placement techniques to determine the optimum number and location. These researches focused on essential issues that should be considered in the design of wireless indoor positioning systems, such as the radio signal coverage requirements [12, 13], the minimization of localization error [14], and the selection of the fitness values for the RN-placement techniques [15].

Second, in the localization phase, the physical information at the target location is measured and then analyzed to estimate its coordinates using localization algorithms. An example of these localization algorithms include the algorithms based on the Triangulation approach, such as Time Difference of Arrival (TDOA) [16] and Angle of Arrival (AOA) [17]. Another example of a localization approach called Scene Analysis utilizes a database of received signal strength (RSS) patterns recorded during the offline phase (i.e., location fingerprint database). The sample vectors of RSS values are matched with the patterns in the database during the online phase to estimate the location of the target [4, 18]. The algorithms used in the Scene Analysis approach are the probabilistic methods [18], the Weighted K-Nearest Neighbor (WKNN) [19, 20], and the neural networks [21].

Based on the localization phase, several recent researches have aimed at investigating the properties of the RSS values inside the building. The findings of these analyses are needed in order to understand the underlying features of RSS characteristics, such as the distribution of RSS values [22] and the effect of the human body on RSS values [23]. Other researches have aimed at the development of the underlying mechanism for the localization algorithms, whereby those researches focused on improving the accuracy and the precision performance [24] and on reducing the computational complexity in location estimation [25].

One of the main challenges in wireless indoor positioning systems is the reliability. RN failures during the online (location determination) phase cause RSS values to be unavailable. This leads to performance degradation of localization accuracy and a lack of system reliability. In the design of indoor positioning systems, reliability is the capability required in order to maintain the functionality of the location determination in uncontrollable environments. In such cases, some RNs in the systems may fail, which could affect the entire operation of the indoor positioning system [26]. Under such unexpected situations, the performance accuracy of the system could drop by almost half [27]. According to the literature reviewed in Section 2, although some existing works have studied the reliable network design problem under node failure [28–30], they only investigated the provisioning of systems designed for telecommunications networks, and these network design solutions cannot be used effectively for indoor positioning systems. Furthermore, the system design for an indoor positioning system that supports the RN-failure scenario during the online phase has not yet been investigated in the existing works. Thus, in this article, we will investigate how to design an RN-placement technique for robust indoor positioning systems that can overcome the problem of when some RNs fail. The major contributions of our article are as follows:

(i) We propose a robust system design for wireless indoor positioning systems based on the location fingerprinting technique. The proposed system design is provisioned to support the robust operation both during a normal situation and when some RNs fail. Our proposed design can be applied to various service area structures ranging from single-floor to multiple-floor environments.

(ii) We have developed a novel mathematical formulation using a Binary Integer Linear Programming (BILP) approach that employs simulated annealing (SA) based on the heuristic solution technique to solve the design problem for wireless indoor positioning systems.

The remainder of this manuscript is organized into six sections as follows. In Section 2, we briefly summarize existing works on the system design for indoor positioning systems. Section 3 provides the problem definition and the problem formulation. Section 4 describes the experimental environment, the wireless transceivers, and the setup parameters used in this work. Section 5 presents the experimental results and discussion. Finally, Section 6 concludes the findings of this article.

2. Related Works

In wireless sensor network (WSN) system design, the positioning of the nodes can affect numerous network performance metrics. The placement of any node will affect overall data collection and must take into account the condition of the physical environment. To ensure usability, it is necessary to propose a reasonable method for the installation of nodes in the network [31]. Moreover, many current applications require the wireless network to continue functioning under unexpected situations and hostile environments, such as node failures. For these reasons, the reliability of the network also requires considerable interest. Recently, several studies have focused on different system designs of reliable and survivable wireless networks to support robust operation in uncontrollable environments. These studies have proposed system design methods that can achieve the reliability and survivability requirements for wireless networks.

In the system design of telecommunications networks, node failures will result in the complete outage of some wireless links in the network. This situation can cause disruption to the network's connectivity and loss of availability. Therefore, several studies have focused on the survivability of the network systems and how to handle the problem of failure situations. Correia et al. [28] presented a fault-tolerance network plan for the wireless optical broadband access network (WOBAN). Their fault-tolerance model can be applied to any scenario of optical wireless failure. Liu et al. [29] presented a reliable mixed-integer programming model for the IP layer network. Their system design model was provisioned to support faulty relay nodes caused by hardware failures or by an overloaded network. Luo et al. [30] presented a gateway placement approach for wireless mesh networks (WMNs). The optimal number and locations of the gateways are provisioned to reduce the gateway interference and provide fault-tolerance assurance. Based on the system design solutions presented in the literature, node failures in a telecommunications network may cause the network either to become disconnected or to have unavailable wireless links in the network. In particular, this may result in complete network outage. To address this problem, several research studies have proposed reliable system designs, whereby those developed systems could still function after some failure of certain network components, usually based on the provision of backup links or bandwidth management schemes.

While the worst-case scenario of node failures in a telecommunications network may result in the network's unavailability, the situations are different for indoor positioning systems. The faulty RNs in indoor positioning systems may decrease the location performance accuracy, although the systems can still provide location estimations. However, the localization accuracy performance of such systems may be less than half of normal if the system is affected by RN failures during the online estimation phase. There is a significant difference between the telecommunications systems and the indoor positioning systems in that the general network design solutions cannot be used effectively for indoor positioning systems.

With regard to the system design for indoor positioning systems, several studies have focused on addressing various design issues in order to meet performance requirements. In [12, 32, 33], the authors proposed a system design for indoor positioning systems based on location fingerprinting techniques. Zhang et al. [12] presented a mathematical formulation of RN placement that aims to minimize the number of RNs, while ensuring that their locations still have sufficient coverage of the service area. Sharma et al. [32] proposed an RN-placement method that aims to minimize the total

number of similar fingerprints so as to achieve better location accuracy. Fang et al. [33] presented a framework for linking the RN placement and the positioning performance. The objective of their algorithm was to choose a suitable set of RN locations so that the signal-to-noise ratio (SNR) was maximized.

Other research works on system design focused on improving the indoor positioning systems based on a triangulation approach [13–15]. Aomumpai et al. [13] proposed a system design approach based on a genetic algorithm to find the optimal solution of RN placement in an RSS-based localization approach. They focused on minimizing the average localization error and maximizing the signal coverage of the service area. Redondi et al. [14] presented an RN-placement method for indoor positioning systems based on the Cramer-Rao lower bound (CRLB) approach. The objective of their system design was to minimize localization errors when operating with a limited number of RNs due to budget constraints are fixed. Merkel et al. [15] presented an optimal RN-placement approach for an indoor positioning system based on distributed range-free localization. They focused on achieving the optimal coverage of a certain area while simultaneously minimizing the necessary number of RNs.

In the literature, existing system designs for indoor positioning systems limit their focuses to the achievement of location accuracy or the provision of signal radio coverage in the service area. Furthermore, the major design concern in the reliability of indoor positioning systems under the faulty RNs during the online phase has not yet been investigated in the exiting works. Unfortunately, existing solutions to survivable network design problems for telecommunications networks cannot be used effectively for indoor positioning systems. Based on these knowledge gaps, the design of RN placement for robust indoor positioning systems is still an open research issue. Therefore, in this work, we propose a robust system design model for wireless indoor positioning systems based on location fingerprinting techniques. Our proposed model aims to place a suitable number of RNs and to determine their locations whereby their placement is provisioned to support robust system operation both during a normal situation and when some RNs have failed.

3. Problem Definition and Formulation

In this section, we describe our mathematical models for the problem of RN placement in indoor multifloor positioning systems. First, in Section 3.1, the problem definition of wireless indoor multifloor positioning systems under RN-failure scenarios is described. Then, in Section 3.2, we explain our mathematical formulation for the robust system design problem. Finally, in Section 3.3, the overall framework of the solution technique is described.

3.1. Definition of System Design for Wireless Indoor Positioning Systems. In wireless indoor positioning system designs, placement of insufficient RNs can lead to accuracy performance degradation. Furthermore, the system may achieve less than half of its intended performance levels during the online estimation phase in the event of RN failures [27]. Under unexpected situations, such as some RNs being faulty during the online phase, unreliable location results are presented because of the unavailability of RSS component(s) during the location estimation process. For example, Figure 1 illustrates an example of the structure of an indoor positioning system in a three-story building (e.g., a smart office building) under the RN-failure scenario in which two RNs have failed. In this diagram, four RNs are installed on each floor. The dashed lines represent the RSS values that a target node receives from all RNs. In Figure 2, one RN on the 1st floor and another RN on the 2nd floor have become unavailable. This may be caused by hardware failures or software errors. In either case, the online scanned RSS values of the target node would not have signals from these RNs and the faulty RNs could reduce the location estimation performance.

To ensure the reliability of indoor positioning systems, a system design model that can handle situations of signal unavailability due to RN failure is required. Furthermore, the RN-placement design model as a major system concern in the reliability of indoor positioning systems in the event of some faulty RNs during the online phase has not been investigated.

3.2. Robust System Design Model. In this section, we describe the robust system design for wireless indoor positioning systems. The proposed system design is called the Robust-Maximum Summation of Max RSSI (R-MSMR), which is enhanced from our previous mathematical model presented in [34]. The task of the proposed R-MSMR is to place a sufficient number of RNs and determine their optimum locations, which may be located on a single floor or on multiple floors. In designing such a system, we focus on the network planning in terms of the wireless network infrastructure in which RNs serve as the wireless nodes in the network. The number of RNs and their locations are determined to ensure that the radio signal coverage is sufficient for the service area. In particular, the robust system design should result in a design solution that supports robust operation both during a normal situation and when some RNs have failed. We formulate the mathematical formulation of the robust system design problem as a Binary Integer Linear Programming (BILP) problem. This deals with models that are the same as linear programming with one additional restriction. The variables of BILP have integer values in which all variables are binary variables [35]. We adopt Simulated Annealing (SA) based on the heuristic solution technique to solve the BILP problem for the R-MSMR because of its simplicity and effectiveness. The SA is a variation of a hill-climber heuristic search approach. The SA allows the search to move to nonimproving solutions with a certain probability. This allows the SA search to avoid being trapped at local optima [36]. The R-MSMR based on BILP consists of two main components: the objective function and the constraints. The notations defined and used in the mathematical formulation of the robust system design problem are summarized in Table 1.

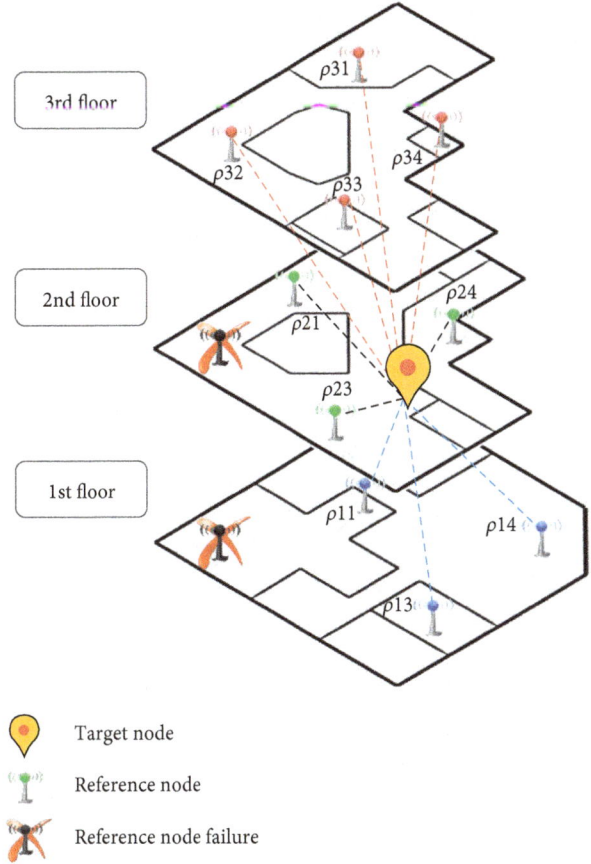

FIGURE 1: Floor determination schematic diagram.

We developed the robust system design in order to solve the RN-failure problem in wireless indoor positioning systems. The proposed system design not only achieves high positioning accuracy during normal situations but also yields reliable location results under unexpected situations such as RN failures. The main objective of the proposed R-MSMR is to place a sufficient number of RNs in optimum locations so that the system can achieve a maximizing summation of the maximum RSS at the signal test points (STPs) received from the RNs installed in the service area as written in the objective function:

$$\text{Maximize} \sum_{\forall i \in T} \max_{\forall j \in B} (S_{ij} P_{ij}). \tag{1}$$

The constraints for the R-MSMR are as follows: constraint (2) states that STP i is under the coverage of RN j if the signal strength received at STP i from RN j (P_{ij}) is greater than the sensitivity threshold P_T. This constraint is used to ensure a high quality of radio signal propagation in the service area. Constraint (3) enforces that each STP must be able to receive a signal from at least a recommended number of RNs for a given accuracy and reliability which are specified with a summation of the accuracy index α and reliability index R. The accuracy index α is a minimum number of radio signals obtained from RNs that should provide a good performance according to Kaemarungsi [8]. A large value (more than the recommended value) will result in higher accuracy and precision while a small value will decrease

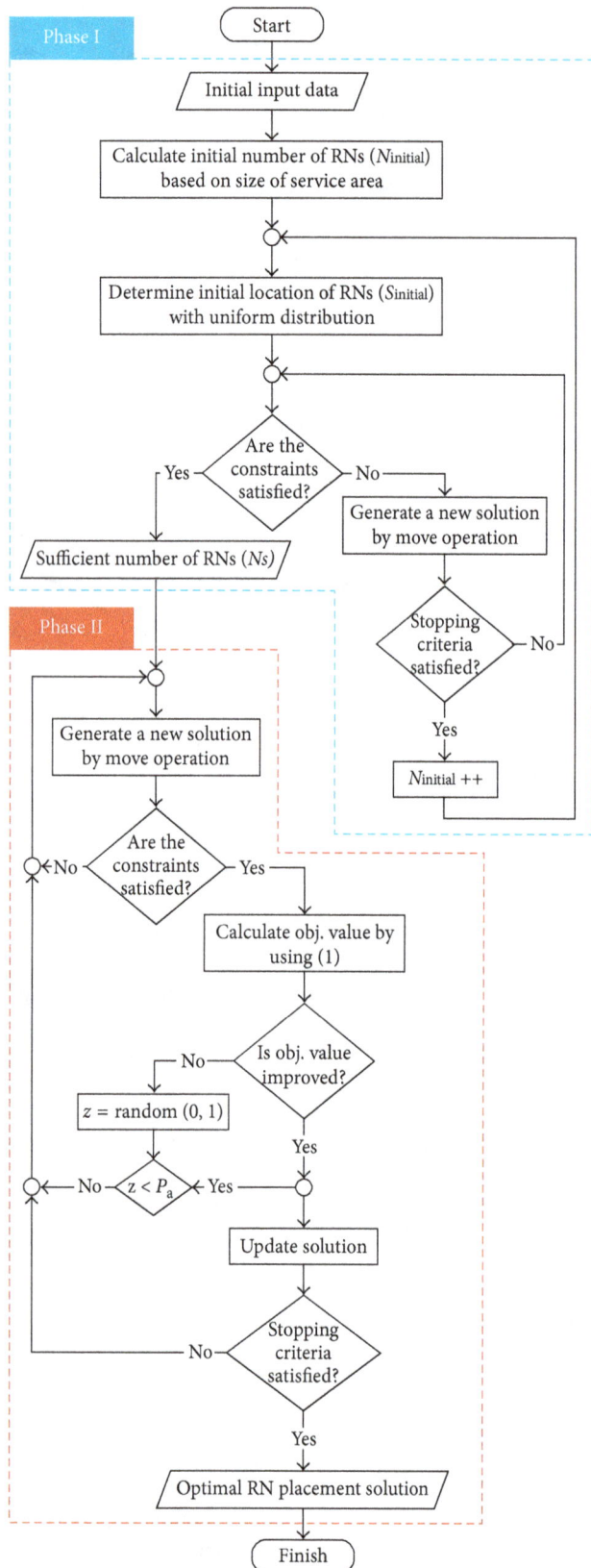

FIGURE 2: Framework of the solution technique.

TABLE 1: Notations.

Sets	
B	A set of candidate sites (CS) that are locations for installing the reference node (RNs)
T	A set of signal test points (STP) that are locations for testing received signal strength
Decision variables	
c_j	A binary {0, 1} variable that equals 1 if the RN is installed at site j, $j \in B$; 0 otherwise
S_{ij}	A binary {0, 1} variable that equals 1 if the STP i is assigned to RN j, $i \in T$ and $j \in B$; 0 otherwise
Constant parameters	
P_{ij}	The RSS that STP i receives from RN j, $i \in T$ and $j \in B$
P_T	The RSS sensitivity threshold
N_s	The sufficient number of RNs installed in the service area
α	The accuracy index or the recommended number of RNs that should provide higher accuracy and precision according to [8]
R	The reliability index or the number of RNs that provide reliability performance

index R is a margin of the number of RNs that is acceptable to fail. Constraint (4) enforces that there is a sufficient number of RNs (N_S) to be utilized in the service area. Constraint (5) specifies that STP i can receive a signal from RN j if RN j is installed.

$$S_{ij}\left(P_{ij} - P_T\right) \geq 0, \quad \forall i \in T, \forall j \in B, \qquad (2)$$

$$\sum_{\forall j \in B} S_{ij} \geq \alpha + R, \quad \forall i \in T, \qquad (3)$$

$$\sum_{\forall j \in B} c_j = N_S, \qquad (4)$$

$$S_{ij} \leq c_j, \quad \forall i \in T, \forall j \in B. \qquad (5)$$

3.3. Solution Technique for R-MSMR. In this section, we describe the overall framework of the solution technique. Figure 2 shows the framework of the solution technique for the R-MSMR. Initial input data of the solution technique are divided into two inputs. The first input involves the physical properties of the service area (i.e., the floor dimension and the total number of floors) and the wireless transceiver information (i.e., the RN-signal coverage area). The second input specifies the parameters for fundamental calculations of the solution technique, including a set of signal test points (STPs), a set of candidate sites for installing RNs (CS), the RSS data of all STPs that have been received from RNs in Watts (Z_{RSS}), the accuracy index, and the reliability index. The output from the solution technique is the optimal number of RNs installed and their locations that ensure the continued effectiveness of the system when some RNs in the system fail. Note that this work considers the discrete candidate sites for installing RNs in order to reduce the computational complexity of the system design.

performance of both accuracy and precision. However, a large value of the accuracy index α will also require a high cost of node infrastructure installation. The reliability

In the proposed framework, the solution technique consists of two phases. Phase I aims to generate a good starting solution that provides a sufficient number of RNs installed (N_S). First, an initial number of RNs (N_{initial}) as a minimum starting number of RNs installed is calculated based on the size of the service area. The starting number is then used to determine the initial location of the RNs (S_{initial}), whereby the RNs are uniformly distributed across the service area. Then, the process is repeated until the set of optimization constraints (2)–(5) are satisfied. In each iteration, move operation based on the SA approach is used to generate the new possible solution (called the neighbor solution), in which specific attributes of the current solution are adjusted [36]. After that, a solution that provides a sufficient number of RNs is obtained.

In Phase II, the SA approach based on the heuristic solution technique is used to determine the optimal location of the RNs (S_o) which is specified with the number of RNs obtained from Phase I. In this phase, the cost of each RN-placement solution is calculated by using the evaluation function as written in (1). Unlike a general heuristic solution technique which only moves to a solution that improves the objective function, SA allows the search to move to nonimproving solutions with the probability of accepting. The acceptance probability for the optimization problem is written as (6), where Δcost is the difference between the cost of a neighbor solution and the cost of the current solution. The control temperature τ is used to control search progresses, in which the temperature is reduced according to the increasing number of iterations in the annealing schedule. When a worse solution is met, a random number z in the ranges [0, 1] is generated and is compared to the probability of accepting. The worse solution is rejected when a random number is lower than the acceptance probability value ($z < P_a$) [36]:

$$P_a = e^{-(\Delta \cos t/\tau)}. \tag{6}$$

Then, the SA process continues until a stopping condition is reached. Finally, the solution that provides the optimal RN placement for indoor positioning systems is obtained. In particular, this optimal RN-placement solution can support the robust operation either during a normal situation or when there is a failure of some RNs. Figure 3 illustrates an example of the RN-placement solution obtained from the proposed R-MSMR, in which the reliability index is equal to two.

4. Experimental Setups

To evaluate the R-MSMR results, the positioning performance of the R-MSMR was analyzed, and the results were compared with other RN-placement designs under a normal situation and when some RNs have failed. In particular, the RN-placement approaches were compared between the proposed R-MSMR and the following three different designs: the coverage and uniform placement (Uniform) model, the Maximize-Sum of Maximum RSS (MSMR) model [34], and Phase I (i.e., sufficient number of RNs obtained from the proposed design) with the uniform placement (PhI-Uni). For example, the objective of

the Uniform is to place the RNs in a service area, in which the whole service area must be able to receive signals from at least one RN. In order to estimate the target location inside the multi-story buildings, the RMoS floor algorithm [37] and the active Euclidean distance technique based on the location fingerprinting approach are used. Instead of matching the RSS patterns obtained from all the RNs in the service area as occurs with the traditional Euclidean distance technique, the active Euclidean distance technique only considers the RSS values that are transmitted from the available RNs (i.e., active RNs) for a matching RSS pattern process. The RMoS floor algorithm is used to determine the floor number of the target, while the active Euclidean distance is used to estimate the target location in coordinates (x, y), in which the matching process of the algorithm considers only the available RSS received from any RN installed. The core of this study can be divided into two objectives:

(1) To compare the positioning performance of indoor positioning systems based on the location fingerprinting technique in which the system is employed by different system designs (which will be discussed in Section 5.2).

(2) To analyze the impact of different RN-failure patterns on the accuracy of indoor multifloor positioning systems (which will be discussed in Section 5.3).

4.1. Experimental Settings. Figure 4 illustrates the floor layouts of a three-story building, which consists of office zones, classrooms, student lounge, and common rooms. The dimension of each floor is approximately 75 m (width) × 75 m (length). We divided the system design experiment of the service area into three scenarios. Scenario 1 considers a single-floor service area in which the service area is located on the first floor. Scenario 2 considers a two-floor service area that covers both the first and the second floor of the building. Scenario 3 considers a three-floor service area. Note that each floor has open space in the center. In each scenario, two parameter sets with the same configuration are used, consisting of the signal test points and the candidate sites for the RN locations as shown in Table 2.

We assign the grid spacing of the fingerprint locations at four meters as shown by the blue cross in Figure 4. A total number of 158, 316, and 474 test points (i.e., target locations) were randomly selected for all three test scenarios, respectively. Note that these numbers of test points were obtained by determining the sample size with confidence intervals [38]. Table 2 shows a summary of the parameters used in our experiments. We conducted four value tests of the reliability index R (i.e., $R = 1, 2, 3$, and 4) to observe how it impacts indoor positioning performance. Note that we define the input parameters of the R-MSMR on SA terms with the cooling ratio $\varphi = 0.9$, the stopping temperature $\tau_{\text{stop}} = 0.01$, and the maximum count of the no-improvement iteration $n_{\text{worse\_max}} = 2000$. These constraints follow the recommendations of Kirkpatrick et al. [36].

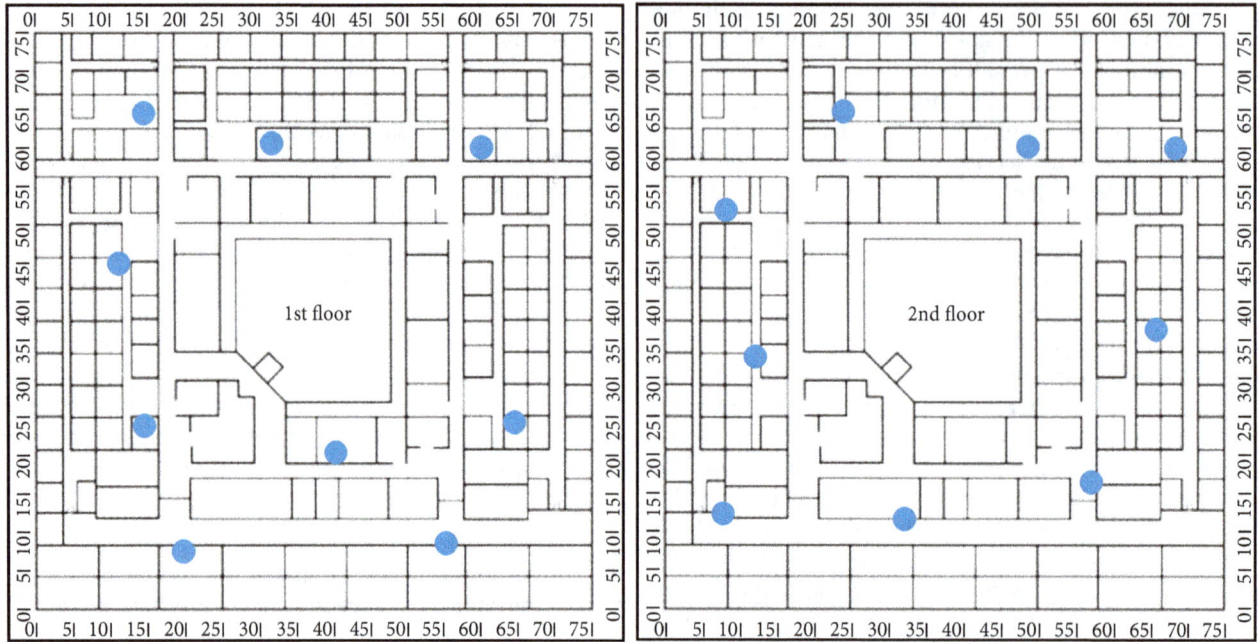

FIGURE 3: Example of the R-MSMR with $R = 2$ for the two-story building.

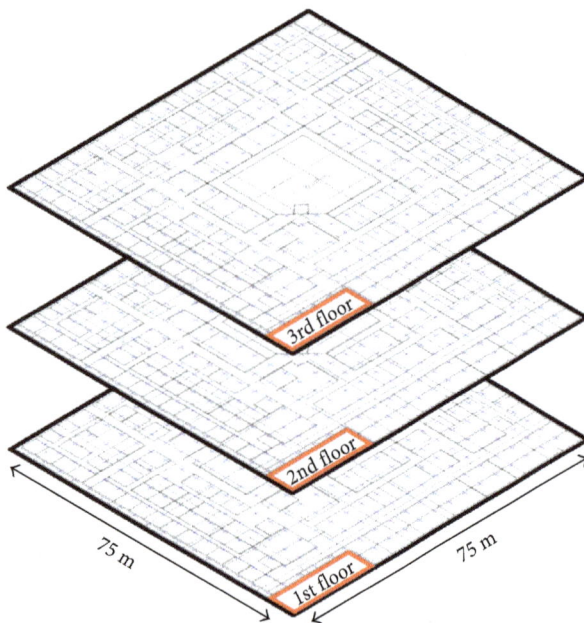

FIGURE 4: The three-story structure with the assigned STPs location.

4.2. Experimental Equipment. Figure 5 illustrates the experimental equipment installed on each floor. This includes RNs which are placed at a height of two meters. The target node is connected to a computer notebook on which the localization technique is executed. The height of the target node is 0.8 meters. IEEE 802.15.4 wireless transceivers were deployed in this work. Each device (RNs and target) has a Freescale MC13224V third-generation chipset with built-in ARM7TDMI processors. The antennas of the wireless transceivers are the inverted F-shape antennas and SMA

antennas [39]. All devices are operated at 2.480 GHz (i.e., channel 26 of IEEE 802.15.4 standard). This is to avoid or minimize the interference from Wi-Fi networks in the area. The transmit power is +3 dBm, while the typical sensitivity of the wireless transceivers is −95 dBm. A typical range (indoors, nonline of sight) is about 30 meters. The target node will gather the RSS values that are transmitted from the RNs in the service area. These measured RSS values are used to estimate the location of the target node. Note that the proposed solution technique calculates the RSS data in linear scale (i.e., Watts). Therefore, we did not use dBm in our calculation because it has limitations in terms of the multiplication and division of log scales. The sampling rate of wireless transceivers in this work is 1 sample for every 3 seconds. Note that we only consider the stationary node in our experimental study. Table 3 summarizes the specifications of the wireless transceivers used in our experiments.

5. Results and Discussion

In this section, several aspects of the numerical results of the RN placement are discussed. First, in Section 5.1, the numerical results of different RN-placement designs for indoor positioning systems are described. Next, in Section 5.2, we provide a comparative performance evaluation of the R-MSMR versus other system designs. Finally, in Section 5.3, the impact of RN-failure patterns on the accuracy performance of indoor positioning systems is analyzed.

5.1. System Design Results. Figures 6–8 illustrate the results of the RN-placement designs for Scenarios 1 to 3, respectively. Each graphic reports the number and the locations of the RNs obtained from four different system designs, which are the Uniform, the MSMR, the R-MSMR with $R = 2$,

TABLE 2: Summary of parameters used in the experiments.

Parameter	Details
Floor dimensions	75 m × 75 m (for 1st, 2nd, and 3rd floor)
RNs placement	Uniform, MSMR, R-MSMR, and PhI-Uni
Number of signal test points (STP) and candidate sites (CS)	328 locations for Scenario 1, 656 locations for Scenario 2, and 984 locations for Scenario 3
Grid spacing of fingerprint locations	4 m × 4 m
Number of test points (i.e., target locations)	158 locations for Scenario 1, 316 locations for Scenario 2, and 474 locations for Scenario 3
The RSS sensitivity threshold (P_T)	0.1 pW (i.e., −100 dBm)
Accuracy index (α)	4 [8]
Reliability index (R)	1, 2, 3, and 4
Floor determination technique	RMoS floor algorithm [37]
Localization technique	Active Euclidean distance

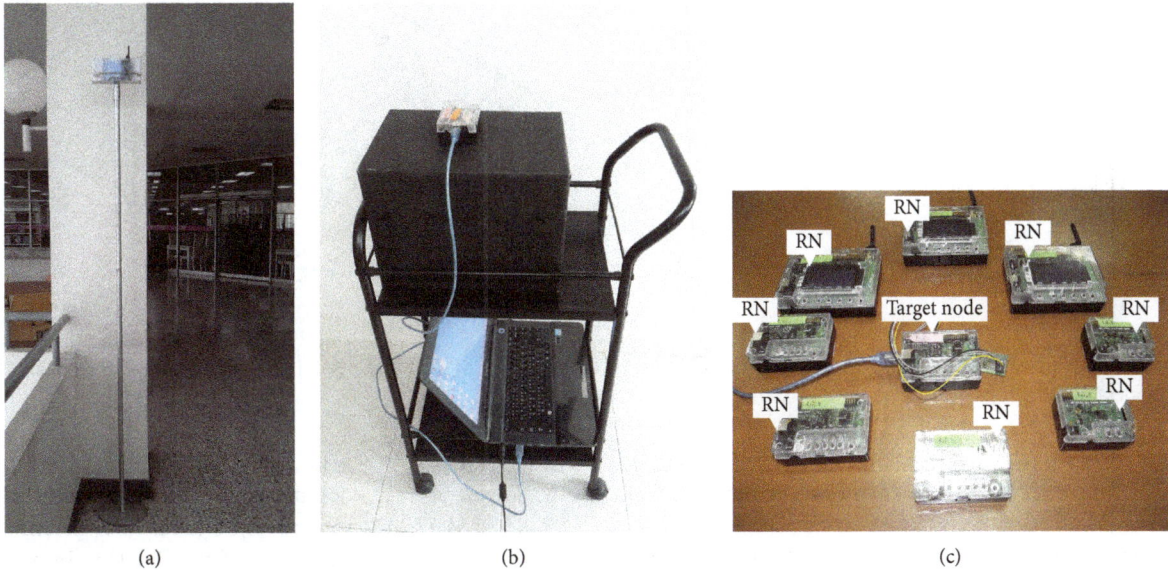

FIGURE 5: Experimental equipment used in this work. (a) Reference node (RN). (b) Target node on a cart with laptop PC. (c) Experimental equipment used in this work.

TABLE 3: Specifications of the wireless transceivers.

Specification	Details
Manufacturer	Freescale
Chipset	MC13224V
Frequency range	2.405 GHz to 2.480 GHz
Operating channel	CH 26 (2.480 GHz)
Rx sensitivity	−95 dBm
Transmit power	+3 dBm
Antenna	Inverted F-antenna and SMA antenna

and the PhI-Uni. In these results, we found that the Uniform has the lowest number of RNs installed in all three scenarios, while the R-MSMR with $R = 2$ has the highest number of RNs installed for all three scenarios. The reason for the R-MSMR requesting a higher number of RNs to be installed than other system designs is that the proposed R-MSMR structure is designed to provide high positioning accuracy during a normal situation and also achieve reliable location results when some RNs in the system fail. Therefore, our

algorithm seeks to distribute the RN locations across the service area while maintaining high quality radio signal coverage.

5.2. Performance Evaluation of the System Designs. In this section, we evaluate the location estimation performance of different system designs. In particular, the proposed R-MSMR was compared with the Uniform, the MSMR, and the PhI-Uni under the same experimental settings as shown in Table 2. The RMoS floor algorithm and the active Euclidean distance technique based on the location fingerprinting approach were used to estimate the target locations in coordinates *x*, *y*, and *floor*. Two different RN scenarios were considered in our experiments. The first scenario is a *fault-free scenario* in which all RNs worked properly. The second scenario is a *3RN-failure per floor scenario* in which three of the RNs on each floor fail. The experimental results of the 3RN-failure per floor were averaged from four faulty patterns. For example, we randomly selected and turned off

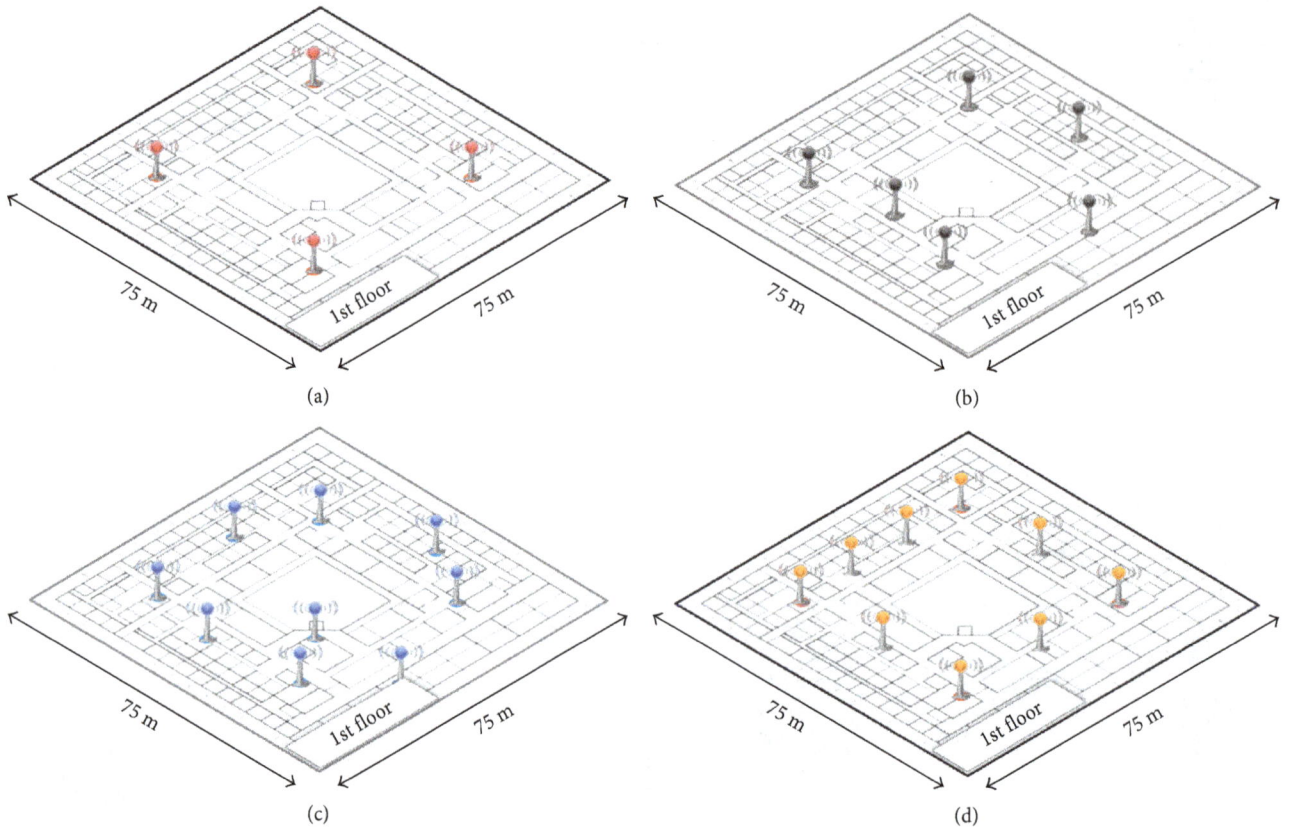

FIGURE 6: The RN placement designed for Scenario 1. (a) Uniform (4 nodes). (b) MSMR (6 nodes). (c) R-MSMR, $R = 2$ (9 nodes). (d) PhI-Uni (9 nodes).

three RNs on each floor to create four patterns of RN failure. Then, an average accuracy performance was computed from those four patterns of RN failure.

The positioning performance results of different system designs were divided into two sections. In Section 5.2.1, the results of the correct floor determination are presented. Then, in Section 5.2.2, the accuracy and the precision of four different indoor positioning systems that employ different system designs are compared.

5.2.1. Correct Floor Determination Results. Under the fault-free scenario, all four indoor multifloor positioning systems that employ different design structures were able to achieve results of 100% correct floor determination in both Scenario 2 and Scenario 3 (i.e., inside the two- and three-floor service area, resp.). Unlike the fault-free scenario, indoor multifloor positioning systems under the RN-failure scenario encountered problems from the missing RSS values during the online phase due to the three faulty RNs on each floor. Figure 9 compares the floor determination accuracy under the 3RN-failure per floor scenario. The red bins, gray bins, blue bins, and yellow bins represent the correct floor determination percentages for the Uniform, the MSMR, the R-MSMR with $R = 2$, and the PhI-Uni, respectively. From these results, we found that the floor determination accuracy of indoor multifloor positioning systems that employ the Uniform, the MSMR, and the PhI-Uni decreased in the event of faulty RNs.

In the results of Scenario 2 and Scenario 3, the Uniform had the lowest correct floor determination percentage of about 92.01% and 62.34%, respectively. Similar floor determination results are seen in the MSMR and the PhI-Uni. In particular, we found that the insufficient RN placement had a significant effect on floor determination performance. The floor determination of the PhI-Uni which used a sufficient number of RNs (N_S) had an incorrect floor number. The PhI-Uni did not provide 100% correct floor determination in either Scenario 2 or Scenario 3 (98.25% and 91%, resp.). These figures are different from the proposed system design, in which the R-MSMR with $R = 2$ was able to achieve the highest correct floor determination percentage of up to 100% in both Scenario 2 and Scenario 3. Moreover, we found that the floor determination accuracy of the Uniform, the MSMR, and the PhI-Uni depended on the number of RNs which failed inside the building; the floor determination accuracy decreased in line with the rise in the number of RN failures.

This observation suggests that a larger number of the RN failures (i.e., considering six and nine RN failures as large numbers in Scenario 2 and Scenario 3, resp.) degrade the floor determination accuracy of the Uniform, the MSMR, and the PhI-Uni. Only the performance of the proposed R-MSMR was not affected by the increasing number of RN failures. It succeeded 100% in correct floor determination in the scenarios of both six and nine RN failures in the building.

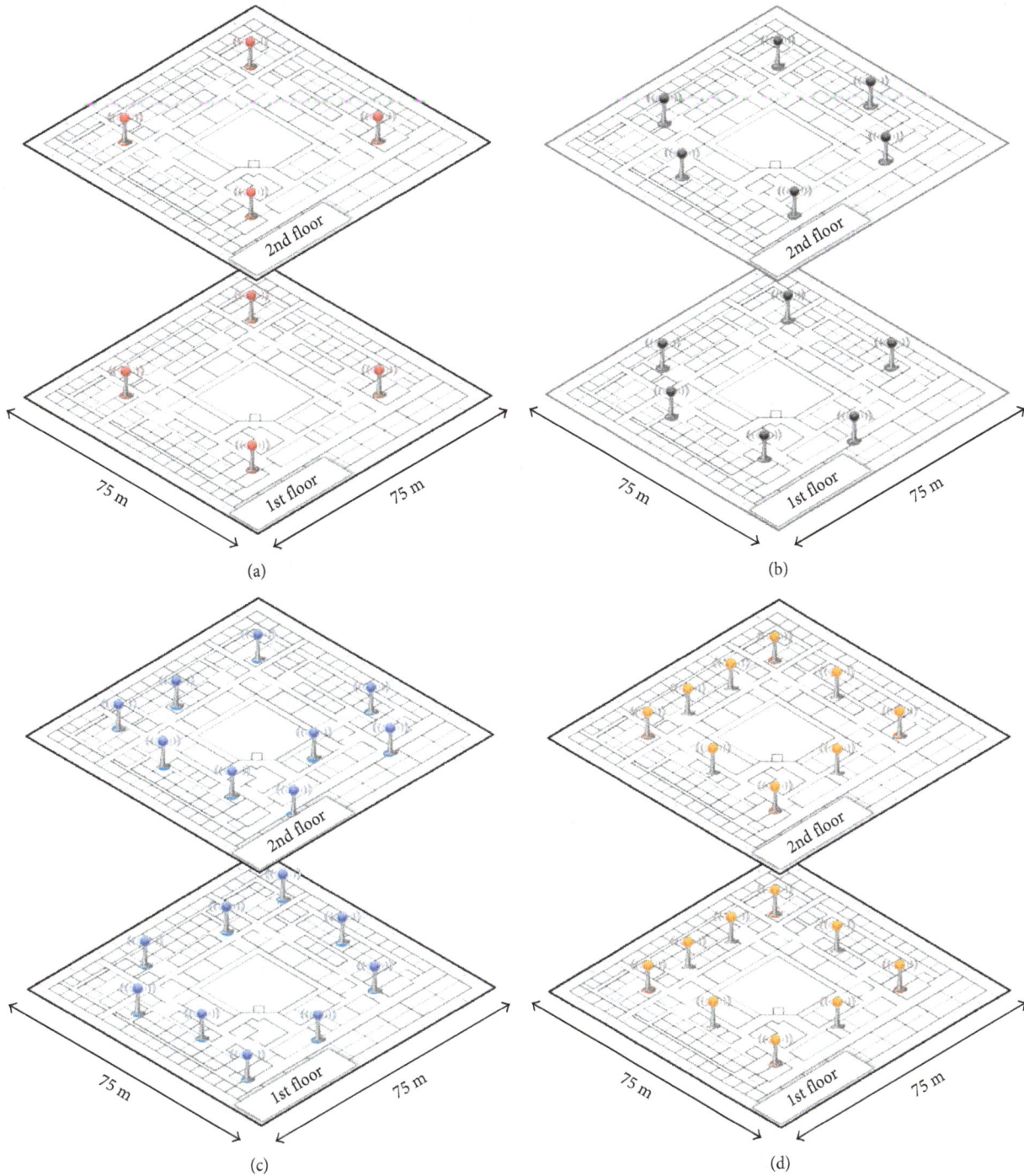

FIGURE 7: The RN placement designed for Scenario 2. (a) Uniform (8 nodes). (b) MSMR (12 nodes). (c) R-MSMR, $R = 2$ (18 nodes). (d) PhI-Uni (18 nodes).

5.2.2. Results of the Positioning Accuracy (x, y). Figures 10–12 report average error distances of indoor positioning systems that employ four different design structures for Scenario 1 to Scenario 3, respectively. We first consider the accuracy performance of the indoor positioning systems under the fault-free scenario as represented by empty bins. The proposed R-MSMR with $R = 2$ shows a better accuracy performance than other system designs in all three scenario areas. For example, Figure 11 illustrates four different

positioning results under the fault-free scenario in the two-story building of Scenario 2. The average error distance of the R-MSMR is 4.42 meters, while the average error distances of the Uniform, the MSMR, and the PhI-Uni are 6.35 meters, 5.20 meters, and 5.13 meters, respectively. R-MSMR has 30.4%, 15.0%, and 13.8% better performance than the Uniform, the MSMR, and the PhI-Uni, respectively. In comparing the accuracy performance of all three service areas under the fault-free scenario, the proposed R-MSMR

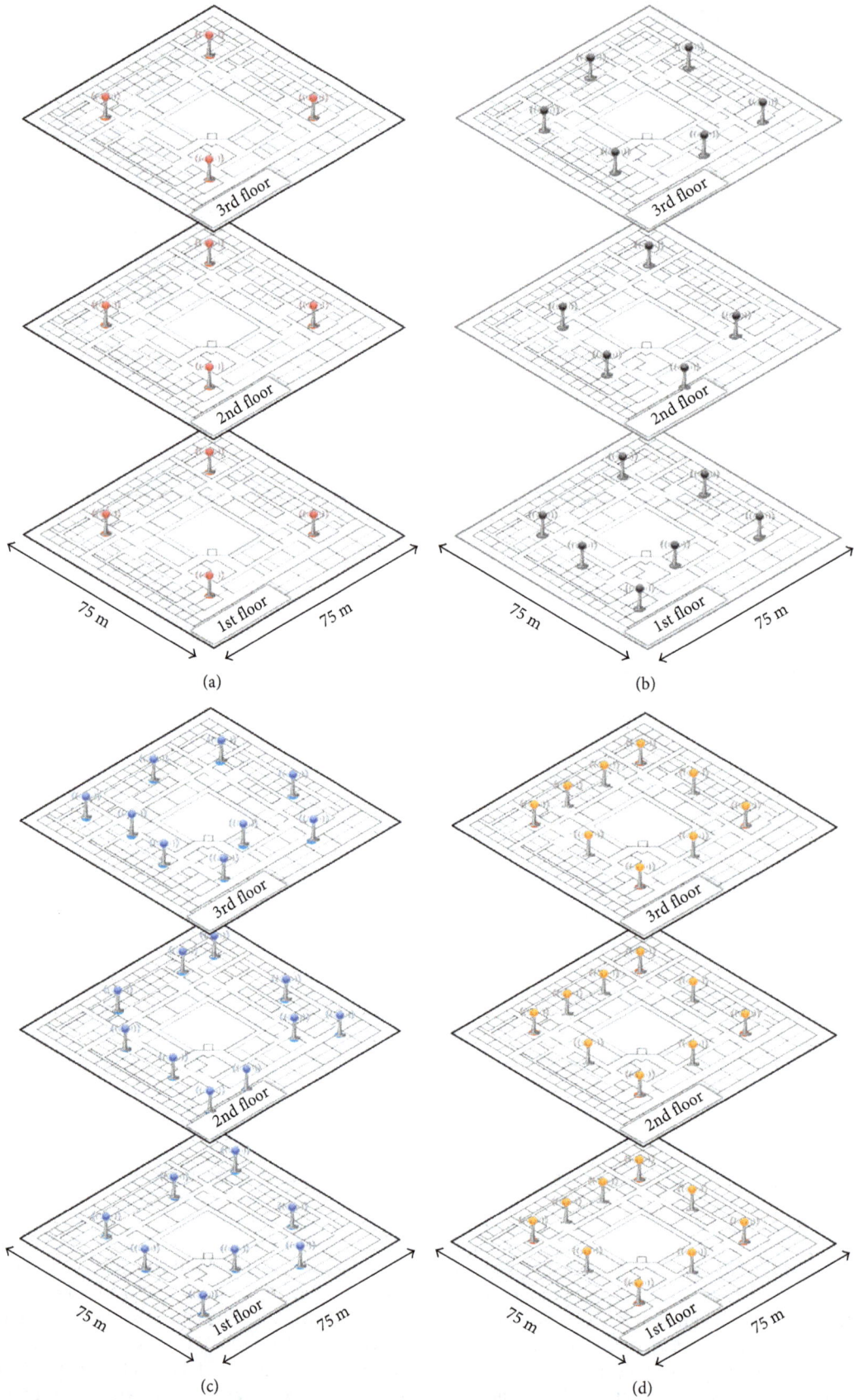

FIGURE 8: The RN placement designed for Scenario 3. (a) Uniform (12 nodes). (b) MSMR (18 nodes). (c) R-MSMR, $R = 2$ (27 nodes). (d) PhI-Uni (27 nodes).

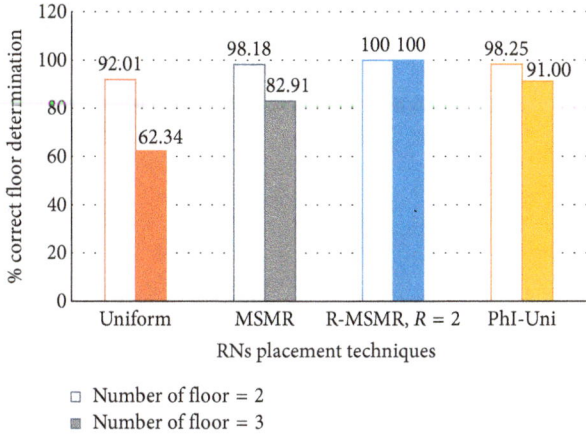

FIGURE 9: Percentage of the correct floor determination under three RN failure per floor.

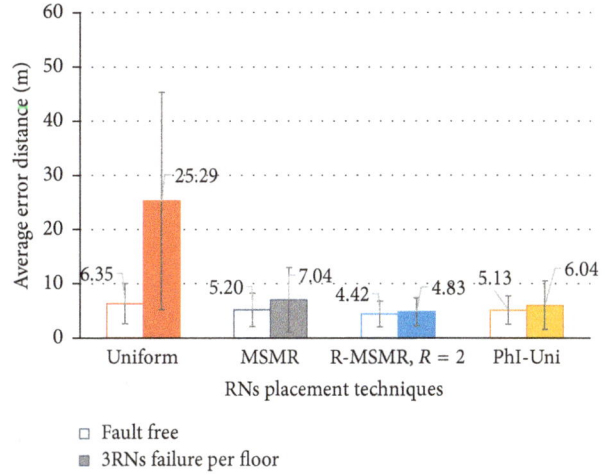

FIGURE 11: An average error distance of wireless indoor positioning systems in Scenario 2 (2 floors).

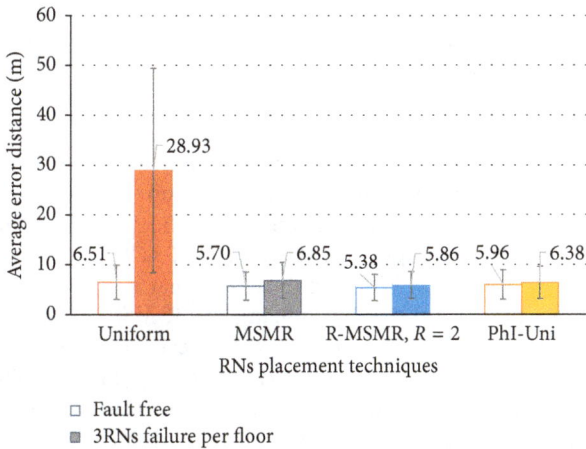

FIGURE 10: An average error distance of wireless indoor positioning systems in Scenario 1 (1 floor).

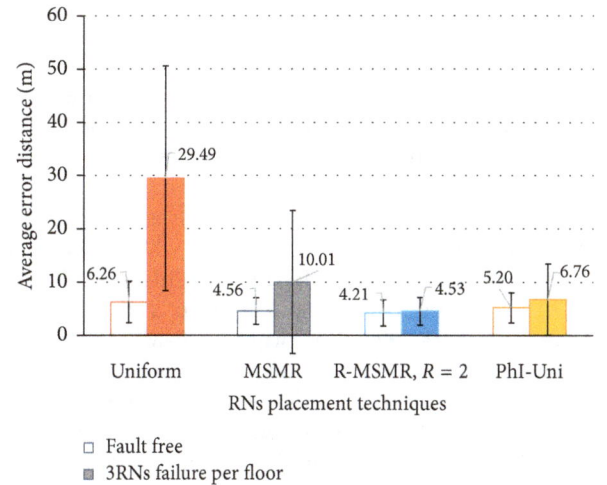

FIGURE 12: An average error distance of wireless indoor positioning systems in Scenario 3 (3 floors).

shows a better performance for location estimation than the Uniform by about 17.4% to 32.8%, the MSMR by about 5.6% to 15.0%, and the PhI-Uni by about 9.8% to 13.8%. Additionally, we found that the accuracy performance trend for all four system designs under the fault-free scenario follows the increasing number of RNs installed in the service areas. A larger number of RNs installed can improve the location accuracy performance. Notice that this follows the same trend as in the performance improvement recommendations by Kaemarungsi [8].

We next consider the performance of the indoor positioning systems under the 3RN failure per floor scenario. Figure 13 shows an example of the cumulative density function (CDF) of the error distance for the two-floor service area of Scenario 2. The red lines, the black lines, the blue lines, and the yellow lines represent the precision performance of the Uniform, the MSMR, the R-MSMR with $R = 2$, and the PhI-Uni, respectively. The results indicate that faulty RNs in the system decrease the positioning performance of all system designs. However, the R-MSMR with $R = 2$ also achieves the highest positioning performance, in which the

performance under the RN-failure scenario is approximately 90% precision within 8.1 meters. This represents an 85.9% better performance than the Uniform (i.e., 90% precision within 57.7 meters), 29.6% better performance than the MSMR (i.e., 90% precision within 11.5 meters), and 11.4% better performance than the PhI-Uni (i.e., 90% precision within 9.14 meters).

Considering the overall accuracy performance under the 3RN-failure per floor scenario, Figures 10–12 show that the R-MSMR with $R = 2$ provides a better accuracy performance than the Uniform by about 79.7% to 84.6%, the MSMR by about 14.5% to 54.7%, and the PhI-Uni by about 8.2% to 32.9%. Furthermore, the fault tolerance of the proposed R-MSMR under all of the RN-failure scenarios is greater than that of the other system designs. For instance, Figure 12 shows the positioning performance under the scenario of nine RN failures inside the three-floor service area. The R-MSMR with $R = 2$ has a percentage of difference between

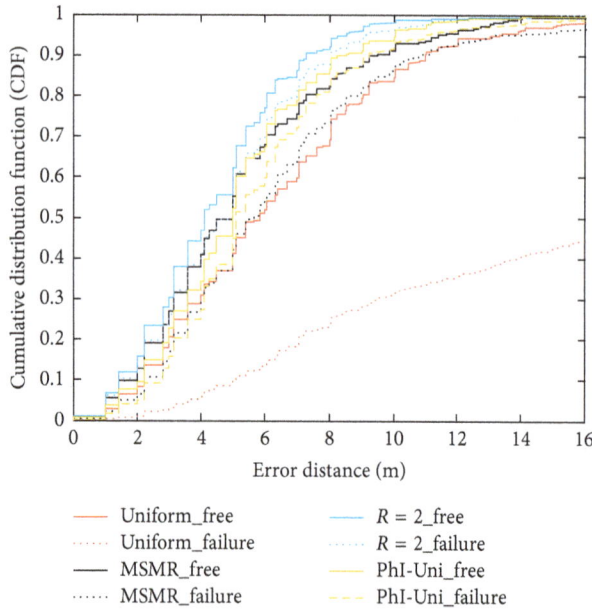

FIGURE 13: CDF of error distance in Scenario 2.

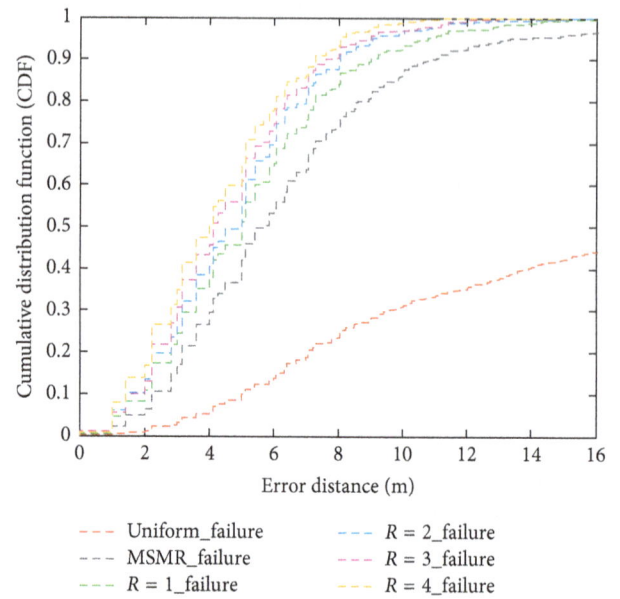

FIGURE 14: Effect of the reliability index on the error distance distribution (CDF) in Scenario 2.

an average error distance in the normal situation and the 3RN-failure per floor scenario of about 7.1%, whereas the Uniform, the MSMR, and the PhI-Uni have percentage of up to 79%, 55%, and 23.1%, respectively. Although the number of RNs installed in the PhI-Uni is the same as the number of RNs installed in the proposed R-MSMR, the indoor positioning performances are different. The PhI-Uni only provides a correct floor determination score of 91% with a fault tolerance of 76.9%, while the proposed R-MSMR achieves a successful correct floor determination score of 100% with a fault tolerance of 92.9%. The reason why the proposed R-MSMR with $R = 2$ can provide better fault tolerance during the online estimation phase than the other system designs is that the proposed R-MSMR is specifically designed to guarantee that the localization area (i.e., all STPs) must be able to receive at least the recommended number of RSS values with a summation of the accuracy index (α) and the reliability index (R). Thus, the structure of the proposed system design can tolerate the missing RSS values during the online estimation phase where some RNs in the system fail. This ensures good positioning accuracy during a normal situation and yields the robustness of the location estimation under the scenario of RN failures. Furthermore, the sufficient number of RNs (N_S) that is obtained from Phase I can also prevent the high cost of node infrastructure installation. Thus, the designers can apply the determining optimal number of RNs obtained from Phase I to place the RNs manually in a small service area or an uncomplicated building.

Figure 14 illustrates the performance improvement of the R-MSMR when the value of the reliability index is increased while the number of faulty RNs is fixed at six nodes (i.e., 3RN-failure per floor inside a two-floor service area). In Figure 14, four of the reliability indices for the R-MSMR are compared, specified as $R = 1$ to $R = 4$. The results show that a larger value for the reliability index can improve the

positioning performance of the location estimation. For example, under the 3RN-failure per floor scenario, the precision performance at 5 meters of the R-MSMR with $R = 1$ to $R = 4$ is approximately 51.1%, 55.6%, 61.1%, and 65.3%, respectively. This demonstrates that the system can handle the faulty RNs by utilizing the R-MSMR with the large value of its reliability index. However, increasing the reliability index for the R-MSMR requires a greater number of RNs to be installed in the system. This leads to higher installation costs, computation time, and memory space requirement for the fingerprint database. These factors will depend on the limitations of the indoor positioning systems chosen, such as the budget for installing the wireless network and the performance requirements of the positioning application.

5.3. *Impact of RN-Failure Patterns on Accuracy Performance.* In this section, we analyze the impact of different RN-failure patterns on accuracy performance. The system design results of the Uniform, the MSMR, the R-MSMR with $R = 2$, and the PhI-Uni under the two-floor service area were considered, in which those RN-placement designs are as shown in Figure 7. The experimental settings in this study were assigned the same configuration as in Scenario 2. The RMoS floor algorithm and the active Euclidean distance technique were used to estimate the target locations. In this study, two different RN-failure patterns were considered. The first pattern was a *similar RN-failure pattern*, in which all RNs on the 1st floor and the 2nd floor fail on the same side of the building. The second pattern involved an *across RN-failure pattern*, in which all faulty RNs on the 1st floor were on the opposite side of the building to the faulty RNs on the 2nd floor. In each RN-failure pattern, the four cases of the faulty RNs inside the building were divided

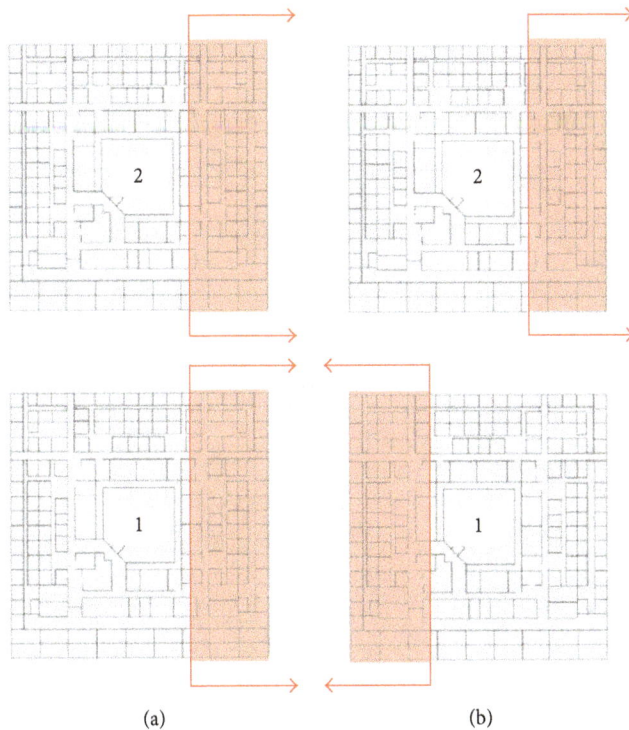

FIGURE 15: Example of the faulty RNs pattern in this experiment. (a) Similar RN-failure pattern. (b) Across RN-failure pattern.

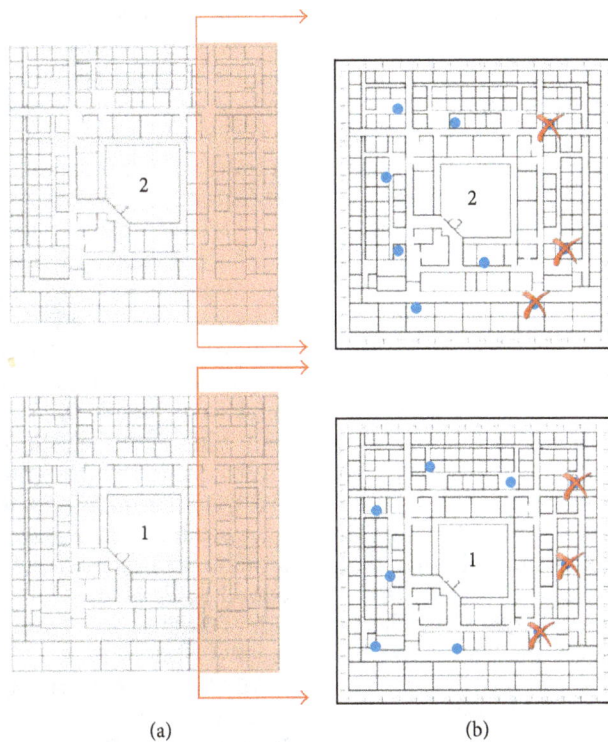

FIGURE 16: Example of the 1st similar RN-failure scenario for the R-MSMR, $R = 2$.

according to the cardinal directions (i.e., north, south, west, and east direction). Figures 15(a) and 15(b) show an example of the 1st similar and the 1st across RN-failure pattern, respectively. In particular, the six cases of the

RN-failure per floor situations, consisting of from zero nodes to five nodes, are compared.

Figure 16 shows an example of indoor positioning systems in the two-story building that employs R-MSMR

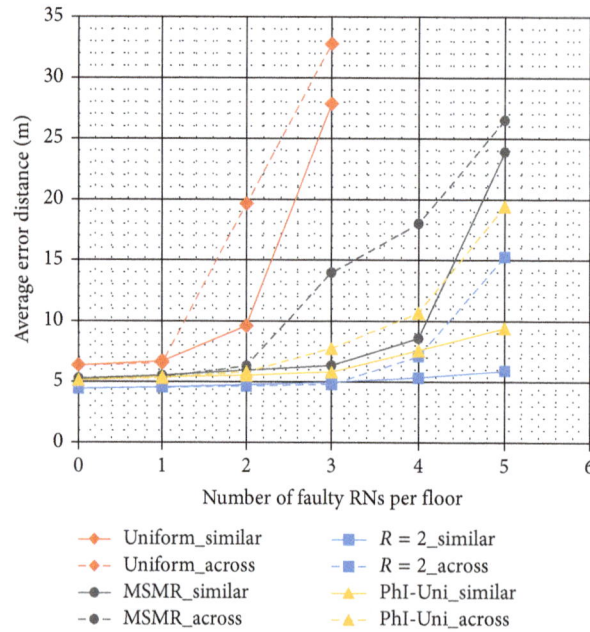

FIGURE 17: Effect of different numbers of RN failures per floor on average error distance.

with $R = 2$. The blue circles represent the location of the RNs installed, while the red crosses represent the location of the RN failures. In this figure, it can be seen that the system encountered the 1st similar RN-failure pattern, in which three RNs on each floor fail. Those faulty RNs on each floor are located in the east section of the service area.

We first consider the impact of the number of RN failures on performance accuracy in Figure 17. The results of each system designs were averaged from the four RN-failure sections (i.e., the four cardinal directions). Clearly, performance accuracy decreases in line with an increase in the number of RN failures. For example, Figure 17 shows the trend of the average error distance when the number of faulty RNs in each floor is increased. The MSMR under the similar RN-failure pattern (the black solid line with circles) has an increasing average error distance of 5.25, 5.50, 5.93, 6.34, 8.59, and 23.88 meters for 0, 1, 2, 3, 4, and 5 failed RNs on each floor, respectively. This indicates that the accuracy performance of the MSMR drops by almost 80% if ten RNs inside the two-floor service area fail. These results are different from those of the proposed R-MSMR with $R = 2$. Our proposed design can provide a better fault tolerance than other system designs. In this case, under the ten RN failure inside the two-floor service area scenario, the average error distance of the R-MSMR with $R = 2$ does not exceed 6 meters.

Moreover, we found that the across RN-failure pattern (the dashed lines) degrades accuracy performance more than the similar RN-failure pattern (the solid lines). For example, the MSMR under the across RN-failure pattern (black dashed line with circles) has an increasing average error distance of 5.25, 5.34, 6.31, 13.95, 18.02, and 26.49 meters for 0, 1, 2, 3, 4, and 5 failed RNs on each floor, respectively. It resulted in a maximum of 52.4% worse location accuracy than the similar RN-failure pattern. This means that the

MSMR may have poor positioning performance if the across RN-failure pattern occurs. Similar results were obtained for the cases of the Uniform, the R-MSMR with $R = 2$, and PhI-Uni design under the across RN-failure pattern.

Next, to investigate how the patterns of RN failure influence accuracy performance, we report the location estimation error performance of the similar and the across RN-failure patterns as shown in Figures 18 and 19, respectively. In each figure, the indoor two-floor positioning systems that employ the Uniform and the R-MSMR with $R = 2$ were compared under the case of the two RN failure per floor scenario. The pink lines denote the error distance that is greater than eight meters. The blue circles represent the actual target locations, while the pink crosses represent the estimated target locations. We first consider indoor positioning systems under the similar RN-failure pattern as shown in Figure 18. Clearly, the Uniform has higher location estimation error than the R-MSMR with $R = 2$. The average error distance of the Uniform is 10.33 meters, while the average error distance of the R-MSMR with $R = 2$ is 4.71 meters. Moreover, Figure 18(a) indicates that most of the location estimation errors for the Uniform are located on the opposite side of the floor to the faulty RNs (i.e., at the west side of the building). These results are different from those of the R-MSMR with $R = 2$. As shown in Figure 18(b), the R-MSMR with $R = 2$ has 54.4% better accuracy performance than the Uniform. Note that both the Uniform and the R-MSMR with $R = 2$ can achieve 100% correct floor determination in the case of the similar RN-failure pattern.

We next consider the indoor positioning systems under the across RN-failure pattern, as shown in Figure 19. The results indicate that the Uniform again has higher location estimation error than the proposed R-MSMR with $R = 2$. Those location estimation errors are mostly located on the opposite side of the floor to the faulty RNs. Besides the issue

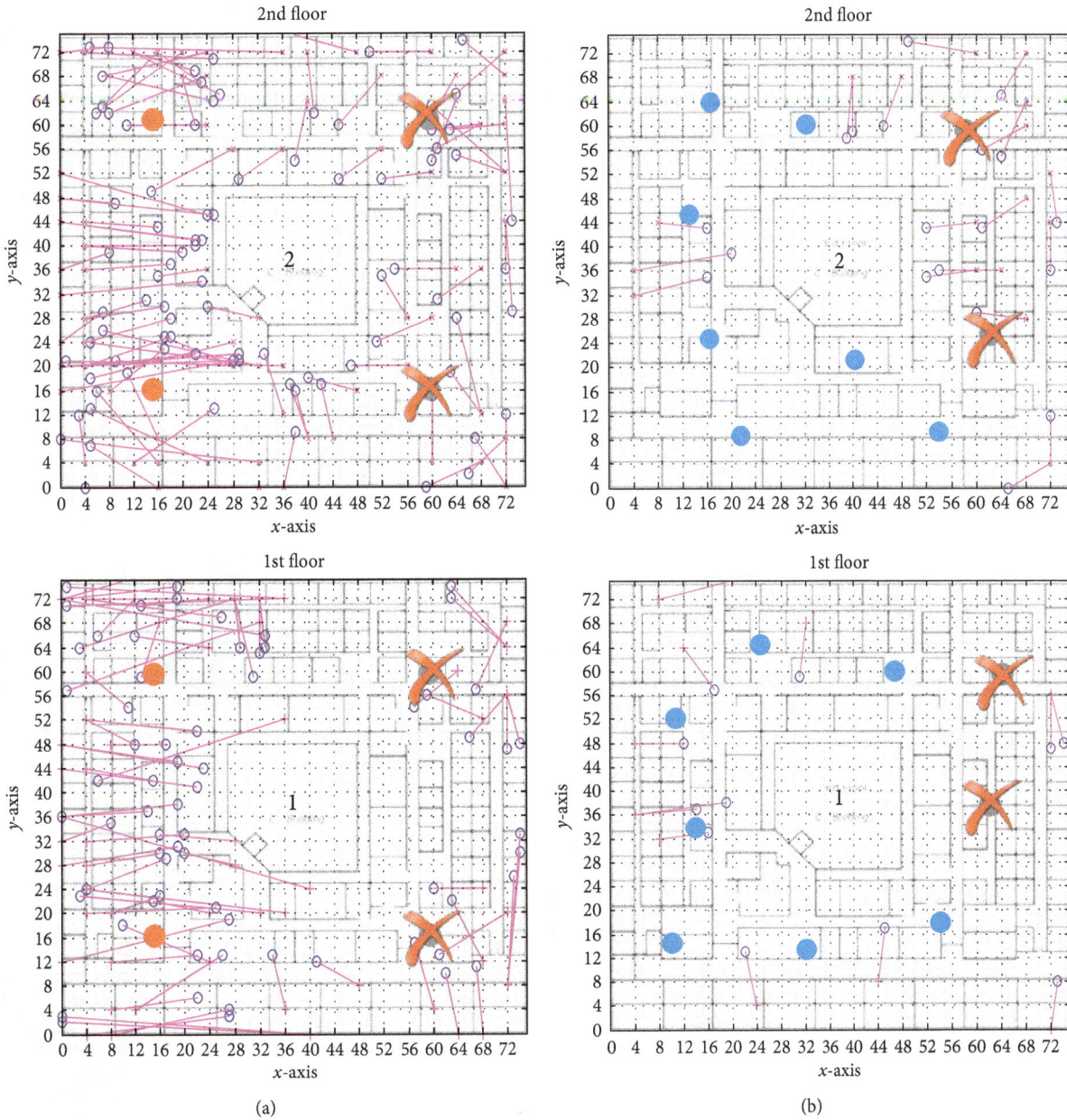

FIGURE 18: The location estimation errors that are greater than 8 meters under the similar RN-failure pattern. (a) The Uniform. (b) The R-MSMR, $R = 2$.

of location estimation errors, incorrect floor determination also occurred under the across RN-failure pattern. In Figure 19(a), the red solid lines represent the results of the error distance when estimating the incorrect floor determination. The Uniform shows the lowest floor accuracy performance under the across RN-failure pattern (with an average error distance of 20.3 meters). The Uniform provides 74.0% correct floor determination, with most of the floor determination errors located on the same side of the floor as the faulty RNs. These results are different from those of the proposed system design. The R-MSMR with $R = 2$ also has an average error distance of 4.72 meters, but it yields 100% correct floor determination. Once again, the overall accuracy performance of the Uniform under the across RN-failure pattern

was up to 49.1% worse than the location accuracy for the similar RN-failure pattern, whereas the overall accuracy performance of the proposed R-MSMR with $R = 2$ was almost the same under both the similar and the across RN-failure patterns.

The results show the advantages of the proposed robust system design, which not only provides the highest performance accuracy during a normal situation, but also yields reliable location results under unexpected situations such as RN failures. The reason for this is that the mathematical formulation of the proposed R-MSMR model considers not only the radio signal coverage requirements but also the accuracy and reliability requirements. Unlike the proposed system design, existing RN-placement designs limit their

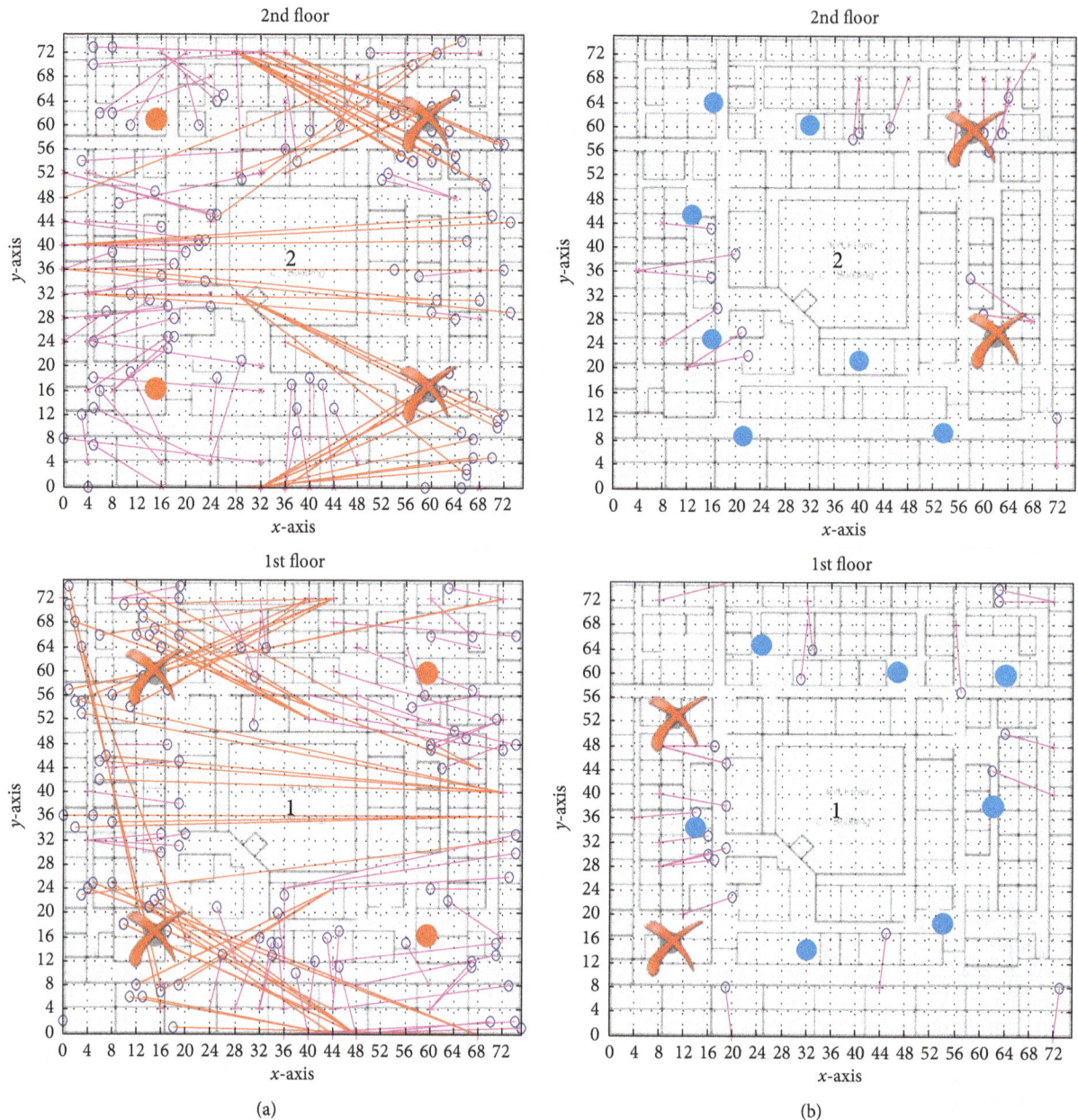

FIGURE 19: The location estimation errors that are greater than 8 meters under the across RN-failure pattern. (a) The Uniform. (b) The R-MSMR, $R = 2$.

focuses to the provision of signal radio coverage in the service area. These are insufficient and inefficient methods for designing indoor positioning systems in unexpected situations. Thus, we can conclude that an indoor positioning systems that employ our proposed R-MSMR is fault tolerant and robust in the RN-failure scenarios considered in our study.

6. Conclusion

In this article, we have considered the problem of wireless indoor positioning system design, for systems based on the location fingerprinting technique. We proposed a novel mathematical formulation using a Binary Integer Linear Programming (BILP) approach that employs Simulated

Annealing (SA) based on the heuristic solution technique. The task of the proposed system design is to determine and place a suitable number of RNs and to find the optimum locations which may be located on a single floor or on multiple floors. In particular, the proposed system design has been developed to support robust operation both during a normal situation and when some RNs have failed. Experimental results reveal that the proposed robust system design outperformed other system designs and was able to achieve the highest location performance accuracy in both *fault-free* and *RN-failure scenarios*. Specifically, the proposed system design delivered 84.6%, 54.7%, and 32.9% better performance accuracy than the Uniform, the MSMR, and the PhI-Uni, respectively.

Our future works will consider a combination of the major components between the robust system design and the robust localization algorithm that may be supported during the normal situation or when some RNs have failed.

Acknowledgments

The authors would like to express their deepest appreciation to the late Dr. Chutima Prommak, Assistant Professor of the School of Telecommunication Engineering at Suranaree University of Technology, Thailand, whose contribution to this work was of great significance. Without her experience, guidance, and patience, it would never have been possible to complete this work successfully.

References

[1] A. Gharaibeh, M. A. Salahuddin, S. J. Hussini et al., "Smart cities: a survey on data management, security and enabling technologies," *IEEE Communications Surveys & Tutorials*, vol. 19, no. 4, pp. 2456–2501, 2017.

[2] C. C. Li, J. Su, and E. T.-H. Chu, "Building/environment data/information enabled location specificity and indoor positioning," *IEEE Internet of Things Journal*, vol. 4, no. 6, pp. 2116–2128, 2017.

[3] A. A. Nazari, A. Yassine, and S. Shirmohammadi, "Equipment location in hospitals using RFID-based positioning system," *IEEE Transactions on Information Technology in Biomedicine*, vol. 16, no. 6, pp. 1058–1069, 2012.

[4] Y.-C. Wang and C.-C. Yang, "3s-cart: a lightweight, interactive sensor-based cart for smart shopping in supermarkets," *IEEE Sensors Journal*, vol. 16, no. 17, pp. 6774–6781, 2016.

[5] K. Pahlavan and P. Krishnamurthy, *Principles of Wireless Access and Localization*, Wiley, Hoboken, NJ, USA, 2013.

[6] D. Dardari, P. Closas, and P. M. Djurić, "Indoor tracking: theory, methods, and technologies," *IEEE Transactions on Vehicular Technology*, vol. 64, no. 4, pp. 1263–1278, 2015.

[7] R. S. Campos, L. L. Marcello, and L. R. Campos, "Wi-Fi multi-floor indoor positioning considering architectural aspects and controlled computational complexity," *Expert Systems with Applications*, vol. 41, no. 14, pp. 6211–6223, 2014.

[8] K. Kaemarungsi, "Efficient design of indoor positioning systems based on location fingerprinting," in *Proceedings of Wireless Networks Communications and Mobile Computing*, pp. 181–186, Lahaina, HI, USA, June 2005.

[9] R. Akl, K. Pasupathy, and M. Haidar, "Anchor nodes placement for effective passive localization," in *Proceedings of International Conference on Selected Topics in Mobile and Wireless Networking (iCOST)*, pp. 127–132, Shanghai, China, October 2011.

[10] O. Baala, Y. Zheng, and A. Caminada, "The impact of AP placement in WLAN-based indoor positioning system," in *Proceedings of ICN '09 Eighth International Conference on Networks*, pp. 12–17, Guadeloupe, France, March 2009.

[11] P. Sadhukhan, P. Sadhukhan, and Z. Pervez, "Impact of beacon coverage on clustering strategies for fingerprinting localization system," in *Proceedings of International Conference on Computing, Networking and Communications (ICNC)*, 5 pages, Maui, Hawaii, USA, March 2017.

[12] R. Zhang, W. Xia, Z. Jia, L. Shen, and J. Guo, "The optimal placement method of anchor nodes toward RSS-based localization systems," in *Proceedings of Sixth International Conference on Wireless Communications and Signal Processing (WCSP)*, pp. 1–6, Hefei, China, October 2014.

[13] S. Aomumpai, K. Kondee, C. Prommak, and K. Kaemarungsi, "Optimal placement of reference nodes for wireless indoor positioning systems," in *Proceedings of 11th International Conference on Electrical Engineering/Electronics, Computer, Telecommunications and Information Technology*, pp. 1–6, Nakhon Ratchasima, Thailand, July 2014.

[14] A. E. C. Redondi and E. Amaldi, "Optimizing the placement of anchor nodes in RSS-based indoor localization systems," in *Proceedings of 12th Annual Mediterranean Ad Hoc Networking Workshop (MED-HOC-NET)*, pp. 8–13, Ajaccio, France, June 2013.

[15] S. Merkel, P. Unger, and H. Schmeck, "Evolutionary algorithm for optimal anchor node placement to localize devices in a mobile ad hoc network during building evacuation," in *Proceedings of Genetic and Evolutionary Computation Conference (GECCO '13)*, pp. 1407–1414, New York, NY, USA, July 2013.

[16] G. Mao, B. Fidanb, and B. D. O. Anderson, "Wireless sensor network localization techniques," *Computer Networks*, vol. 51, no. 10, pp. 2529–2553, 2007.

[17] C. Yang and H.-R. Shao, "WiFi-based indoor positioning," *IEEE Communications Magazine*, vol. 53, no. 3, pp. 150–157, 2015.

[18] S. He and S.-H. G. Chan, "Wi-Fi fingerprint-based indoor positioning: recent advances and comparisons," *IEEE Communications Surveys & Tutorials*, vol. 18, no. 1, pp. 466–490, 2015.

[19] Q. Wang, Y. Feng, X. Zhang, Y. Sun, and X. Lu, "IWKNN: an effective Bluetooth positioning method based on isomap and WKNN," *Mobile Information Systems*, vol. 2016, Article ID 8765874, 11 pages, 2016.

[20] P. Kriz, F. Maly, and T. Kozel, "Improving indoor localization using Bluetooth low energy beacons," *Mobile Information Systems*, vol. 2016, Article ID 2083094, 11 pages, 2016.

[21] S.-H. Fang and T.-N. Lin, "Indoor location system based on discriminant-adaptive neural network in IEEE 802.11 environments," *IEEE Transactions on Neural Networks*, vol. 19, no. 11, pp. 1973–1978, 2008.

[22] K. Kaemarungsi, "Distribution of WLAN received signal strength indication for indoor location determination," in *Proceedings of 1st International Symposium on Wireless Pervasive Computing*, 5 pages, Phuket, Thailand, January 2006.

[23] J. Chen, Y. Ye, and K. Pahlavan, "UWB characteristics of creeping wave for RF localization around the human body," in *Proceedings of IEEE 23rd International Symposium on Personal Indoor and Mobile Radio Communications (PIMRC)*, pp. 1290–294, Sydney, Australia, September 2012.

[24] Q. Li, W. Li, W. Sun, J. Li, and Z. Liu, "Fingerprint and assistant nodes based Wi-Fi localization in complex indoor environment," *IEEE Access*, vol. 4, pp. 2993–3004, 2016.

[25] T. Sathyan and M. Hedley, "Fast and accurate cooperative tracking in wireless networks," *IEEE Transactions on Mobile Computing*, vol. 12, no. 9, pp. 1801–1813, 2012.

[26] Y. Qian, K. Lu, and D. Tipper, "A design for secure and survivable wireless sensor networks," *IEEE Wireless Communications*, vol. 14, no. 5, pp. 30–37, 2007.

[27] C. Khauphung, P. Keeratiwintakorn, and K. Kaemarungsi, "On robustness of centralized-based location determination using WSN," in *Proceedings of 14th Asia-Pacific Conference*

on Communications (APCC), pp. 1–5, Tokyo, Japan, October 2009.

[28] N. Correia, J. Coimbra, and G. Schütz, "Fault-tolerance planning in multiradio hybrid wireless-optical broadband access networks," *IEEE/OSA Journal of Optical Communications and Networking*, vol. 1, no. 7, pp. 645–654, 2009.

[29] T. Liu, W. Yang, and R. Bu, "Reliable telecommunication network design problem under node failure," in *Proceedings of 11th International Symposium on Operations Research and its Applications in Engineering, Technology and Management 2013 (ISORA 2013)*, pp. 1–8, Huangshan, China, August 2014.

[30] J. Luo, W. Wu, and M. Yang, "Interference-aware gateway placement for wireless mesh networks with fault tolerance assurance," in *Proceedings of IEEE International Conference on Systems Man and Cybernetics (SMC)*, pp. 2373–2380, Istanbul, Turkey, October 2010.

[31] Y. Zhang and W. Ye, "Design and placement of light monitoring system in museums based on wireless sensor networks," in *Proceedings of International Symposium on Advanced Control of Industrial Processes (ADCONIP)*, pp. 512–517, Hangzhou, China, May 2011.

[32] C. Sharma, Y. F. Wong, W.-S. Soh, and W.-C. Wong, "Access point placement for fingerprint-based localization," in *Proceedings of IEEE International Conference on Communication Systems (ICCS)*, pp. 238–243, Singapore, November 2011.

[33] S.-H. Fang and T.-N. Lin, "A novel access point placement approach for WLAN-based location systems," in *Proceedings of IEEE Wireless Communications and Networking Conference (WCNC)*, pp. 1–4, Sydney, Australia, April 2010.

[34] K. Kondee, S. Aomumpai, and C. Prommak, "A novel technique for reference node placement in wireless indoor positioning systems based on fingerprint technique," *ECTI Transactions on Computer and Information Technology (ECTI-CIT)*, vol. 9, no. 2, pp. 131–141, 2015.

[35] W. L. Winston and J. B. Goldberg, *Operations Research: Applications and Algorithms*, Thomson Brooks/Cole, Pacific Grove, CA, USA, 2004.

[36] S. Kirkpatrick, C. D. Gelatt, and M. P. Vecchi, "Optimization by simulated annealing," *Science*, vol. 220, no. 4598, pp. 671–680, 1983.

[37] K. Maneerat, K. Kaemarungsi, and C. Prommak, "Robust floor determination algorithm for indoor wireless localization systems under reference node failure," *Mobile Information Systems*, vol. 2016, Article ID 4961565, 12 pages, 2016.

[38] R. Jain, *The Art of Computer Systems Performance Analysis: Techniques for Experimental Design, Measurement, Simulation, and Modeling*, Wiley, Hoboken, NJ, USA, 1st edition, 1991.

[39] *MC13224V: 2.4 GHz 802.15.4 RF and 32-bit ARM7 ™ MCU with 128KB flash, 96KB RAM, NXP Semiconductors*, http://www.nxp.com/products/wireless-connectivity/2.4-ghz-wireless-solutions/2.4-ghz-802.15.4-rf-and-32-bit-arm7-mcu-with-128kb-flash-96kb-ram:MC13224V, 2010.

14

UAV-Assisted Data Dissemination in Delay-Constrained VANETs

Xiying Fan ⓘ,[1,2] **Chuanhe Huang** ⓘ,[1,2] **Bin Fu,**[3] **Shaojie Wen** ⓘ,[1,2] **and Xi Chen**[1,2]

[1]*School of Computer Science, Wuhan University, Wuhan, China*
[2]*Collaborative Innovation Center of Geospatial Technology, Wuhan, China*
[3]*Department of Computer Science, The University of Texas Rio Grande Valley, Edinburg, TX, USA*

Correspondence should be addressed to Chuanhe Huang; huangch@whu.edu.cn

Academic Editor: Nicola Bicocchi

Due to the high mobility of vehicles, the frequent path failures caused by dynamic network topology, and a variety of obstructions, efficient data dissemination with delay constraint in vehicular ad hoc networks (VANETs) is a challenging issue. To address these problems, a novel mobile relaying technique by employing unmanned aerial vehicles (UAVs) is considered to facilitate data dissemination in vehicular environments where the communication infrastructures are not available or the network connectivity is poor. This paper studies and formulates the throughput maximization problem in UAV-assisted VANETs, which aims to achieve high throughput while guarantee the delay constraint of data flows to the vehicles in the area. To maximize the network throughput, the maximization problem tries to find an optimal delivery strategy for data dissemination by optimizing the transmission rate. To solve the problem, the knapsack problem can be reduced to the maximization problem, which is proved NP-hard. A polynomial time approximation scheme is proposed to achieve an approximate solution. Detailed theoretical analysis including time complexity and approximation ratio of the proposed algorithm is presented. Simulation results demonstrate the effectiveness of the proposed algorithm.

1. Introduction

As important components of Intelligent Transportation System (ITS), vehicular ad hoc networks (VANETs) are large-scale mobile ad hoc networks composed of vehicles with communication functions and roadside infrastructures, which aim to provide services for autonomous driving and high-speed information sharing [1, 2]. In VANETs, drivers mainly obtain real-time road conditions and safety information sent by other vehicles through wireless communication technology. In this way, traffic accidents and road congestion can be effectively avoided while travel time and energy consumption can be reduced. Meanwhile, VANETs can provide information services, such as news and entertainment, which can add fun to the boring journey.

However, VANETs have some unique characteristics that other ad hoc networks do not share, such as high vehicle mobility, dynamic network topology, and intermittent network connectivity. These features bring a variety of challenges to data dissemination. To deal with the issues, unmanned aerial vehicles (UAVs) can be utilized to co-operate with VANETs. Compared to traditional terrestrial wireless communications, UAV-enabled communications are significantly less affected by channel impairments such as shadowing and fading and in general possess more reliable air-to-ground channels due to higher possibility of having line-of-sight (LoS) links with ground users [3]. Additionally, in the areas where the infrastructures are difficult or too costly to install and maintain to provide ideal network coverage, UAVs can serve as a viable option, as they can collect information from an area of interest and transmit the information to ground VANETs [4, 5]. They can also act as relays to ground networks when direct multihop communications are not available.

Considering the advantages of UAVs, a cooperative hybrid network framework is proposed, integrating UAVs with ground vehicles for data dissemination in VANETs. In the studied scenario, a vehicle carries a message and aims to transmit the message to a target area where exist a number of vehicles and UAVs. To complete the transmission, the message can be either transmitted over vehicle-to-vehicle (V2V) links, vehicle-to-infrastructure (V2I), or air-to-ground (A2G) communication links. To improve the performance of data dissemination, it should transmit data as much as possible in a specific period, which means to maximize the network throughput. Transmission rate and transmission delay of the links over which data is transmitted are utilized to reflect the throughput. Therefore, to achieve the maximum throughput is equivalent to maximize the sum of transmission rate of selected links on data delivery path. As there may exist more than one path from source to destination, the study aims to select a path with the maximum throughput while satisfying a predefined delay threshold.

Graph theory is applied to abstract the network as a connected graph, then the well-known 0/1 knapsack problem can be reduced to the throughput maximization problem. Due to the property of transmission in VANETs, the problem is regarded as the graph knapsack problem which is one of the classical NP-complete problems [6]. Then, a polynomial time approximation algorithm for the graph knapsack problem is derived based on the approximation scheme for subset sum problem [7]. Since the throughput maximization problem can be reduced from the graph knapsack problem, the proposed approximation algorithm can be applied to the maximization problem and to obtain an end-to-end path with the maximum throughput.

The main contributions of this paper are described as below.

(i) A throughput maximization problem in delay-constrained UAV-assisted VANETs is formulated, which considers the tradeoff of data transmission rate and transmission delay. Then, a multiedge graph knapsack problem is constructed based on 0/1 knapsack problem and reduced to the throughput maximization problem, which is proved to be NP-hard.

(ii) A polynomial time approximation scheme is developed for the multiedge graph knapsack problem to obtain the approximate solution. In the proposed scheme, the edges and vertices are assigned with values to indicate their weight. To select a path with the maximum weight, a trim procedure is applied to remove the unnecessary values. Theoretical analysis proves that the algorithm runs in polynomial time with a bound which is polynomial in the size of the input and $1/\epsilon$, where ϵ denotes the approximation parameter. Additionally, the approximation ratio caused by trimming the unnecessary edges in path selection is also derived as $1 + \epsilon$. The results can be applied to general graph knapsack problem.

(iii) An efficient data dissemination algorithm based on the approximate scheme for the graph knapsack problem is proposed to solve the throughput maximization problem. The values of edges in the graph knapsack problem correspond to the transmission rate and delay of the links. Considering the approximation in the knapsack problem, the proposed algorithm for the maximization problem has a quadratic approximation, which is the combination of the approximation to obtain the optimal transmission rate of links and the approximation to trim the unnecessary edges when selecting the path with the maximum throughput. The time complexity and approximation ratio of the proposed algorithm are also given.

The remainder of the paper is organized as follows. Section 2 overviews the related work. Section 3 describes system model and problem formulation. Section 4 develops a polynomial time approximation scheme for the graph knapsack problem, based on which an algorithm for the throughput maximization problem is proposed. Performance evaluation is presented in Section 5. Section 6 concludes the paper. Finally, Section 7 discusses the tradeoff between the benefit and cost of employing UAVs and gives the direction of future work.

2. Related Work

Lots of research has been done to achieve data dissemination with high efficiency in vehicular networks, most of which is devoted to analyzing the performance of delay, throughput and utility of data dissemination [8]. In this section, data dissemination in ground VANETs and UAV-assisted VANETs is mainly discussed.

2.1. Data Dissemination in Ground VANETs. Tan et al. [9] proposed an analytical model to characterize the downlink average throughput and distribution achieved for each vehicle during the sojourn time by the Markov reward model. Zhang et al. [10] proposed an analytical model to facilitate the real-time data delivery as well as delay-tolerant data delivery, in which the theoretical per-vehicle throughput was derived. Lin et al. [11] developed an analytical model that accurately characterized the maximum throughput rate performance achievable under a prescribed outage probability constraint. As the first study on reliable transmission for bulk or stream-like data in DTNs (delay tolerant networks) [12], Zeng et al. proposed a dynamic segmented network coding scheme to efficiently exploit the transmission opportunity. Xing et al. [13] formulated the multimedia scheduling problem to maximize the utility and designed a heuristic algorithm. As continuous research, the authors [14] investigated multimedia dissemination for large-scale VANETs considering the tradeoff of delivery delay, the quality of service (QoS) of delivered data, and the storage cost.

As an emergent paradigm, some research work has applied SDN (Software Defined Network) to support applications in DTNs while reduce the operating costs [15] as it separates the control and data communication layers to simplify the network management. Liu et al. [16] described the application of SDN concept in VANETs and studied data scheduling problem. Nobre et al. [17] defined an architecture that adapted SDN to battlefield networking (BN), which integrated BN and SDN into dynamic and heterogeneous network-centric environments. Zacarias et al. [18] combined SDN and DTN concepts to address the needs of tactical-operational networks, which could support the diverse range of strict requirements for applications.

2.2. UAV-Assisted Wireless Communications.

To provide wireless communications to a given geographical area, Mozaffari et al. [19] analyzed the deployment of an UAV as a flying base station and derived an analytical framework for the coverage and rate analysis for the device-to-device communication network. Then they investigated the optimal 3D deployment of multiple UAVs [20] to maximize the downlink coverage performance with a minimum transmit power. Orfanus et al. [21] utilized the self-organizing paradigm to design efficient UAV relay networks, to provide robust connections to the devices on the military field. Oubbati et al. [22] proposed a UAV-assisted routing protocol to assist data dissemination and improve the reliability of data delivery by filling the communication gap. Wang et al. [23] studied hybrid VANETs that utilized on-vehicle drones and proposed a distributed location-based routing protocol. Xiao et al. [24] employed UAVs to improve network performance against smart jammers and formulated the interaction between UAVs and jammers as an anti-jamming UAV relay game. Seliem et al. [25] proposed a mathematical framework to obtain the minimum drone density, which was equivalent to the maximum separation distance between two adjacent drones, to limit the worst delay of vehicle-drone packet transmissions. Shilin et al. [26] considered a drone-aided communication network model in an isolated VANET segment to enhance network connectivity. Fawaz et al. [27] developed a mathematical model that utilized drones to evaluate the impact of non-cooperative vehicles on forwarding path availability.

Most of the related work did not consider maximizing the network throughput taking consideration of delay constraint in UAV-aided vehicular networks, which motivates this research.

3. System Model and Problem Formulation

In this section, the system model is presented while the maximization problem is formulated.

3.1. System Model.

To improve the reliability and efficiency of data dissemination in VANETs, UAVs are employed to form a cooperative air-to-ground network. By exploiting the UAV-aided VANETs, UAVs can help ground vehicles explore the area of interest and enhance network connectivity.

As stated in [22], most urban applications that use UAVs like small Quad-Copters do not fly at high altitudes [28]. Thus, this study assumes that UAVs have a low and constant altitude during the flight in order to communicate with vehicles on the ground. IEEE 802.11p MAC protocol is adopted for both V2V and A2G communications. UAVs in the network use a large transmission range (i.e., up to 1000 m [29]) and have a global view of the network. Vehicles and UAVs are equipped with GPS and digital maps to obtain their geographical positions. UAVs can also act as relay nodes to forward data packets when direct multihop V2V links are not available.

The cooperative network architecture of the UAV-assisted VANETs is depicted as Figure 1, which is composed of the UAV network and the ground vehicular network. The scenario includes A2G and V2V communication links, which is a hybrid mode that allows the network to apply both A2G and V2V communications for data dissemination in VANETs.

The network can be abstracted as an edge-weighted graph $G(V, E)$ (see Definition 1), where V is a set of vehicles and UAVs, E is a set of edges to indicate the communication links for data dissemination. The weight of each edge is represented by the transmission condition of the corresponding link.

Definition 1.

Given a weighted network graph $G(V, E)$, where V is the set of vertices and E is the set of edges. Let (w_e, d_e) denotes the value of edge e, where w_e indicates the transmission rate of e, and d_e indicates the transmission delay. The tuple (W_i, D_i) denotes the value of node $i \in G \cdot V$, where W_i and D_i indicates the total transmission rate and transmission delay from source node s to i, respectively.

Note that $G(V, E)$ only considers the edges over which two nodes can communicate, which means no silent edges are included.

3.2. Problem Formulation.

Assume a packet with size K carried by vehicle s needs to be transmitted to a specific area. There may exist more than one end-to-end path from source node s to other nodes in V, which can be denoted by P_s and p indicates a path in P_s. It is important to note that the paths may only exist among vehicles through V2V links or they may contain a hybrid of A2G and V2V communications. For simplicity, the A2G and V2V links are considered as common links with different properties hereinafter. The differences of the links are reflected by their transmission rate and delay.

As the network throughput can be mapped by the transmission rate of the end-to-end path, this study discusses how to optimize the transmission rate of each individual path to achieve the maximum throughput. To guarantee the real-time transmission, the end-to-end delay is limited to a predefined threshold. A continuous convex function $f(r_l) = \log r_l$ with the transmission rate as parameter is utilized to depict the throughput, where l denotes a link on path p and r_l denotes the transmission rate of link l. The reason to consider $f(r_l)$ instead of r_l is that the logarithmic utility function $\log r_l$ can better reflect the transmission rate of the delivery path and guarantee the maximum transmission rate. Meanwhile, the logarithm is concave and, hence, has diminishing returns. Here, it seeks a utility for that naturally achieves the maximum throughput and some level of fairness among the links.

To optimize the transmission rate of links and improve the network throughput, the throughput maximization problem can be formulated as below:

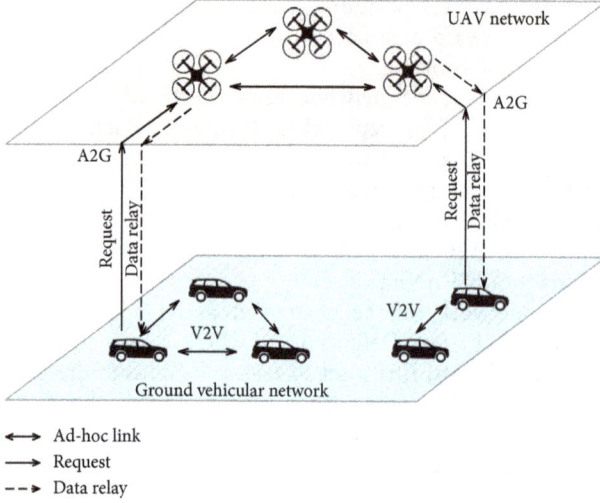

FIGURE 1: An overview of the cooperative air-to-ground network architecture.

$$\begin{aligned} \max \quad & \sum_{l \in L} f(r_l) \cdot x_l, \\ \text{s.t.} \quad & \sum_{l \in L} d_l \cdot x_l \le \delta, \end{aligned} \tag{1}$$

where $r_l \in (1, c_l]$, c_l indicates the maximum capacity of link l, d_l is the transmission delay of link l, and δ is the predefined delay threshold. If link l is selected, x_l is equal to 1, otherwise, x_l is 0. Transmission delay d_l can be calculated by the following equation according to the channel model [30]:

$$d_l = K / (c_l - r_l). \tag{2}$$

The problem can be stated as follows. Given a delay threshold δ and n pairs of positive values (r_l, d_l) to indicate the transmission rate and transmission delay of link l, it aims to select a delivery path which contains a few links to maximize the transmission rate while satisfying the delay constraint. The well-known 0/1 knapsack problem can be reduced to the throughput maximization problem. Then, a polynomial time approximation scheme is proposed to solve the problem.

3.3. An Example. An example is given to illustrate how to derive an approximation solution for the maximization problem. An undirected graph $G(V, E)$ with six vertices is shown in Figure 2. In graph G, each edge has a pair of values (w_e, d_e) and each node has (W_i, D_i) as its weight. The goal is to find a path from source node s to node d, such that the path has the maximum transmission rate W_d while the transmission delay D_d does not exceed δ. The procedure to obtain the path utilizing the approximation method is described.

First, the weight of each edge is given as (w_{sa}, d_{sa}), (w_{sb}, d_{sb}), (w_{ab}, d_{ab}), (w_{ac}, d_{ac}), (w_{be}, d_{be}), (w_{ae}, d_{ae}), (w_{ce}, d_{ce}), (w_{cd}, d_{cd}), and (w_{ed}, d_{ed}). The initial value of node s is $(0, 0)$ and other nodes is $(0, \infty)$. The values of the nodes are recorded, and a list of values for each node will be generated. There might be quite a few values if there are

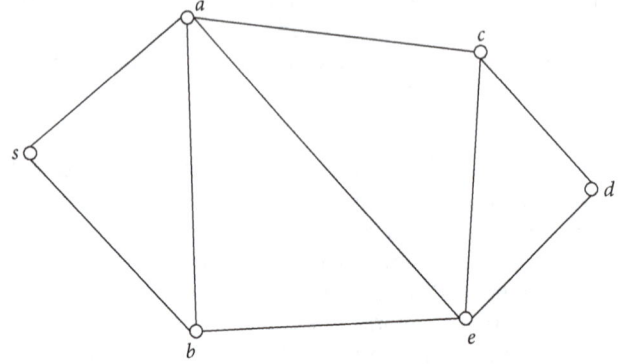

FIGURE 2: An example of an undirected graph $G(V, E)$ with six vertices, each edge connecting the nodes is assigned with weight (w_e, d_e), which helps illustrate how to find a path from s to d.

a large number of nodes. To eliminate redundant values, a trim procedure will be executed if two values in L are close to each other since there is no need to keep both of them. More accurately, a trimming parameter α is utilized such that $0 < \alpha < 1$. When trimming a list by α, remove as many elements as possible, in such a way if L' is the result of trimming L, then for every element y that was removed from L, there is still an element z still in L' that approximates y, that is,

$$\frac{y}{1 + \alpha} \le z \le y. \tag{3}$$

Through the trim procedure, the approximate values (W_d, D_d) of node d can be obtained; thus, an approximate path from s to d will be achieved.

Before explaining the details of the proposed scheme, a list of variables that will be used throughout this research is provided as Table 1.

4. Proposed Solution

In this section, the knapsack problem is reduced to the throughput maximization problem first. Then, a polynomial time approximation scheme is proposed to solve the graph knapsack problem, which can return an approximate solution. Finally, a throughput maximization algorithm is presented based on the approximate scheme for graph knapsack. The approximation ratio and the running time of the proposed algorithm are also analyzed.

4.1. Problem Reduction. To maximize the network throughput, this study optimizes the transmission rate r_l while the sum of d_l does not exceed δ is satisfied. The relation between the transmission rate r_l and transmission delay d_l is presented as $d_l = K / (c_l - r_l)$, subject to $\sum_{l \in L} d_l \le \delta$. From the equation, it can be seen that the transmission delay d_l will increase when r_l increases. There should exist an optimal transmission rate r_l^* with corresponding d_l^*, so the selected path could achieve the maximum throughput while the total delay is within delay constraint. To reduce the complexity of obtaining the optimal values, the approximate values for r_l, d_l are derived.

TABLE 1: Variables used in this paper.

Variable	Definition
$Neighbor[i]$	List of neighbors of node i
δ	Predefined delay threshold
ϵ	A positive real number used for approximation
l	Link between two nodes
L	Set of links on the path for data delivery
r_l	Transmission rate of link l
x_l	$X_l = 1$ means links l is selected otherwise not
s	Source node that carries the information
u, v	Nodes active in the network
W_v	Total weighted transmission rate at node v
D_v	Total transmission delay at node v
w_e	Weighted transmission rate of link between u and v
d_e	Transmission delay of link e between u and v
w_l	Weighted transmission rate of link l, calculated by $f(r_l)$
d_l	Transmission delay of link l
Y	Set of pair values (W_v, D_v) of vehicle v
Y'	Trimmed list of (W_v, D_v) of vehicle v
P^*	Optimal solution of the maximization problem
P	Approximate solution of the maximization problem

According to the equation $d_l = K/(c_l - r_l)$, $r_l = c_l - K/d_l$ holds. When $r_l \longrightarrow 1$, d_l has the minimum value, that is $d_l \longrightarrow K/(c_l - 1)$. Hence, the range of d_l is $(K/(c_l - 1), \delta]$, while the range of r_l is $(1, c_l]$. To achieve the approximate values of r_l and d_l, let r_l increase $1 + \epsilon$ each time until it reaches the largest value, where $0 < \epsilon < 1$ is the parameter used for approximation. Then a list of values for r_l is obtained, which is shown as $\{1, 1 + \epsilon, (1 + \epsilon)^2, \cdots, (1 + \epsilon)^t\}$. According to $(1 + \epsilon)^t \leq c_l$, it has $t \leq \log_{1+\epsilon} c_l$, and t is the largest integer satisfying the inequality, which means $(1 + \epsilon)^{t+1} > c_l$. d_l may have different values according to different r_l calculated by $d_l = K/(c_l - r_l)$, satisfying the condition that $d_l \in (K/(c_l - 1), \delta]$. Therefore, t pairs of (r_l, d_l) can be derived. After $f(r_l)$, t pairs of corresponding values (w_l, d_l) are generated. Consequently, the approximation ratio to obtain the approximate value of r_l is $1 + \epsilon$.

Different pairs of values for each link can be treated as different weights of corresponding edges between two nodes. Accordingly, there may exist multiple edges between two nodes. Then Definition 2 is described.

Definition 2. Given a weighted graph $G(V, E)$, there may exist multiple tuples of values for each edge, which can be treated that there are multiple edges with different values between the corresponding nodes. Accordingly, $G(V, E)$ becomes a multiedge-weighted graph.

As each link is represented by an edge of $G(V, E)$, the graph knapsack problem can be reduced to the throughput maximization problem. The study aims to select a set of links over which the maximum throughput can be achieved while the delay constraint is satisfied.

4.2. Approximation Scheme for the Graph Knapsack Problem. As there may exist more than one path from source node s to node i, node i could have different pairs of weighted values (W_i, D_i). Let Y denote the set of values. Assume Y is sorted

into monotonically increasing order of W_i. A procedure *trim*() (see Algorithm 1) is designed to remove unnecessary values of node i, based on the idea of approximation. The procedure scans the elements of Y in monotonically increasing order. An element is appended onto the returned list Y' only if it is the first element of Y or if it cannot be represented by the most recent values placed into Y'. The output of the procedure *trim*() described as Algorithm 1 is a trimmed, sorted list.

Given the trim procedure, a polynomial time approximation scheme can be constructed for the graph knapsack problem, which is described as Algorithm 2. The approximation procedure takes as input a set of values for node u, $Q = \{(W_{u_1}, D_{u_1}), (W_{u_2}, D_{u_2}), \cdots, (W_{u_n}, D_{u_n})\}$ (in arbitrary order), the delay threshold δ, and the approximation parameter ϵ. Algorithm 2 calls Algorithm 1 to trim the input list. An approximate solution denoted by P within a $1 + \epsilon$ factor of the optimal solution will be returned by the scheme. Lemma 1 is developed to prove that the proposed scheme runs in polynomial time. Meanwhile, Theorem 1 is derived to show that there is a polynomial time approximation algorithm for the multiedge graph knapsack problem.

Lemma 1. *The algorithm for the multiedge graph knapsack problem runs in* $O(3n^2 m \ln W^*/\epsilon)$ *time, where m, n denotes the number of edges and vertices of graph G, respectively.*

Proof. It will show that the running time of the proposed scheme is polynomial in both $1/\epsilon$ and the size of the input. The first part of the algorithm runs in time $O(nm)$, since the initialization in line 1 takes $\Theta(n)$ time, each of the $|V| - 1$ passing over the edges takes $\Theta(m)$ time, where $n = |V|, m = |E|$.

Now the running time of trim process will be analyzed. Assume W^* is the optimal weighted transmission rate of link l and $y.W < y.W^*$. After trimming, successive elements y and Y' of Y have the relationship $Y'.W/y.W > 1 + \epsilon/2n$; that is, they differ by a factor of at least $1 + \epsilon/2n$. Thus, each list contains possibly the value 1 and up to $\log_{1+\epsilon/2n} W^*$ values. It can be deduced that the number of elements in each list Y is at most

$$\lfloor \log_{1+\epsilon/2n} W^* \rfloor + 1 = \frac{\ln W^*}{\ln(1 + \epsilon/2n)} + 1$$

$$\leq \frac{2n(1 + \epsilon/2n)\ln W^*}{\epsilon} + 1 \qquad (4)$$

$$< \frac{3n \ln W^*}{\epsilon} + 1.$$

In summary, the overall running time of the algorithm is $O(3n^2 m \ln W^*/\epsilon)$. This bound is polynomial in the size of the input n and $1/\epsilon$. $\square$

Theorem 1. *The proposed algorithm for the multiedge graph knapsack problem is a polynomial time approximation scheme with an approximation ratio $1 + \epsilon$.*

Input: Y: a list of $(W_i, D_i), \forall i \in V$;
     δ: the predefined delay threshold;
     ϵ: a real number
Output: Y': a trimmed list of Y

(1) Let Y' be empty;
(2) Remove every tuple (W, D) in Y with $D > \delta$;
(3) Partition the tuples of Y into $A_1, \cdots, A_t$ such that for every two tuples $(W, D), (W', D',)$ in the same A_s, they satisfy $W \leq (1 + \epsilon/2n)W'$ and $W' \leq (1 + \epsilon/2n)W$;
(4) **for** $i = 1$ to t **do**
(5)   Select one tuple (W, D) from A_i with the least D;
(6)   Append (W, D) to Y';
(7) **end for**
(8) **return** Y'

ALGORITHM 1: Trim (Y, δ, ϵ).

Input: Q: a list of (W_u, D_u) for node u, every $u \in G.V$;
     δ: the predefined delay threshold
Output: approximation solution P

(1) INITIALIZE $G(V, E)$, set up the value of (w_e, d_e) for corresponding link;
(2) Let $Y_u = \varnothing$ for every $u \in G.V$;
(3) **for** $i = 1$ to $|G \cdot V|$ - 1 **do**
(4)   **for** each edge $e = (u, v) \in G \cdot E$ **do**
(5)   **for** each $(W_u, D_u) \in Y_u$ **do**
(6)   Calculate $(W_v, D_v) = (W_u, D_u) + (w_e, d_e)$;
(7)   Add (W_v, D_v) to Y_v;
(8)   **end for**
(9)   Trim(Y_v);
(10)   **end for**
(11) **end for**
(12) **return** P, which contains the set of links selected

ALGORITHM 2: Polynomial time approximation scheme for graph knapsack.

Proof. Let P^* denote the optimal solution of the problem. From the proposed scheme, it is easily seen that $P \leq P^*$. It needs to show that $P^*/P \leq 1 + \epsilon$.

After trimming, successive tuples y and y' of Y' have the relationship $y \cdot W/y \cdot W > 1 + \epsilon/2n$, where n indicates the number of nodes in G. Scan all the edges, find the path from source node s to destination node u, and let $v_1 v_2 \cdots v_t$ denote the path, $s = v_1, u = v_t$. Due to the trim process executed at the receiver node of each edge, there exists a $(1 + \epsilon/2n)$ factor approximation.

As to v_i, there are $i - 1$ edges between s to v_i; therefore, the approximation ratio should be $(1 + \epsilon/2n)^{i-1}$. Then, v_i reaches v_{i+1} through edge $v_i v_{i+1}$, the approximation ratio at v_i after trimming should be $(1 + \epsilon/2n)^{i-1} \cdot (1 + \epsilon/2n)$, which is equal to $(1 + \epsilon/2n)^i$. Since there are totally n nodes, the number of edges on the path is at most $n - 1$. From the induction of the above procedure, an overall approximation ratio can be expressed as $(1 + \epsilon/2n)^{n-1}$, which can be presented as below:

$$P^*/P \leq (1 + \epsilon/2n)^{n-1}. \tag{5}$$

Now, it needs to show that $P^*/P \leq 1 + \epsilon$, by proving

$$(1 + \epsilon/2n)^{n-1} \leq 1 + \epsilon. \tag{6}$$

Since $\lim_{n \to \infty} (1 + x/n)^n = e^x$, the equation $\lim_{n \to \infty} (1 + \epsilon/2n)^{n-1} = e^{\epsilon/2}$ holds. Since $d/dn(1 + \epsilon/2n)^{n-1} > 0$, function $(1 + \epsilon/2n)^{n-1}$ is monotonically increasing, which means the function increases with n as it approaches the limit $e^{\epsilon/2}$. Thus, the following inequality stands:

$$(1 + \epsilon/2n)^{n-1} \leq e^{\epsilon/2} \leq 1 + \epsilon/2 + (\epsilon/2)^2 \leq 1 + \epsilon. \tag{7}$$

Combine with $P^*/P \leq (1 + \epsilon/2n)^{n-1}$, it has

$$P^*/P \leq (1 + \epsilon), \tag{8}$$

and the analysis of the approximation ratio is completes.

Combined with Lemma 1, it proves that the proposed scheme is a polynomial time approximation scheme. $\square$

4.3. Proposed Algorithm for Throughput Maximization Problem. As the graph knapsack problem is reduced to the throughput maximization problem, a throughput maximization algorithm is proposed based on the approximation scheme for the graph knapsack problem in this section.

Given n items, the ith item is worth w_i and d_i pounds in weight. The 0/1 knapsack problem aims to find a subset of items that the total value is maximum while the total weight is limited to a value. Assume that d_i is at most δ and the items are indexed in monotonically increasing order of their values, that is, $w_1 \le w_2 \le \cdots \le w_n$. Theorem 2 is derived to show that the throughput maximization problem with delay constraint is NP-hard.

Theorem 2. *The throughput maximization problem is NP-hard.*

Proof. Reduce the 0/1 knapsack problem to the throughput maximization problem. Consider Q is a list of n items, denoted by $v_1, v_2, \cdots, v_n$, with corresponding values $\{(w_1, d_1), (w_2, d_2), \cdots, (w_n, d_n)\}$, where (w_i, d_i) indicates the value of the ith item.

Construct a graph $G(V, E)$ with $V = \{s, t, v_1, \cdots, v_n\}$, where s and t denote the source node and destination node, respectively. For every node $v_i \in V$, there is a pair of values (w_i, d_i) for every edge (v, v_i) that goes from v to v_i for any $v \ne v_i$ and $v \ne t$. There is a pair of values $(0, 0)$, for every edge (v, t) that goes from v to t for any $v \ne s$ and $v \ne t$. The knapsack problem aims to select a subset $U \subseteq \{v_1, v_2, \cdots, v_n\}$ of items such that $\sum_{v_i \in U} w_i$ is maximized and the total weight $\sum_{v_i \in U} d_i \le \delta$. A subset U is a feasible solution for the knapsack problem if and only if there is a path that goes from s to the vertex v_i with $v_i \in U$ and then to t. It is easy to see that the time of construction is in polynomial time.

Select items satisfying the required conditions and add them to the knapsack, which is also the way to select the path for the problem. Therefore, if there exists a solution for the knapsack problem, the maximization problem can be solved. Vice Versa, existence of a solution to the maximization problem means there is a solution to the knapsack problem. Thus, the maximization problem is NP-hard.

After reducing the graph knapsack problem to the throughput maximization problem, a Throughput Maximization algoRithm (TMR) is proposed based on Algorithm 2, shown as Algorithm 3. Assume source node s intends to disseminate information to a specific area, an approximate delivery path is desired to achieve the maximum throughput and satisfy the delay constraint.

To be more clearer, a detailed description on how the proposed TMR works on the throughput maximization problem to solve the path-finding issue is presented as follows.

In the initialization process, let $N[v]$ represent a set of v's neighbors. Starting from source node s, execute the following steps to each edge $e \in G \cdot E$.

(1) To find the neighbor vehicles $N[v]$ for node v, exchange information and obtain the corresponding values $(W_{N[i]}, D_{N[i]})$ of the neighbors.

(2) To obtain the links connecting node v with its neighbors, calculate the channel capacity c_l according to the channel condition. Then, get the transmission rate and delay of the corresponding link. Calculate $(W_v, D_v) \in Y_v$ for v by adding the value of its neighbor u, (W_u, D_u) with (w_e, d_e) of the corresponding link e. Therefore, a list Y_v for v can be achieved.

(3) According to the previous step, several pairs of values (W_v, D_v) may exist for node v. First, remove the values with delay that are larger than the delay threshold δ. Then, if there are values with the same delay, keep those with larger transmission rate. Also, remove the values with larger delay and smaller transmission rate. In the following case, such as $Y_v(i) \in Y_v$ with values of $(W_v(i), D_v(i))$, $Y_v(j) \in Y_v$ with $(W_v(j), D_v(j))$, if $W_v(i) > W_v(j)$ and $D_v(i) > D_v(j)$, which means item $Y_v(i)$ is with larger transmission rate but also with larger delay compared with $Y_v(j)$, a trim procedure will be executed to determine whether to remove an element. If $W_v(i) > (1 + \epsilon/2n) \cdot W_v(j)$, append $Y_v(i)$ onto list Y'; otherwise, remove $Y_v(i)$ and append $Y_v(j)$ onto list Y'.

After the iterative operations, paths containing a set of selected links are obtained. If there are more than one path from the source to the destination, choose the one with the largest transmission rate, which is the approximate solution intended to achieve for data delivery.

Theorem 3 is presented to show that the proposed TMR is a polynomial approximation algorithm with an approximation ratio of $1 + \gamma$, where $0 < \gamma < 1$.

Theorem 3. *The throughput maximization algorithm can achieve an approximation ratio $1 + \gamma$ within running time $O(n^2 m \ln C/\gamma)$, where m denotes the number of edges, $C = \sum_{i=1}^{m} c_i$, c_i indicates the transmission capacity of link i.*

Proof. The input of the proposed maximization algorithm is w_l, d_l. As stated in 4.1, the value of w_l, d_l is within a $1 + \epsilon$ factor approximation of the optimal value. Considering the approximation in the multiedge knapsack problem, the throughput maximization problem should have a quadratic approximation. According to theorem 1, the comprehensive approximation ratio is

$$(1 + \epsilon)^2 = 1 + 2\epsilon + \epsilon^2. \tag{9}$$

Assume $\gamma = \epsilon/3$, the inequality $(1 + \epsilon)^2 = 1 + 2\epsilon + \epsilon^2 < 1 + \gamma$ holds.

Hence, an approximation solution within a $1 + \gamma$ factor of the optimal solution can be achieved.

As $r_l \in (1, c_l]$, assume there are m edges on the path. Let $C = \sum_{i \in m} c_i$, then $\forall \ y \in (m, C)$. From Lemma 1, it has

$$\lfloor \log_{1+\epsilon/2n} C \rfloor + 1 < \frac{3n \ln C}{\epsilon} + 1$$

$$= \frac{n \ln C}{\gamma} + 1. \tag{10}$$

Therefore, the total running time is $O(n^2 m \ln C/\gamma)$. $\square$

Input: A list V containing all the vehicles and UAVs in the network
Output: The selected path

(1) Let $N[v]$ denote the neighbor of node v;
(2) $a = 0$, which denotes the number of executions;
(3) **for** each node v **do**
(4)   Send a request to its neighbors;
(5)   Receive the channel information (CI) from its neighbors;
(6)   Calculate c_l according to CI;
(7)   Calculate the approximate values of r_l and d_l according to $d_l = K/(c_l - r_l)$;
(8)   Obtain the transmission rate and delay of the link between each neighbor and node v, denoted as (w_e, d_e);
(9) **end for**
(10) **repeat**
(11) Calculate the transmission rate and delay for each node v in the network, by adding the neighbor's corresponding values to (w_e, d_e), denoted as W_v, D_v;
(12) Apply the trim procedure to remove unnecessary values of node v;
(13) $a = a + 1$;
(14) **until** $(a = |G \cdot V| - 1)$
(15) **return** A path containing a set of selected links

ALGORITHM 3: Throughput Maximization Algorithm.

5. Performance Evaluation

In this section, simulation settings and results are presented and analyzed.

5.1. Simulation Settings. To evaluate the performance of the proposed algorithm, TMR is implemented and compared with other algorithms. In the simulations, the following default settings are used.

The simulations select a 2000 m × 2000 m rectangle street area on the map of Los Angeles and extract the area using openstreetmap [31], the satellite map of which is presented as Figure 3(a). Then, Simulation of Urban Mobility (SUMO) [32] is used to convert the extracted area to the road topology layout, shown in Figure 3(b). The realistic mobility trace of vehicles is generated by the open-source microscopic space-continuous and time-discrete vehicular traffic generator package SUMO. SUMO uses a collision-free car-following model to determine the speeds and the positions of the vehicles. The simulations deploy a number of UAVs which can cooperatively form a full coverage of the simulated area. The speed of UAVs varies from 0 to 15 m/s, and the UAVs maintain a constant altitude that does not exceed 200 m during the flight. The random walk mobility model is applied for the UAVs covering the area. Table 2 gives a list of simulation parameters.

The simulations implement the proposed algorithm and two other algorithms which are UVAR [33] and VBN [11], respectively. Extensive simulations are conducted to thoroughly investigate the efficiency of the proposed algorithm in aspect of delivery ratio, throughput, and number of hops when the number of vehicles varies and the deployed UAVs are set to 20. Additionally, the performance of the proposed algorithm with different UAV densities is evaluated when the number of vehicles is set to 300. A comparison between the proposed solution and optimal solution in terms of throughput is also presented.

5.2. Impact of Number of Vehicles on Delivery Ratio. Delivery ratio is defined as the percentage of packets that are successfully delivered, that is, the ratio of the total number of data packets received by the target destinations to the total number of data packets generated from the sources. A higher delivery ratio means better performance.

In Figure 4, the delivery ratio under different number of vehicles for the compared algorithms is compared. As shown in the figure, the evaluated schemes achieve higher delivery ratio when there are more vehicles in the network. Besides, it can be seen that the proposed scheme and UVAR have better delivery ratio, due to the advantage of UAVs that applied to maintain better network connectivity and guarantee a significant accuracy of path selection. VBN mainly chooses the delivery paths based on cooperation among RSUs and vehicles, which cannot be accurate all the time and may select the paths that are not appropriate for data transmission, resulting in lower delivery ratio.

5.3. Impact of Number of Vehicles on Throughput. An important performance indicator of the algorithms is the throughput of the path from the source to the destination nodes.

In Figure 5, the throughput is plotted versus the number of vehicle nodes. It is observed that the proposed algorithm TMR outperforms the other two algorithms and has the highest network throughput. When the number of vehicles is 50, the throughput of TMR is 1.48 Mbps, higher than that of UVAR and VBN. It also shows that all the compared algorithms achieve higher network throughput as the vehicle density level increases. As the number of vehicles increases to 400, the corresponding throughput of the three schemes increases to 3.45, 3.3, and 3.1 Mbps.

(a) (b)

FIGURE 3: Selected area of Los Angeles, CA, USA.

TABLE 2: Simulation setup.

Parameters	Settings
Simulation area	2000 m × 2000 m
MAC protocol	IEEE 802.11p
Communication range of vehicles	300 m
Communication range of UAVs	1000 m
Vehicle velocity	0–13 m/s
UAV velocity	0–15 m/s
Number of vehicles	50–400
Number of UAVs	10–40

5.4. Impact of Number of Vehicles on Number of Hops.
Number of hops can be obtained from the total number of hops performed by disseminating the message to the target area.

The trend of required hops with the increasing number of vehicle nodes is shown in Figure 6. The proposed algorithms TMR and UVAR perform fewer hops than VBN. This is because that with the help of employed UAVs, available delivery paths from the source node to the destination area can be quickly found by avoiding unnecessary transmissions among vehicle nodes, such that the number of hops consumed is smaller. As the network size becomes larger, the number of hops increases for all the compared schemes. This is mainly because that as the number of vehicles in the target area increases, more hops are needed to deliver the message to the vehicles and complete the dissemination.

5.5. Evaluating the Throughput Maximization Algorithm.
Figure 7 shows the impact of the number of UAVs on data delivery ratio and delivery delay. As more UAVs

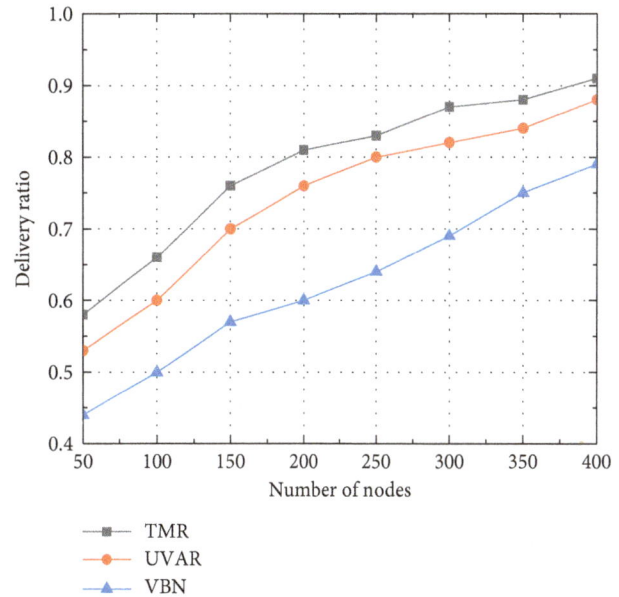

FIGURE 4: Data delivery ratio comparison when the number of nodes increases from 50 to 400.

participate in data transmission, the delivery ratio increases while the delivery delay tends to decrease. This is because that the UAVs can serve relay nodes in data dissemination when there are no available vehicles to carry and forward the data. When more UAVs participate in the communications, the vehicle nodes could select a better UAV relay with a higher probability, which results in the changing data delivery delay and delivery ratio shown in the figure.

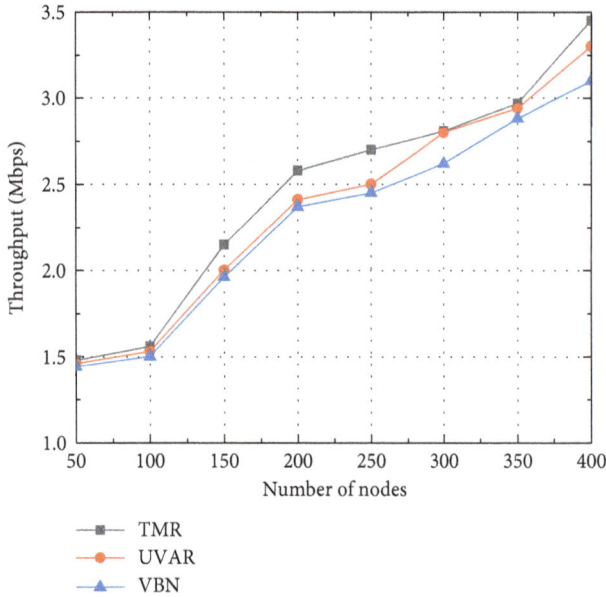

FIGURE 5: Throughput comparison when the number of nodes increases from 50 to 400.

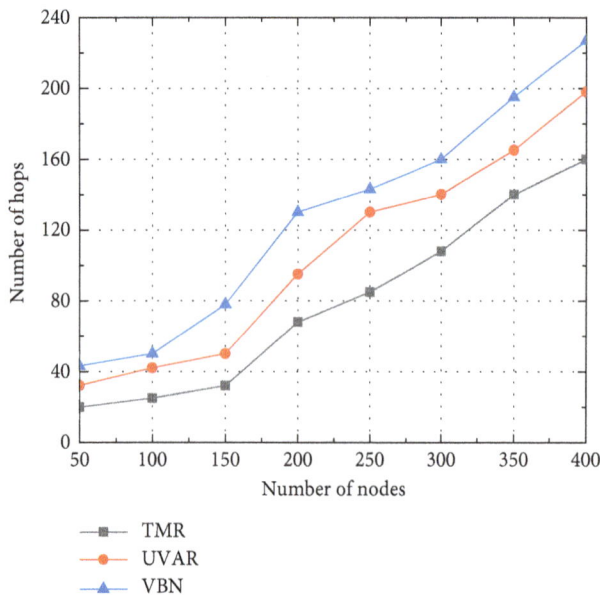

FIGURE 6: Consumed hops comparison when the number of nodes increases from 50 to 400.

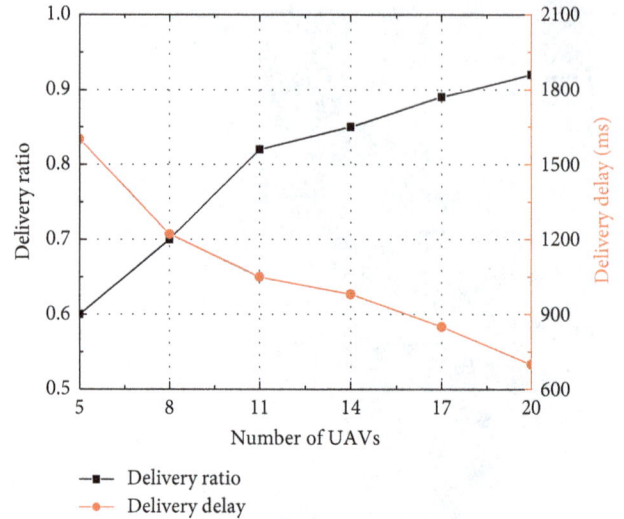

FIGURE 7: Impact of number of UAVs on data delivery efficiency, illustrated by delivery ratio and delivery delay.

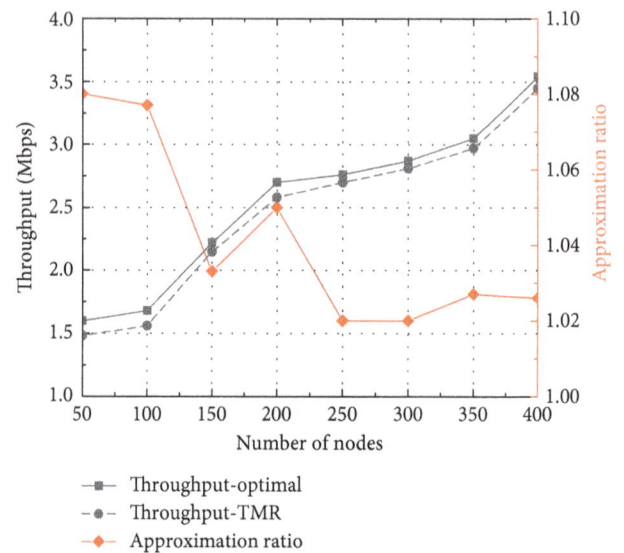

FIGURE 8: Throughput comparison between the proposed solution and the optimal solution, the approximation ratio reflects the throughput difference and the effectiveness of the proposed solution.

6. Conclusion

To show the difference between the proposed solution and the best possible solution, Figure 8 compares the proposed and optimal solutions in terms of throughput, considering the UAV-assisted vehicular environment. Comparing the simulation throughput with the optimal throughput in Figure 8, it can be seen that the simulation result consists with the optimal throughput to a great extent. Meanwhile, the approximation ratio is smaller than 1.1, which also verifies the effectiveness of the proposed algorithm. Observing the changing trends of the throughput, it is easy to find that the system throughput improves when the number of nodes in the network increases.

In this paper, efficient data dissemination in cooperative UAV-assisted VANETs is investigated. To optimize the network throughput, this study formulates a network throughput maximization problem to find the best delivery strategy and select the optimal paths for data delivery, with consideration of the transmission rate of links and the delay constraint for data dissemination. Then reduce the graph knapsack problem to the throughput maximization problem, and a polynomial time approximation scheme is proposed to solve the graph knapsack problem. As to the maximization problem, a throughput maximization algorithm is developed based on the approximation scheme.

Theoretical analysis including the approximation ratio and running time of the proposed solution is provided. Finally, simulations are conducted to evaluate the performance of the proposed algorithm.

7. Discussions and Future Work

While the utilization of UAVs brings significant advantages, it also faces the cost problem. UAV communications are subjected to the additional energy consumption to fly at high altitudes, which is more significant than the communication energy consumption due to signal processing. Nevertheless, the limited on-board energy due to high propulsion energy consumption of UAVs poses critical limits on their communication performance and endurance.

It can be seen that there exists a fundamental tradeoff between the achievable utility benefit and system cost in UAV-assisted communication networks. Using UAVs can increase the network throughput and improve quality of service, which is important to users, especially to the applications with high quality of service requirements. Although the use of UAVs increases the system cost, UAVs have significant advantages over common roadside infrastructure. The tradeoff between the benefit and cost can be achieved by energy-efficient design to enhance the performance of UAV-assisted communication, which is a promising future work direction, such that the deployment and trajectory of UAVs can be carefully designed to save the energy consumption and improve the quality of communications (improved transmission rate and transmission delay).

Despite the contributions presented in this work, many challenges remain to be solved by academia and industry. Future work will focus on the frequent handover problem and interference caused by the high mobility of UAVs and vehicles. Also, energy efficiency of the UAVs remains a relevant topic to be explored to achieve fully utilization of UAVs and improve data dissemination in cooperative network. Further, the integration of the proposed architecture with the concept of SDN and the development of envisaged applications which can adapt to more complicated scenarios might be considered as another future work direction.

Disclosure

An earlier conference version of this paper [34] has been presented in the 13th International Conference on Wireless Algorithms, Systems, and Applications (WASA), 2018.

Conflicts of Interest

The authors declare that they have no conflicts of interest.

Acknowledgments

This work is supported by the the National Natural Science Foundation of China (nos. 61772385, 61373040, and 61572370).

References

[1] R. Ghebleh, "A comparative classification of information dissemination approaches in vehicular ad hoc networks from distinctive viewpoints: a survey," *Computer Networks*, vol. 131, pp. 15–37, 2018.

[2] F. Cunha, L. Villas, A. Boukerche et al., "Data communication in VANETs: protocols, applications and challenges," *Ad Hoc Networks*, vol. 44, pp. 90–103, 2016.

[3] Q. Wu and R. Zhang, "Common throughput maximization in UAV-enabled ofdma systems with delay consideration," 2018, https://arxiv.org/abs/1801.00444.

[4] N. Zhang, S. Zhang, P. Yang et al., "Software defined space-air-ground integrated vehicular networks: challenges and solutions," *IEEE Communication Magazine*, vol. 55, no. 7, pp. 101–109, 2017.

[5] Y. Zhou, N. Cheng, N. Lu, X. Shen, and Sherman, "Multi-UAV-aided networks: aerial-ground cooperative vehicular networking architecture," *IEEE Vehicular Technology Magazine*, vol. 10, no. 4, pp. 36–44, 2015.

[6] M. R. Karp, *Reducibility among Combinatorial Problems*, Springer, New York, NY, USA, 1972.

[7] H. T. Cormen, E. C. Leiserson, L. R. Rivest, and C. Stein, *Introduction to Algorithms*, MIT Press, Cambridge, UK, 3rd edition, 2009.

[8] N. Benamar, K. D. Singh, M. Benamar et al., "Routing protocols in vehicular delay tolerant networks: a comprehensive survey," *Computer Communications*, vol. 48, no. 8, pp. 141–158, 2014.

[9] W. L. Tan, W. C. Lau, O. C. Yue et al., "Analytical models and performance evaluation of drive-thru Internet systems," *IEEE Journal on Selected Areas in Communications*, vol. 29, no. 1, pp. 207–222, 2011.

[10] B. Zhang, X. Jia, K. Yang et al., "Design of analytical model and algorithm for optimal roadside AP placement in VANETs," *IEEE Transactions on Vehicular Technology*, vol. 65, no. 9, pp. 7708–7718, 2016.

[11] Y. Lin and I. Rubin, *Throughput maximization under guaranteed dissemination coverage for VANET systems Information Theory and Applications Workshop*, pp. 313–318, ITA, UK, 2015.

[12] D. Zeng, S. Guo, and J. Hu, "Reliable bulk-data dissemination in delay tolerant networks," *IEEE Transactions on Parallel and Distributed Systems*, vol. 25, no. 8, pp. 2180–2189, 2014.

[13] M. Xing, J. He, and L. Cai, "Maximum-utility scheduling for multimedia transmission in drive-thru Internet," *IEEE Transactions on Vehicular Technology*, vol. 65, no. 4, pp. 2649–2658, 2016.

[14] M. Xing, J. He, and L. Cai, "Utility maximization for multimedia data dissemination in large-scale VANETs," *IEEE Transactions on Mobile Computing*, vol. 16, no. 4, pp. 1188–1198, 2017.

[15] K. Zheng, L. Hou, H. Meng et al., "Soft-defined heterogeneous vehicular network: architecture and challenges," *IEEE Network*, vol. 30, no. 4, pp. 72–80, 2016.

[16] K. Liu, J. K. Y. Ng, V. C. S. Lee et al., "Cooperative data scheduling in hybrid vehicular ad hoc networks: VANET as a software defined network," *IEEE/ACM Transactions on Networking*, vol. 24, no. 3, pp. 1759–1773, 2016.

[17] J. Nobre, D. Rosario, C. Both et al., "Toward software-defined battlefield networking," *IEEE Communications Magazine*, vol. 54, no. 10, pp. 152–157, 2016.

[18] I. Zacarias, L. P. Gaspary, A. Kohl et al., "Combining software-defined and delay-tolerant approaches in last-mile tactical

edge networking," *IEEE Communications Magazine*, vol. 55, no. 10, pp. 22–29, 2017.

[19] M. Mozaffari, W. Saad, M. Bennis, and M. Debbah, "Unmanned aerial vehicle with underlaid device-to-device communications: performance and tradeoffs," *IEEE Transactions on Wireless Communications*, vol. 15, no. 6, pp. 3949–3963, 2016.

[20] M. Mozaffari, W. Saad, M. Bennis, and M. Debbah, "Efficient deployment of multiple unmanned aerial vehicles for optimal wireless coverage," *IEEE Communications Letters*, vol. 20, no. 8, pp. 1647–1650, 2016.

[21] D. Orfanus, P. E. Freitas de, and F. Eliassen, "Self-organization as a supporting paradigm for military UAV relay networks," *IEEE Communications Letters*, vol. 20, no. 4, pp. 804–807, 2016.

[22] O. S. Oubbati, A. Lakas, F. Zhou, M. Gunes, N. Lagraa, and M. Yagoubi, "Intelligent UAV-assisted routing protocol for urban VANETs," *Computer Communications*, vol. 107, pp. 93–111, 2017.

[23] X. Wang, L. Fu, Y. Zhang et al., "VDNet: an infrastructure-less UAV-assisted sparse VANET system with vehicle location prediction," *Wireless Communications & Mobile Computing*, vol. 16, no. 17, pp. 2991–3003, 2016.

[24] L. Xiao, X. Lu, D. Xu et al., "UAV relay in VANETs against smart jamming with reinforcement learning," *IEEE Transactions on Vehicular Technology*, vol. 99, p. 1, 2018.

[25] H. Seliem, M. Ahmed, R. Shahidi et al., "Delay analysis for drone-based vehicular ad-hoc networks," in *Proceedings of 28th IEEE Annual International Symposium on Personal, Indoor, and Mobile Radio Communications*, IEEE, Montreal, QC, Canada, October 2017.

[26] P. Shilin, R. Kirichek, A. Paramonov et al., "Connectivity of VANET segments using UAVs," in *Internet of Things, Smart Spaces, and Next Generation Networks and Systems*, Springer International Publishing, New York, NY, USA, 2016.

[27] W. Fawaz, "Effect of non-cooperative vehicles on path connectivity in vehicular networks: a theoretical analysis and UAV-based remedy," *Vehicular Communications*, vol. 11, pp. 12–19, 2018.

[28] L. Gupta, R. Jain, and G. Vaszkun, "Survey of important issues in UAV communication networks," *IEEE Communications Surveys & Tutorials*, vol. 18, no. 2, pp. 1123–1152, 2016.

[29] W. Fisher, *Development of DSRC/Wave Standards*, IEEE, Annapolis, MD, USA, 2007.

[30] S. Guo, C. Dang, and Y. Yang, "Joint optimal data rate and power allocation in lossy mobile ad hoc networks with delay-constrained traffics," *IEEE Transactions on Computers*, vol. 64, no. 3, pp. 747–762, 2015.

[31] M. Haklay and P. Weber, "OpenStreetMap: user-generated street maps," *IEEE Pervasive Computing*, vol. 7, pp. 12–18, 2008.

[32] SUMO-simulation of urban mobility, http://sumo.sourceforge.net.

[33] O. S. Oubbati, A. Lakas, N. Lagraa, and M. Yagoubi, "UVAR: an intersection UAV-assisted VANET routing protocol," in *Proceedings of the IEEE Wireless Communications and Networking Conference (WCNC)*, IEEE, Doha, Qatar, April 2016.

[34] X. Fan, C. Huang, X. Chen, S. Wen, and B. Fu, "Delay-constrained throughput maximization in UAV-assisted VANETs wireless algorithms, systems, and applications (WASA)," in *Lecture Notes in Computer Science*, Vol. 10874, Springer, Cham, Switzerland, 2018.

15

Trustworthy Event-Information Dissemination in Vehicular Ad Hoc Networks

Rakesh Shrestha and Seung Yeob Nam

Department of Information and Communication Engineering, Yeungnam University, 280 Daehak-Ro, Gyeongsan-si, Gyeongsangbuk-do 712-749, Republic of Korea

Correspondence should be addressed to Seung Yeob Nam; synam@ynu.ac.kr

Academic Editor: Francesco Gringoli

In vehicular networks, trustworthiness of exchanged messages is very important since a fake message might incur catastrophic accidents on the road. In this paper, we propose a new scheme to disseminate trustworthy event information while mitigating message modification attack and fake message generation attack. Our scheme attempts to suppress those attacks by exchanging the trust level information of adjacent vehicles and using a two-step procedure. In the first step, each vehicle attempts to determine the trust level, which is referred to as truth-telling probability, of adjacent vehicles. The truth-telling probability is estimated based on the average of opinions of adjacent vehicles, and we apply a new clustering technique to mitigate the effect of malicious vehicles on this estimation by removing their opinions as outliers. Once the truth-telling probability is determined, the trustworthiness of a given message is determined in the second step by applying a modified threshold random walk (TRW) to the opinions of the majority group obtained in the first step. We compare our scheme with other schemes using simulation for several scenarios. The simulation results show that our proposed scheme has a low false decision probability and can efficiently disseminate trustworthy event information to neighboring vehicles in VANET.

1. Introduction

Vehicular networks are expected to be used for traffic control, accident avoidance, parking management, and so on [1]. Communication security between vehicles needs to be addressed carefully due to the safety requirements of vehicular network applications [2]. There is a lot of ongoing research on security topics, which aims to provide secure communications and verification of data to thwart malicious attackers. One of the major issues in vehicular ad hoc network (VANET) is message trust, which can be used to secure VANET communications. It is essential to periodically evaluate the trustworthiness of event information based on trust metrics. Generally, trust computation in a static network is relatively simple, because the trust level can be calculated based on the behavior of the nodes with sufficient observations [3]. However, message trust computation in VANET is challenging due to the ephemeral nature of the network topology.

The wireless access in vehicular environments (WAVE) protocol is based on the IEEE 802.11p standard and provides the basic radio standard for dedicated short range communication (DSRC) operating in the 5.9 GHz frequency band [4]. Vehicular communications can be achieved in the infrastructure domain for vehicle-to-infrastructure (V2I) communications or in the ad hoc domain for vehicle-to-vehicle (V2V) communications. We mainly focus on V2V communications because road side units (RSUs) [1] may not be available in some parts of the country during the initial stages of deployment of the vehicular communications infrastructure. Vehicles communicate with other vehicles through on-board units (OBUs) forming mobile ad hoc networks that allow communications in a completely distributed manner [5]. We note that some event information (e.g., accident reports) needs to be disseminated quickly and accurately, with minimum delay. Failure in timely and accurate dissemination of such time-critical information might lead to collateral damage to neighboring vehicles.

Some of the issues in vehicular networks include simple routing problems and application-oriented problems like Sybil attacks and false data dissemination [6]. The traditional reputation systems may not work efficiently in vehicular networks [7]. Public key infrastructure (PKI) may not be available everywhere during the initial stages of vehicular network deployment around a country, because some regions may not be covered due to deployment costs or budget issues. Generally, cryptography-based verification of message trustworthiness is computationally expensive. It can protect against a few types of attack from external nodes. However, it will not protect against malicious nodes in the network, which already have the required cryptographic keys, and may not be suitable for V2V ephemeral network communications. Our scheme does not use cryptography and centralized servers, and, thus, it does not have a single point of failure. Most VANET models assume that the system is up and running, where all vehicles have a certain trust score. However, it is not easy to know the trustworthiness of vehicles without having had any interaction with those vehicles. In highly distributed vehicular networks, vehicles can join and leave a network frequently [8, 9]. When a new vehicle joins the network for the first time, there is no information about it. One of the challenges faced by VANET is that the trust model of the VANET should consider the requirement for anonymity of vehicles. The trust model should have minimal overhead in terms of computation complexity, as well as storage. The trust model should be robust to data-centric attacks and be able to detect those attacks [10–12]. VANET security frameworks should be light, scalable, reliable, and secure.

Our proposed scheme investigates the trustworthiness of event information received from adjacent vehicles, which serves as multiple pieces of evidence. We use truth-telling probability as a measure for the trustworthiness of a vehicle. The vehicles communicate through safety messages to report events, such as accident information, safety warnings, information on traffic jams, weather reports, and reports of ice on the road. In our proposed scheme, all vehicles are assumed to have a pseudo identity (PID), which is independent of the node identity. Each vehicle broadcasts an event message to adjacent vehicles from the time it collects information about that event. Every vehicle maintains the trust level of its neighbors in a distributed manner to cope with the propagation of false information. We introduce an enhanced K-means clustering technique to minimize the effect of malicious nodes on trust level calculation. We use a modified threshold random walk algorithm with a single threshold to make a final decision about the occurrence of an event, while supporting real-time decision. We focus on determining the trustworthiness of the event information in the received messages by considering reports from neighboring vehicles differently with a truth-telling probability.

The main contributions of our work can be summarized as follows:

(i) Our proposed scheme can contribute to dissemination of trustworthy information since each vehicle makes a decision on the trustworthiness of information in the received message individually, while dropping packets containing fake information.

(ii) Since all the decisions are made based on the information received from the neighbor vehicles, our proposed scheme can work in an infrastructure-less environment as well.

(iii) Our proposed scheme can make a better decision on the trustworthiness of a given message compared to a simple voting mechanism, since the modified threshold random walk (TRW) can give a higher weight on the opinion of a vehicle, which makes more true statements than false statements.

The remainder of this paper is organized as follows. In Section 2, we discuss the related work. In Section 3, we propose a trustworthy event-information dissemination scheme for VANET. In Section 4, we evaluate the performance of the proposed scheme using simulation. Section 5 concludes the paper with future work.

2. Related Work

Several trust management systems have been proposed for VANET [13–15]. Trust management systems evaluate the trust values of the neighbor nodes to prevent them from interacting with the malicious nodes. The authors in [16] provide a quantitative and systematic review of existing trust management schemes for VANET. They address comprehensive trust model concepts, problems, and solutions related to VANET trust management. There are several works on trust management scheme based on infrastructure framework and cryptography techniques. Trust management schemes can be divided into four categories based on the use of infrastructure and cryptographic measures such as public key infrastructure (PKI) as shown in Table 1. The first category represents the trust management techniques based on infrastructure such as RSU and PKI. In the second category, the nodes rely on infrastructure for trust management without using PKI. In the third category, each node handles the issue of message trustworthiness based on PKI without using any infrastructure. In the fourth category, the nodes are fully decentralized and operate in infrastructure-less environment, and they do not depend on PKI.

In the first category, trust management systems are based on infrastructure such as RSU and PKI and can be effective in identifying malicious nodes with some accuracy [17–23]. However, trust management schemes in this category may not work if infrastructure is not available. The trust management based on PKI is computationally expensive and cannot secure VANET against insider attack, where the malicious nodes have already acquired the cryptographic keys [5, 23]. Some researchers used group signatures (GS) techniques [17–21] to authenticate message sender and guarantee message integrity such as Identity-Based Group Signatures (IBGS) in [18], GSIS in [19], and Identity-based Batch Verification (IBV) in [20]. However, GS schemes are usually based on PKI, and message sender authentication cannot prevent legitimate nodes from sending malicious messages.

TABLE 1: Trust management category based on infrastructure and PKI.

	PKI based	Non-PKI based
Infrastructure based	DTE [5], ESA [17], IBGS [18], GSIS [19], IBV [20], EGSS [21], iTrust [22], PMBP [23]	STRM [13], TRIP [24], RaBTM [25]
Non-infrastructure based	ETM [26], MTM [27], LSOT [28], BTM [29]	CTID [14], ITM [15], RMCV [30], ERM [31], VARS [32], TMEP [33], CRMS [34]

Trust management schemes in the second category require infrastructure including roadside unit or central authority without using PKI [13, 24, 25]. Schemes such as Trust and Reputation Infrastructure-based Proposal (TRIP) [24] and road side unit (RSU) and Beacon-Based Trust Management (RaBTM) [25] may not work if infrastructure is not available. In the third category, the issue of message trustworthiness was investigated based on PKI in an infrastructure-less environment [26–29]. In [27], the authors proposed a multidimensional approach for trust management in decentralized VANET environments using four different types of roles for different types of vehicles. The sender needs to authenticate itself to the receiver node using PKI for verifying each other's role implemented in a distributed manner. However, VANET faces several issues while deploying the PKI scheme due to key distribution in a real world.

In order to overcome the limitation of existing approaches, some researchers investigated trust management without using PKI in an infrastructure-less environment, which corresponds to the fourth category [14, 15, 30–34]. Trust management schemes in this category such as Vehicle Ad Hoc Reputation System (VARS) [32] are more suitable for distributed VANET architecture. Our proposed scheme also belongs to this category, since the decision on the trustworthiness of event information received from neighbor nodes is made in an infrastructure-less environment without using PKI.

The existing trust management systems are established on specific application domain implementing different trust-based models to enhance intervehicular communication. The trust-based models can be classified into three main categories. They are entity based, data-centric based, and hybrid trust models [35]. Entity based trust model deals with the trustworthiness of each node considering the opinions of the peer nodes [26–28, 36]. In [24], the authors proposed a fuzzy approach for the verification of the trustworthiness of the nodes by using feedback from their neighbors. However, the trustworthiness of a message may not always agree with the trustworthiness of the node itself. Thus, this model cannot resolve the issue of message trustworthiness properly.

On the other hand, in data-centric trust model, the trustworthiness of the reported events from the neighbor vehicles is evaluated rather than the trust of the entities or the node itself [5, 30, 31, 35]. In [5], the authors used a Bayesian inference decision module to evaluate the received event reports. But, the inference module uses the prior probability, which is not easy to obtain due to dynamic topology of VANET. In [28], the authors proposed a trust model called a Lightweight Self-Organized Trust (LSOT), which

contains trust certificate-based and recommendation-based trust evaluations. However, it did not distinguish between the trust value of a node and that of the reported message. The trustworthiness of the nodes does not guarantee the trustworthiness of the message as the trustworthy nodes can send fake or faulty messages, if attackers compromise them. In [30], the authors proposed Real-time Message Content Validation (RMCV) scheme in an infrastructure-less mode. This scheme assigns a trust score to a received message based on three metrics, that is, message content similarity, content conflict, and message routing path similarity. The message trustworthiness is based on the maximum value of final trust scores collected from the neighbor nodes. However, this scheme does not consider high mobility of the vehicles and its time complexity is high.

Hence, a hybrid trust model is introduced that combines the entity based and data-centric trust models to evaluate the trustworthiness of a message [32–34]. The authors in [29] proposed a hybrid trust management mechanism called Beacon-based Trust Management (BTM) system, which constructs entity trust from beacon messages and computes data trust from crosschecking the plausibility of event messages and beacon messages. However, their trust model is based on PKI and digital signature, which incurs overhead while signing and authenticating each beacon message before broadcasting.

Thus, we attempt to overcome the limitation of the existing schemes by improving the hybrid trust model for message trustworthiness. As a first step, we use initialization-step enhanced K-means clustering algorithm (IEKA) for clustering of the vehicles into normal and malicious vehicle groups to determine the trustworthiness of each neighbor node. As a second step, we use a modified threshold random walk (TRW) algorithm to decide the trustworthiness of a given message. Thus, our scheme is based on a hybrid trust model. Although RMCV is based on data-centric trust model, it belongs to the fourth category, that is, trust management scheme that requires neither PKI nor infrastructure, according to the classification in Table 1. Thus, we compare our proposed scheme with the RMCV scheme. The detailed comparison and performance evaluation is discussed in Section 4.

3. System Model

Each vehicle collects sufficient information to assess the validity and correctness of a message. Notations explains the parameters and variables used in this paper.

When an event occurs on the road, a vehicle that is near that event sends the safety event message, M_E, to neighboring

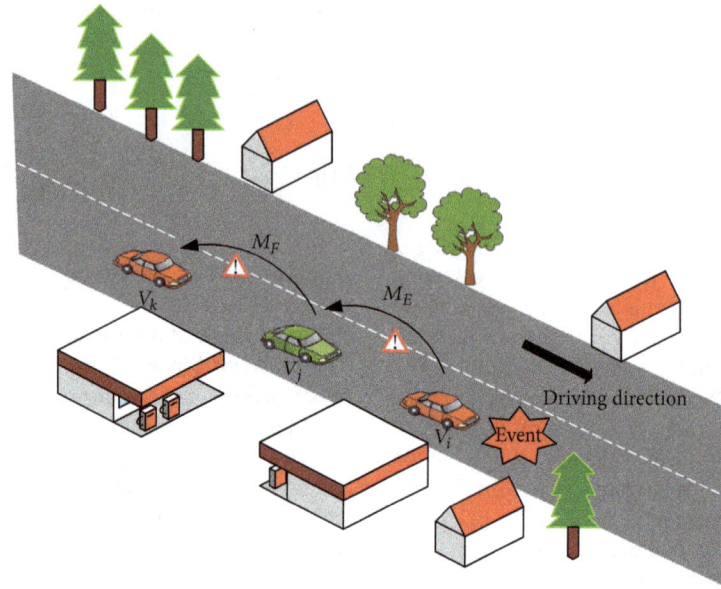

FIGURE 1: Trustworthy message dissemination scheme.

vehicles. Let us suppose that vehicle V_j wants to know the true information about the event reported by vehicle V_i in Figure 1.

The vehicle V_j manages an information pair $(\mathbf{p_i}, \boldsymbol{\theta_i})$ for each neighbor vehicle V_i, where $\mathbf{p_i}$ is the pseudo identity of the ith neighbor vehicle and $\boldsymbol{\theta_i}$ is the trust level, that is, truth-telling probability, of vehicle V_i. We assume that the transportation authority preloads the pseudo ID of the vehicles during vehicle registration, and it should be renewed periodically. To maintain the privacy in VANET, the pseudonym should change over time to achieve unlinkability that protects the vehicle from location tracking. Only privileged authorities are allowed to trace or resolve a pseudonym of the vehicle to a real identity under specific condition [37]. The truth-telling probability $(\boldsymbol{\theta_i})$ is the ratio of the number of true event reports propagated by vehicle V_i to the total number of event reports sent by vehicle V_i over a specific time period.

3.1. Proposed Trustworthy Information Dissemination Scheme. An outline of the proposed scheme to determine the trustworthiness of event information in the received message is shown in Algorithm 1. The vehicle parameters such as pseudo ID (PID) and default trust level are initialized at the beginning. All the vehicles periodically broadcast beacon messages, along with status information such as speed and location to neighboring vehicles. If there is no event triggered, then the vehicles will gather information from the neighboring vehicles. If a vehicle encounters any event by itself, then it broadcasts a safety message along with the trust levels of neighboring vehicles that it knows. Each vehicle accumulates the trust levels of the neighboring vehicles based on the collected safety messages. V_j creates a trust matrix based on the trust level opinions from other vehicles. Thus, the trust matrix manages the trust levels of each neighboring vehicle. Sometimes, vehicles misbehave by

sending false information due to selfish motives like getting easier and faster access to the road, or due to faults. To prevent such false information that can corrupt the trust level of legitimate vehicles, we use a clustering algorithm. Our proposed clustering algorithm attempts to separate the trust level opinions of normal vehicles from the trust level opinions of malicious vehicles. The vehicle will calculate the aggregated trust level of adjacent vehicles belonging to the majority group of normal vehicles from the trust matrix. It will update the trust matrix using the average of trust levels. Then, a modified TRW is applied to know whether the event has actually occurred or not. The modified TRW can provide better decision on the trustworthiness of an event information by giving higher weights on the true event messages. After the trustworthiness of the event information has been verified, the event message is disseminated to other neighboring vehicles along with the updated trust levels. If the event information contained in the message turns out to be untrustworthy, then the message is dropped.

When new vehicles join the VANET, they are not likely to have enough information to infer the trust levels of neighboring vehicles at the beginning. We need a trust level bootstrapping procedure to assign a default trust level for this situation [38]. The trust level, that is, the truth-telling probability, ranges from 0 to 1. If vehicle A does not have any information on vehicle B, then the truth-telling probability of vehicle B is set to 0.5 at vehicle A. We assume that each vehicle sets the truth-telling probability for itself to 1 by default.

We mainly deal with two types of messages: beacon messages and safety messages. The vehicles use beacons to periodically broadcast and advertise status information to neighboring vehicles at intervals of 100 ms. The sender reports its speed, position, and so on to neighboring vehicles with beacon messages via one-hop communications [39]. On the other hand, safety messages support vehicles on the road

```
     //The process is executed by a receiver vehicle upon receiving safety message
     // p_ij: Pseudo ID of ith neighbor vehicle of V_j
     //θ_j: truth-telling probability of V_j
     //θ_ij: estimator of θ_i by V_j
     //Θ_j: trust level opinion generated by V_j
     //θ̂_i: estimator for truth-telling probability of V_i
     Input: Y = {Θ_i} (i = 1, 2, ..., n)
     Output: {θ̂_i}, updated trust matrix
(1)  Information gathering from neighbor vehicles
(2)  If event is triggered then goto step (5)
(3)  Else goto step (2).
(4)  If event source is the V_j itself then goto step (13).
(5)  Else V_j accumulates the trust levels opinions of neighbor vehicles
          Θ_j = ((p_{1j}, θ_{1j}), (p_{2j}, θ_{2j}), ..., (p_{jj}, θ_{jj}), ..., (p_{nj}, θ_{nj}))
(6)  V_j generates a trust matrix based on the trust level opinions.
(7)  Use modified clustering algorithm to separate trust level opinions of normal from malicious vehicles.
(8)  Calculate aggregated trust level of adjacent vehicles belonging to majority group from trust matrix
          θ̂_i = (1/n) Σ_{j=1}^{n} θ_{ij},
(9)  Update the trust matrix
(10) Use modified TRW to know if the event has actually occurred or not.
(11) If we decide that the event message is trustworthy, then goto step (13).
(12) Else drop the message.
(13) Broadcast safety message and trust level to neighboring vehicles.
```

ALGORITHM 1: Determining trustworthiness of event information in the received message.

by delivering time-critical information so that proper action can be taken to prevent accidents and to save people from life-threatening situations. Safety messages include different types of events, E_x, such as road accidents, traffic jams, slippery roads, road constructions, poor visibility due to fog, and emergency vehicle warnings. Vehicles broadcast a safety message to neighboring vehicles when they encounter events on the road [1]. The message payload includes information about the vehicle's position, message sending time, direction, speed, and road events [19]. Each vehicle gathers information about the neighboring vehicles within its communication range.

One advantage of our proposed message dissemination scheme is to avoid a central trusted third party for trust accumulation in a distributed vehicular networking environment. We consider VANET without infrastructure such as RSUs. Vehicles communicate with each other in V2V mode using DSRC [40]. This allows fast data transmission for critical safety applications within a short range of 250 m. A basic safety application contains vehicle safety-related information, such as speed, location, and other parameters, and this information is broadcast to neighboring vehicles [41–43]. Let us consider two vehicles: V_i and V_j. The truth-telling probability of V_i depends on whether vehicle V_i is truthful when relaying event information. According to Velloso et al. [44], the more positive experiences vehicle V_j has with vehicle V_i, the higher the trust vehicle V_j will have towards vehicle V_i.

Let us suppose that vehicle V_i has a pseudo ID p_i and broadcasts a safety warning message M_E, which is defined in (1), when event E_x, where x represents an event type, is detected. If the vehicle itself detects the event, then it broadcasts the safety message along with the trust levels of neighboring vehicles. If a vehicle receives a safety message from other vehicles, it will accumulate the safety message along with trust levels from neighboring vehicles. When vehicle V_j collects event information from vehicle V_i, it finds the type and location of event from the message. Let event message M_E be given by

$$M_E = (p_i, t, L_E, l_i),\qquad(1)$$

where p_i is the pseudo ID of vehicle V_i, t is the message generation time, L_E is the location of event E_x, and l_i is the location of V_i at time t.

In addition to this, each vehicle periodically broadcasts a beacon message defined as $M_B = (p_i, t_i, l_i, s_i)$, where p_i is the pseudo ID of V_i, t_i is the beacon generation time, l_i is the location of V_i, and s_i is the speed of V_i.

Let $θ_i$ be the trust level, that is, truth-telling probability, of vehicle V_i. Truth-telling probability $θ_i$ is defined as the ratio of the number of true events reported by vehicle V_i divided by the total number of events reported by vehicle V_i over a specific period of time. Let m denote the total number of true events reported by V_i and let n denote the total number of events reported by the vehicle up to the current time. Then, the truth-telling probability is

$$θ_i = \frac{m}{n}.\qquad(2)$$

A value for $θ_i$ approaching 1 indicates reliable behavior of the corresponding vehicle, whereas a value close to zero indicates a high tendency towards providing false information [45].

3.2. Calculation of Trust Level of Neighbor Vehicles. When an event occurs, the nearby vehicles broadcast safety messages with additional data, such as pseudo IDs and truth-telling probabilities of other vehicles. Based on the safety messages from the neighboring vehicles, trust matrix $[\theta_{ij}]$ can be obtained, where θ_{ij} is estimation of θ_i by vehicle V_j. The trust matrix manages the truth-telling probability of each neighboring vehicle from the viewpoint of other vehicles. We assume that each vehicle sets its own truth-telling probability to 1. If the trust matrix is constructed, the aggregated trust level, that is, truth-telling probability of vehicle V_i, is calculated from the trust matrix by

$$\hat{\theta}_i = \frac{1}{n}\sum_{j=1}^{n}\theta_{ij}, \tag{3}$$

where $\hat{\theta}_i$ is the estimator for the truth-telling probability of V_i.

3.2.1. Estimation of Truth-Telling Probability Based on the Correctness of Message Information. If we can decide whether specific event information received from a vehicle is correct, this information can be used to estimate the truth-telling probability of the reporting vehicle more accurately. The reliable information about a specific event might be obtained from direct observation of an event spot, or announcement from a public and reliable group.

We explain how the truth-telling probability can be estimated more accurately if we collect more evidence to decide the correctness of messages generated by a given vehicle. We can estimate the truth-telling probability θ_i, defined in (2), based on the correctness of recent N messages from V_i. We introduce a random variable X_n to estimate the number of true reports among the recent N reports from V_i on arrival of the nth report from V_i. Then, the truth-telling probability of V_i can be estimated by X_n/N. We attempt to estimate X_n from X_{n-1} using the following relation:

$$X_n$$
$$= \begin{cases} 1 + \left(1 - \dfrac{1}{N}\right)X_{n-1}, & \text{when the report } n \text{ is correct,} \\ \left(1 - \dfrac{1}{N}\right)X_{n-1}, & \text{otherwise.} \end{cases} \tag{4}$$

Then, we can show that X_n/N approaches the truth-telling probability θ_i of V_i under the assumption that the correctness of one message is independent of the correctness of other messages. By taking expectation on (4), we can obtain $E[X_n]$ as

$$E[X_n] = \Pr[\text{message } n \text{ is correct}]$$
$$\cdot E[X_n \mid \text{message } n \text{ is correct}]$$
$$+ \Pr[\text{message } n \text{ is incorrect}]$$
$$\cdot E[X_n \mid \text{message } n \text{ is incorrect}]$$

$$= \theta_i E\left[1 + \left(1 - \frac{1}{N}\right)X_{n-1}\right] + (1 - \theta_i)$$
$$\cdot E\left[\left(1 - \frac{1}{N}\right)X_{n-1}\right] = \theta_i + \left(1 - \frac{1}{N}\right)E[X_{n-1}]. \tag{5}$$

By solving the recursive relation in (5), we can obtain

$$E[X_n] = \left(1 - \frac{1}{N}\right)^n \{E[X_0] - \theta_i N\} + \theta_i N. \tag{6}$$

Thus, regardless of the initial condition on X_0, we have $\lim_{n\to\infty}E[X_n] = \theta_i N$, and $\lim_{n\to\infty}E[X_n/N] = \theta_i$ from (6). In other words, we can say that X_n/N approaches the truth-telling probability θ_i asymptotically, and we use the estimator of X_n/N and the relation in (4) to update the truth-telling probability of some vehicle whenever we have some evidence to determine the correctness of a message from that vehicle.

3.3. Clustering Algorithm. If there is no evidence to determine the truth of a given message, then the truth-telling probability of vehicle V_i will be calculated using (3). However, malicious vehicles can modify the trust levels of neighboring vehicles to mislead vehicles in a vehicular network. Thus, we need a clustering algorithm that can separate the trust levels of normal vehicles from the trust levels of malicious vehicles. It can reduce the effect of malicious vehicles on the trust levels of normal vehicles. In this subsection, we propose a new clustering algorithm to tackle this issue.

The main goal of our modified clustering algorithm is outlier detection. Our modified clustering algorithm classifies the trust level (truth-telling probability) opinions of the vehicles into two groups, one with the trust level opinions of normal vehicles and the other with the trust level opinions of malicious vehicles. We will select the majority group and neglect the outliers corresponding to the minority group.

Let us assume that an event has occurred on the road and the vehicles near the event location send event messages along with trust level opinions to neighbor vehicles. The vehicle V_j gathers reports about a specific event from neighbor vehicles and manages the trust level opinions of other vehicles as follows. Each vehicle maintains a sorted vehicle list (SVL), which manages pseudo IDs of all the adjacent vehicles in an ascending order, and the vehicle index will be assigned based on the sequence in the sorted list as shown in Table 2. Whenever a vehicle V_j needs to disseminate its own trust level opinion to its neighbors, it sends its trust level opinion Θ_j defined as

$$\Theta_j = \left((p_{1j},\theta_{1j}),(p_{2j},\theta_{2j}),\ldots,(p_{jj},\theta_{jj}),\ldots,(p_{nj},\theta_{nj})\right), \tag{7}$$

where p_{kj} is the pseudo ID of the kth neighbor vehicle of the vehicle j and θ_{jj} is likely to be set to 1 because every node will trust itself.

If V_i receives trust level opinion Θ_j, then V_i updates its own SVL by adding the vehicles that are in Θ_j, but are not in the SVL. After updating SVL, V_i derives $\widetilde{\Theta}_{j'}$ from the received Θ_j as

$$\widetilde{\Theta}_{j'} = \left(\widetilde{\theta}_{1j'},\widetilde{\theta}_{2j'},\ldots,\widetilde{\theta}_{j'j'},\ldots,\widetilde{\theta}_{n'j'}\right), \tag{8}$$

TABLE 2: Example of sorted vehicle list (SVL).

Vehicle index	Pseudo ID
1	1135
2	2056
3	2079
4	2146
5	3012
...	...

where j' is the new index of the vehicle j according to its sequence in the updated SVL and n' is the total number of vehicles in the updated SVL. When p_{kj} in the received Θ_j agrees with the k'th pseudo ID in the updated SVL, $\tilde{\theta}_{k'j'}$ in $\tilde{\Theta}_{j'}$ is updated as

$$\tilde{\theta}_{k'j'} = \theta_{kj}. \tag{9}$$

In this case, n' is always larger than or equal to n since $\tilde{\Theta}_{j'}$ accommodates all the vehicles in Θ_j. If $n' > n$, then it means that there is some pseudo ID that is in the SVL, but not in Θ_j. If an index l corresponds to such a pseudo ID, $\tilde{\theta}_{lj'}$ will be set to 0.5 since the vehicle j' does not know the vehicle l. If $\tilde{\Theta}_{j'}$ is derived, then the trust matrix table is updated by adding the transpose of $\tilde{\Theta}_{j'}$ as the j'th column.

If each vehicle includes the pseudo IDs and the truth-telling probabilities of all the vehicles that it knows in the trust level opinion message defined in (7), then the traffic overhead due to this message can be excessively large. However, we can reduce the message overhead by omitting trivial information. For example, if θ_{kj} in Θ_j is 0.5, this means that the vehicle j does not know the vehicle k since 0.5 is the default value used to initialize the truth-telling probability of a new vehicle. In this case, the vehicle j need not advertise this probability because this default value can be easily filled up by the neighbor vehicles according to the trust matrix updating rule mentioned above with (7), (8), and (9). The vehicles on the road are likely to be ignorant of each other in terms of the trust matrix table, since they need not exchange the trust level opinions if there is no event. Thus, we expect that the policy of omitting trivial information can significantly reduce traffic overhead due to trust level opinion messages.

If V_j collects trust level opinions Θ_i $(i \neq j)$ from other vehicles along with event information, then V_j can construct trust matrix Γ, defined as

$$\Gamma = \begin{bmatrix} \theta_{11} & \theta_{12} & \cdots & \theta_{1n} \\ \theta_{21} & \theta_{22} & \cdots & \theta_{2n} \\ \vdots & \vdots & \ddots & \vdots \\ \theta_{n1} & \theta_{n2} & \cdots & \theta_{nn} \end{bmatrix}. \tag{10}$$

We use a simple example to show our proposed clustering algorithm illustrated by Table 3 and Figure 2. Table 3 shows an example of trust matrix defined in (10), when $n = 3$. Three columns in Table 3 correspond to points A, B, and

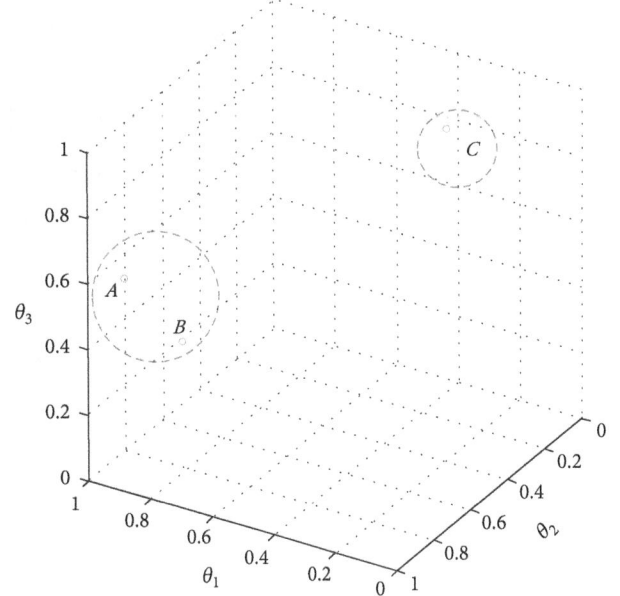

FIGURE 2: Clustering example for the proposed clustering algorithm.

TABLE 3: Trust matrix table of vehicle.

V_j	V_i		
	1	2	3
1	1	0.7	0.2
2	0.8	1	0.4
3	0.5	0.5	1
	A	B	C

C in Figure 2, respectively. The three probability values in the first column of Table 3 correspond to θ_{11}, θ_{21}, and θ_{31}, respectively, and these values are estimation of θ_1, θ_2, and θ_3 by V_i. This tuple of probability values is represented as point A in the three-dimensional space of Figure 2, where each axis represents θ_1, θ_2, and θ_3, respectively. If there is no attacker, that is, all vehicles tell the truth, then all the points will be close to each other. The trust level opinions (Θ_i's) with similar characteristics are likely to form the same cluster. If there is an attacker that tells a lie, then the corresponding point will deviate from the majority group, and this point can be distinguished as an outlier. Even if the attacker tries to change or give higher trust levels by using collusion attack, it can be detected as an outlier. Let us suppose that point C represents an attacker that tells a lie by changing the trust level, as shown in Table 3. The clustering algorithm will separate the trust level opinions into two groups, one group with normal-vehicle trust levels (i.e., A and B) and the other group with malicious vehicle trust levels (i.e., C). Figure 2 describes the outcome of one possible clustering algorithm. The final aggregated trust level is calculated based on the trust level opinions corresponding to the majority group using (3). The final aggregated trust level based on the majority group will be used to update the trust level of the vehicle itself. The resulting trust level is appended to the message during message propagation.

Input: $Y = \{\Theta_i\}$ $(i = 1, 2, \ldots, n)$
Output: $C = \{c_j\}$ $(j = 1, 2)$
Initialize: Calculate unique centroid as initial cluster center,

$$\mu = \frac{\sum_{i=1}^{n} \Theta_i}{|Y|}$$

(1) for each $c_j \in C$ do
(2) $c_j \leftarrow \Theta_i \in Y$
(3) $c_1 = \underset{\Theta_i \in Y}{\arg\max} \|\mu - \Theta_i\|$
(4) $c_2 = \underset{\Theta_i \in Y}{\arg\max} \|c_1 - \Theta_i\|$
(5)   end
(6)   While two centroids are not converged, do
(7)   for each $\Theta_i \in Y$ do
(8)   Assign Θ_i to nearest centroid,
(9) $c_j = \underset{j}{\arg\min} \|c_j - \Theta_i\|^2$
(10)   end
(11)   Update cluster centroid;
(12)   Calculate new centroid c_j as
(13)   for each $c_j \in C$ do
(14) $c_j = (1/|c_j|) \sum_{\Theta_i \in c_j}^{i} \Theta_i,$
(15)   end
(16) end

ALGORITHM 2: Proposed modified clustering algorithm.

We propose a modified K-means clustering algorithm. The main problem with a K-means algorithm lies in the initialization step, so we introduce an enhanced K-means clustering technique by modifying the initialization step, which is called initialization-step enhanced K-means clustering algorithm (IEKA). We use the IEKA to cluster the trust level opinions, while reducing the effect of malicious vehicles on trust levels for other vehicles. Our proposed clustering algorithm can be described in more detail as follows.

After generating a trust matrix, IEKA partitions the trust level opinions into $K(\leq n)$ groups $C = \{c_1, c_2, \ldots, c_k\}$. We designate Y to be a set of the Θ_i vectors; that is, $Y = \{\Theta_1, \Theta_2, \ldots, \Theta_n\}$. We consider only two clusters for our scheme. Initially, we take the mean of all the data points in Y to find a unique centroid, that is, μ.

$$\mu = \frac{\sum_{i=1}^{n} \Theta_i}{|Y|}. \tag{11}$$

We calculate the Euclidean distance between μ and each vector in Y. Choose the point that has the maximum distance from the unique centroid; that is, the selected point is at the farthest distance from the unique centroid. We consider this point as the first centroid, c_1, for the first cluster:

$$c_1 = \underset{\Theta_i \in Y}{\arg\max} \|\mu - \Theta_i\|. \tag{12}$$

Similarly, we compute the Euclidean distance between first centroid c_1 and the remaining points in Y and select the

point with the maximum distance from c_1. Then, this point becomes the second centroid, c_2:

$$c_2 = \underset{\Theta_i \in Y}{\arg\max} \|c_1 - \Theta_i\|. \tag{13}$$

As a next step, we run the conventional K-means clustering algorithm, with c_1 and c_2 being the centroids of two separate groups. Update centroids c_1 and c_2 by calculating the mean value for each group. This gives new centroids c_1 and c_2 and then reassigns each data point to the cluster to which it is closest. We will repeat this process until those two centroids converge. The proposed modified clustering algorithm is given in Algorithm 2.

After clustering using Algorithm 2, which is based on (11), (12), and (13), the aggregated trust level of each neighbor vehicle is calculated based on the trust level opinions belonging to the majority group. We assume that the number of malicious vehicles is less than that of normal vehicles. The aggregated trust level is used for TRW calculation. In the next subsection, we discuss the decision on the event based on hypothesis testing using TRW.

3.4. Event Decision Based on Threshold Random Walk (TRW).
Sequential hypothesis testing is usually used to determine if a specific hypothesis is true or not based on sequential observations [46]. Among the sequential hypothesis testing schemes, threshold random walk has been used to detect scanners with a minimal number of packet observations, while guaranteeing false positives and false negatives [47]. Since we are interested in determining whether a given message is true or not, if true message constitutes one of

Neighbor PID	Trust level (truth-telling prob.)	Event observations (X_i)
p_1	$\hat{\theta}_1$	$X_1 = E_1$
p_2	$\hat{\theta}_2$	$X_2 = \overline{E}_1$
p_3	$\hat{\theta}_3$	$X_3 = \overline{E}_1$
...	...	...
p_n	$\hat{\theta}_n$	$X_n = E_1$

the two hypotheses, then threshold random walk might be applied to this problem. The threshold random walk scheme in [47] uses two thresholds, that is, one upper bound and one lower bound, and the decision is made when a likelihood ratio reaches either threshold. However, in this threshold random walk scheme, we cannot know the number of samples required to reach either threshold in advance. This means that real-time decisions may not be possible if we cannot collect a sufficient number of samples in a short interval. In this paper, we use a modified threshold random walk scheme to determine the validity of a given event, while resolving the issue of real-time decision. We resolve this issue by applying threshold random walk with a single threshold instead of two thresholds. Hereafter, we explain the threshold random walk (TRW) scheme applied to our problem in more detail.

E_1 represents one of the events that can happen on a road. After clustering trust level opinions of neighbor vehicles using IEKA, each vehicle determines the occurrence of event E_1 based on the aggregated trust level table. The aggregated trust level table consists of vehicle PIDs, aggregated trust levels, and event observations, as shown in Table 4.

In Table 4, X_i is the report received from the ith neighbor vehicle about event E_1. Event E_1 represents the occurrence of an event, and $\overline{E}_1$ represents nonoccurrence of event E_1. We need a rule to make a decision about the occurrence of the event. We assume that X_i's are independent of each other among different vehicles. For a given event, suppose random variable Y_i can take only two values (0 and 1); that is,

$$Y_i = \begin{cases} 0; & \text{if } X_i = E_1 \\ 1; & \text{if } X_i = \overline{E}_1. \end{cases} \quad (14)$$

After collecting a sufficient number of reports, we wish to determine whether the event (E_1) has really occurred using sequential analysis [46].

Let us consider two hypotheses: one is null and the other is an alternate hypothesis (i.e., H_0 and H_1), where H_0 is the hypothesis that event E_1 has occurred and H_1 is the hypothesis that event E_1 has not occurred, that is, $\overline{E}_1$. We also assume that conditionals on the hypothesis $Y \mid H_j$ where $j = 0, 1$ are independent. From the definition of the truth-telling probability and (14), we obtain

$$\Pr(Y_i = 0 \mid H_0) = \theta_i,$$

$$\Pr(Y_i = 1 \mid H_0) = 1 - \theta_i,$$

$$\Pr(Y_i = 0 \mid H_1) = 1 - \theta_i,$$

$$\Pr(Y_i = 1 \mid H_1) = \theta_i, \quad (15)$$

where $\Pr(Y = k \mid H_j)$ is the conditional probability that the observation of Y, given hypothesis H_j, is k. Then, $\Pr(Y_i = 0 \mid H_0) = \theta_i$ becomes the truth-telling probability, and $\Pr(Y_i = 1 \mid H_0) = 1 - \theta_i$ becomes the lying probability. In order to make a timely decision, we collect report samples from neighbor vehicles during an interval of fixed duration T. Let N denote the number of report samples collected during this interval. Following the approach of Wald [46], we use collected report samples to calculate the likelihood ratio by

$$\Lambda = \prod_{i=1}^{N} \frac{\Pr(Y_i \mid H_1)}{\Pr(Y_i \mid H_0)}. \quad (16)$$

Although the TRW scheme in [47] makes a decision based on two thresholds, the upper and lower bounds, we use a single threshold to make a decision without the issue of long waiting time. When the threshold is η, the decision rule is as follows:

If $\Lambda \geq \eta$, then accept hypothesis H_1.

If $\Lambda < \eta$, then accept hypothesis H_0.

In this paper, the threshold η will be set to 1, and the truth-telling probability θ_i of an unknown vehicle i will be set to 0.5. When a vehicle receives N report messages, if the Nth report has come from a vehicle with no information on the truth-telling probability, $\Pr(Y_N \mid H_1) = \Pr(Y_N \mid H_0)$ since $\theta_i = 1 - \theta_i$. Thus, the report from the unknown vehicle will not affect the likelihood ratio by (15) and (16). Furthermore, if all the report messages are from the vehicles with no history information, then the likelihood ratio in (16) becomes 1, and, thus, it is fair to put $\eta = 1$, since it is not easy to make a decision in this case.

The advantage of our threshold random walk compared to a simple voting scheme can be described with a simple example as follows. Let us consider a case where an event E_1 is true, and a vehicle receives 5 report messages. Among them, only two report that E_1 is true, and the other three claim that E_1 did not happen. If we make a decision based on a simple voting, then the decision will be $\overline{E}_1$. However, if we apply threshold random walk considering the truth-telling probability of each node, the decision can be different as follows. If the truth-telling probability of the two nodes claiming E_1 is 0.8 and the truth-telling probability of the three nodes claiming $\overline{E}_1$ is 0.6, then the likelihood ratio defined in (16) becomes

$$\Lambda(X) = \frac{0.2}{0.8} \times \frac{0.2}{0.8} \times \frac{0.6}{0.4} \times \frac{0.6}{0.4} \times \frac{0.6}{0.4} = 0.21$$

$$< 1 \, (= \eta). \quad (17)$$

Thus, we will select the hypothesis H_0 according to the decision rule mentioned above, since the likelihood ratio calculated in (17) is less than the threshold η. This means

(a)

(b)

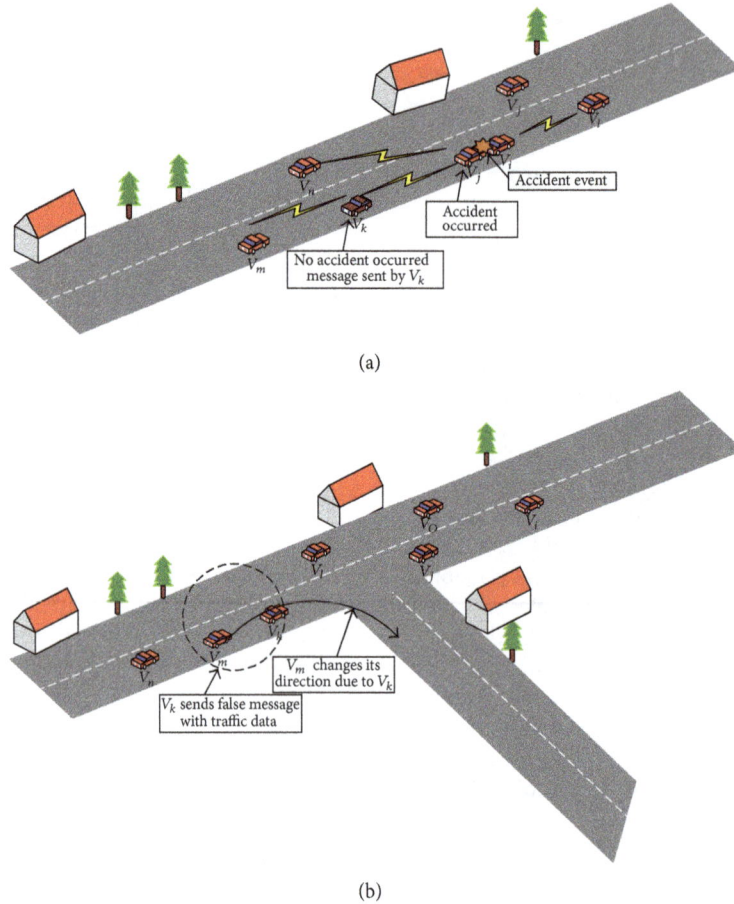

FIGURE 3: Two types of attack patterns considered in this paper: (a) message modification attack and (b) fake message generation attack.

the correct decision of E_1 is made by the proposed threshold random walk. This advantage comes from the fact that the likelihood ratio in (16) gives a higher weight to the opinion of vehicles with a high truth-telling probability.

After the decision on the actual occurrence of the event is made, vehicle j will forward the received message to its neighboring vehicles (with aggregated trust levels) within radio range, which is denoted by

$$M_F = \left(p_j, t, M_E, \Theta_j \right), \tag{18}$$

where p_j is the PID of vehicle j, which forwards the message, t is the time at which M_F was sent, and Θ_j denotes the trust level opinion of vehicle j defined in (7).

3.5. Attack Model. We consider two types of attacks: message modification attack and fake message generation attack in a VANET environment. Figure 3 shows an example of both message modification and fake message generation attack. A malicious vehicle might modify warning messages, either with malicious intent or due to an error in the communications system. In the message modification attack, malicious vehicles can modify message information at any time and falsify the parameters.

In Figure 3(a), when an accident event occurs on the road, the vehicles in an accident or the vehicles which are

close to that accident broadcast the accident event message. After vehicle V_j sends an accident report to other vehicles, a malicious vehicle V_k modifies the message and sends the modified no-accident message as M_F, defined in (18), with the intent to affect decisions taken by other vehicles. Similarly, in a fake message generation attack, malicious vehicles generate a false warning message. For example, in Figure 3(b) [48], a malicious vehicle might send an accident message to neighboring vehicles, even when there is no such event on the road, to clear the route it wants to take. In this case, the malicious vehicle wants to convince other vehicles that an event has occurred. In this scenario, the attacker may have already compromised one or more vehicles and launches attacks by generating a fake message for neighboring vehicles. We assume that the number of malicious vehicles is less than the number of normal vehicles [12]. In simulation, we vary the number of malicious vehicles from 5% to 50% of overall vehicles to evaluate the performance of our proposed scheme in an adversarial environment.

4. Performance Evaluation

4.1. Simulation Setup. The performance of our proposed scheme was evaluated through simulation. We used the Vehicles in Network Simulation (VEINS) framework version

TABLE 5: Simulation parameters.

Parameters	Value
Network simulation package	OMNET++
Vehicular traffic generation tool	SUMO
Wireless protocol	802.11p
Simulation time	300 s
Scenario	Urban/highway
Transmission range	250 m

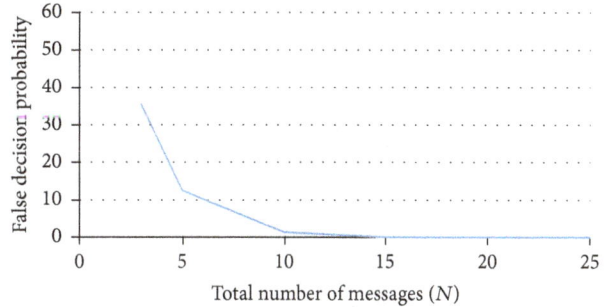

FIGURE 4: False decision probability versus total number of messages (N).

4a2 [49], which is based on both OMNeT++ version 4.6 [50], a discrete event-driven network simulator, and Simulation of Urban Mobility (SUMO) version 22 for road traffic simulation [51]. VEINS connects OMNeT++ and SUMO through Traffic Control Interface (TraCI). VEINS provides realistic models for IEEE 802.11p networks. It provides OMNeT with a set of application programming interfaces to connect the SUMO platform and to dynamically access information about SUMO simulated objects. SUMO allows the creation of scenarios that include realistic mobility patterns, such as vehicle movement and overtaking, as well as lane changing.

We use the default map of Erlangen, Germany, from the VEINS framework with the map size of 2500 m × 2500 m for our simulation. We evaluated our scheme under different traffic densities to consider diverse situations. When the vehicles reach the edge of the road, the vehicles reroute their path and can meet other vehicles multiple times during simulation. The number of vehicles increases linearly with time from 0 s to 300 s. The average vehicle speed changes from 40 km/h in an urban scenario to 110 km/h for highway scenarios. The key parameters considered in our simulation are summarized in Table 5.

We considered two scenarios (urban and highway) by varying parameters such as speed, vehicle density, and percentage of malicious vehicles, as shown in Table 6. The number of malicious vehicles was varied considering the mobility of vehicles in a realistic simulation environment by adjusting vehicle densities and vehicle speeds. We assume that the normal vehicles and the malicious vehicles are uniformly distributed on the roads for each ratio of malicious vehicles [52].

4.2. Simulation Results. In this section, we analyze the simulation results based on OMNet++. The traffic density increases from free-flow traffic (5 vehicles/km^2) to congested traffic (100 vehicles/km^2) where vehicles can meet multiple times. The simulation scenarios are summarized in Table 6. For performance evaluation, we have considered false decision probability and message overhead. We compared our scheme with other schemes under different scenarios. In order to evaluate our proposed scheme, we considered the message modification attack and the fake message generation attack one by one, while increasing the number of malicious vehicles from 5% to 50% in both scenarios. The positions of normal vehicles and the initial distribution of the attackers were randomly determined. We calculated the average false decision probability by averaging the simulation results for

30 simulation runs. A decision is regarded as a false decision when the decision result does not agree with the true status of the event at the time of the decision. In other words, a false decision probability is the ratio of the number of incorrect decisions to the total number of decisions.

In order to update the truth-telling probability of vehicle V_i based on the truth of a given message according to (4), we need to decide the parameter N, that is, the number of recent messages from V_i that will be considered in this estimation. In order to decide N, we run 20 simulations under fake message attack with 30% of malicious vehicles in a highway scenario. We calculate the average false decision probability for various values of N, and Figure 4 shows the result. As the value of N increases, the false decision probability tends to decrease. The false decision probability reaches zero at $N = 15$ and does not change for larger values of N. Thus, N is fixed to 15 hereafter based on this result.

In Figure 4, the false decision probability is high for lower values of N. Let us consider an example to explain worse performance for lower values of N. Let us take an extreme case of $N = 1$. Then, this means that when V_j receives a message from V_i, it decides the truth-telling probability of V_i only based on the last message, since $N = 1$. Thus, if V_j finds that the last message from V_i was false, then V_j will think that the truth-telling probability of V_i is 0, according to the updating rule described in (4). On the other hand, if V_j finds that the last message from V_i was true, then V_j will think that the truth-telling probability of V_i is 1. Thus, the estimated truth-telling probability of each vehicle is either 0 or 1. However, if the truth-telling probability of a given vehicle is different from 0 or 1, then this updating rule (with $N = 1$) will never find the accurate truth-telling probability, since the truth-telling probability is always 0 or 1 according to the updating rule. Hence, the truth-telling probability can be significantly different from the correct truth-telling probability for lower values of N, especially when $N = 1$.

For our modified TRW scheme, we need to determine an optimal value of message collection time T to achieve a good decision accuracy. We run several simulations under fake message generation attack with 30% of malicious vehicle in highway scenario. We calculated the average false decision probability against the message collection time under the simulation parameters given in Table 5. In the sparse network, we received report message as low as five messages

TABLE 6: Simulation scenarios.

Scenario number	Type	Total vehicles	Malicious vehicle ratio	Average speed	Std. of speed
(1)	Urban	100	0~50%	40 km/h	4.76
(2)	Highway	100	0~50%	110 km/h	7.52

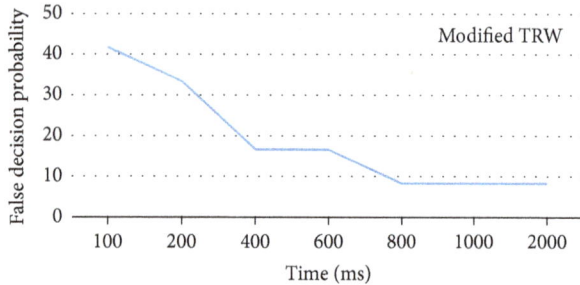

FIGURE 5: False decision probability for various values of message collection time (T).

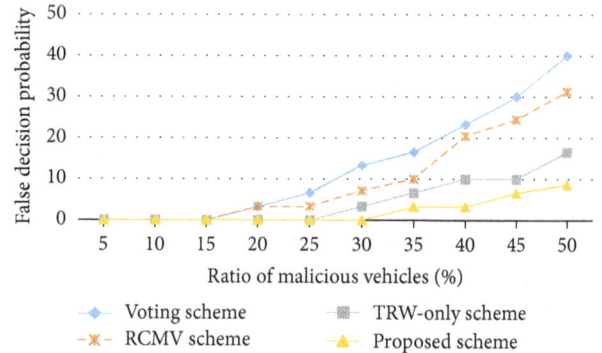

FIGURE 6: Comparison of the proposed scheme with other schemes under a message modification attack in a highway scenario.

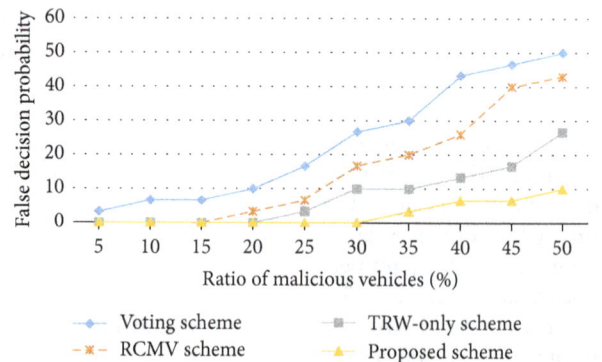

FIGURE 7: Comparison of the proposed scheme with other schemes under a fake message generation attack in a highway scenario.

with report collection time less than 100 ms which results in high false decision probability. The simulation result is shown in Figure 5. As the report collection time interval increases, the false decision probability decreases and, from 800 ms, the false decision probability does not decrease anymore. Based on this, we set the value of T to 1 sec, and this value will be used for T hereafter.

We now compare our proposed scheme with other schemes: RMCV scheme, a simple voting scheme, and TRW-only scheme. RMCV is an information oriented trust model and the outcome of the scheme is a trustworthiness value associated with each received message. In RMCV scheme, we consider the message trustworthiness based on the content similarity. The message trustworthiness is likely to increase as the message contents are similar among different vehicles. In the TRW-only scheme, we used the modified threshold random walk to make a decision about the event in the warning message without applying our proposed clustering algorithm. Several voting methods have been proposed to estimate the trustworthiness of each report message [53–55]. In the simple voting mechanism, each vehicle collects a fixed number of warning messages from the neighboring vehicles regarding an event and makes a decision by following the opinion of the majority group [55]. For the voting scheme, we collected 15 messages to make a decision, as this was the optimal number according to our simulation.

We compare our proposed scheme with other schemes in terms of false decision probability for various ratios of malicious vehicles under the message modification attack in a highway scenario as shown in Figure 6. We can see that our proposed scheme yields a lower false decision probability compared to the other mechanisms, even when the number of malicious vehicles increases. The simple voting mechanism performs worst among the four schemes. The performance of the RMCV scheme is close to TRW-only scheme when the malicious vehicle ratio is low. However, it degrades significantly compared to our proposed scheme as the malicious vehicle ratio increases. Our proposed scheme

has a false decision probability of 0% when the ratio of malicious vehicles is 30% in highway scenario.

We compare our proposed scheme with other schemes in terms of false decision probability under a fake message attack in a highway scenario as shown in Figure 7. We consider a case where the attacker generates messages about a fake event. Our proposed scheme yields better performance compared to the RMCV, simple voting, and TRW-only schemes with a low false decision probability of less than 10%. The false decision probability of RMCV and voting scheme exceed 40% when the ratio of malicious vehicles increased to 50%. In Figure 7, the false decision probability increases as the ratio of malicious vehicles increases, with a tendency similar to Figure 6.

We compare our proposed scheme with other schemes in terms of false decision probability under a message modification attack in an urban scenario in Figure 8. Our proposed scheme exhibits better performance compared to other schemes. The false decision probability does not exceed

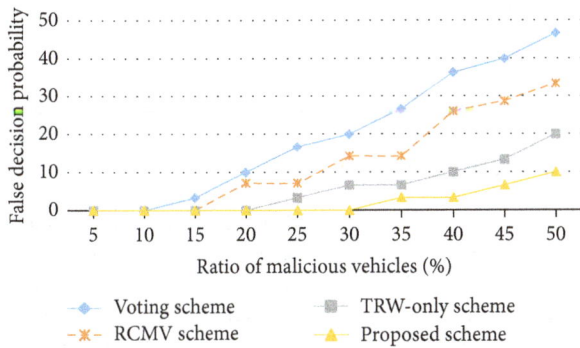

FIGURE 8: Comparison of the proposed scheme with other schemes under a message modification attack in an urban scenario.

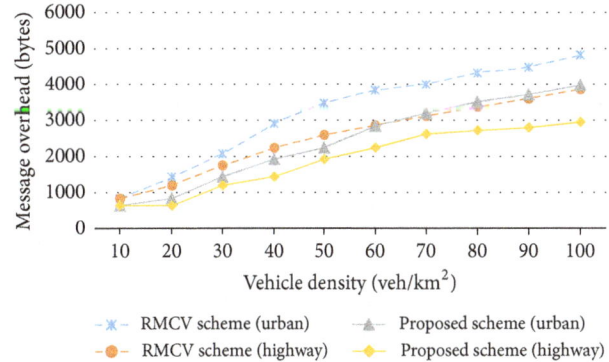

FIGURE 10: Message overhead for various values of vehicle densities under no malicious vehicle.

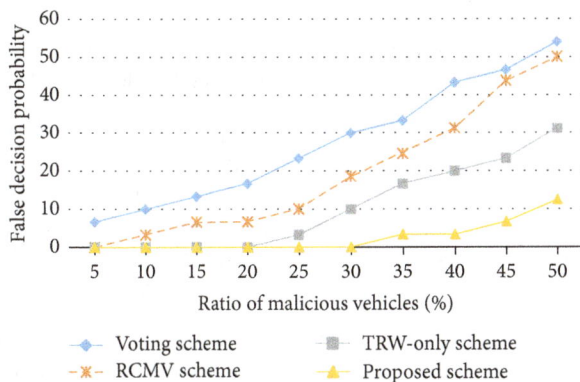

FIGURE 9: Comparison of the proposed scheme with other schemes under a fake message generation attack in an urban scenario.

message overhead is the cost incurred due to the extra message that is exchanged with neighboring vehicles. In terms of message overhead, as the vehicle density increases, the message overhead also increases in both scenarios, as shown in Figure 10. In the beginning, when the vehicle density is low, our proposed scheme has low message overhead as the scheme does not advertise the default trust level of new neighbor vehicles; however the pseudo IDs in the trust level opinion pair cause some message overhead. We can see that the message overhead is higher for the RMCV as compared to our scheme in both scenarios because in their scheme the vehicle nodes send query messages to the neighboring vehicles and then receive response messages regarding the accident event. On the contrary, there is no query message, but only one-way report messages are sent in our scheme. The message overhead in the urban scenario for both schemes is slightly higher than the highway scenario, because the speed of the vehicles in the urban scenario is less than that of highway scenario, and, thus, each vehicle accumulates more messages, compared to the highway scenario.

10% for our proposed scheme. However, it reaches 20% for the TRW-only scheme. The RMCV and the simple voting schemes exhibit much higher false decision probabilities compared to our proposed scheme.

We compare our proposed scheme with other schemes in terms of false decision probability under a fake message attack in an urban scenario in Figure 9. The false decision probability of the proposed scheme increases when the density of the malicious vehicles generating the false message increases, resulting in a false decision probability slightly greater than 10%. In an urban scenario, the high density of vehicles and low speeds help the propagation of false event messages generated by attackers. Thus, the false decision probability in this case is slightly higher than that for the highway scenario. In this scenario, the RMCV and the simple voting schemes exhibit higher false decision probabilities compared to the proposed scheme, with a tendency similar to Figure 8.

We now compare our scheme with the RMCV scheme in terms of message overhead. We considered a situation where there is an actual accident without malicious vehicles. In Figure 10, we present the simulation results of the message overhead with respect to the varying density of vehicles per square kilometer in both urban and highway scenarios. The

We also compare our scheme with RMCV in terms of message overhead in the presence of malicious vehicles under fake message attack. The average vehicle density increases from 1 vehicle per km^2 to 100 vehicles per km^2 throughout the simulation time in urban and highway scenarios. We run several simulations by increasing the ratio of the malicious vehicles from 5% to 50% in both scenarios. Our scheme collects warning messages from neighboring vehicles to detect the trustworthiness of event information contained in the received messages. In both schemes, the message overhead increases as the ratio of the malicious vehicles increases because the vehicles accumulate more messages due to the presence of the malicious vehicles. The message overhead for different ratios of malicious vehicles is shown in Figure 11. Our scheme has a lower message overhead compared to the RMCV scheme in both scenarios.

5. Conclusion and Future Work

In this paper, we proposed a trustworthy event-information dissemination scheme in VANET. We determine and disseminate only the trustworthy event messages to neighbor

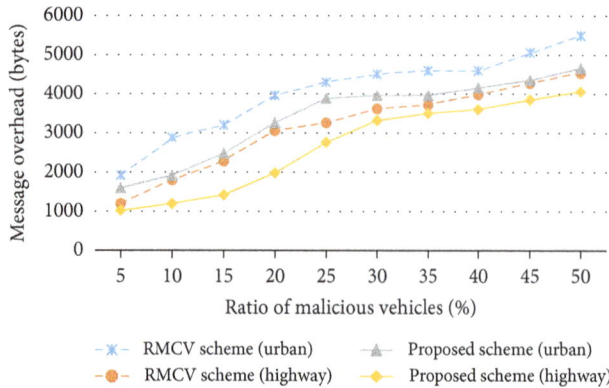

FIGURE 11: Message overhead for various ratios of malicious vehicles under fake message generation attack.

vehicles. We introduced a modified K-means clustering algorithm to reduce the effect of malicious vehicles on the trust levels (i.e., the truth-telling probabilities) of other vehicles. In other words, the issue of node trustworthiness is resolved through a modified K-means clustering algorithm in our proposed scheme. In the next step, the issue of message trustworthiness is resolved by applying a modified TRW to the report messages received from neighbor vehicles along with the information on node trustworthiness. We compared our proposed scheme with RMCV, simple voting, and TRW-only schemes through simulation. The simulation results show that our proposed scheme has a lower false decision probability compared to other schemes as well as low message overhead compared to the RMCV scheme. The simulation results also show that our proposed scheme can effectively cope with message modification attack and fake message generation attack as long as the number of benign vehicles is larger than the number of malicious vehicles. Our scheme has an additional advantage that the decision on the trustworthiness of a given message is made in an infrastructure-less environment without using PKI.

In this paper, we assumed that the malicious vehicles are uniformly distributed on the roads. However, this assumption may not be valid if colluding malicious vehicles move as a group to increase their influence on the nearby vehicles. Such a complicated issue will be studied in more detail in our future work.

Notations

p_i: Pseudo ID of vehicle V_i
θ_i: Trust level (truth-telling prob.) of vehicle V_i
E_x: Event of type x
M_E: Event message
M_B: Beacon message
M_F: Forwarded message
L_E: Location of event E_x
l_i: Location of vehicle V_i
s_i: Speed of vehicle V_i
θ_{ij}:   Estimation of trust level θ_i for vehicle i by vehicle j.

Conflicts of Interest

The authors declare that there are no conflicts of interest regarding the publication of this paper.

Acknowledgments

This research was supported in part by Basic Science Research Program through the National Research Foundation of Korea (NRF) funded by the Ministry of Education (2013R1A1A2012006 and 2015R1D1A1A01058595), and by the MSIT (Ministry of Science and ICT), Korea, under the ITRC (Information Technology Research Center) support program (IITP-2017-2016-0-00313) supervised by the IITP (Institute for Information & communications Technology Promotion).

References

[1] H. Hartenstein and L. Kenneth, *VANET: Vehicular Applications and Inter-Networking Technologies*, Wiley, New Jersey, NJ, USA, 1st edition, 2009.

[2] P. Papadimitratos, L. Buttyan, T. Holczer et al., "Secure vehicular communication systems: design and architecture," *IEEE Communications Magazine*, vol. 46, no. 11, pp. 100–109, 2008.

[3] K. Govindan and P. Mohapatra, "Trust computations and trust dynamics in mobile adhoc networks: a survey," *IEEE Communications Surveys & Tutorials*, vol. 14, no. 2, pp. 279–298, 2012.

[4] Y. L. Morgan, "Notes on DSRC & WAVE standards suite: its architecture, design, and characteristics," *IEEE Communications Surveys & Tutorials*, vol. 12, no. 4, pp. 504–518, 2010.

[5] M. Raya, P. Papadimitratos, V. D. Gligor, and J.-P. Hubaux, "On data-centric trust establishment in ephemeral ad hoc networks," in *Proceedings of the 27th IEEE Communications Society Conference on Computer Communications (INFOCOM '08)*, pp. 1912–1920, Arizona, Ariz, USA, April 2008.

[6] R. Shrestha, S. Djuraev, and S. Y. Nam, "Sybil attack detection in vehicular network based on received signal strength," in *Proceedings of the 3rd International Conference on Connected Vehicles and Expo, ICCVE '14*, pp. 745-746, November 2014.

[7] J. Doucer, "The Sybil Attack," in *Proceedings of the IPTPS '01 Revised Papers from the First International Workshop on Peer-to-Peer Systems*, 2002.

[8] G. Martuscelli, A. Boukerche, and P. Bellavista, "Discovering traffic congestion along routes of interest using VANETs," in *Proceedings of the 2013 IEEE Global Communications Conference, GLOBECOM '13*, IEEE, Atlanta, GA, USA, December 2013.

[9] Z. Huang, S. Ruj, M. Cavenaghi, and A. Nayak, "Limitations of trust management schemes in VANET and countermeasures," in *Proceedings of the IEEE 22nd International Symposium on Personal, Indoor and Mobile Radio Communications (PIMRC '11)*, pp. 1228–1232, IEEE, Toronto, Canada, September 2011.

[10] M. Raya, P. Papadimitratos, and J.-P. Hubaux, "Securing vehicular communications," *IEEE Wireless Communications Magazine*, vol. 13, no. 5, pp. 8–15, 2006.

[11] Z. Li and C. T. Chigan, "On joint privacy and reputation assurance for vehicular ad hoc networks," *IEEE Transactions on Mobile Computing*, vol. 13, no. 10, pp. 2334–2344, 2014.

[12] F. Ishmanov, S. W. Kim, and S. Y. Nam, "A secure trust establishment scheme for wireless sensor networks," *Sensors*, vol. 14, no. 1, pp. 1877–1897, 2014.

[13] N. Yang, "A similarity based trust and reputation management framework for vanets," *International Journal of Future Generation Communication and Networking*, vol. 6, no. 2, pp. 25–34, 2013.

[14] K. Rostamzadeh, H. Nicanfar, N. Torabi, S. Gopalakrishnan, and V. C. M. Leung, "A context-aware trust-based information dissemination framework for vehicular networks," *IEEE Internet of Things Journal*, vol. 2, no. 2, pp. 121–132, 2015.

[15] R. A. Shaikh and A. S. Alzahrani, "Intrusion-aware trust model for vehicular ad hoc networks," *Security and Communication Networks*, vol. 7, no. 11, pp. 1652–1669, 2014.

[16] S. A. Soleymani, A. H. Abdullah, W. H. Hassan et al., "Trust management in vehicular ad hoc network: a systematic review," *EURASIP Journal on Wireless Communications and Networking*, vol. 2015, no. 1, article 146, 2015.

[17] M. Raya, A. Aziz, and J. P. Hubaux, "Efficient secure aggregation in VANETs," in *Proceedings of the 3rd ACM International Workshop on Vehicular Ad Hoc Networks (VANET '06)*, pp. 67–75, ACM, New York, NY, USA, September 2006.

[18] B. Qin, Q. Wu, J. Domingo-Ferrer, and L. Zhang, "Preserving security and privacy in large-scale VANETs," in *Proceedings of the 13th International Conference Information and communications security (ICICS '11)*, vol. 7043 of *Lecture Notes in Computer Science*, pp. 121–135, Springer Berlin Heidelberg, Berlin, Germany, 2011.

[19] X. Lin, X. Sun, P.-H. Ho, and X. Shen, "GSIS: a secure and privacy-preserving protocol for vehicular communications," *IEEE Transactions on Vehicular Technology*, vol. 56, no. 6 I, pp. 3442–3456, 2007.

[20] C. Zhang, R. Lu, X. Lin, P.-H. Ho, and X. Shen, "An efficient identity-based batch verification scheme for vehicular sensor networks," in *Proceedings of the 27th IEEE Communications Society Conference on Computer Communications (INFOCOM '08)*, pp. 816–824, IEEE INFOCOM, Arizona, Ariz, USA, April 2008.

[21] A. Wasef and X. Shen, "Efficient group signature scheme supporting batch verification for securing vehicular networks," in *Proceedings of the IEEE International Conference on Communications (ICC '10)*, pp. 1–5, Cape Town, South Africa, May 2010.

[22] H. Zhu, S. Du, Z. Gao, M. Dong, and Z. Cao, "A probabilistic misbehavior detection scheme toward efficient trust establishment in delay-tolerant networks," *IEEE Transactions on Parallel and Distributed Systems*, vol. 25, no. 1, pp. 22–32, 2014.

[23] D. Tian, Y. Wang, H. Liu, and X. Zhang, "A trusted multi-hop broadcasting protocol for vehicular ad hoc networks," in *Proceedings of the 2012 1st International Conference on Connected Vehicles and Expo, ICCVE '12*, pp. 18–22, December 2012.

[24] F. Gómez Mármol and G. Martínez Pérez, "TRIP, a trust and reputation infrastructure-based proposal for vehicular ad hoc networks," *Journal of Network and Computer Applications*, vol. 35, no. 3, pp. 934–941, 2012.

[25] Y.-C. Wei and Y.-M. Chen, "An efficient trust management system for balancing the safety and location privacy in VANETs," in *Proceedings of the 11th IEEE International Conference on Trust, Security and Privacy in Computing and Communications (TrustCom '12)*, pp. 393–400, Liverpool, UK, June 2012.

[26] U. F. Minhas, J. Zhang, T. Tran, and R. Cohen, "Towards expanded trust management for agents in vehicular ad-hoc networks," *International Journal of Computational Intelligence: Theory and Practice*, vol. 5, no. 1, pp. 3–15, 2010.

[27] U. F. Minhas, J. Zhang, T. Tran, and R. Cohen, "A multifaceted approach to modeling agent trust for effective communication in the application of mobile Ad Hoc vehicular networks," *IEEE Transactions on Systems, Man, and Cybernetics, Part C: Applications and Reviews*, vol. 41, no. 3, pp. 407–420, 2011.

[28] Z. Liu, J. Ma, Z. Jiang, H. Zhu, and Y. Miao, "LSOT: a lightweight self-organized trust model in VANETs," *Mobile Information Systems*, vol. 2016, no. 1, Article ID 7628231, pp. 1–15, 2016.

[29] Y.-M. Chen and Y.-C. Wei, "A beacon-based trust management system for enhancing user centric location privacy in VANETs," *Journal of Communications and Networks*, vol. 15, no. 2, Article ID 6512239, pp. 153–163, 2013.

[30] S. Gurung, D. Lin, A. Squicciarini, and E. Bertino, "Information-oriented trustworthiness evaluation in vehicular ad-hoc networks," *Lecture Notes in Computer Science (including subseries Lecture Notes in Artificial Intelligence and Lecture Notes in Bioinformatics): Preface*, vol. 7873, pp. 94–108, 2013.

[31] Q. Ding, X. Li, M. Jiang, and X. Zhou, "Reputation-based trust model in vehicular ad-hoc networks," in *Proceedings of the International Conference on Wireless Communications And Signal Processing (WCSP)*, 2010.

[32] F. Dötzer, L. Fischer, and P. Magiera, "VARS: a vehicle ad-hoc network reputation system," in *Proceedings of the 6th IEEE International Symposium on a World of Wireless Mobile and Multimedia Networks, WoWMoM '05*, pp. 454–456, June 2005.

[33] C. Chen, J. Zhang, R. Cohen, and P.-H. Ho, "A trust modeling framework for message propagation and evaluation in VANETs," in *Proceedings of the 2nd International Conference on Information Technology Convergence and Services (ITCS '10)*, IEEE, Cebu, Phillippines, August 2010.

[34] A. Patwardhan, A. Joshi, T. Finin, and Y. Yesha, "A data intensive reputation management scheme for vehicular ad hoc networks," in *Proceedings of the 2006 3rd Annual International Conference on Mobile and Ubiquitous Systems: Networking and Services, MobiQuitous*, July 2006.

[35] J. Zhang, "A survey on trust management for VANETs," in *Proceedings of the 25th IEEE International Conference on Advanced Information Networking and Applications (AINA '11)*, pp. 105–112, Biopolis, Singapore, March 2011.

[36] P. Wex, J. Breuer, A. Held, T. Leinmüller, and L. Delgrossi, "Trust issues for vehicular ad hoc networks," in *Proceedings of the IEEE 67th Vehicular Technology Conference-Spring (VTC '08)*, pp. 2800–2804, Singapore, May 2008.

[37] J. Petit, F. Schaub, M. Feiri, and F. Kargl, "Pseudonym schemes in vehicular networks: a survey," *IEEE Communications Surveys & Tutorials*, vol. 17, no. 1, pp. 228–255, 2015.

[38] Z. Malik and A. Bouguettaya, "Reputation bootstrapping for trust establishment among web services," *IEEE Internet Computing*, vol. 13, no. 1, pp. 40–47, 2009.

[39] W. R. J. Joo and D. S. Han, "An enhanced broadcasting scheme for IEEE 802.11p according to lane traffic density," in *Proceedings of the 20th International Conference on Software, Telecommunications and Computer Networks (SoftCOM '12)*, September 2012.

[40] W. Fehr, *Security system design for cooperative vehicleto-vehicle crash avoidance applications using 5.9 GHz Dedicated Short Range Communications (DSRC) wireless communications*, 2012.

[41] H. Xiong, K. Beznosov, Z. Qin, and M. Ripeanu, "Efficient and spontaneous privacy-preserving protocol for secure vehicular communication," in *Proceedings of the 2010 IEEE International Conference on Communications, ICC 2010*, May 2010.

[42] F. Jiménez, J. E. Naranjo, and Ó. Gómez, "Autonomous manoeuvring systems for collision avoidance on single carriageway roads," *Sensors*, vol. 12, no. 12, pp. 16498–16521, 2012.

[43] X. Tang, D. Hong, and W. Chen, "Data Acquisition Based on Stable Matching of Bipartite Graph in Cooperative Vehicle–Infrastructure Systems," *Sensors*, vol. 17, no. 6, p. 1327, 2017.

[44] P. B. Velloso, R. P. Laufer, D. D. O. O. Cunha, O. C. M. B. Duarte, and G. Pujolle, "Trust management in mobile ad hoc networks using a scalable maturity-based model," *IEEE Transactions on Network and Service Management*, vol. 7, no. 3, pp. 172–185, 2010.

[45] F. Bao, I. Chen, M. Chang, and J. Cho, "Hierarchical trust management for wireless sensor networks and its applications to trust-based routing and intrusion detection," *IEEE Transactions on Network and Service Management*, vol. 9, no. 2, pp. 169–183, 2012.

[46] A. Wald, *Sequential Analysis*, John wiley ad Sons, New York, NY, USA, 7th edition, 1965.

[47] J. Jung, V. Paxson, A. W. Berger, and H. Balakrishnan, "Fast portscan detection using sequential hypothesis testing," in *Proceedings of the IEEE Symposium on Security and Privacy*, IEEE, California, Calif, USA, 2004.

[48] G. Peter, S. Zsolt, and A. Szilard, "Highly automated Vehilce Systems," Mechatronics Engineer MSc Curriculum Development, BME MOGI, 2014.

[49] C. Sommer, R. German, and F. Dressler, "Bidirectionally coupled network and road traffic simulation for improved IVC analysis," *IEEE Transactions on Mobile Computing*, vol. 10, no. 1, pp. 3–15, 2011.

[50] A. Varga and R. Hornig, "An overview of the OMNeT++ simulation environment," in *Proceedings of the 1st International ICST Conference on Simulation Tools and Techniques for Communications, Networks and Systems (SIMUTools '08)*, March 2008.

[51] D. Krajzewicz, J. Erdmann, M. Behrisch, and L. Bieker, "Recent Development and Applications of SUMO - Simulation of Urban MObility," in *Proceedings of the Recent Development and Applications of SUMO - Simulation of Urban MObility*, vol. 5, pp. 128–138, December 2012.

[52] W. Ben Jaballah, M. Conti, M. Mosbah, and C. E. Palazzi, "The impact of malicious nodes positioning on vehicular alert messaging system," *Ad Hoc Networks*, vol. 52, pp. 3–16, 2016.

[53] A. Tajeddine, A. Kayssi, and A. Chehab, "A privacy-preserving trust model for VANETs," in *Proceedings of the 10th IEEE International Conference on Computer and Information Technology, CIT-2010, 7th IEEE International Conference on Embedded Software and Systems, ICESS-2010, 10th IEEE Int. Conf. Scalable Computing and Communications, ScalCom '10*, pp. 832–837, July 2010.

[54] J. Petit and Z. Mammeri, "Dynamic consensus for secured vehicular ad hoc networks," in *Proceedings of the 2011 IEEE 7th International Conference on Wireless and Mobile Computing, Networking and Communications, WiMob '11*, pp. 1–8, October 2011.

[55] B. Ostermaier, F. Dötzer, and M. Strassberger, "Enhancing the security of local danger warnings in VANETs - A simulative analysis of voting schemes," in *Proceedings of the 2nd International Conference on Availability, Reliability and Security, ARES '07*, pp. 422–431, April 2007.

16

A New Vehicle Localization Scheme based on Combined Optical Camera Communication and Photogrammetry

Md. Tanvir Hossan ⓘ, **Mostafa Zaman Chowdhury** ⓘ, **Moh. Khalid Hasan** ⓘ,
Md. Shahjalal ⓘ, **Trang Nguyen**, **Nam Tuan Le** ⓘ, **and Yeong Min Jang** ⓘ

Department of Electronics Engineering, Kookmin University, Seoul, Republic of Korea

Correspondence should be addressed to Yeong Min Jang; yjang@kookmin.ac.kr

Academic Editor: Jeongyeup Paek

The demand for autonomous vehicles is increasing gradually owing to their enormous potential benefits. However, several challenges, such as vehicle localization, are involved in the development of autonomous vehicles. A simple and secure algorithm for vehicle positioning is proposed herein without massively modifying the existing transportation infrastructure. For vehicle localization, vehicles on the road are classified into two categories: host vehicles (HVs) are the ones used to estimate other vehicles' positions and forwarding vehicles (FVs) are the ones that move in front of the HVs. The FV transmits modulated data from the tail (or back) light, and the camera of the HV receives that signal using optical camera communication (OCC). In addition, the streetlight (SL) data are considered to ensure the position accuracy of the HV. Determining the HV position minimizes the relative position variation between the HV and FV. Using photogrammetry, the distance between FV or SL and the camera of the HV is calculated by measuring the occupied image area on the image sensor. Comparing the change in distance between HV and SLs with the change in distance between HV and FV, the positions of FVs are determined. The performance of the proposed technique is analyzed, and the results indicate a significant improvement in performance. The experimental distance measurement validated the feasibility of the proposed scheme.

1. Introduction

Localization refers to the process of identifying the location (x and y coordinates in two-dimensional (2D) space and x, y, and z coordinates in three-dimensional (3D) space) of an object in a certain point in space at a specific time. Several studies are contributing to the development of accurate localization schemes owing to increased demand for Internet of Things (IoT) applications. The necessity of a localization scheme is integrated within the requirement of IoT. IoT relies on an enormous number of physical objects (e.g., sensor nodes and sensor networks) that are connected via the Internet [1]. These objects can be interconnected to each other either via wire or wireless mediums. A localization scheme is an important concern for connecting sensor nodes in remote location. A node cannot access or wirelessly communicate with other nodes without accurately positioning itself. The characteristics of localization schemes vary with the features of indoor and outdoor environments [2].

It is well known that localizing sensor nodes indoor can be a crucial obligation for modern businesses and commerce. However, issues related to outdoor localization, particularly vehicle localization, are prioritized over indoor localization. Recently, following road traffic safety [3] has become important owing to the increasing number of fatal road accidents. World Health Organization statistics [4] show that traffic-related accidents worldwide resulted in 1.3 million deaths of people between 15 and 29 years, and the number of nonlethal injuries is 15–40 times greater (between 20 and 50 million). Thus, traffic fatalities rank among the 10 top causes of death, comparable to suicide, HIV/AIDS, homicide, and other diseases. The most common cause of traffic fatalities (around 60%) is high vehicle speeds (above 80 km/h) on the road [5]. Autonomous vehicles can help minimize traffic deaths. Meanwhile, the demand for autonomous vehicles has been rising dramatically to avoid accidents [6]. Furthermore, outdoor localization is of prime importance in the transportation domain, particularly, for autonomous vehicles, which requires localizing other vehicles from the host vehicle (HV) in road environments such as highways. For autonomous vehicles, the features of localization are classified as active and passive. Active features include setting region of interest (ROI) and measuring the possibility of communicating with other vehicles and maintaining safe distance from other vehicles to avoid unwanted collisions by measuring spatial and temporal scenarios [7]. Passive features include obtaining localization information from individual vehicles, which can then be accumulated by a traffic control center and utilizing in effective way to mitigate traffic congestion.

1.1. Existing Solutions, Limitations, and Current Trends in Vehicle Localization. Global positioning system (GPS) is considered as the most prominent solution for outdoor localization scheme. GPS provides a line of sight vehicle localization solution using the sensor information from [8–10] and data from a satellite orbiting at an altitude of approximately 20,000 km. GPS uses the radio frequency (RF) band for positioning the HV on road. However, the HV cannot measure its own distance from other vehicle, such as the forwarding vehicle (FV), via GPS; it offers only the current location of the HV. Moreover, this localization scheme is fraught with several challenges, such as GPS signals being blocked by obstacles such as buildings, subway, tunnels, and trees. Localization using GPS can generate a localization error of up to 1 m within 10 s [11]. A wireless network standard for vehicle states, called IEEE 802.11p [12, 13], is available; this is referred to as wireless access in vehicular environments (WAVE) [14]. This standard is used to maintain a communication network among vehicles within vehicular ad hoc networks (VANETs) [15] and to support intelligent transport system applications. RF signals in VANET systems are used for communication and vehicle localization [16]. Owing to various environmental effects and the multipath nature of the network, non-Gaussian noise is included with the transmitted signal, whose strength shows nonlinear characteristics over distance. The WAVE standard uses a license-free RF band (i.e., 2.4 GHz) [17], which is open to interference from other signal sources, thus making the entire network vulnerable from the viewpoints of both communication and localization. Other existing technologies for vehicle localization include light detection and ranging (LiDAR) [18–20] and the time-of-flight (ToF) camera technique [21–24]. Light-emitting diodes (LEDs) and cameras or photodiodes are embedded in LiDAR and ToF system infrastructures; however, they are used only for detection and ranging. These equipment are not useful for vehicle-to-vehicle or vehicle-to-infrastructure communications [25–29] and are expensive to be used in a vehicular environment.

Optical wireless communication (OWC) is an emerging and promising technology [30] that is viable for handling scenarios wherein RF faces challenges. OWC is not intended to replace RF; however, the coexistence of both can provide a better solution [31] for communication and localization. Optical camera communication (OCC) [32] is a subarea of OWC that uses a camera as a receiver to decode signals from a modulated light source, for example, LEDs, by varying the state of the light source to transmit binary data via optical channels. It is a secure, safe, reliable, and fast method for communication as well as localization [33]. A unique feature of OCC is that the camera used for vehicle localization can simultaneously be used to communicate with other vehicles that transmit signals using modulated lights. With little modifications, LEDs in existing infrastructure, that is, vehicles and streetlights (SLs), can be used for communication (e.g., bidirectional communication between two vehicles or between vehicles and infrastructure) [34–40].

To better communicate in outdoor environments, vehicles around the HV must be localized precisely. More importantly, multiple-input and multiple-output (MIMO) features of OCC [41] should allow the HV to simultaneously communicate with more than one vehicle. In [42], author presents a received signal strength-based visible light communication localization scheme, but it could not improve localization performance such as more complex models of the environment or additional hardware are required for localization. The localization of multiple vehicles would require incorporating OCC and photogrammetry technologies [43]. Photogrammetry [44, 45] deals with a branch of geometry wherein an image sensor (IS) is used to measure an object by quantifying the photon intensities of different wavelengths of light incident on an area, that is, a unit pixel of a camera. Photogrammetry helps accumulate information on semantic and geometric properties and variation of relative distances of objects, which refers to vehicles in this context. This vehicle location information can be shared with following vehicles with the help of OCC and rear-facing LED lights. Figure 1 shows a vehicle localization scheme combining OCC and photogrammetry.

A vehicle localization technique, wherein each FV broadcasts its identity (ID) to the HV as FV-ID, is proposed herein. After extracting the unique ID from the received signal, the HV can distinguish an FV from other FVs. Since the HV and FV simultaneously change their positions over time, location of the HV should be normalized based on the location of a fixed object, for instance, an SL. Comparing the locations of more than one SL relative to the HV, a virtual location of the HV can be temporarily generated. This HV location information acts as the origin of a Cartesian coordinate system that allows determining the FV location relative to the HV.

For autonomous vehicle localization, infrared LED array can be attached at the SLs and back side of the FVs to clarify the area of the LED array. Though the near-infrared (NIR) source is visible to the camera, it is possible to receive data from the FVs and SLs. Detecting light intensity for the nearest light sources is much higher on the IS of the camera of HV rather than far distance light source. Compared with visible light-based communication, NIR-based communication is influenced by the optical channel. Under daylight, it is challenging to receive data from an NIR-based transmitter. Importantly, a recent development of high-dynamic

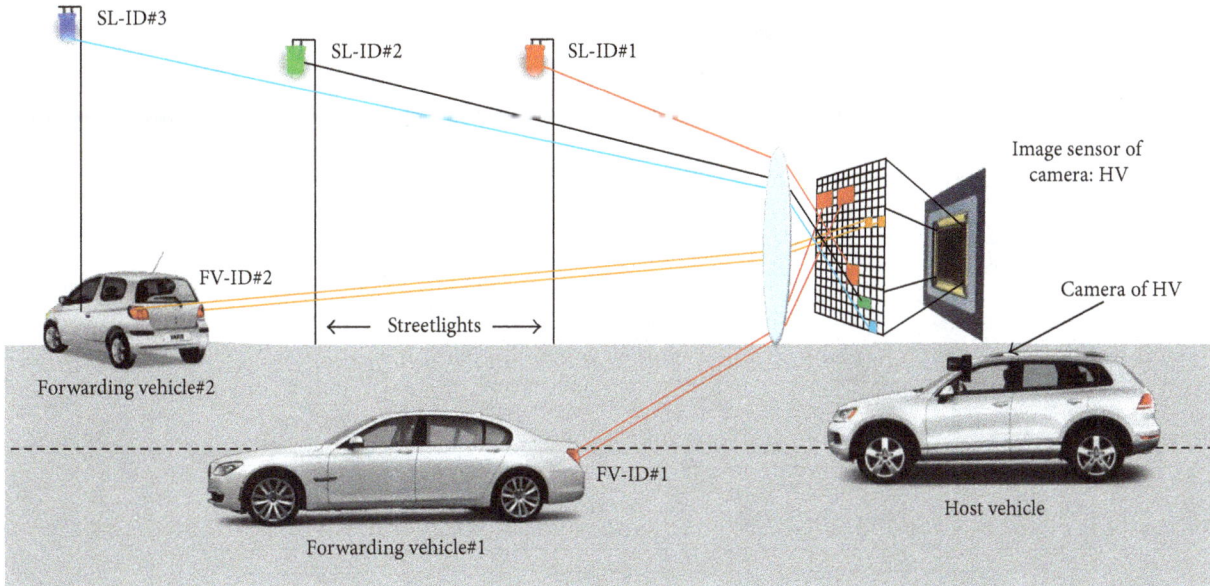

FIGURE 1: OCC and photogrammetry-based vehicles localizing by comparing relative position with the help of streetlights.

range imaging technique reduces noise and enhances the image quality under daylight [46, 47]. Therefore, it is expected that ambient light no longer poses a problem for OCC, even when the transmitter possesses an NIR-optical band. Simulation results show vehicle localization accuracy with considering the impact of several parameters including signal-to-interference-plus-noise ratio (SINR), IS resolution, camera exposure time, and the distance between two SLs.

The remainder of this paper is organized as follows: Section 2 explains a detailed theoretical and mathematical model of our proposed scheme. Experimental setup for distance measurement is shown in Section 3. In Section 4, the simulation results associated with vehicle localization studies are presented. Finally, Section 5 presents a summary of lessons learned and concludes this study.

2. Development of Proposed Scheme

Almost every vehicle produced in recent years is equipped with a camera (i.e., less than 30 frames/sec) that is used to monitor the outdoor scenarios and to assist the drivers by providing a view of their blind spot. Herein, the HV communicates with the FV and measures the distance between vehicles using such a camera mounted in front of the vehicle. Using OCC, this camera detects transmitted signal IDs, such as FV-ID and SL-ID from each FV and SL simultaneously. A pair of taillight on FVs transmits ID in different phases to modulate the data using a modulation scheme called spatial two phase-shift keying (S2-PSK) [48] to the HV's camera. These LEDs transmit at a constant clock rate (e.g., 125 or 200 Hz) to send a flicker-free signal. The SLs use the same modulation scheme as the FV for transmitting the SL-ID. MIMO is a distinctive functionality of a camera that helps distinguishing FV-ID from SL-ID. These IDs are required to determine the ROI for vehicle localization. The ROI specifies the camera's viewing region within an image and helps minimize the scope of false-position results from the main event. On the road, an FV can move side to side or change its direct distance with respect to the HV, which we stated as horizontal shift and vertical shift, respectively. These position shifts lead to a change in image size that can be measured from the IS. Both the FV and HV move simultaneously; therefore, it is not always possible to localize the position of the FV relative to the HV. However, if the position of the HV is known, the relative positions of the FV and HV can be easily compared. The position of SL is fixed relative to every vehicle on the road; therefore, it is necessary to receive SL-IDs from the SLs to determine the HV position. Figure 2 shows a flowchart of the proposed localization scheme wherein the FV location information is compared with the current HV location to identify special and temporal cases. After receiving IDs, algorithm will move ahead if the size of detecting image area of FV is greater or equal to unit pixel area. In decision symbol of the algorithm, the threshold value indicates the minimum distance between HV and FVs to avoid collision.

2.1. LED-ID from SL and FV. In Figure 3, the two LED pairs fixed on the back of the FV transmit a modulated FV-ID [48]. The SL transmits the SL-ID using the same modulated signal (i.e., S2-PSK) by dividing a single LED array into two pairs of LED arrays. Depending on the input bit sequence, the transmitting signal phases of the LED array pairs can differ. The scheme uses a symmetric Manchester symbol to map each LED symbol. Using a spatial under-sampling approach, the LED pairs transmit in the same phase for bit 0 and in different phases for bit 1. The bit interval $s_1(t)$ for one of the tail LEDs is as follows:

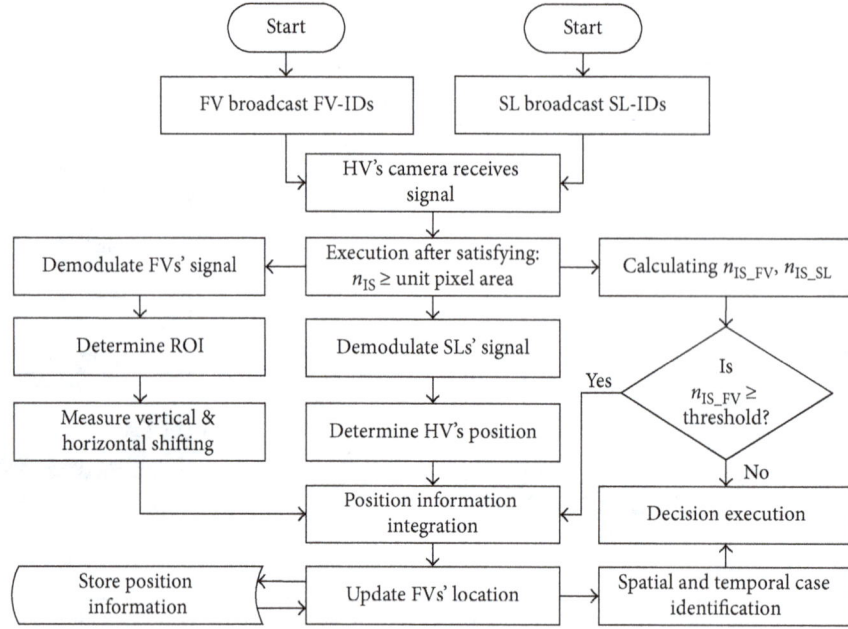

FIGURE 2: Flow chart for a vehicle localization technique based on OCC and photogrammetry.

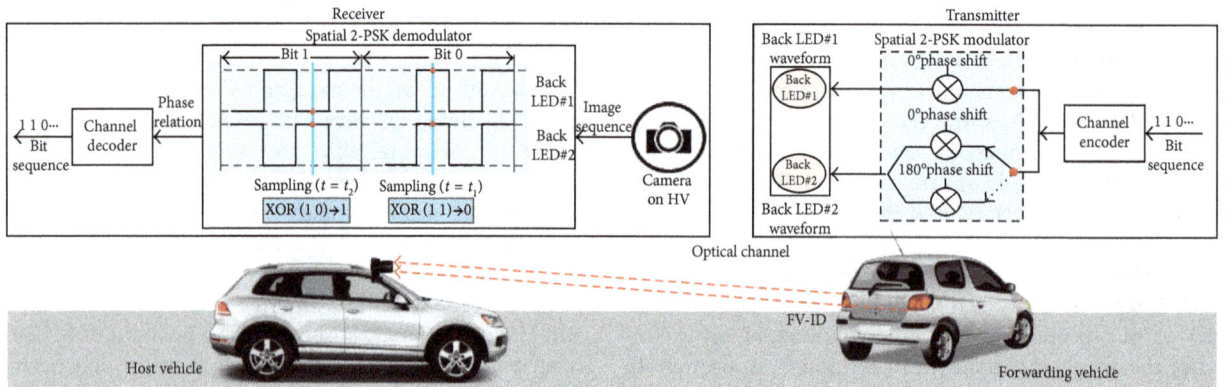

FIGURE 3: Coding and decoding LED-ID using OCC.

$$s_1(t) = \sum_{k=0}^{N} s_1(t_k + kT)$$

$$\text{where} \quad 0 \le t < T_{\text{bit}}, \; s_1(t_k) = \begin{cases} 1, & 0 \le t_k < \dfrac{T}{2} \\ 0, & \dfrac{T}{2} \le t_k < 0, \end{cases} \tag{1}$$

where k is an unsigned integer of N bit-interval cycles, T_{bit} is a bit interval, and T is the cyclic interval of signal.

The bit interval $s_2(t)$ for other of tail LED is as follows:

$$s_2(t) = \sum_{k=0}^{N} s_2(t_k + kT)$$

$$\text{where} \quad 0 \le t < T_{\text{bit}}, \; s_2(t_k) = \begin{cases} 1, & \overline{s_1(t_k)} \\ 0, & s_1(t_k). \end{cases} \tag{2}$$

From the same camera image, the S2-PSK demodulates a bit from a pair of states of two different LEDs. At sampling time t_s, the same states of two tail LEDs on the same image resemble bit 0, otherwise bit is 1. XOR operation determines the value of bit captured in the same image as follows:

$$\text{bit} = s_1(t_s) \oplus s_2(t_s), \tag{3}$$

where $s_1(t_s)$ and $s_2(t_s)$ are the states of two LEDs at sampling time t_s.

Compared with other modulation schemes (e.g., undersampled phase shift on-off keying) [49], this demodulation can gain a lower bit error rate (BER) within an image. A nonlinear XOR classifier can remove the remaining BER. The BER performance of this modulation scheme [41] is stated as follows:

$$P_{e,\text{S2–PSK}} = 2\alpha p_e(1 - \alpha p_e), \tag{4}$$

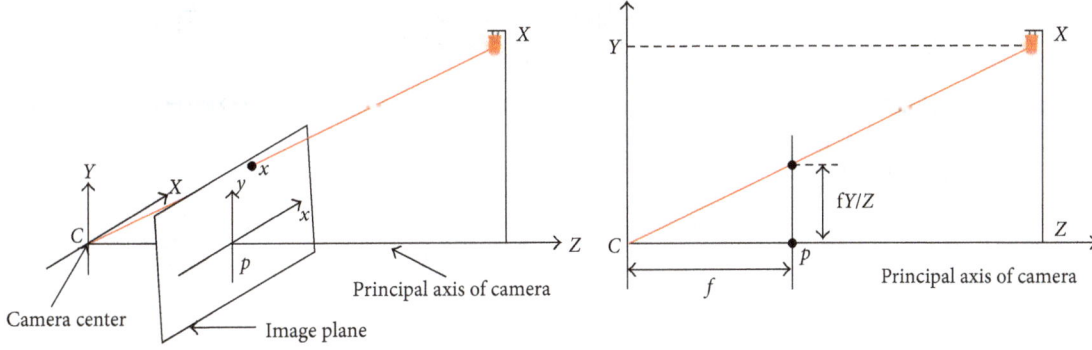

FIGURE 4: Camera calibration for vehicle localization.

where p_e is the bit error probability of the LED state and α is the error rate enhancement.

Considering environmental effects, the SINR [50] is expressed as follows:

$$\text{SINR} = \frac{\left(\kappa P_{\text{opt}} H\right)^2}{\iota^2 N_0 B + \sum \left(\kappa P_{\text{opt}} H_{\text{else}}\right)^2}, \tag{5}$$

where κ is the optical-to-electrical conversion efficiency at the camera, N_0 is the noise power spectral density, B is the modulation bandwidth, H is the optical channel gain, H_{else} is the channel gain for interfering light sources, P_{opt} average optical power, and ι is the conversion between average electrical power P_{elec} and average optical power P_{opt}.

Meanwhile, the optical channel gain is expressed as follows:

$$H = \begin{cases} \dfrac{(m+1)A_c}{2\pi D^2} g(\theta) T_s(\theta) \cos^m(\phi) \cos(\theta), \end{cases} \tag{6}$$

where m is the Lambertian index, A_c is the physical area of IS, D is the distance between transmitter and receiver, θ is the angle of incidence, and ϕ is the angle of irradiation.

2.2. Camera Calibration and Photogrammetry.

In computer vision applications, camera calibration is essential to determine real-world coordinates from simple 2D images. The simplest camera calibration method involves using a pinhole camera model to provide a perfect perspective transformation [51]. In a Euclidean coordinate system, the origin of the projected object coordinates is shifted from the principal point of the camera's image plane, as shown in Figure 4. Mapping an object's Euclidean three-space $\mathbb{R}^3$ coordinates $(X, Y, Z)^T$ to the Euclidean two-space $\mathbb{R}^2$ allows for the mapping of an object from the real world to image coordinates as follows:

$$(X, Y, Z)^T \rightarrow \left(\frac{FX}{Z} + p_x, \frac{FY}{Z} + p_y\right)^T, \tag{7}$$

where F is the focal length of the camera and $(p_x, p_y)^T$ are the principal point coordinates of the camera.

Homogeneous vectors allow us to map the coordinates of the real world, and an image in terms of matrix multiplication as follows:

$$\begin{pmatrix} fX + Zp_x \\ fY + Zp_y \\ Z \end{pmatrix} = \begin{bmatrix} R & -RC \\ 0 & 1 \end{bmatrix} \begin{bmatrix} F_x & s & p_x & 0 \\ & F_y & p_y & 0 \\ & & 1 & 0 \end{bmatrix} \begin{pmatrix} X \\ Y \\ Z \\ 1 \end{pmatrix}, \tag{8}$$

where $F_x (= Fm_x)$ and $F_y (= Fm_y)$ represent the focal length of the camera in terms of pixel area along the x and y directions, respectively; s is the skew parameter, and it is normally zero; R is the camera's orientation relative to real-world coordinates; and C denotes the camera's coordinates. Here, m_x and m_y denote the number of pixels per unit distance, expressed as image coordinates in the x and y directions, respectively. Equation (8) can be expressed succinctly as follows:

$$x = RK[I|-C]X, \tag{9}$$

where K is the calibration matrix of the camera, X is the coordinate matrix in a world coordinate frame, and I is an identity matrix.

Let the distance from the LED to the camera lens be D, and the distance from the focal point of the camera to the projected image on the IS be e. Then, the ratio of LED distance and image distance is distance as follows:

$$\frac{D}{e} = \frac{D}{F} - 1. \tag{10}$$

Image area calculation performance depends on F and D, and it must satisfy the condition $F << D$. Therefore, $(D - F)$ is equivalent to D. The ratio of height and width of an LED (l, r) and the same ratio of the projectile image (l_i, r_i) are known as the magnification of camera lens. This ratio is similar to the ratio of LED distance D and image distance e, which is described as follows:

$$l_i r_i = \frac{F^2 l r}{D^2}. \tag{11}$$

The number of pixel on IS for a particular object n_{IS} is the ratio of projected image area to the unit pixel area of the IS. In an IS, the unit pixel length is ρ, unit pixel area is ρ^2, and A is the area of the LED light source. Thus, the following equation can be stated from (11) as follows:

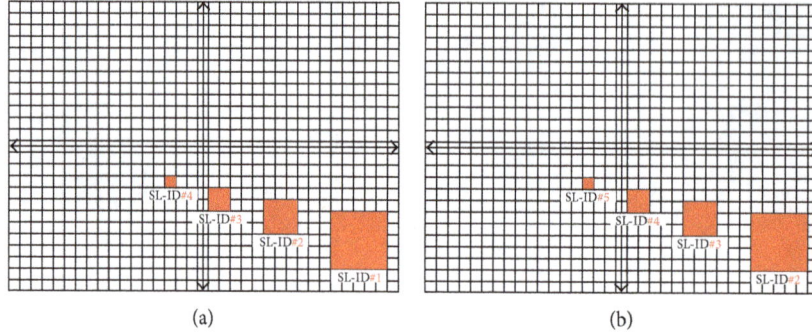

(a)   (b)

FIGURE 5: SL-IDs change with the change of HV's position from (a) at time t to (b) at time $(t+1)$.

$$D = \frac{F}{\rho} \sqrt{\frac{A}{n_{\text{IS}}}}. \qquad (12)$$

Distance is always an absolute value; therefore, the negative sign in (12) can be discarded. If the camera focal length F and unit pixel length ρ are maintained constant for a certain camera, the distance of the LED is kept proportional with respect to the square root of the LED's area and disproportional with respect to the square root of pixel area of that LED on the IS [39].

2.3. Determining HV Position. The origin in coordinate systems, such as the Cartesian and polar coordinate systems, is required to determine the position of one or more objects in either 2D or 3D space. In an outdoor environment, nearly every vehicle frequently changes its location; the measurement distances from HV to FV are not always accurate because of subsequent variations in their relative positions over time. Therefore, the location of FV from the origin or any stable location cannot be measured, and it is better if the HV location is known throughout this period. The shift in the FV's location can be measured by comparing is location with the current location of the HV.

In our proposed localization scheme measures, the HV's location by comparing it with the location of SLs. SLs' location is always fixed with respect to other vehicles within this mobile scenario. This distance comparison yields location information for the HV, which also represents its virtual coordinates. To ensure the accuracy of this measurement system, location information from the onboard diagnostic II (OBD II) system and SLs is combined by the HV. The SL-ID should contain unique information that helps to distinguish this ID from other transmitted signals, such as FV-ID. The header of the SL-ID indicates that this ID belongs to a specific SL. In addition, other information, such as height of the SL from the ground and distance between two SLs on the same road, can be added after the header of the ID. At the same direction, there is a similarity among all SL-IDs of the SLs and a unique value within the IDs increasing or decreasing gradually.

After selecting the ROI, the distance between the camera and LED of the SL is measured using photogrammetry. Figure 5 shows the change in getting an SL-ID

within the field of view (FOV) of the camera owing to a change in the HV position. The two axes are used to show the midpoint of IS. In Figure 5(a), the ID from the SL shows *SL-ID#1~SL-ID#4* at time t. These IDs vary *from SL-ID#2~SL-ID#5* at time $(t+1)$, which is shown in Figure 5 (b). The size of the projectile image area of the nearest SL occupies a greater area on the IS compared with other SLs. The direct distance is calculated using (12); the distance for *SL-ID#1* is shorter compared with that for *SL-ID#4*, as shown in Figure 5(a).

Using OCC, the camera decodes the SL-IDs of the SLs. Figure 6(a) shows that the SL's height is (SL_h_n), and the constant distance between two SLs is d_n, where nth $(n = 1, 2, \ldots, \mathbb{N})$ is related to the number of SL. Using photogrammetry, $D_{\text{SL}_j-\text{HV}}$ is determined as the measured direct distance between the camera and SL's LED, where jth $(j = 1, 2, \ldots, \mathbb{N})$ states the number of iteration sequence over a period. The horizontal distance between the camera and SL is a_j. These horizontal distances are calculated by applying Pythagorean theorem on a right triangle where $D_{\text{SL}_j-\text{HV}}$ and SL's height are the remaining two sides of that right triangle.

From the top geometric view, a few triangulations can be generated after decoding this distance information. At a certain time, applying Pythagorean theorem again to triangles CSL_1H_t and CSL_1SL_2, we obtain

$$a_1^2 = c^2 + h^2, \qquad (13)$$

$$a_2^2 = (d + c)^2 + h^2, \qquad (14)$$

where h is the horizontal distance from the camera to the pavement, c is the distance between the cross point of the horizontal line and the shortest distance from the cross point to the SL, a_1 is the horizontal distance for SL_1, and a_2 is for SL_2 shown in Figure 6(b). In all cases, $c < d$. We combine (13) and (14) as

$$c = \frac{(a_2^2 - a_1^2) - d^2}{2d}. \qquad (15)$$

The horizontal distance is determined from the camera to the pavement by combining (13) and (15) as follows:

$$h = \sqrt{a_1^2 - \left\{ \frac{(a_2^2 - a_1^2) - d^2}{2d} \right\}^2}. \qquad (16)$$

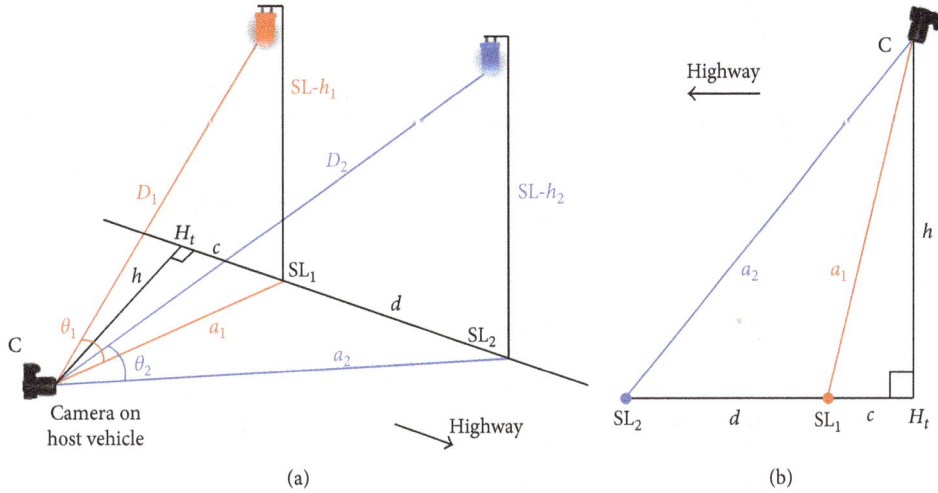

FIGURE 6: Obtaining virtual coordinates from (a) road scenario (geometric view (side)) and (b) measuring vertical distance from the HV to pavement (geometric view (top)).

The position of HV is a function that varies with the horizontal position h_j, which is always positive; angular position $\theta_{\text{SL}_j-\text{HV}}$ of the SL relative to the HV; SL's LED image area is $n_{\text{IS\_SL}}$ on IS; and velocity of HV is V_{HV}. When the HV moves, the parameters related to the HV's position change. If the initial position is recorded at time t; after Δt, the position of HV states as follows:

$$P_{\text{HV}}(t + \Delta t) : \left\{ h_j \pm \Delta h; \Delta\theta_{\text{SL}_j-\text{HV}}; n_{\text{IS\_SL}} \geq \rho^2; V_{\text{HV}}\left(\frac{\Delta[c_j + d_n]}{\Delta t}\right) \right\}. \tag{17}$$

The horizontal distance between the HV and the pavement is a function of the horizontal direct distance a_j, distance between two SLs, that is, d_n; and distance between the cross point of the horizontal line and the shortest distance from the cross point to the SL, that is, c_j, where all these values are also changed according to the change of the angular position $\theta_{\text{SL}_j-\text{HV}}$.

$$h(a_j, c_j, d_n) : \left\{ \Delta\theta_{\text{SL}_j-\text{HV}} \right\}. \tag{18}$$

The direct distance between the SL and HV depends on the area of the SL's LED $n_{\text{IS\_SL}}$ on the IS and the angular position $\theta_{\text{SL}_j-\text{HV}}$. If the value $n_{\text{IS\_SL}}$ is less than the unit pixel area ρ^2 (this will happen when the position of the SL is too far from the HV); the value of image area is ignored from the calculation. On the other hand, angular position $\theta_{\text{SL}_j-\text{HV}}$ changes with the bending of the road (or the edge of a road) and $D_{\text{SL}_j-\text{HV}}$ changes accordingly. Therefore, the following expression is stated for measuring direct distance:

$$D_{\text{SL}_j-\text{HV}}\left(n_{\text{IS\_SL}}, \Delta\theta_{\text{SL}_j-\text{HV}}\right) : \left\{ n_{\text{IS\_SL}} \geq \rho^2; \Delta\theta_{\text{SL}_j-\text{HV}} \right\}. \tag{19}$$

From initial time t to Δt, the changes of angular position is found by simply comparing current angular position with previously recorded value as follows:

$$\Delta\theta_{\text{SL}-\text{HV}}(t + \Delta t) : \left\{ \theta_{\text{SL}_1-\text{HV}} \sim \theta_{\text{SL}_2-\text{HV}} \right\}. \tag{20}$$

The horizontal distance h_j is set as the x coordinate and c_j is set as the y coordinate for the HV with respect to the nearest SL. Therefore, when the HV moves, this distance information is updated. Figure 7 shows the flowchart for calculating and correcting horizontal position information. From (20), a considerably initial angular position $\theta_{\text{SL}_1-\text{HV}}$ compared with the conjugate angular position $\theta_{\text{SL}_2-\text{HV}}$ indicates the presence of a ground curvature. In such cases, infrastructure effect optimization is required; otherwise, the horizontal distance can be easily updated.

2.4. Determining the Position of FVs using the Position of HV. Each vehicle has a pair of headlight and taillight. Using OCC, taillights of FVs transmit modulated signals to a receiver, that is, the camera of the following vehicle. Using this modulated signal, the FV transmits emergency information along with some basic vehicle information, for example, the area of the single light from the rear of the vehicle. This transmitted signal from one pair of taillights is noted as FV-ID, and one ID is unique compared to other vehicles' IDs.

For a proper communication among vehicles (i.e., FVs and HV) on the road, the signal transmitted from both taillights must be received by the camera. There are scenarios in which signal interruption can occur; for instance, the HV may monitor two vehicles from an angle wherein one of the

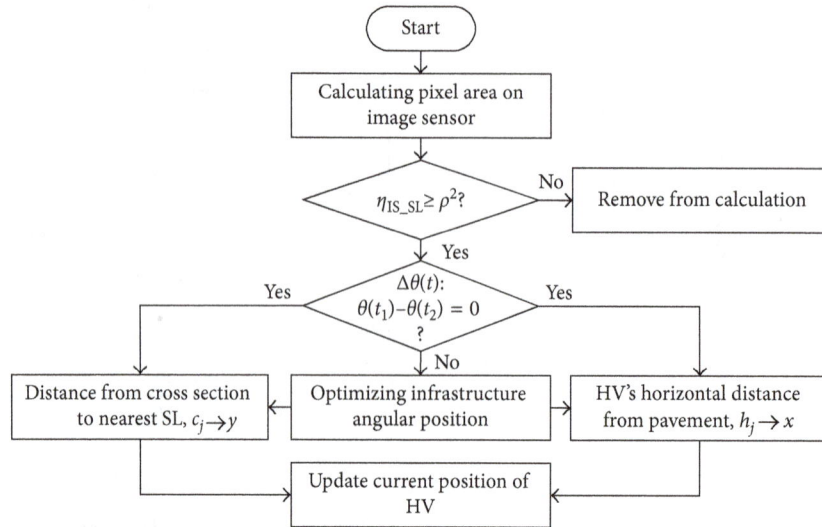

FIGURE 7: Flowchart of detailed development of the HV's location information.

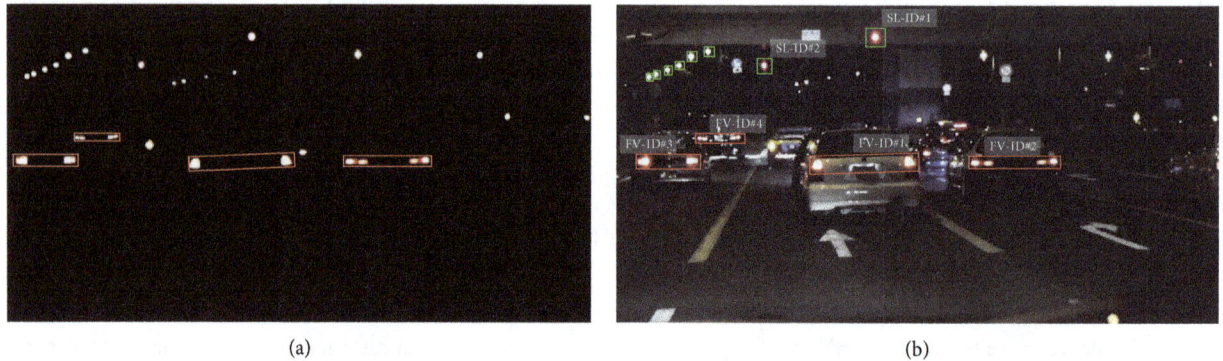

(a)

(b)

FIGURE 8: Using OCC, (a) selecting region of interest and (b) receiving IDs.

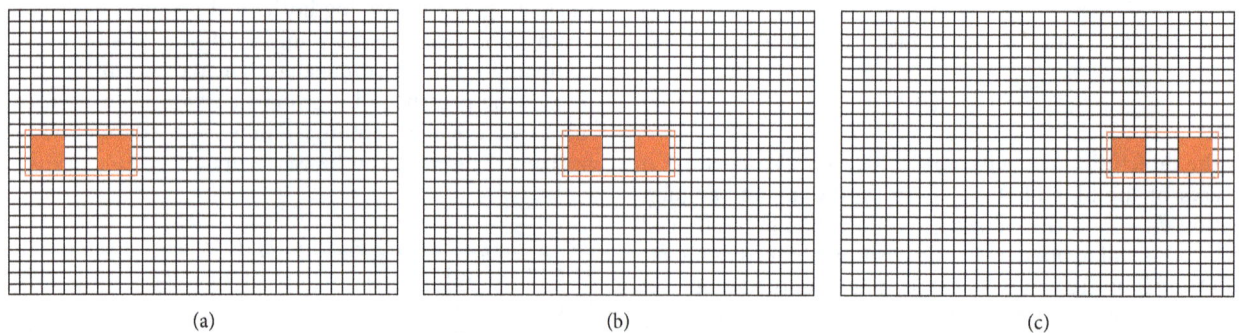

(a)

(b)

(c)

FIGURE 9: A pair of FV's taillights moves from the left to the middle and finally to the right on the image sensor; implying that the FV is moving right to left with respect to the HV. (a) FV at the right side of the HV. (b) FV stays straight of the HV. (c) FV at the left side of the HV.

lights from a single vehicle is covered by the other vehicle. In this case, data extraction is not possible although a single light signal is received by the HV's camera. Moreover, two vehicles can be differentiated using their FV-IDs even if they are moving in parallel. The advantages of LED-ID-based vehicle identification make it possible to fix the ROI, which is the preliminary condition for successful communication and localization. Figure 8 shows every FV broadcasting FV-IDs

along with the SL as SL-IDs. In Figure 8(a), the background is turned black by controlling the shutter speed of the camera, which is mounted in the HV. After demodulation and decoding of the transmitted signal, all IDs are accumulated, as shown in Figure 8(b).

The area of taillight LED arrays on the IS changes relative to changes in the distance between the HV and FV. By calculating the area of these images on the IS, two types of

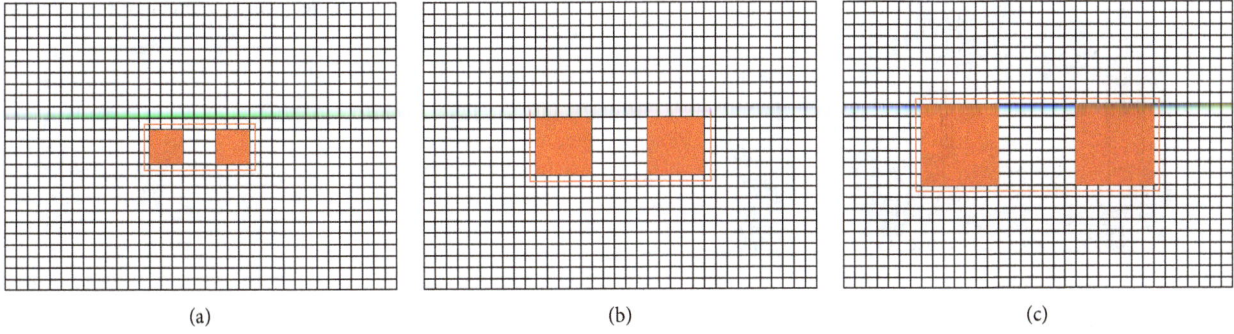

FIGURE 10: A pair of FV's taillight increases in size gradually from left, middle, and right on the image sensor; implying that the FV is moving closer to the HV. (a) FV keeps safe distance from HV. (b) FV stays min. distance from HV. (c) Critical distance between FV-HV.

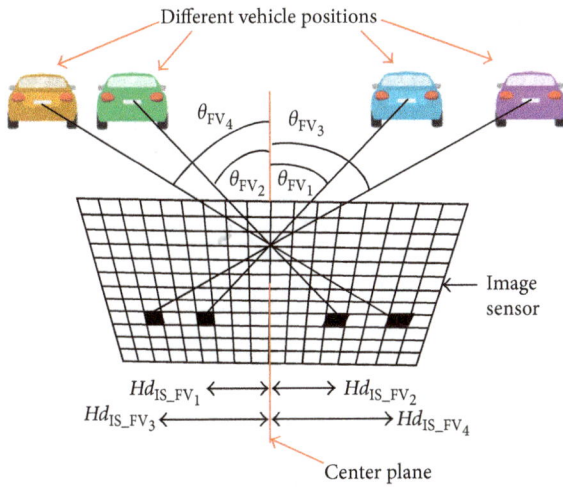

FIGURE 11: Measurement of vehicular angular position from horizontal displacement on the image sensor.

FV position shifts can be determined with respect to the HV, namely, horizontal shifting and vertical shifting. Horizontal FV shift will be visible if the vehicle changes its position from side to side. Concurrently, the vehicle can slow down, changing the direct distance between the FV and HV, which is defined as vertical shift. Figure 9(a) shows an image of the taillight LED that is projected on the left side of the image after being refracted by the camera lens. The image of the original light source is on the HV's right side. Furthermore, Figures 9(a) and 9(b) show that the vehicle is moving from the right to the middle and later to the left with respect to the HV. By contrast, in Figure 10(a), the area of the projected image is smaller than that of the other two Figures 10(b) and 10(c), which shows that the FV was initially far from the HV and that this distance decreased gradually.

Generally, an IS consists of a 2D pixel array of photodetectors and transistors, vertical and horizontal access circuitry, and readout circuitry. Each and every pixel is accessed by the access circuitry, and readout circuitry helps to read the signal value in the pixel. In dense traffic scenarios, the angular position of the FV from HV helps to alleviate the position measurement error. Therefore, at the middle of the IS, a plate is considered as the center plane as in Figure 11

which vertically separates the IS into two. With respect to this plane both angular displacement θ_{FV_k} of FV and horizontal displacement $Hd_{IS\_FV_k}$ on the IS for corresponding FV can be measured. Here, $k\text{th}\,(k = 1, 2, \ldots, \mathbb{N})$ is the number of received FV-ID by the HV's camera. In Figure 11, different image colors on the IS distinguish one FV-ID from other FV-IDs. The angular displacement θ_{FV_k} of FV is always zero when the FV locates at center plane. Otherwise, the numerical value of angular displacement helps us to mitigate some challenges from FV positioning, for example, depth estimation, lane changing information, and position estimation error mitigation for left-right side of the road. The calculated horizontal displacement $Hd_{IS\_FV_k}$ on the IS for corresponding FV is the function of FV's taillight image area n_{IS_FV} and depends on angular displacement θ_{FV_k} of FV as follows:

$$Hd_{IS\_FV_k}\left(n_{IS\_FV} \ge \rho^2\right) : \left\{\theta_{FV_k}\right\}. \tag{21}$$

FV's position can be determined by comparing with the position P_{HV} of HV and taillight image area n_{IS_FV} of FV, horizontal displacement $Hd_{IS\_FV_k}$ on the IS for corresponding FV, and the speed of FV, that is, ΔV_{FV}. Overtime, these parameters will change, and consequently, the position of the FV will change. If t is the initial time, then possible position of FV at Δt is as follows:

$$P_{FV}(t + \Delta t) : \left\{P_{HV}(t + \Delta t); n_{IS\_FV} \ge \rho^2; Hd_{IS\_FV_k}; \Delta V_{FV}\right\}. \tag{22}$$

3. Experimental Distance Measurement

Distance measurement using a camera is one of the important steps in the proposed scheme. Figure 12 shows the experimental setup and distance measurement procedure performed using our existing facilities under an ambient light environment. A circular LED light was used to transmit the signal. A smartphone camera was used as the receiver. With movement of the smartphone, the observed distance changed. Figure 13 shows the results of experimental distance measurement. The result shows the percentage error in measurement with respect to the actual distance. The error resolution seems to remain within 1% for most distance measurements. Although the experiment could not be

(a)

(b)

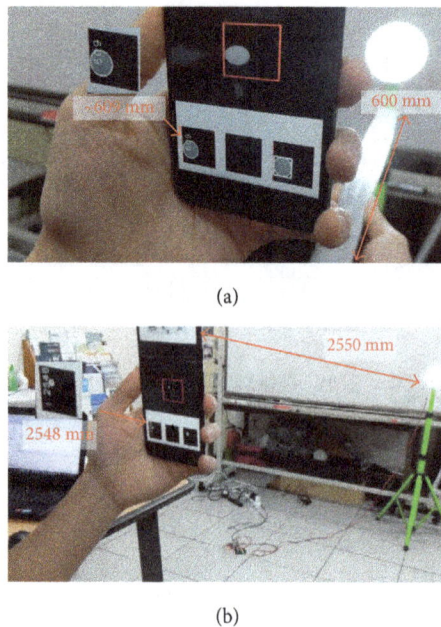

Figure 12: Experimental setup for distance measurement: (a) LED light at the distance of 600 mm and (b) LED light at the distance of 2550 mm from the camera.

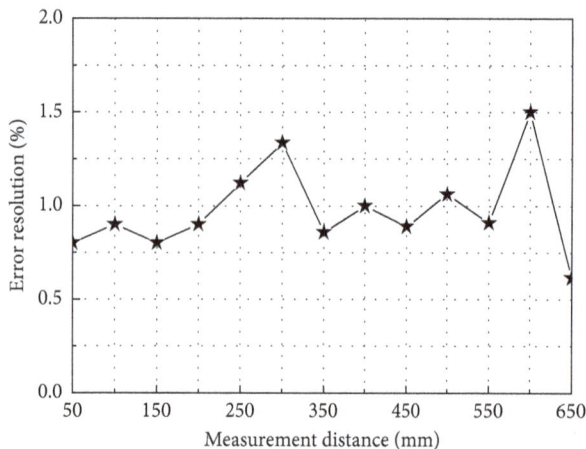

Table 1: Transmitter and receiver parameters for simulation results.

Parameter	Value
Parameters for transmitter	
Size of LED panel	$10 \times 10 \, \text{cm}^2$
Modulation method	S2-PSK
Encoding method	Manchester coding
Data rate	15 bps
Parameters for receiver	
Detection distance	30–200 m
Horizontal FOV	90°–120°
Image resolution	1–10 megapixels
Sensor physical size	$36 \times 24 \, \text{mm}^2$
Frame rate	30 fps
Focal length	16–25 mm
Pixel size	2.5–4 μm
Lens aperture	4
Exposure time	1/2000 to 1/15 sec
Height of SL	7 m
Interdistance between SL	25 m
Lane width	10 m
Vehicle speed	0–110 km/h

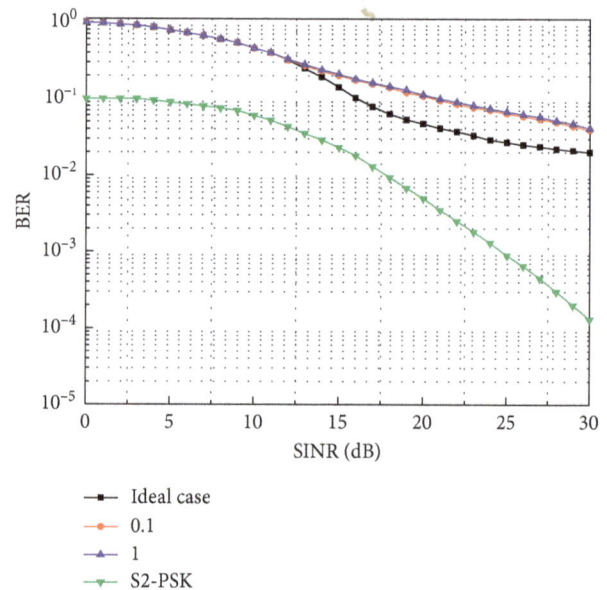

Figure 13: Experimental measured distance value versus error resolution.

Figure 14: SINR versus BER.

performed in a real vehicle environment owing to lack of facilities, this distance measurement experiment validated the feasibility of the proposed scheme.

4. Simulation Results

Several factors and environmental impacts must be considered for achieving localization accuracy. We considered a smooth surface to ignore the turbulence caused by vehicle movements and the impacts of other bad weather conditions (e.g., fog, snow, and rain) for generating the simulation results. The effect of a single parameter on vehicle localization accuracy was considered, whereas the other parameters were maintained constant. Table 1 lists the transmitter

parameters for transmitting and summarizes the specifications of the receiver (i.e., camera) and the optical channel environment.

A low-pass filter-like Gaussian filter is used to estimate the BER performance of the OCC system with respect to the SINR as a blurring filter for image processing. In this case, the variance σ_c^2 ($= 0.5$) for channel filtering is considered zero in ideal state. The curves for the case of estimated variances $\sigma_c = 0.1$ and 1.0 of the Gaussian filter are plotted in Figure 14 to evaluate the influence of the estimation error of the channel filter. In this regard, S2-PSK modulation technique-based OCC system shows better BER performance with respect to SINR.

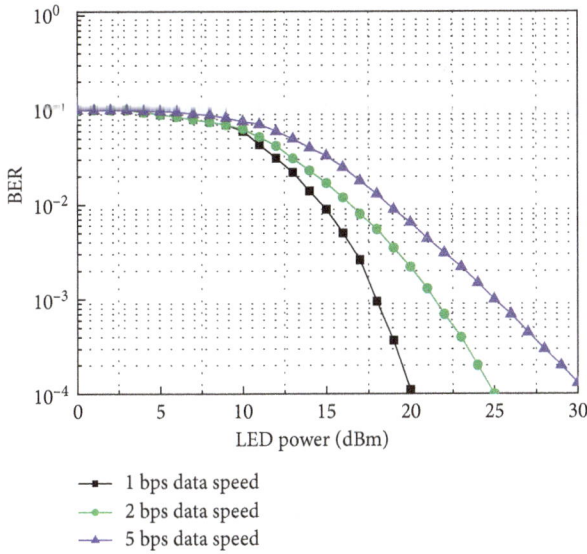

FIGURE 15: LED power versus BER.

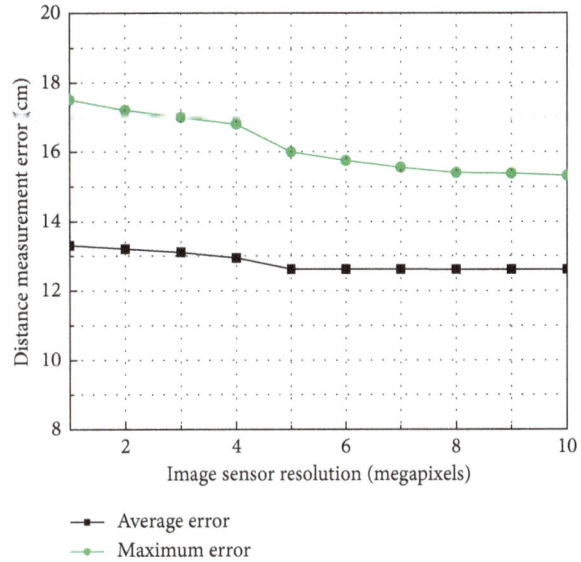

FIGURE 16: Image sensor resolution of the camera with respect to the distance measurement error.

Data rate is depending on the camera frame rate. It is possible to detect one-bit data from one camera frame. For instant, 30-bit data can be received from the camera which frame per second (fps) is 30. In S2-PSK, the Manchester coding is used for data encoding. Therefore, half of total bits per second (bps) generates from 30 fps camera, that is, 15 bps. Figure 15 shows BER performance of the camera receiver with varying data speed. For simulation, result is formulated for 1 bps, 2 bps, 5 bps data speed and required LED power is increased accordingly.

The distance error occurs when there is a discrepancy between actual and measured values of distance. The systematic error caused by environmental facts, surveillance approaches, and tool leads to this mismeasurement in such dynamic vehicular environment and needs to be minimized to achieve better positioning accuracy. The average error takes from series of repeated measurements, whereas the maximum error generates from single measurement. The association of the distance error with different camera parameters in the point of average and maximum errors helps us to improve the performance of distance measurement approach. IS resolution is an important camera parameter which is defined by the number of total pixels and has an impact on distance calculation. It is possible to calculate the area of LED array more precisely if the camera resolution is higher. Higher resolution provides the detail about the detected LEDs to measure their area on the IS. Lower resolution will cause more errors in distance measurement. In Figure 16, at 1 megapixel, both maximum and average distance calculation errors are higher, that is, 17.5 cm and 13.3 cm, respectively. From 5 to 10 megapixels, the maximum distance measurement error is varying linearly, whereas the average error is fixed.

In our proposed scheme, the camera should receive signal from LEDs at very high speed moving scenario. During this dynamic scenarios, the IS should completely expose under the illumination with every detail of the targeted LEDs, that is, streetlights and taillight of forwarding vehicles. The exposure time (or shutter speed) of the camera will ensure a period when the amount of light will be exposed on the IS. In the high-speed vehicular case, large exposure time of the camera will cause a blurred image and short exposure time will allow us to capture detailed flashes of light from a target object. Due to the dependency on the received image quality at IS, exposure time has an impact on evaluating the performance of distance calculation. Both average and maximum distance measurement errors show equivalent evolvements with the exposure time of camera IS in Figure 17. Localization accuracy is better at lower exposure time (i.e., 1/2000). However, when the exposure time is 1/15 in a second, the distance measurement error is maximum (i.e., 18.8 cm) at maximum error case.

In the mobile environment, speed and position shift; these are two important factors that cause effect on vehicle distance measurement. We are considering zero shifting of FV with respect to HV to simulate the effect of speed of FV on the distance calculation error. The speed of FV is varying from 0 to 110 km/h within 200 m distance, whereas the speed of HV is considered constant, that is, 30 km/h, during simulation period which is plotted in Figure 18. Therefore, at the very beginning, the distance measurement error will occur due to the speed of HV with respect to the FV. With the increase of the FV's speed, both average and maximum distance measurement errors are increased gradually up to 110 km/h speed. This distance measurement error occurs due to the execution time required for position calculation by the HV.

At constant vehicular speed (i.e., 50 km/h), position accuracy of HV is measured wherein the distance between SLs varied from 10 to 150 m. From Figure 19, it can be seen that as the internal distance between the SLs increases, the accuracy decreases. Moreover, at the beginning, the

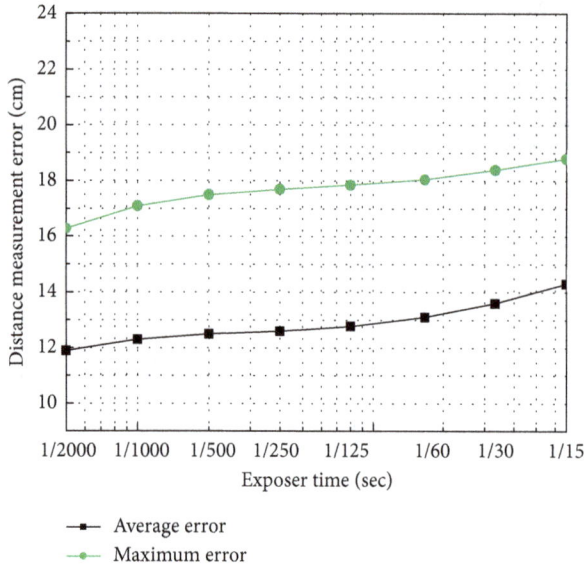

FIGURE 17: Camera exposure time with respect to the distance measurement error.

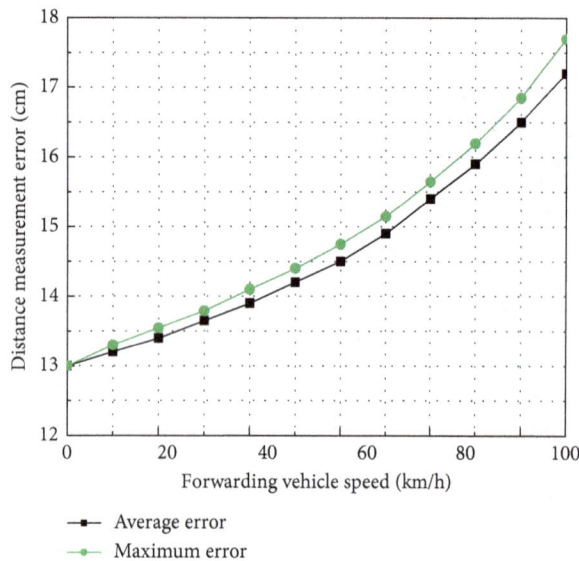

FIGURE 19: Recording the effect of distance between street lights on the measurement accuracy of HV's position.

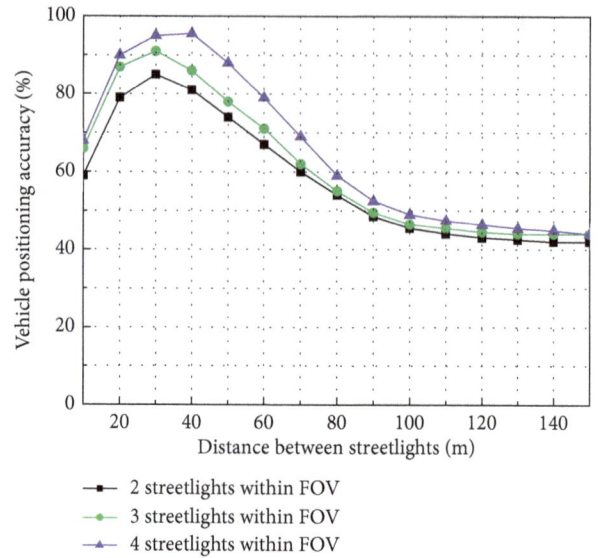

FIGURE 18: Varying speed of FV with respect to the distance measurement error when its position shift maintains at zero value.

simulation results show that the distance measurement accuracy is relatively lower owing to the speed of the HV and data extraction from the SLs. At 50 km/h, the distance between two conjugate SLs is very small, that is, 10 m. Therefore, within a very short period, the number of SLs crossed is greater compared to the case wherein the distance between two SLs is 40 m. At the highest point of the graph, it simplifies that the distance between SLs and receiving SL-IDs well execute to get better distance measurement accuracy, that is, nearly about 90%. In addition, the number of SLs influences the performance calculation. Performance improves as the number of SLs increases. This ensures the possibility of obtaining a greater number of SL-IDs at the

same time and calculating the position of the HV more precisely. Furthermore, as the distance between SLs increases, the chance of comparing the location information of the SLs for accurate HV positioning is minimized. As a result, the slope of distance measurement accuracy moves downward. The variation of the lines maintains a constant margin of up to 150 m because SL-IDs have not been obtained yet.

5. Conclusions

A vehicle localization technique in an outdoor environment is proposed herein. The technique employs photogrammetry, which is a novel idea for localization. Implanted OCC with photogrammetry improved vehicle localization performance. The proposed technique was used to measure the distance between HVs and FVs by calculating the image area on the IS. Beforehand, the HV receives FV-IDs from each FV and uses OCC to decode these IDs. The HV's current location information helps mitigate the possibilities of relative position shifts among the HV and FV. The SL communicates with the HV in the same way as the FVs. Location information of the HV is accumulated by comparing the location of SLs with the HV's OBD II system. Experimental distance measurement confirmed the feasibility of the proposed scheme. Overall distance measurement errors were within 12–20 cm, wherein a change in one parameter was considered. The sizes of the tail LEDs of FVs are different; recognition of such LEDs is out of the scope of this study. A deep learning-based algorithm will be required to boost the performance of this single camera to overcome all challenges related to vehicle detection and localization.

Acknowledgments

This research was supported by the MSIT (Ministry of Science and ICT), Korea, under the Global IT Talent support program (IITP-2017-0-01806) supervised by the IITP (Institute for Information and Communication Technology Promotion).

References

[1] A. Al-Fuqaha, M. Guizani, M. Mohammadi, M. Aledhari, and M. Ayyash, "Internet of things: a survey on enabling technologies, protocols, and applications," *IEEE Communications Surveys & Tutorials*, vol. 17, no. 4, pp. 2347–2376, 2015.

[2] J. Wang, R. K. Ghosh, and S. K. Das, "A survey on sensor localization," *Journal of Control Theory and Applications*, vol. 8, no. 1, pp. 2–11, 2010.

[3] World Health Organization, *Global Status Report on Road Safety: Time for Action*, WHO Press, Geneva, Switzerland, 2009.

[4] World Health Organization, *10 Facts on Global Road Safety*, WHO Press, Geneva, Switzerland, 2017, http://www.who.int/features/factfiles/roadsafety/en/.

[5] World Health Organization, *Global Status Report on Road Safety 2015*, WHO Press, Geneva, Switzerland, 2015, http://apps.who.int/iris/bitstream/10665/189242/1/9789241565066_eng.pdf.

[6] E. Moradi-Pari, A. Tahmasbi-Sarvestani, and Y. P. Fallah, "A hybrid systems approach to modeling real-time situation-awareness component of networked crash avoidance systems," *IEEE Systems Journal*, vol. 10, no. 1, pp. 169–178, 2016.

[7] A. Islam, M. T. Hossan, T. Nguyen, and Y. M. Jang, "Adaptive spatial-temporal resolution optical vehicular communication system using image sensor," *International Journal of Distributed Sensor Networks*, vol. 13, no. 11, 2017.

[8] E. Abbott and D. Powell, "Land-vehicle navigation using GPS," *Proceedings of the IEEE*, vol. 87, no. 1, pp. 145–162, 1999.

[9] H. Fang, M. Yang, R. Yang, and C. Wang, "Ground-texture-based localization for intelligent vehicles," *IEEE Transactions on Intelligent Transportation Systems*, vol. 10, no. 3, pp. 463–468, 2009.

[10] Y. Cui and S. S. Ge, "Autonomous vehicle positioning with GPS in urban canyon environments," *IEEE Transactions on Robotics and Automation*, vol. 19, no. 1, pp. 15–25, 2003.

[11] E. A. Olsen, C.-W. Park, and J. P. How, "3D formation flight using differential carrier-phase GPS sensors," *Journal of the Institute of Navigation*, vol. 46, no. 1, pp. 35–48, 1999.

[12] B. Li, G. J. Sutton, B. Hu, R. P. Liu, and S. Chen, "Modeling and QoS analysis of the IEEE 802.11p broadcast scheme in vehicular ad hoc networks," *Journal of Communications and Networks*, vol. 19, no. 2, pp. 169–179, 2017.

[13] Y. Yang, D. Fei, and S. Dang, "Inter-vehicle cooperation channel estimation for IEEE 802.11p V2I communications," *Journal of Communications and Networks*, vol. 19, no. 3, pp. 227–238, 2017.

[14] A. M. S. Abdelgader and W. Lenan, "The physical layer of the IEEE 802.11p WAVE communication standard: the specifications and challenges," in *Proceedings of the World Congress on Engineering and Computer Science (WCECS)*, San Francisco, CA, USA, October 2014.

[15] S. A. Hussain, M. Iqbal, A. Saeed et al., "An efficient channel access scheme for vehicular ad hoc networks," *Mobile Information Systems*, vol. 2017, Article ID 8246050, 10 pages, 2017.

[16] E. K. Lee, S. Yang, S. Y. Oh, and M. Gerla, "RF-GPS: RFID assisted localization in VANETs," in *Proceedings of 2009 IEEE 6th International Conference on Mobile Adhoc and Sensor Systems*, pp. 621–626, Macau, China, October 2009.

[17] A. M. Ladd, K. E. Bekris, A. P. Rudys, D. S. Wallach, and L. E. Kavraki, "On the feasibility of using wireless Ethernet for indoor localization," *IEEE Transactions on Robotics and Automation*, vol. 20, no. 3, pp. 555–559, 2004.

[18] K. Takagi, K. Morikawa, T. Ogawa, and M. Saburi, "Road environment recognition using on-vehicle LiDAR," in *Proceedings of the IEEE Intelligent Vehicles Symposium*, pp. 120–125, Tokyo, Japan, 2006.

[19] P. Lindner and G. Wanielik, "3D LIDAR processing for vehicle safety and environment recognition," in *Proceedings of the IEEE Workshop on Computational Intelligence in Vehicles and Vehicular Systems*, pp. 66–71, Nashville, TN, USA, May 2009.

[20] LeddarTech, *LiDAR-Enabled Optical Detection Systems*, LeddarTech, Quebec City, QC, USA, 2017, http://www.roadtraffic-technology.com/contractors/detection/leddartech/.

[21] L. Li, *Time-of-Flight Camera–An Introduction*, Technical White Paper, Texas Instruments, Dallas, TX, USA, 2014, http://www.ti.com/lit/wp/sloa190b/sloa190b.pdf.

[22] S. Lanzisera, D. Zats, and K. S. J. Pister, "Radio frequency time-of-flight distance measurement for low-cost wireless sensor localization," *IEEE Sensors Journal*, vol. 11, no. 3, pp. 837–845, 2011.

[23] S. May, D. Droeschel, D. Holz et al., "Three-dimensional mapping with time-of-flight cameras," *Journal of Field Robotics*, vol. 26, no. 11-12, pp. 935–965, 2009.

[24] Y. Cui, S. Schuon, D. Chan, S. Thrun, and C. Theobalt, "3D shape scanning with a time-of-flight camera," in *Proceedings of the IEEE Computer Society Conference on Computer Vision and Pattern Recognition*, pp. 1173–1180, San Francisco, CA, USA, June 2010.

[25] T. Schenk and B. Csathó, "Fusion of LIDAR data and aerial imagery for a more complete surface description," *International Archives of Photogrammetry Remote Sensing and Spatial Information Sciences*, vol. 34, no. 3, pp. 310–317, 2002.

[26] S. Foix, G. Alenya, and C. Torras, "Lock-in time-of-flight (ToF) cameras: a survey," *IEEE Sensors Journal*, vol. 11, no. 9, pp. 1917–1926, 2011.

[27] P. Fürsattel, S. Placht, M. Balda et al., "A comparative error analysis of current time-of-flight sensors," *IEEE Transactions on Computational Imaging*, vol. 2, no. 1, pp. 27–41, 2016.

[28] J. Seiter, M. Hofbauer, M. Davidovic, S. Schidl, and H. Zimmermann, "Correction of the temperature induced error of the illumination source in a time-of-flight distance measurement setup," in *IEEE Sensors Applications Symposium Proceedings*, pp. 84–87, Glassboro, NJ, USA, April 2013.

[29] S. Fuchs, "Multipath interference compensation in time-of-flight camera images," in *Proceedings of the 20th International Conference on Pattern Recognition*, pp. 3583–3586, Istanbul, Turkey, October 2010.

[30] L. Zeng, D. O'Brien, H. Minh et al., "High data rate multiple input multiple output (MIMO) optical wireless communications using white led lighting," *IEEE Journal on Selected Areas in Communications*, vol. 27, no. 9, pp. 1654–1662, 2009.

[31] M. Ayyash, H. Elgala, A. Khreishah et al., "Coexistence of WiFi and LiFi toward 5G: concepts, opportunities, and challenges," *IEEE Communications Magazine*, vol. 54, no. 2, pp. 64–71, 2016.

[32] M. Uysal, C. Capsoni, Z. Ghassemlooy, A. Boucouvalas, and E. Udvary, *Optical Wireless Communications Signals and Communication Technology*, Springer, Cham, Switzerland, 2016.

[33] M. Z. Chowdhury, M. T. Hossan, A. Islam, and Y. M. Jang, "A comparative survey of optical wireless technologies: architectures and applications," *IEEE Access*, vol. 6, pp. 9819–9840, 2018.

[34] C. Danakis, M. Afgani, G. Povey, I. Underwood, and H. Haas, "Using a CMOS camera sensor for visible light communication," in *Proceedings of the IEEE Globecom Workshops*, pp. 1244–1248, Anaheim, CA, USA, Devember 2012.

[35] I. Takai, S. Ito, K. Yasutomi, K. Kagawa, M. Andoh, and S. Kawahito, "LED and CMOS image sensor based optical wireless communication system for automotive applications," *IEEE Photonics Journal*, vol. 5, no. 5, p. 6801418, 2013.

[36] M. Rezaei, M. Terauchi, and R. Klette, "Robust vehicle detection and distance estimation under challenging lighting conditions," *IEEE Transactions on Intelligent Transportation Systems*, vol. 16, no. 5, pp. 2723–2743, 2015.

[37] I. Takai, T. Harada, M. Andoh, K. Yasutomi, K. Kagawa, and S. Kawahito, "Optical vehicle-to-vehicle communication system using LED transmitter and camera receiver," *IEEE Photonics Journal*, vol. 6, no. 5, pp. 1–14, 2014.

[38] T. Yamazato, I. Takai, H. Okada et al., "Image-sensor-based visible light communication for automotive applications," *IEEE Communications Magazine*, vol. 52, no. 7, pp. 88–97, 2014.

[39] Y. Goto, I. Takai, T. Yamazato et al., "A new automotive VLC system using optical communication image sensor," *IEEE Photonics Journal*, vol. 8, no. 3, pp. 1–17, 2016.

[40] C. M. Silva, B. M. Masini, G. Ferrari, and I. Thibault, "A survey on infrastructure-based vehicular networks," *Mobile Information Systems*, vol. 2017, Article ID 6123868, 28 pages, 2017.

[41] T. Nguyen, A. Islam, M. T. Hossan, and Y. M. Jang, "Current status and performance analysis of optical camera communication technologies for 5G networks," *IEEE Access*, vol. 5, pp. 4574–4594, 2017.

[42] N. Wu, L. Feng, and A. Yang, "Localization accuracy improvement of a visible light positioning system based on the linear illumination of LED sources," *IEEE Photonics Journal*, vol. 9, no. 5, pp. 1–11, 2017.

[43] M. T. Hossan, M. Z. Chowdhury, A. Islam, and Y. M. Jang, "A novel indoor mobile localization system based on optical camera communication," *Wireless Communications and Mobile Computing*, vol. 2018, Article ID 9353428, 17 pages, 2018.

[44] J. Baqersad, P. Poozesh, C. Niezrecki, and P. Avitabile, "Photogrammetry and optical methods in structural dynamics—A review," *Mechanical Systems and Signal Processing*, vol. 86, pp. 17–34, 2017.

[45] W. Förstner and B. P. Wrobel, *Photogrammetric Computer Vision*, vol. 11, Springer International Publishing, Cham, Switzerland, 1st edition, 2016.

[46] Z. Li, Z. Wei, C. Wen, and J. Zheng, "Detail-enhanced multiscale exposure fusion," *IEEE Transactions on Image Processing*, vol. 26, no. 3, pp. 1243–1252, 2017.

[47] Y. Huo and F. Yang, "High-dynamic range image generation from single low-dynamic range image," *IET Image Processing*, vol. 10, no. 3, pp. 198–205, 2016.

[48] T. Nguyen, A. Islam, and Y. M. Jang, "Region-of-interest signaling vehicular system using optical camera communications," *IEEE Photonics Journal*, vol. 9, no. 1, pp. 1–20, 2017.

[49] P. Luo, M. Zhang, Z. Ghassemlooy, and D. Han, "Experimental demonstration of RGB LED-based optical camera communications," *IEEE Photonics Journal*, vol. 7, no. 5, pp. 1–12, 2015.

[50] Y. Wang and H. Haas, "Dynamic load balancing with handover in hybrid Li-Fi and Wi-Fi networks," *Journal of Lightwave Technology*, vol. 33, no. 22, pp. 4671–4682, 2015.

[51] R. Hartley and A. Zisserman, *Multiple View Geometry in Computer Vision*, Cambridge University Press, Cambridge, UK, 2nd edition, 2004.

17

Efficient and Reliable Cluster-Based Data Transmission for Vehicular Ad Hoc Networks

Xiang Ji [iD],[1,2] Huiqun Yu [iD],[1] Guisheng Fan [iD],[1] Huaiying Sun,[1] and Liqiong Chen [iD][3]

[1]*Department of Computer Science and Engineering, East China University of Science and Technology, Shanghai 200237, China*
[2]*Shanghai Key Laboratory of Computer Software Evaluating and Testing, Shanghai 201112, China*
[3]*Department of Computer Science and Information Engineering, Shanghai Institute of Technology, Shanghai 201418, China*

Correspondence should be addressed to Huiqun Yu; yhq@ecust.edu.cn

Academic Editor: Alessandro Bazzi

Vehicular ad hoc network (VANET) is an emerging technology for the future intelligent transportation systems (ITSs). The current researches are intensely focusing on the problems of routing protocol reliability and scalability across the urban VANETs. Vehicle clustering is testified to be a promising approach to improve routing reliability and scalability by grouping vehicles together to serve as the foundation for ITS applications. However, some prominent characteristics, like high mobility and uneven spatial distribution of vehicles, may affect the clustering performance. Therefore, how to establish and maintain stable clusters has become a challenging problem in VANETs. This paper proposes a link reliability-based clustering algorithm (LRCA) to provide efficient and reliable data transmission in VANETs. Before clustering, a novel link lifetime-based (LLT-based) neighbor sampling strategy is put forward to filter out the redundant unstable neighbors. The proposed clustering scheme mainly composes of three parts: cluster head selection, cluster formation, and cluster maintenance. Furthermore, we propose a routing protocol of LRCA to serve the infotainment applications in VANET. To make routing decisions appropriate, we nominate special nodes at intersections to evaluate the network condition by assigning weights to the road segments. Routes with the lowest weights are then selected as the optimal data forwarding paths. We evaluate clustering stability and routing performance of the proposed approach by comparing with some existing schemes. The extensive simulation results show that our approach outperforms in both cluster stability and data transmission.

1. Introduction

Vehicular ad hoc network (VANET) is the foundation for the intelligent transportation systems (ITSs), which aims at achieving seamless Internet connectivity between vehicles on the road [1]. With the developments of intelligent vehicle and new generation wireless communication techniques, vehicles equipped with wireless interfaces are able to provide ITS services [2] such as traffic monitoring, vehicle navigation [3], nearby information services, and mobile vehicular cloud computing.

Therefore, the creation of a stable network and communication management is the most challenging task due to the high mobility and uneven spatial distribution of vehicles in VANETs. The clustering technology is testified to be a promising solution to improve routing reliability and scalability by organizing similar vehicles into several virtual groups, called clusters [4]. Each cluster has a capital vehicle, named cluster head, which is responsible for managing communication in the cluster. Vehicles of a cluster can communicate directly via an intracluster communication, while vehicles in different clusters may achieve intercluster communication through cluster heads.

The originally notable clustering algorithms were designed for mobile ad hoc networks (MANETs) [4], such as the popular lowest identifier (LID) [5] and Mobility Clustering (MOBIC) [6]. Later, several other algorithms were designed for clustering in MANET. Recently, some of those algorithms were implemented in VANETs. However, due to the characteristic mobility and channel conditions of VANET, these approaches should be adapted according to the unique properties.

Clustering algorithms proposed for VANETs are used in communication networks to partition similar vehicles into clusters [7]. Therefore, clustering techniques can effectively limit the channel contention among cluster members to ensure fair channel access. Moreover, under the management of the cluster head, clustering algorithms can provide spatial reuse of resources such as bandwidth [8]. Given the high mobility of VANETs, how to select the cluster head and how to improve cluster stability become tough challenges.

Based on VANETs technology, numerous applications have been developed for the ITS. A typical kind of application is to disseminate safety messages among vehicles, including accident warning and congestion information [9, 10]. Another kind of application, infotainment, is also important for successful VANET deployment [11]. Infotainment services provide more pleasurable experience for both drivers and passengers with various applications, such as nearby information access and multimedia application [12].

To serve the infotainment services in urban VANET, this paper proposes a new LREL-based clustering scheme with the purpose of establishing a stable virtual network for data transmission. In order to form stable clusters, we propose an LLT-based neighbor sampling scheme to filter out unstable neighbor vehicles. Different from previous clustering schemes which focus on vehicular mobility, we propose new metric link reliability (LREL) for cluster head selection. The cluster heads are selected in a distributed way. In addition, we propose a routing protocol by using the proposed clustering architecture. We select bridge nodes at intersections to connect clusters in street scenarios. The bridge node acts as the routing path decision maker by monitoring the delay to incur for data transmission over road segments.

The contributions of this paper are mainly listed as follows:

(i) We propose an LLT-based neighbor sampling scheme to filter out the redundant unstable neighbors. A stable neighbor set is selected as a basis of clustering procedures.

(ii) We propose a new LREL-based clustering approach with the purpose of establishing a stable virtual network for efficient and reliable data transmission in urban VANET.

(iii) We propose a routing protocol for urban VANET by utilizing the structure constructed by LRCA. The protocol constructs routing path via the bridge node at intersection.

(iv) We analyze the performance of the proposed scheme by considering clustering performance metrics and routing metrics, including packet delivery ratio, end-to-end delay, and control overhead and clustering stability, by comparing with some existing cluster-based schemes.

The rest of the paper is organized as follows: Section 2 gives a brief description of related works. Section 3 describes the system model. Section 4 presents the proposed link reliability-based clustering scheme. Section 5 shows the data transmission utilizing the proposed clustering architecture. Next, Section 6 shows simulation results of the proposed scheme. Finally, conclusions are presented in Section 7.

2. Related Work

The original clustering algorithms are proposed in the late 1980s. Since then, a large amount of cluster-oriented researches have been introduced to MANETs in general and VANETs in particular [4]. Vehicle clustering is a potential approach to improve the scalability of networking protocols for VANET scenarios. For cluster-based routing protocols, cluster heads take responsibilities for the discovery and maintenance of routing paths, which reduce the control overhead to a great extent [13]. Due to the high-speed mobility of vehicles, network topology changes frequently [14]. Under this circumstance, the cluster maintenance cost increases significantly. Therefore, how to form the stable clusters and maintain their stability during communication are a vital issue in clustering techniques for VANETs.

Many clustering algorithms designed for VANETs have been proposed based on mobility metrics for cluster formation mechanisms. The mobility features, including speed, direction, and location of vehicles, are very important for VANET clustering procedures. Kayis and Acarman [15] proposed a passive clustering algorithm based on predefined speed intervals. They organize the vehicles within the same speed interval into groups. However, the speed interval is not a good metric for assessment because two vehicles with very similar speed around the interval gap might be divided into different clusters. Chen et al. [16] used the distance-based criteria in the cluster construction algorithms. Furthermore, they employ a central server to manage the cluster merging and splitting events. Shea et al. [17] proposed a distributed mobility-based clustering algorithm based on a data clustering technique called affinity propagation. They use the metric of vehicular position and mobility in cluster creation procedure by combining the current and future positions.

Some other clustering mechanisms have been proposed for VANET based on a sum of weighted values. Wang et al. [18] proposed a priority-based clustering approach. The priority is calculated according to the estimated travel duration and speed deviation. Almalag Mohammad et al. [19] presented a lane-based clustering algorithm which selects the vehicle as the cluster head with the highest cluster head level (CHL). CHL is a hybrid metric combining the condition of traffic flow, relative speed, and relative position of the vehicle. Morales et al. [20] proposed the clustering algorithm based on the destination of vehicles. According to their mechanism, vehicles with similar destinations are more likely to form a cluster. The weighted metric is computed as the combination of the current position, relative speed, relative destination, and final destination of vehicles.

Clustering algorithms for VANET can be categorized into two classes of one-hop clustering and multihop clustering. The aforementioned algorithms [16, 17, 19, 20] are based on single-hop clusters in which the cluster members are one-hop away from the CH. Vehicles only join the clusters where the CH is in its local vicinities. One-hop cluster topology can reduce the cluster re-affiliation and decrease cluster maintenance overhead because fewer information exchanges are required [21]. However, the transmission range and density of vehicles affect the size of cluster. In a high vehicular density, data collision may happen in the clusters. On the contrary, a vehicle may fail to detect neighbors in very low density. Recently, plenty of works have been proposed for multihop clustering algorithms. Wolny [22] proposed MDMAC, which is a modification of DMAC. MDMAC is able to form k-hop clusters by introducing the TTL (time-to-live) parameter in message delivery. Zhang et al. [23] proposed a multihop clustering scheme for VANETs. The multihop clusters are constructed based on the relative mobility between vehicles in multihop distance. Ucar et al. [9] proposed a novel multihop clustering scheme, known as VMaSC, which selects CHs based on relative mobility with respect to its neighbors. VMaSC reduces overhead during cluster formation by introducing a direct connection to the neighbor which is already a cluster head or a cluster member, instead of connecting to the CH multihops away. Further, the VMaSC claims to be the first multihop clustering algorithm which is analyzed under a realistic scenario. Ziagham and Noorimehr [24] proposed a single-hop clustering approach named MOSIC based on the changes of relative vehicular mobility. It uses the Gauss–Markov mobility (GMM) model for mobility predication and makes vehicle be able to prognosticate its mobility relative to its neighbors.

In recent years, some researchers build up semiclusters for VANET scenarios. Zhang et al. [25] proposed a novel variant of cluster, which called the microtopology (MT). The MT acts as a basic component of routing paths which consists of vehicles and wireless links among vehicles along the street. Togou et al. [26] proposed SCRP, which is an approximate cluster-based routing protocol based on connected dominating set (CDS). SCRP selects a small number of vehicles as dominating vehicles to form a virtual backbone in the network. Lin et al. [27] designed a moving-zone-based architecture for data delivery in VANETs. Similar to cluster formation, the moving zone is self-organized by vehicles which have similar movement patterns. Rivoirard et al. [28] proposed the chain-branch-leaf (CBL) clustering scheme which combines the information on road configuration, vehicle mobility, and link quality. CBL builds a stable vehicular network infrastructure by selecting vehicles with lower speed in the same traffic direction to form a stable backbone of branch nodes named Chain.

Data transmission over vehicular networks poses a number of challenges and has been widely studied. He et al. [29] proposed a SDN-based wireless communication solution to manage the network resources, which can schedule different network resources and minimize communication cost. Zeng et al. [30] proposed a channel prediction-based scheduling strategy for cooperative data dissemination in VANET, which reduces communication overhead and the data dissemination delay. Zhu et al. [31] proposed a distributed data replication algorithm with the idea of letting the data carrier distribute the data dissemination tasks to multiple nodes to speed up the dissemination process.

In the literature, clustering algorithms have been proposed for the purpose of load balancing, quality-of-service

support, and data transmission in VANET scenario [2]. The built up clusters can serve as a hierarchical infrastructure-like overlay on top of an underlying ad hoc network, which can be used to route packets [4]. There have been some routing-oriented clustering algorithms which include both clustering and routing algorithms. For example, Song et al. [32] proposed a cluster-based directional routing protocol for VANETs which considers moving directions for cluster head selection. Ohta et al. [33] formed the clusters using position and direction information of vehicles. Unlike these researches, we consider the reliability of links between vehicles. We put up a new metric, named LREL, for cluster head selection. In addition, we propose a LLT-based neighbor sampling scheme to filter out unstable neighbors, which can reduce unnecessary message exchanges.

3. System Model

The prominent characteristics of VANET, including the high mobility and the uneven spatial distribution of vehicles, lead to frequent changes in the topologies and disconnections of the network. To solve these problems, we propose a clustered VANET structure to provide reliable connectivity for a group of vehicles. Figure 1 shows the general system architecture for vehicular service scenarios where vehicles are grouped into multiple moving clusters. Each cluster contains a capital vehicle (cluster head) which is responsible for managing information about the cluster members as well as data transmission.

In this paper, we propose a comprehensive cluster-based data transmission approach with pure vehicle-to-vehicle (V2V) communication type being considered. We assume that each vehicle has a unique identity and is equipped with an onboard unit (OBU). The GPS service is available for obtaining basic information, including vehicle's current location, velocity, and moving direction. Vehicles exchange their information with one another through beacon messages. The beacon message is broadcasted and collected at every beacon interval, which includes vehicle's identifier, current position, current velocity, moving direction, vehicle's current state, and cluster head's identifier if it is a cluster member.

In the proposed cluster algorithm, each vehicle may be in one of the following four states:

(i) *Initial node (IN)*: Initial state of the vehicles which do not belong to any cluster.

(ii) *Cluster head (CH)*: The state in which the vehicle acts as the leader of a cluster.

(iii) *Cluster member (CM)*: The state in which the vehicle is attached to an existing CH.

(iv) *Candidate cluster member (CCM)*: The state in which the vehicle intends to be a CM of an existing cluster before receiving a confirmation message.

The transitions among these states are triggered by different events. The details of the state transition process are presented in the following section. The notation used in this paper is presented in Table 1.

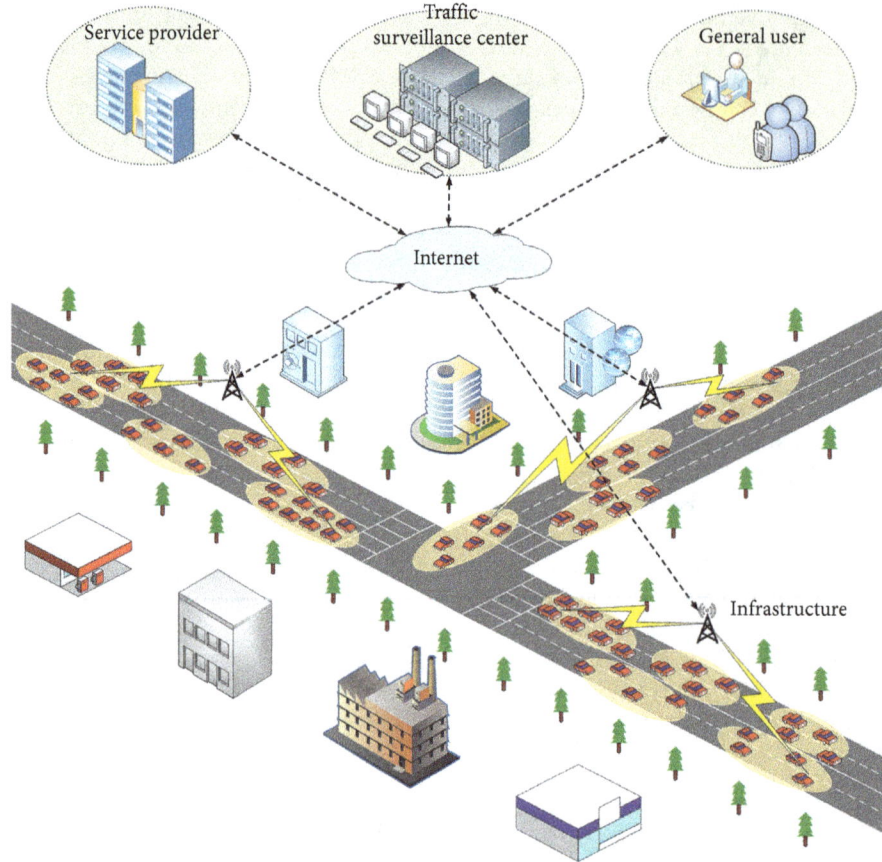

FIGURE 1: Cluster-based VANET architecture for vehicular service scenarios.

4. Vehicle Clustering

This paper proposes a link reliability-based clustering algorithm (LRCA) for urban VANETs. The LRCA is designed to provide efficient and reliable data transmission across the urban VANETs. Before clustering, we propose a novel LLT-based neighbor sampling strategy to filter out the redundant unstable neighbors. The proposed clustering scheme mainly composes of three parts: cluster head selection, cluster formation, and cluster maintenance. The general procedural flow of our proposed clustering algorithm is presented in Figure 2.

4.1. LLT-Based Neighbor Sampling. In VANET, vehicles exchange and collect information about their one-hop vicinities through periodic beacon messages. After receiving the beacons, each vehicle constructs the potential neighbor (PN) set [34]. However, not all the vehicles in PN are ideal for clustering. In order to improve the stability of clusters, this paper introduces an LLT-based neighbor sampling process to select a stable neighbor (SN) set from PN.

Link lifetime (LLT) [35], also called link expiration time (LET), represents the predicted duration time that two adjacent vehicles remain connected. Equation (1) defines LLT calculation. Δv_{ij} and Δd_{ij} represent the difference of velocity and distance between V_i and V_j, respectively. R denotes the transmission range of a vehicle. The LLT is

introduced to evaluate the link sustainability. The link is more sustainable with a larger LLT value.

$$\text{LLT}_{ij} = \frac{\left|\Delta v_{ij}\right| \cdot R - \Delta v_{ij} \cdot \Delta d_{ij}}{\left(\Delta v_{ij}\right)^2}. \tag{1}$$

The detail of the sampling process is presented in Algorithm 1. At the beginning, the system starts a timer (IN_TIMER) for gathering information. The vehicle in SN is defined as the neighboring vehicle which is going to have a constant connection for a predetermined time threshold δ_s. For each vehicle V_i, it maintains a set of stable neighbors SN_i, which contains several entries $\text{SN}_i(j)$ of vehicle V_j.

4.2. Link Reliability Metric. The link reliability model for communication links between two vehicles in urban VANET is given in [36]. The network connectivity status is mainly determined by the velocity distribution over the vehicular traffic flow. The basic link reliability model is defined as the conditional probability in the following equation which describes the probability of the continuous link connectivity between two vehicles over a specified time duration:

$$r(l) = P\{l \text{ continues to } t + \text{LLT} \mid l \text{ is available at } t\}, \tag{2}$$

where $r(l)$ represents the reliability of the link and l is the particular link between two vehicles. LLT indicates the

TABLE 1: Notations.

Notation	Description
IN	Initial node
CH	Cluster head
CM	Cluster member
CCM	Candidate cluster member
R	Communication range
V_i	Vehicle i; i is the ID of vehicle
LLT_{ij}	Link lifetime between V_i and V_j
$LREL_i$	Link reliability metric of V_i
PN_i	Potential neighbor set of V_i
SN_i	Stable neighbor set of V_i
$SN_i(j)$	Entry of V_j in SN_i
$State(V_i)$	State of V_i
IN_TIMER	Initial timer for gathering information
JOIN_TIMER	Join response timer
CH_TIMER	Cluster head selection timer
CH_ACK	Cluster head acknowledgement message
JOIN_REQ	Join request message
JOIN_RESP	Join response message
MERGE_REQ	Cluster merging request message
MERGE_RESP	Cluster merging response message
MERGE_ACK	Cluster merging acknowledgment message
CM_LIST_i	Cluster member list in V_i
$NUM\_CM(V_i)$	Number of vehicles in CM_LIST_i
MAX_CM	Maximum number of vehicles a CH can serve
N_{pm}	Potential merged cluster size
δ_s	Time threshold for neighbor sampling
δ_m	Time threshold for cluster merging
δ_i	Time threshold for isolated CH

```
Input: PN_i;
Output: SN_i;
(1)  while IN_TIMER > 0 do
(2)    for each vehicle V_j ∈ PN_i do
(3)      if V_i receives a Beacon message from V_j then
(4)        V_i calculates LLT_ij;
(5)      end if
(6)      if LLT_ij > δ_s then
(7)        if V_j ∈ SN_i then
(8)          V_i updates SN_i(j)
(9)        else
(10)         V_i adds a new SN_i(j) to SN_i
(11)       end if
(12)     end if
(13)   end for
(14) end while
```

ALGORITHM 1: LLT-based neighbor sampling.

prediction interval of link duration from time t. Equation (2) shows that if the link l is available at time t, the link will be available until $t + LLT$.

The velocity of vehicles is the main parameter to calculate the link reliability. In this paper, we assume that the velocity satisfies a *normal distribution* [37]. Then the probability density function of the velocity $g(v)$ can be calculated as

$$g(v) = \frac{1}{\sigma\sqrt{2\pi}}e^{-((v-\mu)^2/2\sigma^2)}, \tag{3}$$

where μ and σ denote the mean and standard variation, respectively. Let Δv_{ij} be the relative velocity between vehicles V_i and V_j, that is, $\Delta v_{ij} = |v_i - v_j|$. Since v_i and v_j satisfy the normal distribution, Δv_{ij} should also obey the law of normal distribution. Let $f(T)$ be the probability density function of the communication duration T. $f(T)$ can be calculated as

$$f(T) = \frac{4R}{\sigma_{\Delta v_{ij}}\sqrt{2\pi}}\frac{1}{T^2}e^{\left(\left(2R/T - \mu_{\Delta v_{ij}}\right)^2/\left(2\sigma_{\Delta v_{ij}}^2\right)\right)}, \quad \text{for } T \geq 0, \tag{4}$$

where $\mu_{\Delta v_{ij}}$ and $\sigma_{\Delta v_{ij}}$ denote the mean and the standard variation of relative velocity between v_i and v_j, respectively. Then, we can obtain the link reliability value $r(l_{ij})$ by integrating $f(T)$ in (4) from time t to $t + T$, as shown in the following equation:

$$r_t\left(l_{ij}\right) = \begin{cases} \int_t^{t+LLT} f(T)\,dT, & \text{if } LLT > 0, \\ 0, & \text{otherwise.} \end{cases} \tag{5}$$

By using the Gauss error function Erf [38], the integral in (5) can be obtained as follows:

$$r_t\left(l_{ij}\right) = \text{Erf}\left(\frac{((2R)/t) - \mu_{\Delta v_{ij}}}{\sigma_{\Delta v_{ij}}\sqrt{2}}\right)$$
$$- \text{Erf}\left(\frac{((2R)/(t + LLT)) - \mu_{\Delta v_{ij}}}{\sigma_{\Delta v_{ij}}\sqrt{2}}\right), \quad \text{when } LLT > 0. \tag{6}$$

The Erf function is calculated as follows:

$$\text{Erf}(x) = \frac{2}{\sqrt{\pi}}\int_0^x e^{-\eta^2}\,d\eta, \quad -\infty < x < +\infty. \tag{7}$$

For a particular vehicle, it may have connections with multiple vehicles in the surrounding zone. The network connectivity could be diverse due to the complex traffic conditions. Therefore, we only consider those vehicles in the SN while defining the metric LREL:

$$LREL_i(t) = \sum_{v_j \in SN_i} r_t\left(l_{ij}\right). \tag{8}$$

4.3. Cluster Head Selection. This section describes the method whereby a CH is selected. We assess the fitness of a vehicle to act as a cluster head based on link reliability. The calculation of LREL has been presented in the previous section.

As shown in Algorithm 2, a vehicle V_i in the IN state firstly tries to join an existing cluster by listening to the CH_ACK message or beacon message from a CH during the time period CH_TIMER. If V_i fails to join an existing cluster when CH_TIMER expires, V_i calculates a weighted metric,

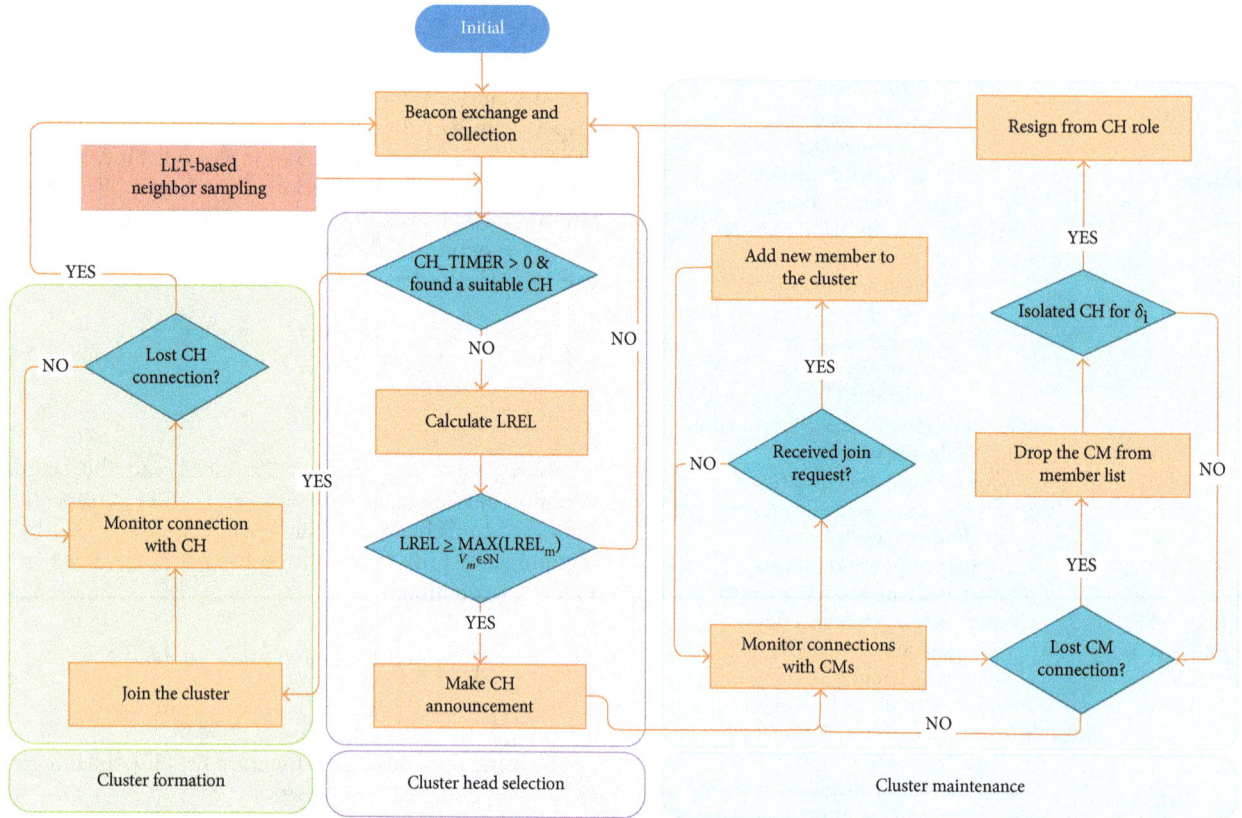

FIGURE 2: The general procedural flow of LRCA.

Input: Set of IN
Output: Set of CCM, CH
(1) **for** each vehicle V_i where State(V_i) = IN **do**
(2) V_i starts CH_TIMER
(3)   **while** CH_TIMER > 0 **do**
(4)   **if** V_i receives CH_ACK or Beacon from CH$_j$ **then**
(5)   State$(V_i) \leftarrow$ CCM
(6)   **goto Cluster Formation**
(7)   **end if**
(8)   **end while**
(9)   **if** V_i does not receive CH_ACK **then**
(10) V_i calculates LREL$_i$
(11)   **if** LREL$_i \geq$ MAX$_{V_m \in SN_i}$(LREL$_m$) **then**
(12)   State$(V_i) \leftarrow$ CH
(13) V_i broadcasts CH_ACK
(14)   **end if**
(15)   **end if**
(16) **end for**

ALGORITHM 2: Cluster head selection.

LREL$_i$, according to (8). Then V_i compares LREL$_i$ with the neighbors in SN$_i$. If it turns out that V_i has the highest value of LREL, V_i broadcasts CH_ACK information to claim itself to be a CH.

4.4. Cluster Formation. As shown in Algorithm 3, the vehicle in the IN state will join a cluster if it receives the CH_ACK

message or beacon message from a CH. The vehicle first transfers its state to CCM, then sends a JOIN_REQ message to the corresponding CH, and starts a join response timer (JOIN_TIMER). On reception of the JOIN_RESP message, the vehicle changes the state from CCM to CM. If the vehicle does not receive any response message, the vehicle resets its state to IN. A special case is that a vehicle V_i in the IN state receives messages from multiple CHs. In this case, the vehicle selects CH$_j$ which has the highest LLT$_{ij}$.

The vehicle which is in the CH state maintains a CM_LIST to store the information of CMs. When the CH receives a JOIN_REQ message from the surroundings, the CH first checks the total number of members in the CM_LIST. If the number of CMs is less than a maximum number of members allowed (MAX_CM), the CH generates a JOIN_RESP message and unicasts it to the vehicle from which the JOIN_REQ message is received. At the meantime, the CH builds an entry of the vehicle and adds it to the CM_LIST.

4.5. Cluster Maintenance. Because of the high mobility of vehicles in VANET, the role of vehicles may keep changing frequently, which brings extra maintenance overhead. In our proposed scheme, the CH resigns from CH role and transfers to the IN state when losing all of its CMs. Otherwise, it remains as CH until the cluster merging process happens. Therefore, we only consider cluster merging (Algorithm 4) and vehicle leaving events in the cluster maintenance procedure.

```
Input: Set of IN, CCM
Output: Set of CM
 (1) for each vehicle V_i where State(V_i) = IN do
 (2)    if V_i receives CH_ACK or Beacon from CH_j then
 (3)       State(V_i) = CCM
 (4)       V_i unicasts JOIN_REQ to V_j
 (5)       V_i starts JOIN_TIMER
 (6)       while JOIN_TIMER > 0 do
 (7)          if V_i receives JOIN_RESP then
 (8)             State(V_i) ← CM
 (9)          end if
(10)       end while
(11)       if V_i does not receive JOIN_RESP then
(12)          State(V_i) ← IN
(13)       end if
(14)    end if
(15) end for
(16) for each vehicle V_j where State(V_j) = CH do
(17)    if V_j receives JOIN_REQ from V_i then
(18)       if NUM_CM(V_j) < MAX_CM then
(19)          V_j adds V_i into CM_LIST_j
(20)          V_j unicasts JOIN_RESP to V_i
(21)       end if
(22)    end if
(23) end for
```

ALGORITHM 3: Cluster formation.

```
Input: Two subclusters
Output: The merged cluster
 (1) if LLT_ij ≥ δ_m then
 (2)    if LREL_i ≥ LREL_j then
 (3)       CH_j sends MERGE_REQ to CH_i
 (4)       if CH_i receives MERGE_REQ then
 (5)          CH_i estimates N_pm
 (6)          if N_pm ≤ MAX_CM then
 (7)             CH_i sends MERGE_RESP to CH_j
 (8)          end if
 (9)       end if
(10)       if CH_j receives MERGE_RESP then
(11)          CH_j broadcasts MERGE_ACK
(12)          State(V_j) ← CM
(13)       end if
(14)    end if
(15) end if
```

ALGORITHM 4: Cluster merging.

4.5.1. Cluster Merging.

As time passes, clusters moving on the road may overlap with one another. When two moving clusters get closer to one another, the overlapping area of these two clusters becomes larger. Heavily overlapped moving clusters introduce redundant in intracluster management and communication overhead. Under the circumstances, the two clusters have potential to be merged. Instead of starting the cluster merging procedure immediately, the merging procedure begins if the two CHs (CH$_i$ and CH$_j$) detect that they will stay neighbors for a certain time threshold δ_m(LLT$_{ij} \geq \delta_m$). This is because if two clusters just passing by one another quickly, the overlap of the two clusters is temporary and would not affect the overall performance in the long run. Once the cluster merging process begins, the two CHs share their cluster information and the CH with lower LREL (CH$_j$) sends MERGE_REQ to the higher CH (CH$_i$).

Upon reception of MERGE_REQ, CH$_i$ estimates the potential merged cluster size (N_{pm}). If $N_{pm} \leq$ MAX_CM, cluster merging is permitted and CH$_i$ then sends MERGE_RESP to CH$_j$. If CH$_j$ receives MERGE_RESP, CH$_j$ gives up the leadership and broadcasts MERGE_ACK to inform its CMs about the merge operation. Otherwise, the CHs remain the role as CHs. On reception of MERGE_ACK, CMs which also in the SN$_i$ of CH$_i$ automatically become cluster members of CH$_i$. The remaining vehicles then search new clusters and join in.

4.5.2. Leaving a Cluster.

During every beacon period, each CH monitors the connections with its CMs dynamically. If a CH does not receive the beacon message from its CM for at least two beacon interval, the CH is considered to loss the connection with the CM. Every time when a CH receives a beacon message from its CM, it updates the CMs related information in its CM_LIST. When the disconnection occurs, CH deletes the entry of the CM from the CM_LIST. The vehicle which lost the connection with CH then transfers its state to IN and tries to find a new cluster. When a CH losses all of its CMs to become an isolated CH for a certain time δ_i, the CH resigns from CH role and turns to IN.

5. Routing Protocol of LRCA

The goal of our proposed LRCA architecture is to delivery data packets to a specified destination in the vehicular networks. For example, if a vehicle is heading for a particular shopping district, this vehicle can obtain the parking information nearby or promotion information from merchants in advance by sending inquires to the vehicles around the shopping district.

In LRCA, clusters span over every road segment. In order to connect them in street scenarios, we propose a bridge node selection scheme to nominate special nodes at intersections. Moreover, these selected bridge nodes are responsible for assessing the network condition over each road segment connected to the intersection.

5.1. Bridge Node Selection at Intersections.

For each intersection, the vehicle which will stay longer at the intersection zone is preferred to be selected as the bridge node. In this case, those vehicles stopped by the red light will be the ideal candidates. If there is only one vehicle, we nominate the vehicle as the bridge node directly; if there are more vehicles, we randomly select a vehicle as the bridge node for the sake of simplicity; if there is no vehicle stopping at the intersection, we will select the vehicle which is approaching the intersection center with the lowest velocity.

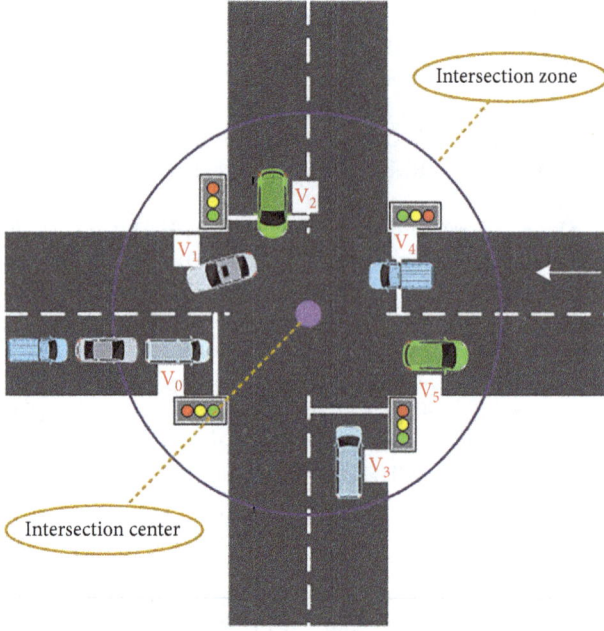

Figure 3: Bridge node selection at intersections.

For example, as shown in Figure 3, V_0 will be selected as the bridge node because it is stopping at the intersection. In other case, assume that the vehicle V_0 does not exist. The rest of the vehicles within the intersection zone are all candidates (i.e., $\zeta = \{V_1, V_2, V_3, V_4, V_5\}$). Among them, V_1 and V_5 are going past the intersection center while V_2, V_3, and V_4 are approaching the intersection center. We discard V_1 and V_5 from ζ because they are on the way leaving the intersection zone. Afterwards, the bridge node is selected from the remaining candidates $\zeta = \{V_2, V_3, V_4\}$. The vehicle in ζ which has the minimum velocity is then chosen as the bridge node.

When the bridge node is about to drive out of the intersection zone, a new bridge node will be selected to guarantee the connectivity in intersection scenarios.

5.2. Road Segment Evaluation. When calculating the routing path, we cannot simply adopt traditional shortest path algorithms (e.g., *Dijkstra's* algorithm) because when the packet carrier arrives at an intersection, it is not guaranteed that it can meet another vehicle moving towards the most preferred direction [39]. In order to evaluate the network condition of road segments, we put forward a distributed procedure named road segment evaluation (RSE) which is dynamically initiated by aforementioned bridge nodes.

RSE is triggered when a bridge node is selected at an intersection (I_i) by sending a light-weight control packet (P_{RSE}) to the adjacent intersection (I_j). Thereafter, P_{RSE} transverses the road segment (RS_{ij}) via relaying forwarders and gathers information regarding connectivity, delay, and hop count at each intermediate forwarder. When P_{RSE} reaches the target intersection, the bridge node at that junction calculates the delivery delay (d_p) as follows:

$$d_p = t_{\text{rc}} - T_{\text{RSE}}, \qquad (9)$$

where t_{rc} and T_{RSE} designate the received time and the generation time of P_{RSE}, respectively. d_p can indicate the network condition in the road because it experiences similar transmission and queuing delay in addition to interference and fading conditions in R_{ij}. Then, d_p is compared to a threshold $T_{M_{ij}}$ which is calculated as below:

$$T_{M_{ij}} = 2 \cdot t_{d_M} \cdot \left\lceil \frac{|\text{RS}_{ij}|}{R} \right\rceil + t_{\text{cf}}, \qquad (10)$$

where t_{d_M} is a constant parameter representing the maximum acceptable delay per forwarder, including the transmission delay and queuing delay, $|\text{RS}_{ij}|$ denotes the length of RS_{ij}, and t_{cf} is a constant parameter representing the maximum tolerable time using carry-and-forward. For a disconnected road segment, P_{RSE} is dropped when a forwarder detects that d_p is larger than $T_{M_{ij}}$. In case a bridge node has not received the P_{RSE} for a certain time, that is, $d_p > T_{M_{ij}}$, the bridge node deduces that R_{ij} is disconnected. Afterwards, the bridge node assigns weight to RS_{ij}:

$$w_{ij} = \begin{cases} d_p/T_{M_{ij}}, & d_p \le T_{M_{ij}}, \\ \infty, & \text{otherwise.} \end{cases} \qquad (11)$$

5.3. Route Construction. The protocol is designed for transmitting data with the lowest delivery delay and the highest stability in terms of connectivity. In particular, we assume that a vehicle generates a DATA_PACKET in the form of $<V_{\text{src}}, M, L_{\text{dest}}>$, where V_{src} denotes the identity of the sender vehicle, M is the message, and L_{dest} is the location of message destination.

As shown in Algorithm 5, the protocol consists of two phases, that is, inter- and intrasegment phases. The intersegment phase deals with the routing path decision at intersections, while the intrasegment phase focuses on packet forwarding within a road segment. When the DATA_PACKET is generated, V_{src} forwards it to the nearest bridge node.

For the intersegment phase, when the current forwarder arrives at an intersection I_i, it delivers DATA_PACKET to the bridge node at that intersection. Once received, the bridge node obtains the destination and calculates the routing metric as below:

$$M_{ij} = \frac{|\text{SP}_j|}{|\text{SP}_i|} \cdot w_{ij}, \qquad (12)$$

where M_{ij} is combination of both geographic information and routing path delay. In (12), $|\text{SP}|_j/|\text{SP}|_i$ denotes the geographical process where SP_i represents the shortest path between current intersection and L_{dest}, and SP_j represents the shortest path between the next candidate intersection and L_{dest}. The road segment RS_{ij} with the minimum M_{ij} is considered to be the optimal routing path. Then, the bridge node delivers the DATA_PACKET to a vehicle on the selected road.

> (1) **Initialization**: DATA_PACKET $\leq V_{\text{src}}, M, L_{\text{dest}} >$
> (2) **if** the current forwarder V_c arrives at an intersection **then**
> (3) $\quad V_c$ delivers DATA_PACKET to the bridge node V_{B_i}
> (4) $\quad V_{B_i}$ calculates M_{ij} and selects the RS_{ij} with the minimum M_{ij}
> (5) $\quad V_{B_i}$ delivers DATA_PACKET to a vehicle on RS_{ij}
> (6) **else**
> (7) $\quad$ **if** V_c is a CH **then**
> (8) $\quad\quad$ **if** L_{dest} inside the cluster range **then**
> (9) $\quad\quad\quad$ Multicast DATA_PACKET to its CMs and **Done**
> (10) $\quad\quad$ **else**
> (11) $\quad\quad\quad$ Unicast DATA_PACKET to a CM closest to the target intersection I_j
> (12) $\quad\quad$ **end if**
> (13) $\quad$ **else**
> (14) $\quad\quad$ **if** DATA_PACKET is from its CH **then**
> (15) $\quad\quad\quad$ Unicast DATA_PACKET to a CM or CH closest to I_j
> (16) $\quad\quad$ **else**
> (17) $\quad\quad\quad$ Unicast DATA_PACKET to its CH
> (18) $\quad\quad$ **end if**
> (19) $\quad$ **end if**
> (20) **end if**

ALGORITHM 5: Routing protocol.

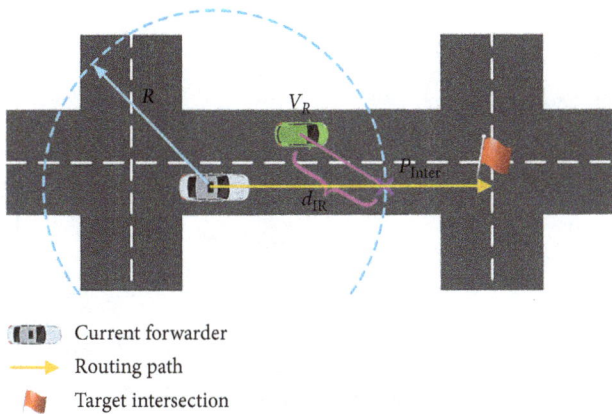

Current forwarder

Routing path

Target intersection

FIGURE 4: Computation of a good relay node.

For the intrasegment phase, if the current forwarder is a CH (Steps 6–11), it checks whether L_{dest} is inside its cluster range. If so, the CH will broadcast the DATA_PACKET to its CMs directly. If not, the CH will look for a good relay vehicle for the message propagation. The CH aims at finding a CM which is closest to the target intersection. The CH first computes the intersection point (P_{Inter}) of the route path and communication range as shown in Figure 4. Then, the CM which is closest to P_{Inter} and moves towards the target intersection is considered to be a good relay vehicle (V_R) for the message propagation. To find such a vehicle, the CH calculates the distance (d_{IR}) between P_{Inter} and its CMs. Afterwards, CH selects the CM with minimum d_{IR} as the relay vehicle and then unicasts the DATA_PACKET to this CM.

In another case, if the current forwarder is a CM (Steps 12–17), it checks the source of the packet. If the packet does not come from its CH, the CM will unicast the packet to its CH directly. If the packet comes from its CH, the CM will be responsible for sending the packet to vehicles in nearby clusters. The CM first checks the neighbor list to find a neighbor CH which is moving towards the target intersection. If there are multiple such CHs, the CM selects the CH with the shortest distance to the target intersection. Then, the DATA_PACKET will be delivered to the selected CH. If the CM fails to find a CH in the neighbor list, the CM will select a one-hop neighbor which is closest to the next intersection as the next propagation vehicle.

6. Performance Evaluation

The proposed approach is compared with three previously proposed clustering-based schemes NHop [23], VMaSC [9], and MOSIC [24], and a nonclustering-based approach GeoSVR [40]. Among them, NHop and VMaSC are the two most cited multihop clustering algorithms and MOSIC is a latest single-hop clustering approach which has been introduced in Section 2. Since our proposed LRCA and the MOSIC are both single-hop clustering algorithms, the one-hop NHop and VMaSC are implemented in the simulation. GeoSVR is a high-cited nonclustering-based routing protocol which combines node location with the digital map. Meanwhile, it selects the routing path based on vehicular density to avoid local maximum and sparse connectivity.

The simulations are implemented in the Network Simulator NS-2 (v-2.35) [41] with the mobility of vehicles generated by SUMO [42]. As shown in Figure 5, the simulation scenario is a 5100m × 4800m ordinary urban environment which is extracted from the OpenStreetMap [43] of Shanghai China. The number of simulated vehicles is set to 1500, and we run the simulation for 100 seconds to let all the injected vehicles move around the map for a while. After 100 seconds, the simulation runs for another 500 seconds to evaluate the total performance metrics. In the simulation, we evaluate the metrics at the transmission ranges of 200 and 500 m. Meanwhile, we vary the maximum allowable velocity

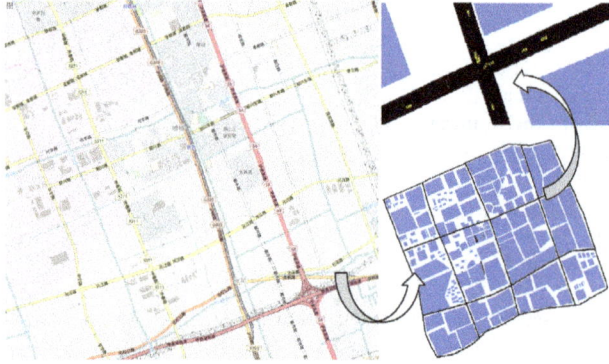

FIGURE 5: Simulation scenario of Shanghai China with SUMO.

of vehicles from 10 to 30 m/s and the maximum acceleration and deceleration are set to 5 m^2/s. The reported result is the average of 10 times repeated run. The details of simulation parameters and values are listed in Table 2.

6.1. Clustering Performance. To evaluate the performance of the proposed clustering algorithm, we focus on the stability of cluster, where cluster stability means lower changes in the CHs and lower changes in the CMs. Therefore, good clustering algorithm should be designed to minimize the rate of CH change and gain long term of CH duration, as well as cluster member duration.

Consequently, the following performance metrics are used for comparison:

(i) *Cluster head duration*: It is defined as the average time from a vehicle becoming a CH to transferring to another state.

(ii) *Cluster member duration*: It is defined as the average time from a vehicle joining a cluster to leaving the cluster.

(iii) *Cluster head change rate*: It is defined as $1 - (|S_{CH}^t \cap S_{CH}^{t-1}|/|S_{CH}^t \cup S_{CH}^{t-1}|)$, where S_{CH}^t and S_{CH}^{t-1} represent the current CH set and previous CH set, respectively. $|s|$ denotes the number of elements in the corresponding set.

The results in Figures 6–8 evaluate the clustering stability for different maximum allowable velocities and transmission ranges by comparing the proposed LRCA with VMaSC, NHop, and MOSIC, from the aspects of the average CH lifetime, the average CM lifetime, and the CH change rate.

Figures 6 and 7 show the performance of CH and CM duration for different maximum allowable velocities and transmission ranges, respectively. Results show that the average CH and CM duration will decrease when the maximum allowable velocity of vehicles increases. This is because when the vehicles move faster, the vehicular network topology becomes more dynamic and eventually makes it difficult for vehicles to maintain a relatively stable condition with their neighbor vehicles for a long period. VMaSC-1hop acquires the longest CH lifetime and CM lifetime when the maximum velocity is 10 m/s. But both the CH duration and CM duration of VMaSC-1hop decrease

TABLE 2: Simulation parameters.

Notation	Description
Simulation area	5100 m × 4800 m
Maximum velocity	10, 15, 20, 25, 30 m/s
Maximum acceleration	5 m/s^2
Transmission range	200, 500 m
MAC protocol	IEEE 802.11p
Data rate	2 Mbps
MAX_CM	10
HELLO_PACKET period	200 ms
HELLO_PACKET size	64 bytes
DATA_PACKET generation rate	10, 20, 30, 40, 50 packet/s
DATA_PACKET size	1024 bytes
IN_TIMER	1 s
CH_TIMER	2 s
JOIN_TIMER	2 s
δ_s	2 s
δ_m	2 s
δ_i	1 s

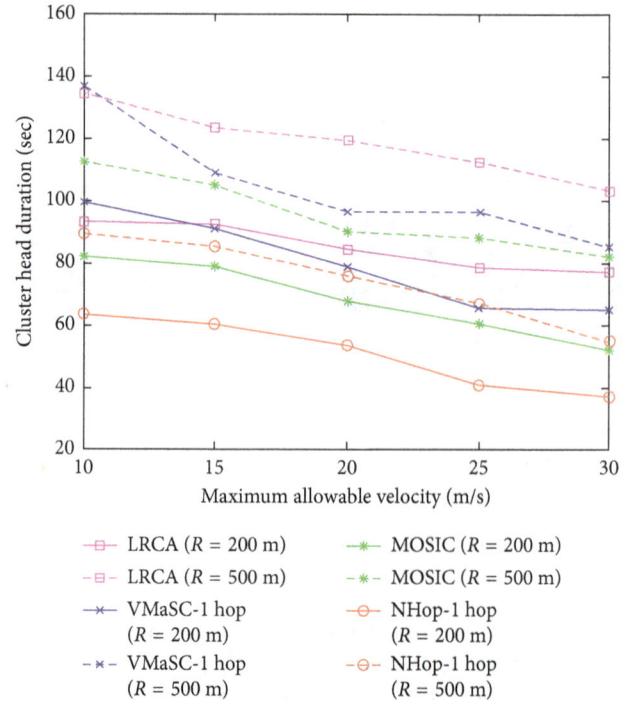

FIGURE 6: CH duration for different maximum allowable velocities and transmission ranges.

rapidly when the maximum allowable velocity increases. When the maximum velocity is bigger than 15 m/s, LRCA performs better against VMaSC (VMaSC-1hop), NHop (NHop-1hop), and MOSIC in terms of both CH and CM duration. This is because in our scheme, we construct stable clusters by employing LLT-based neighbor sampling scheme to pick out stable neighbors. In addition, the CH duration and CM duration are larger at a high-transmission range. The main reason is that the vehicles can communicate with more neighbor nodes and create higher correlation of connectivity behavior when the transmission range is higher. As the transmission range increases from 200 to 500 m, the

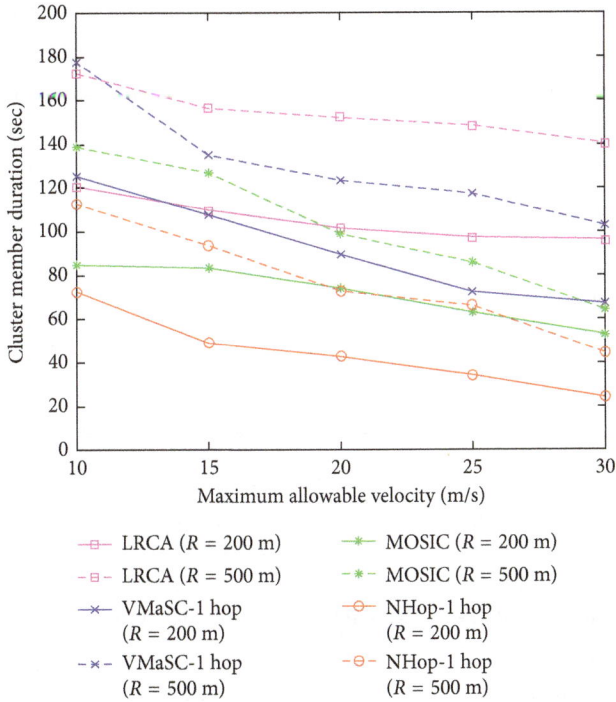

FIGURE 7: CM duration for different maximum allowable velocities and transmission ranges.

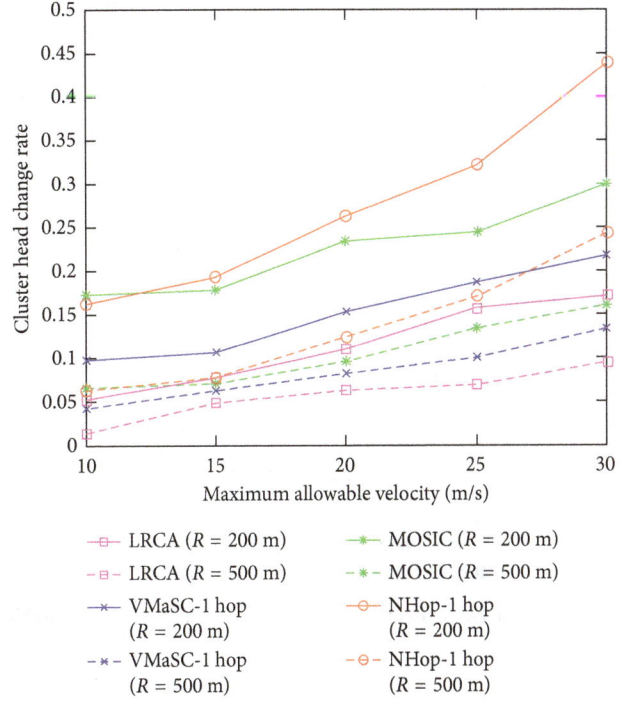

FIGURE 8: CH change rate for different maximum allowable velocities and transmission ranges.

link duration increases for all the simulated clustering algorithms.

Figure 8 shows the performance of the CH change rate for different maximum allowable velocities and transmission ranges. From the results, we can observe that the CH change rate increases with the increment of maximum allowable velocity and decreases as the transmission range increases. The reason is that the increment of maximum allowable velocity accelerates the change of network topology. Afterwards, some CMs may move out of the cluster or cluster merging may happen, which may result in cluster head changes. On the contrary, the higher transmission range provides much more connectivity of the vehicles within a cluster, which reduces the changes in the cluster head. The CH change rate for NHop-1hop is closer to MOSIC at both low- and high-transmission range when the maximum allowable velocity is between 10 and 15 m/s. However, the CH change rate for NHop-1hop increases rapidly as the velocity increases. Our proposed LRCA acquires the lowest CH change rate against VMaSC-1hop, NHop-1hop, and MOSIC in both cases of transmission range. We reduce the CH change rate by changing the CH state to other clustering state only when cluster merging happens or it losses all the CMs to become an isolated CH. In addition, to avoid unnecessary state transitions, we put up two time thresholds δ_m and δ_i before merging two clusters and before transferring an isolated CH to the IN state.

As shown above, mostly, the LRCA gains longer duration of both CH and CM and performs the lowest CH change rate at both low- and high-transmission range. Therefore, the proposed LRCA indicates the better performance of

clustering stability than that of VMaSC (VMaSC-1hop), NHop (NHop-1hop), and MOSIC.

6.2. Routing Performance. The routing performance of our proposed LRCA is compared with three cluster-based routing mechanisms including NHop, VMaSC, and MOSIC, which have been discussed in the related work above, and a non-clustering approach called GeoSVR. We aim at achieving efficient and reliable data delivery with high packet delivery ratio and low network latency.

From the simulation results above, we can learn that clusters acquire more stability at a high-transmission range. Hence, we set the transmission range to 500 m when evaluating the routing performance here. To evaluate the effect of different parameters on the routing performance, we vary the parameters of the data packet generation rate and maximum allowable velocity in the simulation. The default value of the maximum allowable velocity and the data packet generation rate are set to 20 m/s and 30 packet/s, respectively.

Furthermore, we evaluate the routing performance using the following performance metrics:

(i) *Packet delivery ratio (PDR)*: This metric is defined as the ratio of the average number of data packets successfully received by destinations, compared to the total number of generated packets.

(ii) *End-to-end delay (E2ED)*: This metric represents the average delay between the time a packet generated by the source and the time of this packet reached to the destination.

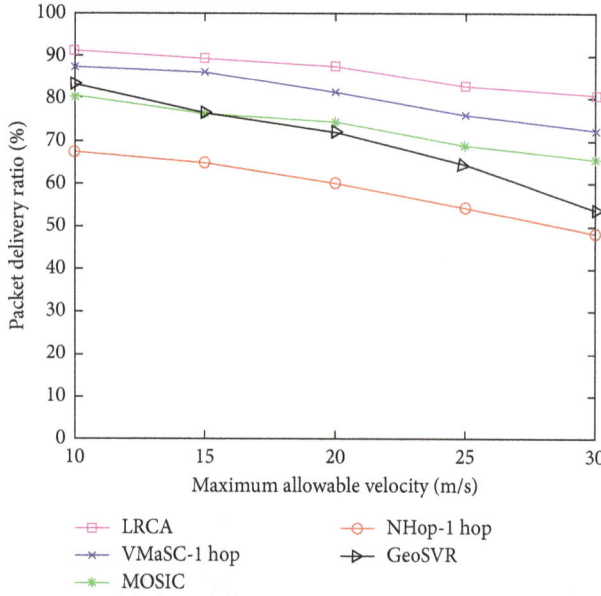

FIGURE 9: Packet delivery ratio for varying maximum allowable velocity.

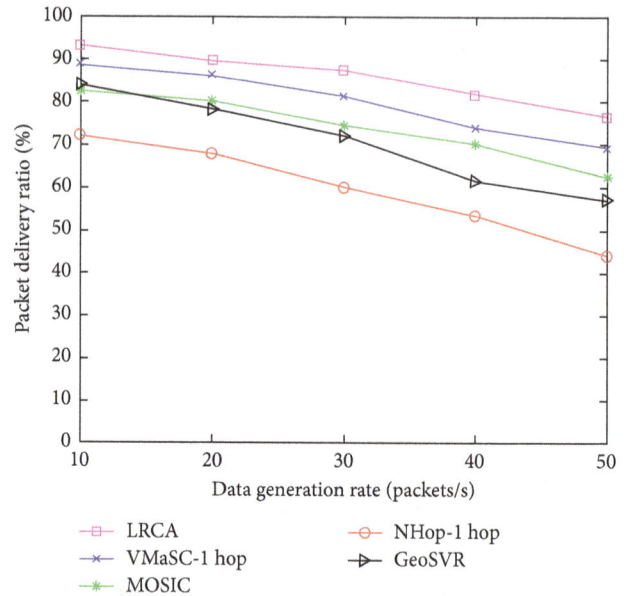

FIGURE 10: Packet delivery ratio for varying data generation rate.

(iii) *Normalized routing overhead (NRO)*: This metric is calculated as the ratio of the size of total generated packets to the size of the data packets successfully received by the destinations.

Figure 9 shows the performance of the packet delivery ratio for varying maximum allowable velocity. It is observed that LRCA achieves higher packet delivery ratio than other schemes. By selecting stable neighbors according to link lifetime (LLT), LRCA builds up a stable clustered virtual network which can provide the stable connections between cluster members, and meanwhile increase the bandwidth availability and reduce data collision. Moreover, the results show that the packet delivery ratio of all protocols decreases when lifting upper limitation of allowable velocity. Because the maximum allowable velocity increases, the network topology changes rapidly, leading to an increment of packet loss ratio. The packet delivery ratio of GeoSVR is very close to that of MOSIC in the low-speed scenario. However, the packet delivery ratio of GeoSVR decreases significantly when the maximum allowable velocity is greater than 20 m/s, which indicates that the clustering-based routing approaches achieve more stability than the nonclustering-based ones, especially in the high-dynamic scenario.

Figure 10 shows the performance of the packet delivery ratio for varying data generation rate. Obviously, the packet delivery ratio of all the simulated protocols decreases when the data generation rate increases. This is because the vehicles need to store and carry the data packets when encountering network partitions. However, the size of packet buffer is limited, leading to subsequent packets being dropped when the buffer is full. Results show that our proposed LRCA achieves better performance of the packet delivery ratio than the other protocols. In LRCA, the bridge node selects the optimal routing path for data delivering considering the latency of each road segment, which decreases the cases of packet carrying by

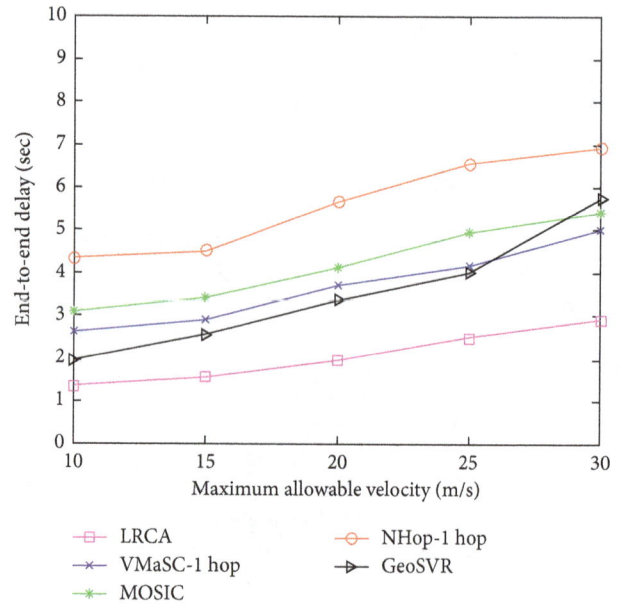

FIGURE 11: End-to-end delay for varying maximum allowable velocity.

setting a time threshold parameter for the carry-and-forward mode as shown in (10).

Figure 11 shows the performance of average end-to-end delay for varying maximum allowable velocity. The increment of velocity results in frequent changes of network topology. Thus, the results show that the end-to-end delay increases in high-speed scenario. The proposed scheme achieves significant reduction of end-to-end delay in comparison with other schemes. This is because LRCA builds up stable clusters which can guarantee the sufficient connectivity and reliable linking. Therefore, the retransmission times and transmission delay are reduced, which results in reduction of end-to-end delay. Another reason is that with the help of the stable connected

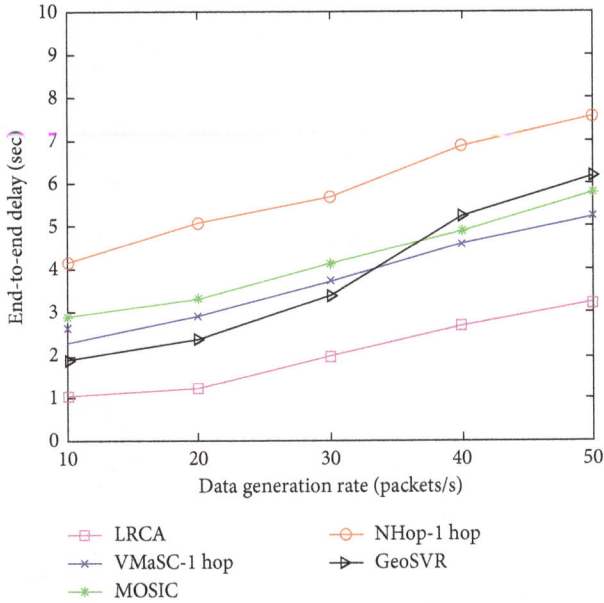

FIGURE 12: End-to-end delay for varying data generation rate.

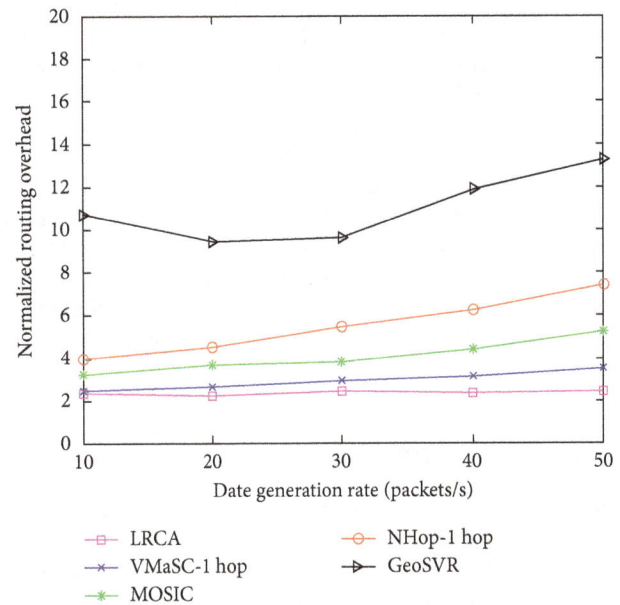

FIGURE 13: Normalized routing overhead for varying maximum allowable velocity.

clusters, packets can be delivered to the next hop with short MAC layer contention, which leads to short network latency. GeoSVR achieves lower delivery delay against VMaSC, MOSIC, and NHop when the maximum speed is less than 25 m/s. The GeoSVR obtains a better result by using the optimal forwarding path and the restricted forwarding algorithm. However, the end-to-end delay of GeoSVR increases significantly when the maximum velocity comes to 30 m/s. This shows that the GeoSVR may not be quite suitable for high-speed VANET scenario.

Figure 12 shows the performance of average end-to-end delay for varying data generation rate. When varying the data generation rate from 10 to 50 packet/s, the queuing delay in the buffer with relay nodes raises, which eventually affects the end-to-end delay. Consequently, the results show that the end-to-end delay of all the simulated protocols tends to increase with the increase of the data generation rate. In LRCA, the routing decision is made based on the estimated delay information over each road segment, which includes both delays due to packet carrying and packet relaying on links. LRCA acquires the lowest delivery delay by selecting the route paths with the lowest delay and highest geographical process as shown in (12). GeoSVR selects road segments with high vehicular density as routing paths to reduce the probability of store-and-carry events. That is why GeoSVR achieves lower end-to-end delay than VMaSC, MOSIC, and NHop when the data generation rate is less than 30 packet/s. But the GeoSVR suffers from data congestion when the data generation rate is high due to the strategy forwarding data over dense roads. Therefore, the end-to-end delay is much higher when the data generation rate is high.

Figure 13 shows the performance of normalized routing overhead for varying maximum allowable velocity. The proposed scheme shows the lowest overhead in comparison with others. Besides, the overhead of clustering schemes is much lower than that of the nonclustering scheme GeoSVR. This is

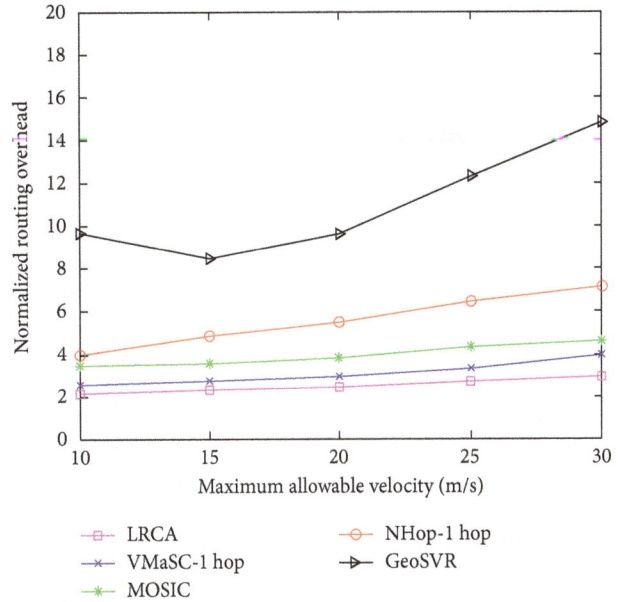

FIGURE 14: Normalized routing overhead for varying data generation rate.

because the clustering approaches can reduce intervehicle communications. The experiment results also show that the overhead of NHop-1hop increases when the vehicles move faster, while the overhead of LRCA does not significantly increase when the maximum velocity increases. This is benefit from the neighbor sampling strategy to filter out unstable neighbors and reduce unnecessary message exchanges. Moreover, the LRCA provides the stable connections for data delivery which can reduce the number of packet retransmissions.

Figure 14 shows the performance of normalized routing overhead for varying data generation rate. Results show that

LRCA achieves the lowest overhead and the clustering-based protocols acquire much lower overhead than the non-clustering scheme GeoSVR. The overhead of our proposed LRCA barely increases when raising the data generation rate, while other clustering methods more or less increase. LRCA reduces redundant message exchanges by selecting stable neighbor vehicles and reduces data packet retransmissions by selecting the route path with more connectivity.

7. Conclusion

In this paper, we propose a new link reliability-based clustering algorithm (LRCA) to provide efficient and reliable data transmission in urban VANET. In LRCA, vehicles are grouped with stable neighbor vehicles which are selected by the LLT-based neighbor sampling scheme. Further, we introduce a cluster-based routing protocol to provide the efficient and reliable data transmission in vehicular networks. In contrast to those protocols proposed for safety critical applications, the routing approach in this paper is designed to support infotainment services in VANET which are not stringent in delay constraints. To transmit data packets through stable paths, we introduce bridge nodes at intersections to make routing decisions. The bridge node evaluates the network condition over road segments and assigns weights to them. Then the road segment with minimum weight is selected to construct the overall routing path. The simulation results show that the proposed LRCA acquires better clustering stability in terms of long cluster head duration, long cluster member duration, and low rate of cluster head change. The proposed routing protocol performs better than the previous proposed schemes, which demonstrates the advantages of the proposed LRCA. In the future work, we will further improve the route strategy to minimize the end-to-end delay and satisfy the requirements of real-time applications in VANET.

Acknowledgments

This work was partially supported by the NSF of China under Grant nos. 61702334 and 61772200, Shanghai Pujiang Talent Program under Grant no. 17PJ1401900, Shanghai Municipal Natural Science Foundation under Grant nos. 17ZR1406900 and 17ZR1429700, Action Plan for Innovation on Science and Technology Projects of Shanghai under Grant no. 16511101000, Collaborative Innovation Foundation of Shanghai Institute of Technology under Grant no. XTCX2016-20, and Educational Research Fund of ECUST under Grant no. ZH1726108.

References

[1] B. T. Sharef, R. A. Alsaqour, and M. Ismail, "Vehicular communication ad hoc routing protocols: a survey," *Journal of Network & Computer Applications*, vol. 40, no. 1, pp. 363–396, 2014.

[2] R. S. Bali, N. Kumar, and J. J. P. C. Rodrigues, "Clustering in vehicular ad hoc networks: taxonomy, challenges and solutions," *Vehicular Communications*, vol. 1, no. 3, pp. 134–152, 2014.

[3] J. Jeong, H. Jeong, E. Lee, T. Oh, and D. H. C. Du, "SAINT: self-adaptive interactive navigation tool for cloud-based vehicular traffic optimization," *IEEE Transactions on Vehicular Technology*, vol. 65, no. 6, pp. 4053–4067, 2016.

[4] C. Cooper, D. Franklin, M. Ros, F. Safaei, and M. Abolhasan, "A comparative survey of VANET clustering techniques," *IEEE Communications Surveys & Tutorials*, vol. 99, p. 1, 2017.

[5] C. R. Lin and M. Gerla, "Adaptive clustering for mobile wireless networks," *IEEE J Selected Areas in Communications*, vol. 15, no. 7, pp. 1265–1275, 1997.

[6] P. Basu, N. Khan, and T. D. C. Little, "A mobility based metric for clustering in mobile ad hoc networks," in *Proceedings of the International Conference on Distributed Computing Systems*, , Mesa, AZ, USA, April 2001.

[7] M. Fathian and A. R. Jafarian-Moghaddam, "New clustering algorithms for vehicular ad-hoc network in a highway communication environment," *Wireless Networks*, vol. 21, no. 8, pp. 2765–2780, 2015.

[8] G. V. Rossi, F. Zhong, W. H. Chin, and K. K. Leung, "Stable clustering for ad-hoc vehicle networking," in *Proceedings of the Wireless Communications and Networking Conference*, pp. 1–6, Edinburgh, UK, December 2017.

[9] S. Ucar, S. C. Ergen, and O. Ozkasap, "Multihop-cluster-based IEEE 802.11p and LTE hybrid architecture for VANET safety message dissemination," *IEEE Transactions on Vehicular Technology*, vol. 65, no. 4, pp. 2621–2636, 2016.

[10] B. Liu, D. Jia, J. Wang, K. Lu, and L. Wu, "Cloud-assisted safety message dissemination in VANET-cellular heterogeneous wireless network," *IEEE Systems Journal*, vol. 99, pp. 1–12, 2017.

[11] P. Salvo, M. D. Felice, F. Cuomo, and A. Baiocchi, "Infotainment traffic flow dissemination in an urban VANET," in *Proceedings of the Global Communications Conference*, pp. 67–72, Atlanta, GA, USA, December 2013.

[12] M. Oche, R. Md Noor, and A. Jalooli, "Quality of service management for IPTV services support in VANETs: a performance evaluation study," *Wireless Networks*, vol. 21, no. 1, pp. 315–328, 2015.

[13] K. Abboud and W. Zhuang, "Stochastic modeling of single-hop cluster stability in vehicular ad hoc networks," *IEEE Transactions on Vehicular Technology*, vol. 65, no. 1, pp. 226–240, 2016.

[14] X. Ji, H. Q. Yu, G. S. Fan, and W. H. Fu, "SDGR: an SDN-based geographic routing protocol for VANET," in *Proceedings of the IEEE International Conference on Internet of Things*, pp. 276–281, Exeter, UK, June 2017.

[15] O. Kayis and T. Acarman, "Clustering formation for inter-vehicle communication," in *Proceedings of the Intelligent Transportation Systems Conference (ITSC)*, pp. 636–641, Hong Kong, China, October 2007.

[16] J. Chen, C. Lai, X. Meng, J. Xu, and H. Hu, "Clustering moving objects in spatial networks," in *Proceedings of the International Conference on Database Systems for Advanced Applications*, pp. 611–623, Bangkok, Thailand, April 2007.

[17] C. Shea, B. Hassanabadi, and S. Valaee, "Mobility-based clustering in VANETs using affinity propagation," in *Proceedings of the Global Telecommunications Conference*, pp. 1–6, FL, USA, December 2010.

[18] Z. Wang, L. Liu, M. C. Zhou, and N. Ansari, "A position-based clustering technique for ad hoc intervehicle communication," *IEEE Transactions on Systems Man & Cybernetics Part C*, vol. 38, no. 1, pp. 201–208, 2008.

[19] S. Almalag Mohammad and C. Weigle Michele, "Using traffic flow for cluster formation in vehicular ad-hoc networks," in *Proceedings of the Local Computer Networks*, pp. 631–636, Denver, CO, USA, October 2010.

[20] M. M. C. Morales, C. S. Hong, and Y. C. Bang, "An adaptable mobility-aware clustering algorithm in vehicular networks," in *Proceedings of the Network Operations and Management Symposium*, pp. 1–6, Maui, MI, USA, April 2011.

[21] M. Ren, L. Khoukhi, H. Labiod, J. Zhang, and V. Vèque, "A mobility-based scheme for dynamic clustering in vehicular ad-hoc networks (VANETs)," *Vehicular Communications*, vol. 9, pp. 233–241, 2016.

[22] G. Wolny, "Modified DMAC clustering algorithm for VANETs," in *Proceedings of the International Conference on Systems and Networks Communications*, pp. 268–273, Sliema, Malta, October 2008.

[23] Z. Zhang, A. Boukerche, and R. Pazzi, "A novel multi-hop clustering scheme for vehicular ad-hoc networks," in *Proceedings of the ACM International Workshop on Mobility Management & Wireless Access (MobiWac)*, pp. 19–26, Miami Beach, Fl, USA, October-November 2011.

[24] A. Ziagham and M. R. Noorimehr, "MOSIC: mobility-aware single-hop clustering scheme for vehicular ad hoc networks on highways," *International Journal of Advanced Computer Science & Applications*, vol. 7, no. 9, 2016.

[25] X. M. Zhang, K. H. Chen, X. L. Cao, and K. S. Dan, "A street-centric routing protocol based on microtopology in vehicular ad hoc networks," *IEEE Transactions on Vehicular Technology*, vol. 65, no. 7, pp. 5680–5694, 2016.

[26] M. A. Togou, A. Hafid, and L. Khoukhi, "SCRP: stable CDS-based routing protocol for urban vehicular ad hoc networks," *IEEE Transactions on Intelligent Transportation Systems*, vol. 17, no. 5, pp. 1298–1307, 2016.

[27] D. Lin, J. Kang, A. Squicciarini, Y. Wu, S. Gurung, and O. Tonguz, "MoZo: a moving zone based routing protocol using pure V2V communication in VANETs," *IEEE Transactions on Mobile Computing*, vol. 16, no. 5, pp. 1357–1370, 2017.

[28] L. Rivoirard, M. Wahl, P. Sondi, M. Berbineau, and D. Gruyer, "Chain-branch-leaf: a clustering scheme for vehicular networks using only V2V communications," *Ad Hoc Networks*, vol. 68, pp. 70–84, 2018.

[29] Z. He, D. Zhang, and J. Liang, "Cost-efficient sensory data transmission in heterogeneous software-defined vehicular networks," *IEEE Sensors Journal*, vol. 16, no. 20, pp. 7342–7354, 2016.

[30] F. Zeng, R. Zhang, X. Cheng, and L. Yang, "Channel prediction based scheduling for data dissemination in VANETs," *IEEE Communications Letters*, vol. 21, no. 6, pp. 1409–1412, 2017.

[31] J. Zhu, C. Huang, X. Fan, S. Guo, and B. Fu, "EDDA: an efficient distributed data replication algorithm in VANETs," *Sensors*, vol. 18, no. 2, p. 547, 2018.

[32] T. Song, W. Xia, T. Song, and L. Shen, "A cluster-based directional routing protocol in VANET," in *Proceedings of the IEEE International Conference on Communication Technology*, pp. 1172–1175, Jinan, China, September 2011.

[33] Y. Ohta, T. Ohta, and Y. Kakuda, "An autonomous clustering-based data transfer scheme using positions and moving direction of vehicles for VANETs," in *Proceedings of the Wireless Communications and Networking Conference*, Paris, France, April 2012.

[34] M. Ren, J. Zhang, L. Khoukhi, H. Labiod, and V. Vèque, "A unified framework of clustering approach in vehicular ad hoc networks," *IEEE Transactions on Intelligent Transportation Systems*, vol. 99, pp. 1–14, 2017.

[35] S. S. Wang and Y. S. Lin, "PassCAR: a passive clustering aided routing protocol for vehicular ad hoc networks," *Computer Communications*, vol. 36, no. 2, pp. 170–179, 2013.

[36] M. H. Eiza and Q. Ni, "An evolving graph-based reliable routing scheme for VANETs," *IEEE Transactions on Vehicular Technology*, vol. 62, no. 4, pp. 1493–1504, 2013.

[37] Z. Niu, W. Yao, Q. Ni, and Y. Song, "Link reliability model for vehicle ad hoc networks," in *Proceedings of the London Communication Symposium*, pp. 1–4, London, UK, 2006.

[38] L. C. Andrews, *Special Functions of Mathematics for Engineers*, McGraw-Hill, New York, NY, USA, 2nd edition, 1991.

[39] J. Jeong, S. Guo, Y. Gu, T. He, and D. H. C. Du, "Trajectory-based data forwarding for light-traffic vehicular ad hoc networks," *IEEE Transactions on Parallel & Distributed Systems*, vol. 22, no. 5, pp. 743–757, 2011.

[40] Y. Xiang, Z. Liu, R. Liu, W. Sun, and W. Wang, "GeoSVR: a map-based stateless VANET routing," *Ad Hoc Networks*, vol. 11, no. 7, pp. 2125–2135, 2013.

[41] *The Network Simulator: NS2*, 2015, http://www.isi.edu/nsnam/ns/.

[42] *SUMO: Simulation of Urban Mobility*, 2015, http://sumo.sourceforge.net.

[43] *OpenStreetMap*, 2017, http://www.openstreetmap.org/.

Permissions

List of Contributors

Jiaqi Lei, Hongbin Chen and Feng Zhao
Key Laboratory of Cognitive Radio and Information Processing, Guilin University of Electronic Technology, Guilin 541004, China

Tao Zhang, Lin Xiao and Dingcheng Yang
Information Engineering School, Nanchang University, Nanchang, China

Laurie Cuthbert
Information Systems Research Centre, Macau Polytechnic Institute, Rua de Luis Gonzaga Gomes, Macau

Jinyi Wen, Qin Yang and Sang-Jo Yoo
School of Information and Communication Engineering, Inha University, 253 Yonghyun-dong, Nam-gu, Incheon 402-751, Republic of Korea

Yin Wu and Bowen Li
Department of Information Science and Technology, Nanjing Forestry University, No. 159, Long Pan Road, Nanjing 210037, China

Yongjun Zhu and Wenbo Liu
College of Automation, Nanjing University of Aeronautics and Astronautics, No. 29, Jiangjun Avenue, Nanjing 211106, China

Chanhyuk Cho and Sanghyun Ahn
Department of Computer Science and Engineering, University of Seoul, Seoul, Republic of Korea

Taemin Ahn
Samsung Electronics Co., Ltd., Yong-tong-gu, Suwon, Gyung-gi-do, Republic of Korea

Jihoon Seok
SK Hynix Co., Ltd., 2091 Gyeongchung-daero, Bubal-eup, Icheon-si, Gyeonggi-do, Republic of Korea

Inbok Lee
Department of Software, Korea Aerospace University, 6 Hanggongdaehang-ro, Deokyang-gu, Goyang-si, Gyeonggi-do 412-791, Republic of Korea

Junghee Han
Department of Telecommunication and Computer Engineering, Korea Aerospace University, 76 Hanggongdaehang-ro, Deokyang-gu, Goyang-si, Gyeonggi-do 412-791, Republic of Korea

Yongjun Ren, Yepeng Liu and Sai Ji
School of Computer and Software, Nanjing University of Information Science & Technology, Nanjing, China Jiangsu Collaborative Innovation Center of Atmospheric Environment and Equipment Technology (CICAEET), Nanjing University of Information Science & Technology, Nanjing, China

Arun Kumar Sangaiah
School of Computing Science and Engineering, Vellore Institute of Technology (VIT), Vellore, India

Jin Wang
School of Computer & Communication Engineering, Changsha University of Science & Technology, Changsha, China

Yu Xu, Lin Xiao and Dingcheng Yang
Information Engineering School, Nanchang University, Nanchang 330031, China

Laurie Cuthbert
Information Systems Research Centre, Macao Polytechnic Institute, Rua de Luis Gonzaga Gomes, Macao SAR, China

Yapeng Wang
MPI-QMUL Information Systems Research Centre, Macao Polytechnic Institute, Macao SAR, China

Kang Liu, Qi Zhu and Ying Wang
Department of Telecommunication and Information Engineering, Nanjing University of Posts and Telecommunications, Jiangsu 210003, China

Muhammad Ahmad and Adil Khan
Institute of Robotics, Innopolis University, Innopolis, 420500 Kazan, Tatarstan, Russia

Muhammad Ahmad and Salvatore Distefano
University of Messina, Messina, Italy

Mohammed A. Alqarni and Sajjad Hussain Chauhdary
Faculty of Computing and Information Technology, University of Jeddah, Saudi Arabia

Asad Khan
Graphic and Computing Lab, School of Computer Science, South China Normal University, Guangzhou, China

Manuel Mazzara
Director of Institute of Technologies and Software Development, Head of Service Science and Engineering Lab, Innopolis University, Innopolis, 420500 Kazan, Tatarstan, Russia

Tariq Umer
Department of Computer Science, COMSATS University, Wah Campus, Islamabad, Pakistan

Sergio Caro-Alvaro, Eva Garcia-Lopez, Antonio Garcia-Cabot, Luis de-Marcos and Jose-Javier Martinez-Herraiz
Computer Science Department, University of Alcala, Alcalá de Henares, Spain

Xujie Li, Lingjie Zhou, Xing Chen, Ailin Qi and Chenming Li
College of Computer and Information, Hohai University, Nanjing 211100, China

Yanli Xu
College of Information Engineering, Shanghai Maritime University, Shanghai 201306, China

Chao Li and Huimei Lu
School of Computer Science and Technology, Beijing Institute of Technology, Beijing, China

Yong Xiang
Department of Computer Science and Technology, Tsinghua University, Beijing, China

Zhuoqun Liu
Neoclub Information Technology Company Limited, Shanghai, China

Wanli Yang
The First Research Institute of the Ministry of Public Security, Beijing, China

Ruilin Liu
Department of Computer Science, Rutgers University, Piscataway, NJ, USA

Jetmir Haxhibeqiri, Elnaz Alizadeh Jarchlo, Ingrid Moerman and Jeroen Hoebeke
Imec, IDLab, Department of Information Technology, Ghent University, iGent Tower, Technologiepark-Zwijnaarde, B-9052 Ghent, Belgium

Halikul Lenando
Faculty of Computer Science and Information Technology, Universiti Malaysia Sarawak, Kota Samarahan, Malaysia

Mohamad Alrfaay
Faculty of Computer and Information Sciences, Jouf University, Jouf, Saudi Arabia

Kriangkrai Maneerat and Kamol Kaemarungsi
National Electronics and Computer Technology Center, NSTDA, Pathumthani, Thailand

Md. Tanvir Hossan, Mostafa Zaman Chowdhury, Moh. Khalid Hasan, Md. Shahjalal, Trang Nguyen, Nam Tuan Le and Yeong Min Jang
Department of Electronics Engineering, Kookmin University, Seoul, Republic of Korea

Xiang Ji, Huiqun Yu, Guisheng Fan and Huaiying Sun
Department of Computer Science and Engineering, East China University of Science and Technology, Shanghai 200237, China

Xiang Ji
Shanghai Key Laboratory of Computer Software Evaluating and Testing, Shanghai 201112, China

Liqiong Chen
Department of Computer Science and Information Engineering, Shanghai Institute of Technology, Shanghai 201418, China

Index